# THE INORGANIC CHEMISTRY OF MATERIALS

## How to Make Things out of Elements

# THE INORGANIC CHEMISTRY OF MATERIALS

## How to Make Things out of Elements

**Paul J. van der Put**

*Delft University of Technology*
*Delft, The Netherlands*

**Plenum Press • New York and London**

Library of Congress Cataloging-in-Publication Data

Put, Paul J. van der.
The inorganic chemistry of materials : how to make things out of elements / Paul J. van der Put.
p. cm.
Includes bibliographical references and index.
ISBN 0-306-45731-8
1. Chemistry, Inorganic. 2. Inorganic compounds--Synthesis.
3. Materials. I. Title.
QD151.2.P89 1998
546--dc21 98-33993
CIP

ISBN 0-306-45731-8


A Division of Plenum Publishing Corporation
233 Spring Street, New York, N.Y. 10013

http://www.plenum.com



10 9 8 7 6 5 4 3 2 1



Printed in the United States of America

# PREFACE

*The devil may write chemical textbooks because every few years the whole thing changes.*

BERZELIUS

The basis of all technology involving materials is chemistry and physics, and technologists who deal with matter need skills in both fields. The technology of each class of material depends on physics and chemistry and the degree of need for each of them is different from one type of material to the next. It can roughly be said that metallurgy is mainly physical, a ceramicist needs more (inorganic) chemistry, and polymer science combines organic chemistry with continuum physics. Chemistry remains underrepresented in materials science curricula, although it has a considerable impact on materials technology. This book is an attempt to fill that gap in materials education.

Inorganic chemistry of materials includes those parts of inorganic chemistry or the chemistry of elements that can be used to make products. It is not one single subject but consists of several widely different disciplines, such as structural chemistry, coordination chemistry, and solid state ionics, to name but a few. These subjects have much chemistry in common and to show their function for materials technology a text on materials chemistry should integrate these different parts of the chemistry of the elements.

There exist excellent introductions to inorganic chemistry* and monographs on all subjects in inorganic chemistry. A selection of monographs might be used in a thorough course. However, in the initial stages of study it is not easy to grasp the significance of the different chemistries from monographs alone. Moreover those who lack time for a thorough training in inorganic chemistry have no patience for what they see as scientific luxuries in academic treatises. When technologists are hunting for facts they need them fast. They have developed the habit when reading of skipping whatever is not of immediate use, and monographs necessarily have too much of that. Moreover, introductory chapters on chemistry addressed to students

*M. Silberberg. *Chemistry: The Molecular Nature of Matter and Change*. Mosby, St. Louis (1996). D. F. Shriver, P. W. Atkins, and C. H. Langford. *Inorganic Chemistry*. Oxford University Press, Oxford (1990). L. L. Zumdahl. *Chemistry*. D. C. Heath, Lexington (1989).

| **Chemical Technology: production of matter** materials and ***consumables*** | | | |
|---|---|---|---|
| **Organic technology** | | **Inorganic technology** | |
| **Biotechnology** | **Synthetic organics** | **Nonmetals** | **Metals** |
| Wood ***Food***<br>Leather ***Pharmaca***<br>Fibers<br>Paper | Plastics ***Pharmaca***<br>Fibers ***Pesticides***<br>***Soap***<br>***Fuel*** | Single crystals ***Fertilizer***<br>Ceramics ***Resources***<br>Glasses ***Chemicals*** | Steel<br>Nonferro |

A rough grouping of the four types of matter in materials for durable products and matter that is consumed (in italics).

in materials engineering* are limited to structural chemistry and omit the subjects that may be of the most use in practice, such as solid state and surface chemistry.

This book on the inorganic chemistry of materials collects those parts of chemistry that are basic for practicing technologists whose job is to adapt matter for a purpose. It is directed to users who are not primarily after scientific insight for its own sake but who want hard chemical facts in order to make a material. It does not provide extensive discussions of the subjects that are covered in detail in orthodox textbooks (quantum chemistry, molecular reaction mechanisms, equilibrium thermodynamics) but instead includes essential topics that other books on inorganic chemistry neglect or ignore (solid state kinetics, design, and morphogenesis).

Roughly speaking there are two types of matter that concern the technologist: consumables, or matter for consumption, and materials, or matter from which to make durable products (see the accompaning illustration). Consumables are mostly molecular substances and most of them are organic compounds such as food, drugs, pesticides, soap, and fuel oil products. Inorganic consumables are fertilizers and mineral resources. The other part of matter that concerns the technologists is materials or matter for durable use. Most but not all of the solid compounds in this category are inorganic. The chemistry of bulk inorganic chemicals for consumption† is not discussed here, nor are organic materials.

Synthesis is basic for materials chemistry. Teaching synthesis raises the issues of the specific versus the general and technology versus science. Directions for synthesis are very matter-specific and relations with generalities are rarely given, because they do not seem to be very helpful from the point of view of the laboratory worker. That might call for a major revision of chemical theory some day. However, the importance of science for technology is generally overrated. One observes and the other makes. Making artifacts with the desired properties (i.e., synthesis) and their

*C. W. Callister. *Materials Science and Engineering: An Introduction.* Wiley, New York (1994). C. Newey and G. Weaver. *Materials Principles and Practice.* Butterworth and Heinemann, Milton Keynes (1991).
†W. Buechner, R. Schliebs, G. Winter, K. H. Buechel. *Industrial Inorganic Chemistry.* VCH, Weinheim (1989). R. Thompson (ed). *Industrial Inorganic Chemicals: Production and Uses.* Royal Chemical Society, Cambridge (1995).

design distinguishes the professions from the sciences. Ironically, engineering schools have gradually become schools of scientific physics and mathematics. The reason for this remarkable neglect of needs is that academic respectability calls for subject matter that is intellectually tough, analytic, formalizable, and teachable, while design and technology is in large measure intellectually soft, intuitive, informal, and cookbooky.* This poses a dilemma for any textbook on technology. Creativity cannot be taught and the directions in Chapter 9 should not be expected to aim for that.



This book is addressed to four groups of users who work in different fields but share a need for inorganic synthesis, a key subject for all of them:

*1. Students of materials science and technology*: There is more to materials science than a physical description of crystal structure and the microstructure of solids. Materials scientists are appreciated mainly for their skill in making matter do what is wanted of it. They have to be able to put atoms in their place and that is doing chemistry. The synthetic chemistry of functional and structural materials is central to this textbook. Nonmetallic inorganic materials get much less coverage in textbooks than metals and plastics although they are increasingly discussed in the biannual conferences of the Materials Research Society. The programs of those conferences indicate where the attention is now: some 4% of the subjects are devoted to metals, 16% to plastics, 50% to inorganic nonmetallic materials, and 30% to general characterization. This shift in focus is reflected in this book.

*2. Students of chemical engineering.* The range of synthetic processes is not restricted to the unit operations that are usually taught to the chemical engineer (basically applied physics of fluids). Synthesis of inorganic materials also includes methods such as forging, casting, sintering, hot-pressing, crystal growth, and lithography for making integrated circuits in semiconductor technology. The world is not a soup of organic molecules subject only to the established rules of transport phenomena. An area that is rightly ignored by chemical engineers is the current academic model for the chemical bond. However, by doing so they miss opportunities: good models of the chemical bond are reliable in their predictions and engineers involved in creating or modifying materials need them. Some of them are collected in this book.

*3. Product designers in need of novel materials.* Product designers often concentrate on shaping and they usually choose their materials from a list of what is available. However, available materials exist because they have been optimized for other purposes than those aimed for in the new application, and materials designers should be aware of that. Designing a product is an activity that includes development of materials because the functions wanted are strongly related to the fabrication process for materials. This book tries to show how the design of products extends to creating materials and property combinations that do not yet exist.

*4. Students of chemistry.* This book has not much to add to subjects that are readily available in other books on chemistry and usually the individual subjects are much better presented there. Yet this nonfundamental and pragmatic text should be useful to academic chemists at the beginning of their careers. After graduation, chemists are predominantly employed in the professions. In their job thc applications of chemistry prevail rather than its fundamentals and that means that some reschooling for the real world remains necessary. This book shows which parts of

*M. Diani (ed.). *The Immaterial Society*. Prentice Hall, Englewood Cliffs (1992).

academic chemistry are applicable in technology. There is also the need for less specialization. Graduates in inorganic chemistry know much about inorganic molecules or about solids but they might profit from being conversant with both. Much practical inorganic chemistry is also developed by materials technologists outside the chemical institutes and that work is less accessible for chemists because it is not published in the journals that are most familiar to them. Even the practicing organic chemist needs familiarity with inorganic compounds, particularly solids, as most organic chemicals are made with reactions involving inorganic catalysts.

Inorganic chemistry of materials is a vast subject and for the sake of being comprehensive the style of this text is inevitably terse. Many items are mentioned but not elaborated. For a more relaxed treatment of them the various references cited should be browsed. This textbook on the inorganic chemistry of materials is not a collection of recipes but supplies a chemical basis. It differs from other treatments on the following issues: (1) It is on synthesis rather than on characterization. (2) It collects those parts of academic chemistry that can be used for materials and parts of chemistry developed in other disciplines. (3) It includes subjects that other textbooks on materials science and technology neglect, such as the chemical bond, morphogenesis, and design.

Finally I would like to mention many others who contributed to this book. First of all, Evelyn Grossberg, for a critical reading of the final manuscript. The book grew out of discussions with graduate students in the Faculty of Chemical Engineering and Materials Science at the University of Technology in Delft, The Netherlands. Also many consultants and engineers working in industry, in development departments, and in production as well as scientists and technologists at uncountable conferences taught me the essentials of our trade. I am pleased to acknowledge my debt to them all.

P. J. van der Put

*Delft*

# CONTENTS

## Chapter 3
## Inorganic Molecules . . . 87



## Chapter 4
## Structural Solid State Chemistry . . . 111



## Chapter 5
## Solid State Reactions . . . 167

*Chapter 1*

# INTRODUCTION

*... nanotechnology implies that we can essentially make anything out of air, earth, fire and water.*

*Tom Abeles, PMC, Internet, 1995*

## 1.1. The Technology of Materials

This introductory chapter discusses atoms and bonds — the basics of the inorganic chemistry of materials — and also sketch out the ways of dealing with these building blocks in different disciplines. Materials chemistry is not a sharply bounded and institutionalized subject but a combination of many crafts and sciences. Some awareness of the norms in relevant areas is necessary for materials designers, who must get their information from those fields.

There are four groups — two organic and two inorganic (Figure 1.1) — into which materials are conveniently classified because each of these groups has its own characteristic formation processes. The organic materials are either synthetic (plastics) or of biological origin (wood, leather, paper, cotton), and are not our concern here. Inorganic materials involve almost the entire periodic system comprising groups of metals and nonmetallic inorganic materials. New technologies demand new materials designed to meet the new needs. Creating all these materials out of the 75 elements available and exploiting their potential is a chemical challenge. Inorganic chemistry of materials is the discipline devoted to the fabrication of inorganic materials from resources derived from the earth's crust[1] and is the basis for all technology involving matter. A large application area is in the field of nonmetallic inorganic materials, ceramics, glasses, and single crystals. Metallurgy, on the other hand, is primarily a branch of applied physics, and after reduction of their ores metals are processed mainly with nonchemical methods. Hybrid composites that combine metals with ceramics are fabricated using both chemical and physical techniques.

It is clear, then, that some knowledge of the chemistry of thc elements is a prerequisite for being able to use them. Societal demands for alternative resources, dematerialization, and durability are also compelling reasons for achieving a thorough knowledge of the chemical behavior of matter:

*Alternatives* are cheap, easily processed, and abundantly available replacements for expensive, scarce, and strategic elements, e.g., cobalt and chromium, necessary in

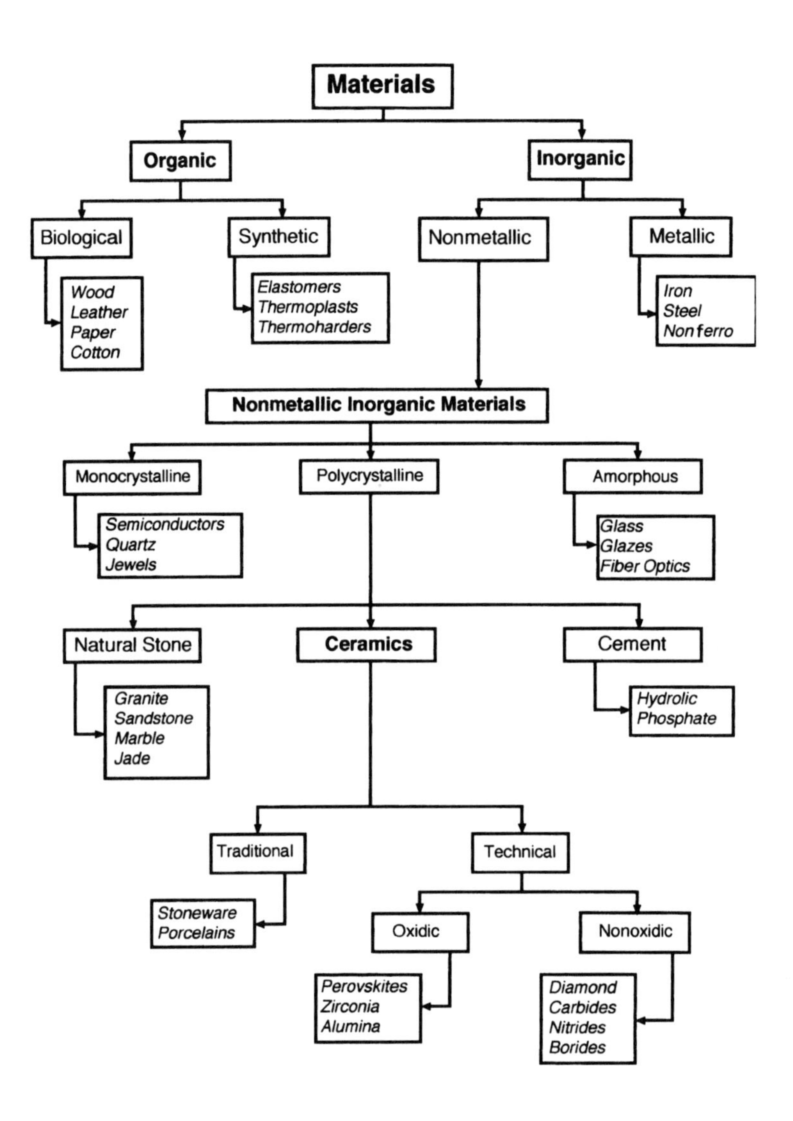


**Figure 1.1.** The family tree of materials with examples.

key industrial alloys. If alternatives can be found it will directly attenuate economic and societal tensions generated by the scarcity of these resources. The stockpiling of strategic metals in rich countries is the clearest indication that the possibilities of alternatives have not been fully realized. Resources are vast and scarcity problems are, in reality, man-made and nontechnical issues, which can be solved by chemical technology. "Never do by law what you can do by engineering, particularly if the law doesn't work."[2] The concept of critical or strategic resources becomes virtually obsolete if all the possibilities represented by the elements in the periodic table are exploited. For example, there are many instances in which expensive, scarce materials can be replaced by abundant and inexpensive ones if the latter are protected by cheap ceramic coatings. In gas turbines silicon nitride replaces superalloys that contain a large amount of cobalt. Basically, anything goes in materials, and what remains to be done is to achieve cost reduction in the fabrication processes and improve the performance of the replacements.

*Dematerialization* is a recent trend in society: we want to do more with less and do not want matter to restrict the range of our fantasy, and novel materials are being developed in materials chemistry to increase our possibilities. Their realization requires highly detailed information on matter while the contribution of material resources themselves to the cost of new products tends to become much less than the cost of the technological development. While virtually 100% of the cost of a brick is the cost of the clay and energy, only some 30% of the cost of a car is of material origin, the balance being the price of invested knowledge, and more than 99% of the cost of integrated circuits and fiber optics is for technology rather than for the constituents. This trend in products parallels the evolution of materials. Clay, obsidian, glass, gold, copper, and lead require little processing, while steel production demands more skills, and use of the more recent single crystals, alloys, technical ceramics, solid state devices, and electronics requires even more. Thus the prices of products fabricated from the last group of materials show an increasingly higher content of invested chemical know-how. The better we know how to use matter, the less we need of it.

*Sustainability* is a corollary to dematerialization, and it requires some change in our habits in dealing with matter. Processes have to be designed so that there is no waste and products need to have longer life and should be designed to be easily deconstructable, repaired, and reused, all of which needs novel materials, optimized for these requirements. At present commercial considerations mitigate against sustainability because durability decreases product turnover. This market pressure against durability translates into the absence of any serious constraint in the use of materials for consumer products and they remain in large part disposables. However, environmental forces that mandate sustainability may prevail in the end. Further, low corrosion and wear-resistance are stronger and stronger selling assets for capital goods, such as tools for production and machine parts, so, owing to customer demand, durability and sustainability in this market stand a better chance.

There is one basic question in materials chemistry: how to make what we want? Synthesis is the central issue addressed in this book. Several questions arise:

1. The playing field for this game is the periodic table. What are the properties of the elements? This question covers the entire field of inorganic chemistry.[3,4] The applicable elements and their behaviors are selected here. Key periodicities in the periodic table are noted in the next section, and they are basic for the use of the elements in compounds. Consideration of the properties of the elements is inherent in all the individual subjects dealt with in this volume.
2. Elements are linked to each other to make compounds. Chemical bonds are central to chemistry. What sort of compounds and bonds are possible? A short introduction to the behavior of valence electrons in bonds is given in Secton 1.3 and Chapter 2 is entirely devoted to a discussion of the chemical bond.

3. Closely related to the issue of bonds are: what types of compounds and materials are there,[5–8] what are their peculiarities, and how can we manipulate them chemically? Processes for molecules differ from those for atomic solids, as described in Section 1.3. Materials are atomic solids and Chapters 4 and 5 address structural chemistry and reactions in solids, respectively. Solids have surfaces, where almost all their reactions occur, and Chapter 6 describes their remarkable chemistry. Synthesis often involves reacting molecules, which is the reason for Chapter 3 on molecules. A description of molecules is somewhat unusual in a textbook on materials science but it is essential for understanding the synthesis of many inorganic materials.
4. How can chemistry be used to relate performance, properties, and microstructure of materials? The nature of solid products, which includes their composition and their microstructure, is fixed during their synthesis. The chemical formation of the microstructure, or morphogenesis, is the subject of Chapter 7. Morphogenesis is also hard to find in textbooks on chemistry or in those on materials science. The design of materials is not well documented either because it is impossible to give hard rules for designing. Materials design is more an art than a science. Chapter 9 introduces the chemical aspects of this new field.
5. Can synthesis be derived from general principles or is it an unstructured collection of recipes? Synthesis is not arrived at from theoretical considerations; it is developed in the laboratory. Recipes are matter-specific and always left that way. It would not be helpful to force them into some generalized scheme that shows their consistency with basic chemical theory. Yet models can sometimes be helpful in their design. This book combines generalities with specific behavior, as inorganic materials chemistry needs both. Chapter 8 is on details of synthesis. Often a material is made and only after it is made is it given the shape that is needed for the product, e.g., metals and plastics. For ceramics and composites the material is made after shaping one of the solid precursors. A novel trend in materials processing is to shape a material simultaneously with its chemical preparation. Examples of shaping during synthesis are the so-called powderless processing of ceramics, reaction-bonded ceramics, and chemical vapor deposition.

## 1.2. The Periodic Table

The periodic table as was noted above is the playing field for the materials technologist. A systematic understanding of the chemical and physical behavior of the elements and their compounds would be invaluable but an adequate text would consist of many volumes. Thus, in place of full descriptions of the elements, some periodicities in the table are included to acquaint readers with the elements and indicate their peculiarities.[3,9]

The elements in the periodic table are arranged according to the numbers of electrons in the different atomic shells (Figure 1.2). The larger the shell, the more electrons it can have, but the maximum number of electrons in each shell is 2, 8, and 18 (the outer shell of the elements in the period with the rare-earth metals can have

Group:

| 1 | 2 | 3 | 4 | 5 | 6 | 7 | 8 | 9 | 10 | 11 | 12 | 13 | 14 | 15 | 16 | 17 | 18 |
|---|---|---|---|---|---|---|---|---|---|---|---|---|---|---|---|---|---|
| 1 H | | | | | | | | | | | | | | | | | 2 He |
| 3 Li | 4 Be | | | | | | | | | | | 5 B | 6 C | 7 N | 8 O | 9 F | 10 Ne |
| 11 Na | 12 Mg | | | | | | | | | | | 13 Al | 14 Si | 15 P | 16 S | 17 Cl | 18 Ar |
| 19 K | 20 Ca | 21 Sc | 22 Ti | 23 V | 24 Cr | 25 Mn | 26 Fe | 27 Co | 28 Ni | 29 Cu | 30 Zn | 31 Ga | 32 Ge | 33 As | 34 Se | 35 Br | 36 Kr |
| 37 Rb | 38 Sr | 39 Y | 40 Zr | 41 Nb | 42 Mo | 43 Tc | 44 Ru | 45 Rh | 46 Pd | 47 Ag | 48 Cd | 49 In | 50 Sn | 51 Sb | 52 Te | 53 I | 54 Xe |
| 55 Cs | 56 Ba | *Ln* | 72 Hf | 73 Ta | 74 W | 75 Re | 76 Os | 77 Ir | 78 Pt | 79 Au | 80 Hg | 81 Tl | 81 Pb | 82 Bi | 83 Po | 84 At | 86 Rn |

| *Ln:* | 57 La | 58 Ce | 59 Pr | 60 Nd | 61 Pm | 62 Sm | 63 Eu | 64 Gd | 65 Tb | 66 Dy | 67 Ho | 68 Er | 69 Tm | 70 Yb | 71 Lu |
|---|---|---|---|---|---|---|---|---|---|---|---|---|---|---|---|

| | | | | | | | | | | | | | | | | | |
|---|---|---|---|---|---|---|---|---|---|---|---|---|---|---|---|---|---|---|
| | H | | | | | | | | | | | | | | | | He |
| 1s: | 1 | | | | | | | | | | | | | | | | 2 |
| | Li | Be | | | | | | | | | | B | C | N | O | F | Ne |
| 2s: | 1 | 2 | | | | | | | | | | 2 | 2 | 2 | 2 | 2 | 2 |
| 2p: | | | | | | | | | | | | 1 | 2 | 3 | 4 | 5 | 6 |
| | Na | Mg | | | | | | | | | | Al | Si | P | S | Cl | Ar |
| 3s: | 1 | 2 | | | | | | | | | | 2 | 2 | 2 | 2 | 2 | 2 |
| 3p: | | | | | | | | | | | | 1 | 2 | 3 | 4 | 5 | 6 |
| | K | Ca | Sc | Ti | V | Cr | Mn | Fe | Co | Ni | Cu | Zn | Ga | Ge | As | Se | Br | Kr |
| 3d: | | | 1 | 2 | 3 | 5 | 5 | 6 | 7 | 8 | 10 | 10 | 10 | 10 | 10 | 10 | 10 | 10 |
| 4s: | 1 | 2 | 2 | 2 | 3 | 1 | 2 | 2 | 2 | 2 | 1 | 2 | 2 | 2 | 2 | 2 | 2 | 2 |
| 4p: | | | | | | | | | | | | | 1 | 2 | 3 | 4 | 5 | 6 |
| | Rb | Sr | Y | Zr | Nb | Mo | Tc | Ru | Rh | Pd | Ag | Cd | In | Sn | Sb | Te | I | Xe |
| 4d: | | | 1 | 2 | 4 | 5 | 5 | 7 | 8 | 10 | 10 | 10 | 10 | 10 | 10 | 10 | 10 | 10 |
| 5s: | 1 | 2 | 2 | 2 | 1 | 1 | 2 | 1 | 1 | 0 | 1 | 2 | 2 | 2 | 2 | 2 | 2 | 2 |
| 5p: | | | | | | | | | | | | | 1 | 2 | 3 | 4 | 5 | 6 |
| | Cs | Ba | Ln | Hf | Ta | W | Re | Os | Ir | Pt | Au | Hg | Tl | Pb | Bi | Po | At | Rn |
| 4f: | | | 1- | 14 | 14 | 14 | 14 | 14 | 14 | 14 | 14 | 14 | 14 | 14 | 14 | 14 | 14 | 14 |
| 5d: | | | | 2 | 3 | 4 | 5 | 6 | 7 | 9 | 10 | 10 | 10 | 10 | 10 | 10 | 10 | 10 |
| 6s: | 1 | 2 | | 2 | 2 | 2 | 2 | 2 | 2 | 2 | 1 | 1 | 2 | 2 | 2 | 2 | 2 | 2 |
| 6p: | | | | | | | | | | | | | 1 | 2 | 3 | 4 | 5 | 6 |

**Figure 1.2.** The periodic table of the elements used in materials. The orbital occupation of valence electrons parallels group properties.

32 electrons). These numbers are accounted for by the orbital model.[10] As the valence electrons are strong determinants of most properties, the periodicity in electron occupation parallels the properties of the elements. The periodic table is based on those periodicities of the elements or their compounds.

## *Metals and Nonmetals*

The first division of the elements is into the groups of metals and nonmetals. The diagonal going from beryllum to polonium forms the boundary of the metals groups, but that boundary is not sharp.

To the left of the Be–Po diagonal the metallic elements form intermetallic solid compounds or alloys with each other. Metallic atoms are characterized by a small number of valence electrons and low electronegativity. These compounds and alloys are themselves usually metals but not always, e.g., CsAu is a semiconductor. The stoichiometries in intermetallics (these are the numbers $n$ and $m$ in the compound $A_nB_m$) do not usually correspond to atomic valencies, as they would in ionic and covalent compounds. The existence regions of intermetallic compounds may be wide, which means that a compound such as $A_nB_m$ may have continuous ranges of $n$ and $m$ in the same phase. In other words, intermetallics can be berthollides.

To the right of the Be–Po diagonal are the nonmetals. Compounds that contain at least one nonmetal ($TiC_x$, $NbN_x$, $ZrO_{2-x}$) are the nonmetallic inorganic (NMI) compounds. Ceramics are polycrystalline materials made of such atomic (nonmolecular) NMI compounds and consist of phases that combine metallic with nonmetallic atoms. When the nonmetallic elements react with each other they can form molecules such as $NH_3$, $SF_6$, $PCl_5$, and $CO_2$ or insulating solids such as BP, $NI_3$, BN, SiC, and $SiO_2$. In contrast with intermetallics they usually have the stoichiometries their valencies predict. The compounds of the elements in one group (column in the periodic table) are to some extent similar because they have the same number of valence electrons. They behave differently from one another because they are of different sizes.

### Atomic and Ionic Size

The atomic or ionic size varies with the position of the element, its coordination number, and its charge. Atomic sizes are derived from interatomic distances in compounds, and are shown in Figure 1.3 for neutral atoms and for ions. The atomic radii are used to estimate structures and densities of compounds. The periodicity in size with increasing atomic number is clear from the figure. The alkaline metals (group 1) and the noble gases (group 18) have large atoms while the atoms around the middle of the transition metal periods are relatively small.

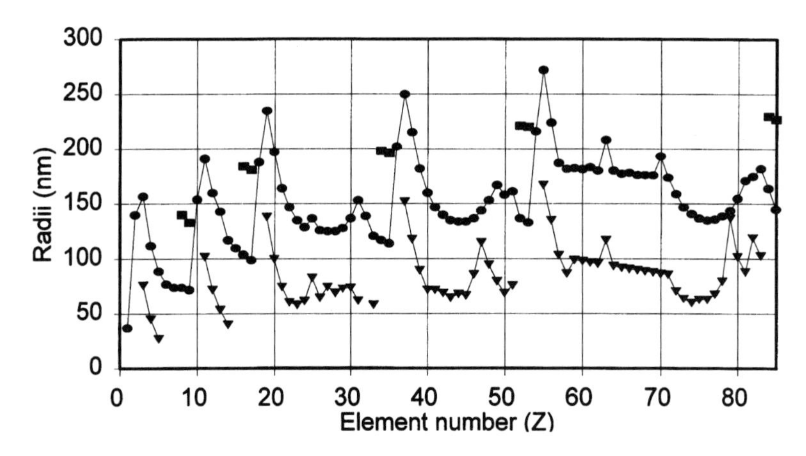


**Figure 1.3.** The size of atoms and ions of the elements. Neutral atoms have covalent sizes marked with circles, cations with triangles, and anions with squares.

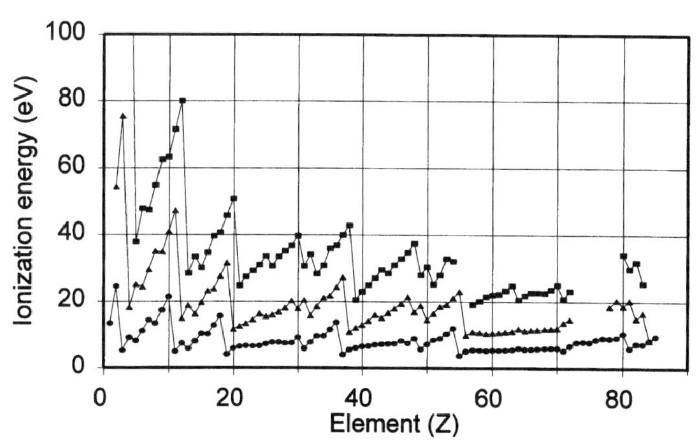


**Figure 1.4.** The first three ionization potentials of the elements. Circles indicate the values of neutral atoms. Higher valencies have higher ionization energies.

### *Ionization Energy, Electron Affinity, and Electronegativity*

Ionization energy, electron affinity, and electronegativity which are more fully discussed in Section 2.7, are used to describe the chemical bond in certain models and form the basis for estimating the properties of compounds.

The ionization potential is the minimum energy required to ionize an atom or an ion and is derived from the atomic spectrum. The periodicities of the first three ionization potentials are shown in Figure 1.4. Closed shells are difficult to ionize and an electron that is alone in its shell is easily removed.

The electron affinity is the energy that the neutral atom uses to bond a free extra electron, i.e., the ionization potential of the mononegative ion. Electron affinities are more difficult to measure than ionization potentials. Their experimental values are plotted in Figure 1.5.

Electronegativity is the tendency to attract charge from neighboring atoms that are bonded. There are many scales for electronegativity but they are all correlated. Electronegativity is a basic feature of chemical bonding.

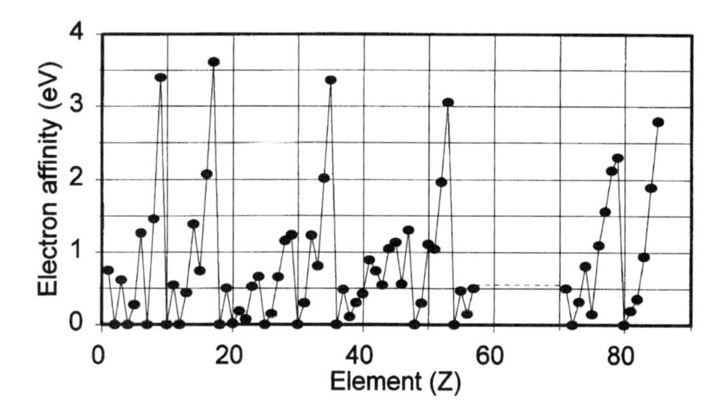


**Figure 1.5.** The periodicity in electron affinities of the elements.

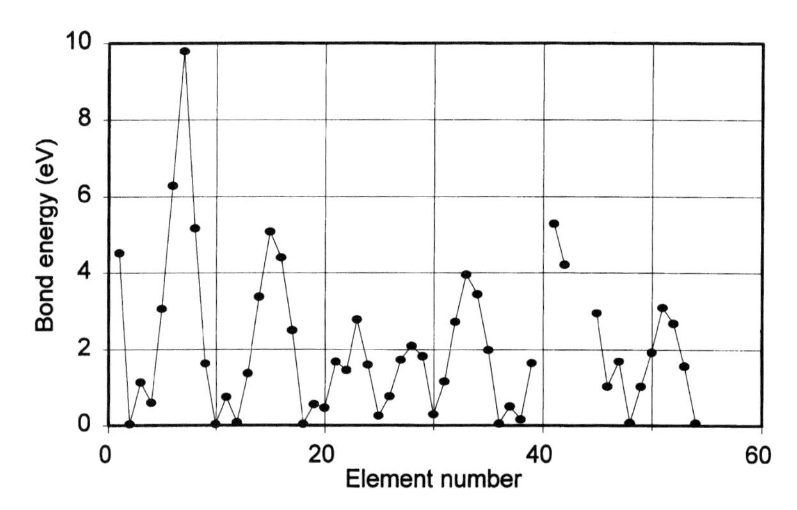


**Figure 1.6.** Periodicity in the bond energies of homopolar dimers of the elements.

### *Homopolar Bonding Strength*

The energy of atomization of the elements is related to the bond energies of the elements, which are shown in Figure 1.6. Electrons in half-filled shells especially those in a *d*-shell, are tightly bound to the atom, and most atoms having half-filled outer shells are less reactive than atoms that do not bind their valence electrons so well. High bonding strengths are correlated with small atoms, as can be seen by comparing Figures 1.3 and 1.6. Atoms that have an $(ns)^2$ valence electron configuration have a helium-type filled outer electron shell and therefore such atoms are weakly bound to atoms of their own type (groups 2 and 12).

### *Nature of the Hydrides*

Almost all elements form hydrides, i.e., compounds with hydrogen (Figure 1.7). Hydrides occur in three of the basic compound types, saltlike lattices, metallic solids, and covalent volatile molecules. The very electropositive elements form ionic hydrides (LiH, $CaH_2$); the transition metals can dissolve hydrogen and form metallic hydrides; the hydrides of the nonmetals are molecules ($CH_4$, $H_2S$).

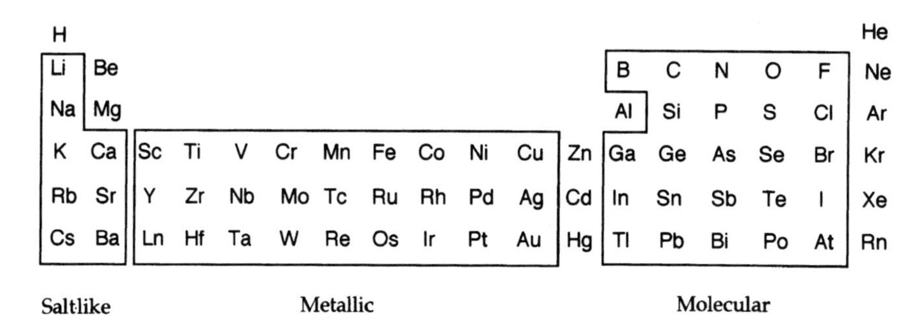


**Figure 1.7.** Nature of the hydrides of the elements in the periodic table.

Although arranged in groups (columns) with similar properties and periods (rows of the periodic table), the elements are all unique. The periodic table groups them roughly according to the properties that they share but the similarities among them should not be exaggerated. Sodium may be thought not to differ much from potassium, but a chemist who confuses them in his synthesis might blow up his laboratory. Sodium chloride and potassium chloride might be quite similar for a physicist but a crystal grower or a cook would disagree. Three single elements illustrate the great variability of properties shown by the elements. Many others will be discussed in the following chapters.

*Carbon.* According to the organic chemists carbon is the center of the periodic table or at least the center of the part of the table that concerns them. The two main forms of carbon—diamond and graphite—are clear examples of the strong influence of structure on properties, as the same atoms arranged in different lattices result in completely different materials. Glassy carbon (an amorphous modification) has properties intermediate between graphite and diamond. Molecular forms of carbon discovered recently ($C_{60}$ and similar structures) have been synthesized in arc discharges between carbon electrodes rather than with organic molecular methods. Carbon reacts with many other elements. It forms metallic compounds in transition metal carbides and covalent solids with elements such as boron, silicon, and nitrogen. In solids it does not form mononuclear ions but the $C_2^{2-}$ ion exists in the salt calcium carbide. Like hydrogen, carbon forms molecular covalent compounds with most other nonmetals.

*Silicon.* Silicon is in the same group as carbon and is another center of the periodic table, situated between the metals and the nonmetals. As a semiconducting element it is the material basis of digitized society. Bound to some metals, it is a good electric conductor and is also chemically inert. $TiSi_2$, e.g., is used for electrical contacts in microelectronics, and $MoSi_2$ is the best corrosion-resistant conductor known at high temperatures in air. Bound to oxygen, silicon constitutes the primary component of the vast class of traditional ceramics and silicones or siloxanes, which are well-known inorganic polymers.

*Boron.* Boron is structurally the most bizarre element in the periodic table. Simple bonding rules that are applicable to the other elements have to be bent considerably in order to accommodate the behavior of boron. Boron should be a metal, but it is a semiconductor with unique structures and anomalous physical properties. Combined with metals it participates in metallic bonding but boron atoms are simultaneously covalently bonded to other boron atoms in the higher metal borides. Similar to carbon it does not form mononuclear ions, its halides are molecules, and its other compounds with nonmetals are solids.

## 1.3. Types of Matter: Structure and Bonding

There are several ways to categorize materials. One way is to group them according to their use or function. Thus structural materials, such as metals, plastics, concrete, and building ceramics, used in construction are chosen for their mechanical

**Table 1.1. Functional Properties of Materials**

| Property | Type of material | Application | Examples |
|---|---|---|---|
| Electric | Insulators | Fire plugs | $Al_2O_3$ |
| | Dielectrics | Active materials | PLZT |
| | Electronic conductors | Chips | $TiSi_2$ |
| | Semiconductors | Chips, sensors | Si |
| | Ionic conductors | Fuel cell electrolytes | PSZ |
| | Mixed conductors | Fuel cell electrodes | $LnMnO_3$ |
| Magnetic | Ferrimagnetics | Information recording | $\alpha$-$Fe_2O_2$ |
| Thermal | Thermal conductors | Heat exchangers | SiC |
| | Thermal insulators | Refractories | MgO |
| Optical | Chromophores | Pigments, ion lasers | Cr:$Al_2O_3$ |
| | Translucent materials | Fiber optics | Ge:$SiO_2$ |
| | Nonlinear optics | Switches | $LiNbO_3$ |
| | Photovoltaics | Solar cells | Si, BP |
| | Transparent conductors | Displays | In:$SnO_2$ |
| Chemical | Active materials | Catalysts | Cermets |
| | Inert materials | Corrosion resistors | Oxide ceramics |
| | Biocompatibles | Prostheses | C/SiC, diamond |

properties—high strength, low weight, good fatigue behavior, and high hardness. Functional materials are not used in constructions but have electrical, optical, thermal, or chemical applications and most functional materials are of the nonmetallic inorganic type. Some functional properties and examples are summarized in Table 1.1.

Materials can also be grouped according to the way their atoms are bound. Molecules have strong intramolecular and weak intermolecular bonds while atomic solids have strong interatomic bonds in extended lattices. A materials engineer is interested in the crystal structure of a material because the (intrinsic) properties depend on it. The structures of some molecules and crystal lattices are shown in several figures further on in this chapter following a brief review of chemical bonds.

The structures in molecules and solid lattices strongly depend on the kind of bond between the atoms. There are five pure types of bonding between atoms, three of which are strong and two weak: covalent, ionic, metallic, hydrogen, and dispersive bonds. Covalent and ionic bonds, which are strong, occur in molecules and in many atomic solids. Metallic bonding, which is also rather strong, is found only in lattices and not in molecules. The weaker bonding types, hydrogen bonding and the dispersive or van der Waals–London forces, account for the interactions between molecules (in liquids or in molecular solids) and for cohesion in polymers. These bonds with their characteristics are shown in Table 1.2.

The covalent bond has localized electrons, and has little or no net charge transfer from one atom to the other. Silicon and diamond are prototypes of solids having covalent bonds, and the bonds in hydrogen and nitrogen molecules are also purely covalent. These bonds are oriented in space with respect to each other. Four-coordination (four nearest-neighbor atoms) often indicates covalency. Examples of covalent bonds in binary compounds are found in the solid silicon carbide (SiC) and nitrogen chloride molecules ($NCl_3$).

**Table 1.2. Types of Solids and Their Chemical Bonds**

| Type | Lattice units | Properties | Examples | Lattice energy (eV/mol) |
|---|---|---|---|---|
| Covalent | Atoms | Semiconductors, or insulators, hard, low mobility | SiC | 10.5 |
| | | | Diamond | 7.4 |
| Ionic | Cations and anions | Electron insulators, ionic conductors, brittle | LiF | 10.4 |
| | | | CsI | 6.0 |
| | | | MgO | 40.5 |
| Metallic | Cations and interstitial electrons | Electron conductors, plastic, tough | Mg | 1.5 |
| | | | Fe | 4.3 |
| | | | Mo | 6.8 |
| Hydrogen | Atoms and protons | Insulators low melting point | Ice | 0.5 |
| | | | HF | 0.3 |
| Dispersive and dipolar | Molecules | Very low melting point | $SiH_4$ | 0.1 |

The ionic bond exists in salts and in complexes (molecules having a central metal ion) between atoms that have very different electronegativities. There is charge transfer (in the extreme form), no directed bonding, and strong localization of the electrons on the atoms. The lattice consists of ions. Ionic bonding leads to structures roughly obeying the Pauling rules for stacking hard charged spheres. These rules prescribe the relationship between the ionic radius and the coordination number.

The metallic bond has delocalized valence electrons, which have left the (electropositive) atoms and fill the interstices among the stacked atoms in the metal lattice. There is no charge transfer from an atom to its neighbors in the lattice, and bonding forces are not directed. A metal can be described as a stack of cations embedded in a gas of free electrons in the interstitials. Three-quarters of the elements in the periodic table from lithium to bismuth form metallic bonds in their lattices. Metallic bonds are not found in molecules (except in small metal clusters). The arrangement of atoms that are bonded with metallic bonds depends on the number of valence electrons and interstitial holes in the stacked atoms of the structure. Metal bonding can often be recognized from the stoichiometry of the compounds, which differs widely from the accepted valencies of the atoms bonded in the lattice: the intermetallic compounds $TiAl_3$, $TiAl_2$, TiAl, and $Ti_2Al$ may all be somewhat nonstoichiometric. Examples of ceramic compounds that have metallic bonds are TiN and TaC.

The hydrogen bond is a weaker bond between atoms than the covalent, ionic and metallic bonds. It occurs if a proton bridge can be formed between two atoms, e.g., in the interaction among water molecules in liquid water and ice and among many organic groups. Boranes are examples of molecules that have an intramolecular hydrogen bond.

The van der Waals or dispersive bond is the weakest. It accounts for intermolecular forces, which represent only a small contribution to bonds in materials except for glues and liquid crystals.

In general, the actual bond between two given atoms is not purely of one type but is a mixture of covalent, metallic, and/or ionic character, each to a variable

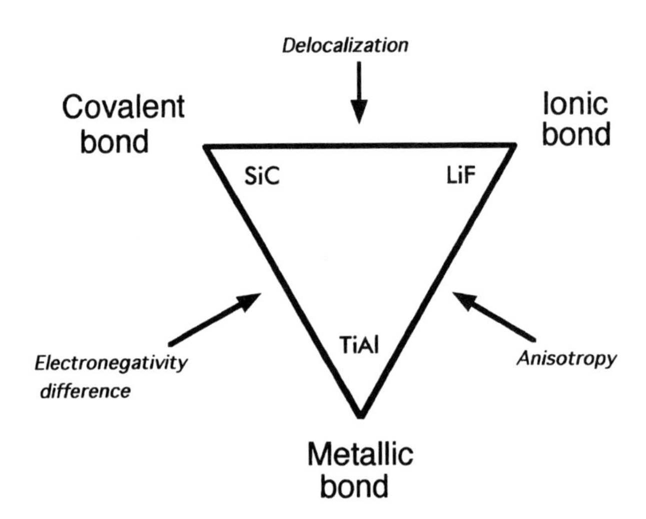


**Figure 1.8.** The character of three types of chemical bonds. Each binary compound has a position in the triangle according to the contribution of each of the three bond types to the actual bond between the atoms. The three compounds indicated near the corners have a chemical bond in which one bond type dominates.

extent, depending on the nature of the particular atoms that are bound to each other. The weight of the three strong bonding types in a given bond can in principle be plotted in a triangular graph in the same way as the phase diagrams of three-component systems or the composition of a triaxial ceramic body are represented. In this graph the place of a chemical bond in a given binary compound is determined by the value of each of the three properties (the coordinates in Figure 1.8), which can be a number between 0 and 1. The sum of the three numbers equals 1. Each of the three pure bonding types (covalent, ionic, and metallic) maximizes one of the three characteristics: the electronegativity difference, the directedness of the chemical bond, and the degree of delocalization of the valence electrons. Compounds with a predominantly ionic bond (salts) are found in the corner of the triangular diagram where the electronegativity difference of the bonded atoms is large. Intermetallics can be found in the corner where the degree of delocalization is high while covalent compounds have highly directed bonds. Usually a bond has characteristics of all three extreme types. The values of the contribution could in principle be estimated but they depend on the calculation scheme and are never given in tables of binary compounds.

Owing to the fact that valence electrons determine bonds, the electrical properties of a material are related to the bond type. In conductors such as metals, alloys, and intermetallics, the atoms are bound to each other primarily by metallic bonds, and metals such as tungsten or aluminum are good conductors of electrons or heat. Covalent bonds occur in insulators such as diamond and silicon carbide and in semiconductors such as silicon or gallium arsenide. Complexes and salts have ions that are bound with electrostatic forces. Ionic conductors can be used as solid electrolytes for fuel cells because solids with ionic bonds may have mobile ions. Most polymers have covalent bonds in their chains but the mechanical properties of plastics are mainly the result of the intermolecular, rather weak, bonding forces.

Materials are synthesized in many ways: in solid state reactions, from melts, solutions, or vapors, assisted by energy sources such as heat, plasmas, light, or mechanical forces. There is some connection between the type of synthesis and the type of bonding but it is not very close. Ionic and metallic solids can be formed by solid state diffusion reactions. Solids are created at crystallite surfaces from liquids or gases, and the atoms in solid surfaces can show all types of bonds. Ionic compounds are often made by mixing the ions in solution and precipitating the solid. If that is done slowly, single crystals may form. Solvent molecules ($NH_3$, $H_2O$) or ions in molten salts used as solvents can compete with reactant ligands in molecules or ionic lattices; if solvation is a problem in synthesis, very acidic solvents are used. Metallic compounds are made simply by mixing the reactants in a melt or by sintering solid particles. Covalent solids are made by radical reactions at the solid/gas interface. Polymers and colloids are usually formed in solutions out of molecular reactants with radical or ion intermediates.

Materials technologists generally do not bother much about bonding. They take covalent, ionic, and metallic bonds for granted and discussion about them in their textbooks is reduced to a minimum. Semiconductor technologists employ a simple band model, enough to systematize properties of bulk crystals, while solid state ionics recognizes only fully charged ions and the Coulomb forces between them. This approach to the chemical bond in materials technology is paradoxical because bonds are known to strongly affect the properties of compounds as well as molecular and crystal structures. The applications of solids and molecules depend on the behavior of the valence electrons in their chemical bonds. Yet convenient bonding models are rarely used, if taught at all, perhaps because the prevailing orbital model of the chemical bond is virtually useless in daily practice and is therefore ignored. The neglect, however, is unnecessary, as there are simple bonding models (Chapter 2) that lead to reliable predictions and are easy to use. Some familiarity with the effects of bonding on properties and synthesis is essential for the professional materials technologist.

The inorganic molecules that are of interest for materials scientists, apart from the very simple ones, such as hydrogen, oxygen, water, and ammonia, are those containing metal atoms, the coordination compounds or complexes, and organometallic compounds. Coordination compounds are used in the synthesis of materials from solvents; organometallic compounds are used to bring metals into the vapor phase at relatively low temperatures and for making solids. The chemistry of these molecules is discussed in Chapter 3, but a few examples are given here to point out frequently occurring structures of molecules and lattices.

Coordination compounds or complexes have a central metal atom bound to surrounding ligands, ions, or molecules. The metal–ligand bonds are ionic in most complexes and the molecular structures have high symmetries: $HgCl_2$ is linear like $CO_2$, $BF_3$ is trigonal planar, tetrachlorosilane ($SiCl_4$) is tetrahedral like methane, $PCl_5$ is trigonal bipyramidal, and $SF_6$ is octahedral. The coordination number of a complex is the number of ligand atoms that surround the central ion and are directly bound to it. The ligands in coordination compounds need not be simple ions or atoms but can be fairly large molecules that are either molecular ions or have a dipole moment sufficiently large for electrostatic bonding to the central metal ion.

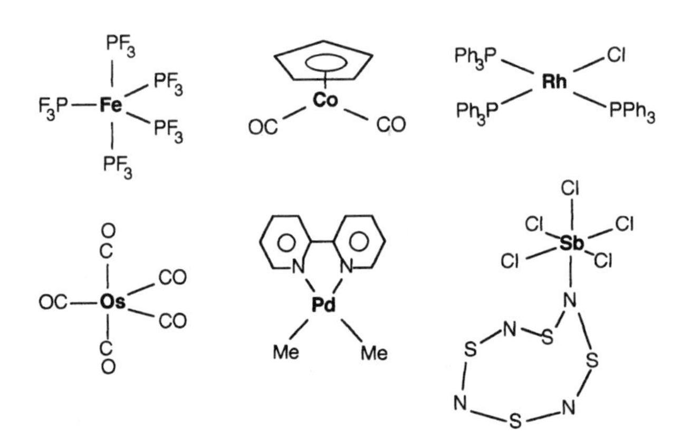


**Figure 1.9.** Structure of some metal complexes with inorganic or organic ligands.

Figure 1.9 shows examples of complexes with ligands that are inorganic or organic molecules. If the organic ligands are attached to the metal center through one of their carbon atoms the complex is an organometallic compound. A complex can be a neutral molecule that is made up of ions. It is usually a central cation linked to surrounding anions, small molecules, or both. Complexes can also be ions themselves (when they have a net positive or negative charge) and have structures like pieces of ionic lattices as shown in Figure 1.10. This analogy between ionic lattices and molecular complexes is actually used to interpret absorption spectra of solutions and solids.

Silatrane (Figure 1.11) is a remarkable molecule—the silicon center is five-coordinate while silicon is usually four-coordinate—and it is created through a reaction between substituted orthosilicate esters and triethanolamine:

$$\text{Z-Si(OR)}_3 + (\text{HOCH}_2\text{CH}_2)_3\text{N} \rightarrow 3\text{ROH} + \text{Z-Si(OCH}_2\text{CH}_2)_3\text{N}$$

where $Z$ and $R$ are organic groups. Drugs are generally organic but silatrane although inorganic seems to have extraordinary biological effects. Silatranes are reported to be useful against a wide range of ills from baldness to cancer and are known to accelerate wound healing and tissue regeneration. Other inorganic drugs

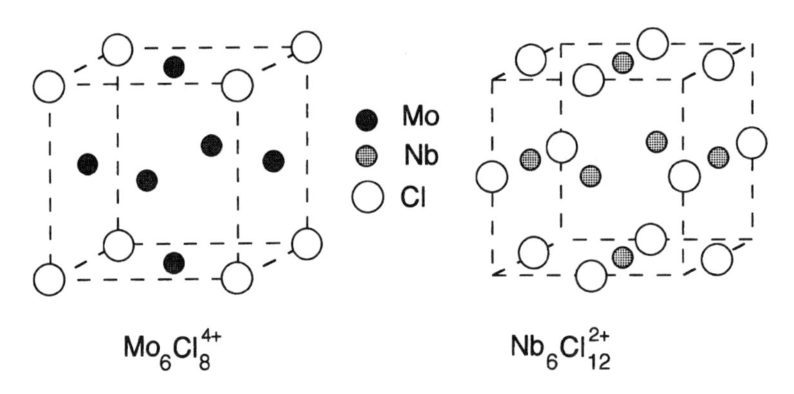


**Figure 1.10.** Structures of octahedral metal chloride cluster cations.

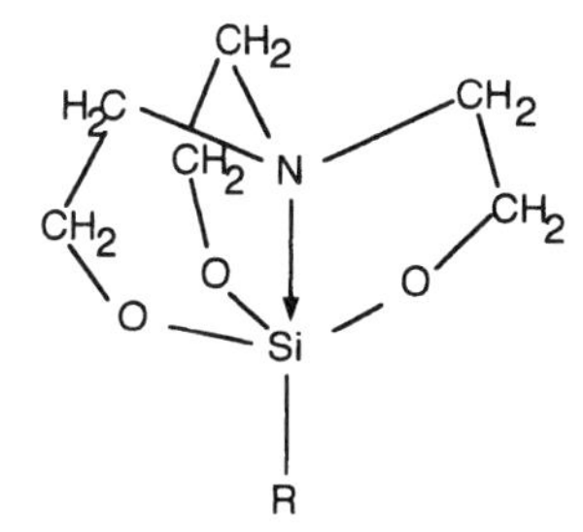


**Figure 1.11.** Basic structure of Voronkov's silatranes. The physiological activity changes with group R: R = methyl: sedative; R = phenyl: local anesthetic; R = alkoxy or alkyl: antitumor activity; R = halomethyl or alkoxy: increased hair growth and wound healing; R = EtO > MeO > Me > Et > aldehyde > ketone: anticoagulants.

are certain gold and platinum complexes, which are used against arthritis and cancer. Coordination chemistry is of considerable interest for biochemistry, as many enzymes are organic ligands that function with metal ions.

Organic and inorganic molecules or radicals can act as bridging ligands as shown in Figure 1.12. In these molecules the methyl group is directly bonded to several metal atoms simultaneously in the same way that hydrogen acts as a bridge in many boranes, where it is bonded to two boron atoms simultaneously. Metal hydroxide polymers are precursors for ceramics made with the sol-gel process. Almost all metal hydroxides can be precipitated from a solution of their salts by growth of the initial hydroxide clusters to form small colloidal particles. Silicates, which form when mixed hydroxides containing silica are calcined, have a wide range of structures in which the $SiO_4$ tetrahedra are linked to each other by sharing one or more oxygen atoms to form oligomers, linear chains (pyroxenes, $SiO_3^{2-}$), double bands (amphiboles $Si_4O_{11}^{6-}$, or $Si_2O_5^{2-}$), branched chains, planar nets, or three-dimensional nets. Countercharges are cations between the nets. The crystal struc-

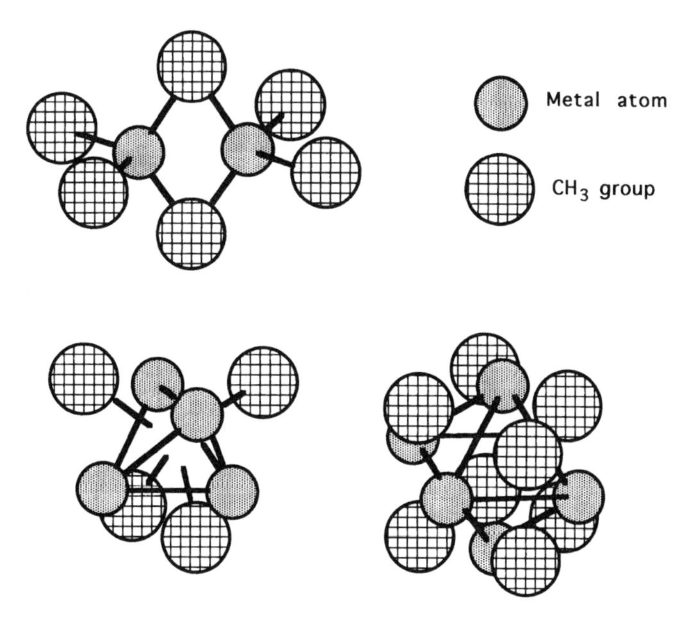


**Figure 1.12.** The molecular structure of beryllium and aluminum alkyls. In some cases the alkyl groups are not bonded to one metal atom only but to three simultaneously.

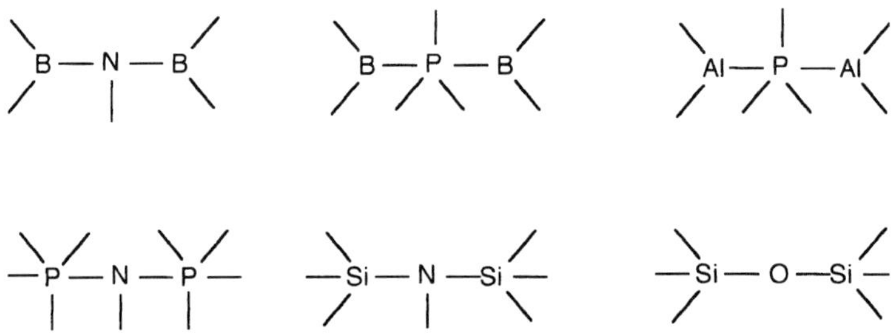

**Figure 1.13.** Backbone patterns in a few inorganic polymers. Side groups can be organic or inorganic.

tures of a silicate can be read from the stoichiometry of the compounds, and the nature of the silica nets endows the silicates with asbestos or micalike properties.

Typical inorganic polymers do not have any carbon in their chains but the side groups attached to the central chain may be organic. Figure 1.13 illustrates several possible inorganic chains. Inorganic polymers, of which silicones and polyphosphazenes are the best known, have different properties than organic polymers, some of which may be better (glass point, fire resistance), and they are also more expensive. Polymers are intermediate between molecules and solids but are discussed in Chapter 3 on molecules.

The chemistry of solids is more complicated than that of molecules. The properties of solid products depend on the way they are made, which is not true for molecules. Orthodox inorganic chemistry is mainly concerned with molecules. Materials chemistry, on the other hand, deals with solids, surfaces, interfaces, molecules, and atoms. The rates of chemical processes involving solids depend strongly on the spatial arrangement or dimension of the reactant, product, or reaction space, as will be described in Chapter 7. Morphogenesis, the formation of shapes and interconnection between shapes on a micro- and nanoscale, is irrelevant for molecules. For materials technology the microstructure is as important as the atomic composition and crystal structure. The properties of matter depend on the dimension of the atomic distribution of the surfaces or the interfaces—in short on its morphology.

The effect of morphology on properties is evident in silicon. A single crystal of silicon has a bandgap of 1.1 eV, nanocrystalline silicon particles have a size-dependent and direct bandgap, which for amorphous silicon is 1.7 eV, and silylenes are polymers with a bandgap around 3 eV.

Most surfaces are fractal and their chemistry differs from that of flat crystal faces. The chapters comprising this volume are concerned with the chemistry of molecules, solids, surfaces, and fractal matter. Not only the morphology of materials controls many of their properties but the fractal dimension of the matter distribution in materials also strongly affects the result of chemical processes in or on those distributions of atoms. Dimensions of interfaces pervade materials properties and it is difficult to ignore them when optimizing polycrystalline composites.

As every material is unique, especially in its chemistry, it is virtually impossible to find generally valid guidelines for making particular structures. There is only one vague general rule: If structures of solids are stable under the conditions in which they are used, one could try to make them under equilibrium conditions provided that the temperatures required are not excessively high. Examples of solids made under near-equilibrium conditions are the high-critical-temperature superconductors and glass ceramics. However, many materials are not intended for use in their equilibrium state, and generally the rule for making them is: determine the reaction kinetics and make use of them. Improve the necessary transport rates, which are often low in solids. Soft chemistry and diamond synthesis at low pressures and temperatures are examples of making inorganic nonequilibrium structures by using kinetics.[11] These are discussed more fully below.

## 1.4. Chemical Change

Materials chemists have to control the reactions when making materials. Their basic problem revolves around when and how much to increase reaction rates and when to slow reactions or prevent them from taking place at all. For this they need a thorough appreciation of what makes chemical reactions go. It is sometimes stated in introductory textbooks on chemistry that one or both of two conditions are necessary for chemical changes to occur: during a reaction the energy of the system decreases or the entropy increases, or both.[12] One should realize that changes in both enthalpy and entropy contribute to the *driving force* for a reaction. However, while a certain thermodynamic driving force is necessary for chemical reactions to occur, the more fundamental factor is the possibility of atomic transport and of reaction between the colliding atoms. The reactants must have a sufficiently high mobility and a low enough activation energy for the reaction. Both of these conditions appear to be so obvious that it is tacitly assumed that they always hold in chemistry: for chemical reactions atoms should be able to move and to react. For two reasons this is not as evident as is usually assumed:

1. It is the differences in mobilities or in reactivity rather than the differences in driving forces that determine the outcome of a chemical process. Resistance against reaction or diffusion not only determines the reaction rates of intermediate steps but also affects the composition and the morphology of the solid product. This is not generally recognized and too often the strength of the driving forces is taken to be the sole determinant of reactions. Actually it has been shown that when driving forces for reactions are very strong, they do not directly determine the process rates at all.[13] Moreover, the end product of a reaction is not always the thermodynamically lowest level of the mixture of elements.
2. It may happen that we do not want the atoms to move or to react. For example, if we need passive materials we want no chemical reactivity. Removing the driving forces for the reaction does not work: very stable complexes are known that can exchange their ligands very fast and noble metals are known to be seriously modified in certain atmospheres in spite of their stability. In other words, just because a compound has a high stability does not necessarily mean that it does not react. Passivation is realized by chemically fixing the atoms (raising the activation barrier for reactions) and taking away their mobility.

Apart from the kinetic requirements of high atomic mobility and low activation barriers, chemical processes also need favorable driving forces. In thermodynamic terms this means that the reaction occurs in the direction of increasing total entropy of the system plus environment or a lower Gibbs free energy of the system itself.[14] The strength of the driving forces for a reaction can be estimated from experimentally determined thermodynamic properties of the reactants and products, the state energies of these can be estimated from calculations using bonding models.

Sufficiently high driving forces can be achieved easily enough by choosing the right conditions or reactants, and in most cases it is not the driving force that becomes a bottleneck in synthesis. If a reaction is thermodynamically "impossible" because the driving force is in the reverse direction ($\Delta G > 0$), activation methods such as the introduction of plasmas can often be found to make the same products from the same reactants as in the thermodynamically impossible processes. In the fabrication of materials thermodynamics become a constraint only in cases where equilibrium compounds are being made. Phase diagrams illustrate this: they generally show the existence of stable compounds under equilibrium conditions because they are made up of data collected after so long a wait that the system is in equilibrium. Occasionally, however metastable phases are also noted in phase diagrams but only if they are readily found and exist for some time. The alumina–silica phase diagram shows only mullite as a possible compound and not others, such as zeolites, that are present but unstable.

Chemical kinetics describe the dependence of the reaction rates on the reaction conditions, i.e., on temperature, pressure, power intensity for activation, and partial concentrations of the reactants. Reaction rates can be measured *in situ* during synthesis e.g., by thermogravimetry or by optically monitoring the amount of product formed during the reaction. Kinetic data are essential for designing production equipment and for controlling the nature of the process product, such as its morphology. Conversely, the morphology of a solid product indicates which reactions have contributed to its formation. A scanning electron micrograph (SEM) of a polycrystalline reaction product shows which growth mechanisms have been dominant during synthesis as clearly as an image of mineral geological deposits shows the processes that have occurred during formation of the sediments.

The relation between driving forces and reaction resistances is schematized in Figure 1.14. All reactions in materials synthesis are complex with many intermediate steps, i.e., reactions between intermediate products and diffusion processes. These steps have their own rate constants, which can have a wide range of values. The overall reaction rate that is measured is determined by the rate-determining step. The rate-determining step in a set of sequential reaction steps is the reaction that has the highest chemical resistance; the steps before or after the rate-determining step with much higher rate constants do not affect the kinetics. For a set of parallel steps the reaction that is easiest is the rate-determining step, and the more sluggish parallel steps have no influence on the overall rate.

This is analogous to the electrical current in a set of resistors subject to a voltage. The lowest resistance among a number of parallel resistors takes the largest current and determines the overall circuit resistance if the others have much higher resistances. If the resistors are in series, the highest resistance in the set is the dominant one and determines the overall resistance if the others are low in comparison. In a set of intermediate reaction steps the rate coefficient $k$ varies with

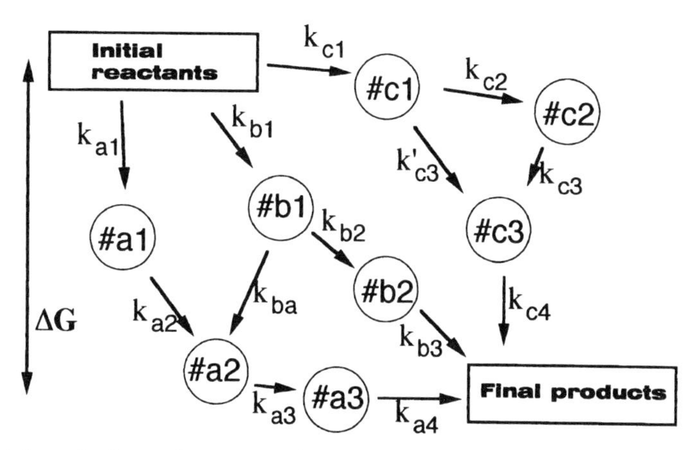


**Figure 1.14.** Schematic of steps that are part of an overall reaction from initial reactants to final products via intermediate species. # indicates a reaction intermediate and the different $k$-values indicate that there are many resistances for processes in parallel and in series. A chemical reaction is analogous to a network of resistors in circuits: a chemical conversion rate is comparable to a current. The overall kinetics are determined by the lowest $k$-value in the reaction series that has the highest rate of production of end products compared to other reaction series running in parallel. The scheme presupposes a small driving force $\Delta G$.

the reciprocal reaction "resistance" $R$, the chemical mass conversion rate $\phi$ as "current" $i$, and the driving force $\Delta G$ as the "voltage" $V$. For the reaction, the chemical analogy of Ohm's law would read $\phi = \Delta G/k$ ($i = V/R$).

The analogy between electrical circuits and sets of reactions (that occur near equilibrium) is useful to single out very sluggish or easy reactions as rate-determining in series or parallel sets of reactions steps. However, there is an essential difference between chemical kinetics and Ohm's law: In electrical circuits the current depends linearly on the voltage if the resistance does not depend on current and potential difference, which is usually the case. In kinetics, the reaction rate coefficient (the reverse resistivity against conversion) is not constant: if the potential difference is changed the resistance does not as a rule remain constant. The activation energy and with it the reaction rate coefficient (reciprocal resistance) then also changes. In that case the "current" is not proportional to the chemical potential as it would be in electrical circuits.

It is well known that if a reaction has a negative $\Delta G$ and is therefore thermodynamically allowed, it does not take place if the reactants are inert (high activation energy). In practice, however, this is sometimes ignored when chemists attempt to increase or decrease the reaction rates by increasing or decreasing the driving forces. Sometimes this works and sometimes it does not. The overwhelming effect of kinetics is dramatically illustrated by the case of oxidation reactions of aluminum. The Ellingham diagram for oxides, which is reproduced in Section 8.1, plots the Gibbs free energy against temperature. This diagram can be used to calculate the partial gas pressure oxygen would have when in equilibrium with both the metal and the oxide simultancously at each temperature. It also shows which metals are thermodynamically able to reduce other oxides (take over the oxygen and liberate the other metal from its oxide). The less noble the metal, the further down on the plot is the Gibbs energy of its oxidation reaction. Gold and platinum have lines above $\Delta G = 0$.

According to the Ellingham diagram the affinity of aluminum for oxygen is as high as that of lithium and much higher than that of sodium. The driving force for the formation of alumina is in fact high enough to make explosives out of aluminum and water for use in ammunition (there is a patent on that). If thermodynamics determined corrosion, and would prevail over kinetics, it would be as senseless to make objects that are going to be used in water or air out of aluminum as to make them of lithium. Ships, cars, and aeroplanes would then have to be made out of gold or porcelain, not out of the metals that are actually used.

We know that corrosion is a kinetic phenomenon. Aluminum is a fine material in water and in oxygen thanks to its corrosion kinetics and notwithstanding its instability; and lead, a material that is known to be inert in oxygen, is pyrophoric in air when finely divided. Thus when one is asked to develop a material that is corrosion-resistant, it is useless to consult the Ellingham diagrams, to look for low solubilities, to find the redox potentials, or to consult other thermodynamic data. Kinetics is the key to processes, and this is true not only for corrosion, but also for other chemical reactions (except those near equilibrium). Salient examples in chemical engineering are crystal growth and catalysis.

## 1.5. Relations to Other Monodisciplines

Inorganic materials chemistry consists of different subjects and it has a considerable overlap with other disciplines, such as chemical technology, materials science, ceramics, solid state physics, physical chemistry, chemical physics, analytical chem-

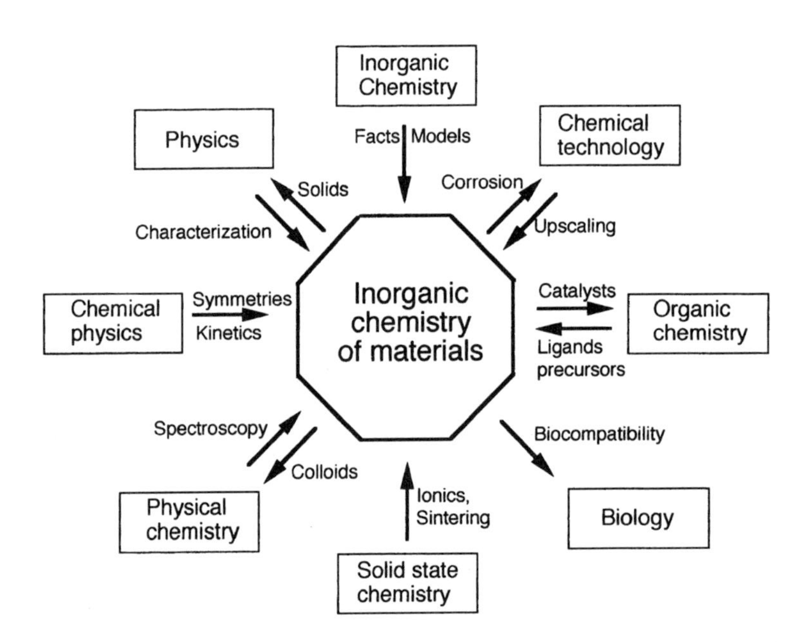


**Figure 1.15.** The relationships between the inorganic chemistry of materials and other sciences.

istry, biochemistry, and organic chemistry. How inorganic materials chemistry relates to these areas and in what sense the materials designer can make use of them will be sketched here. Figure 1.15 summarizes the related fields. Specialists in these related subjects have all developed their own habits, concepts, idiosyncrasies, and rituals, and when borrowing tricks from others one must take this into account. What works in one context may not work in the same way in another. Materials chemists in particular have to deal with colleagues from very different spheres.

### *Physics*

Physics in this context deals with generalities in heat, light, sound, forces, and electric currents. It is a science that reduces problems by simplifying them and removing so many features that are called nonessentials that only one cause is left over and that cause is then found and described. This procedure is both its strength and its weakness. Physicists have raised the standards of measuring properties to olympic levels and are often imitated for that. Anyone thinking of applying experimental physical techniques to measure properties of materials should not hesitate to call in a physicist.

On the other hand, extreme physical approximations are often taken too seriously by outsiders, who try to emulate the physicist's approach even in areas where it is not applicable. Simplifications still pervade all the sciences, even those dealing with complex systems, such as chemistry, biology, materials, and meterorology. However, certain practices that are successful in physics are not transferable to disciplines that cannot simplify their subject because it is too complex. Examples of such methods are quantum-mechanical band calculations of materials more complicated than single crystals, simulations in biophysics, predicting the weather with the Navier–Stokes equations, and applying the Fokker–Planck equation to nonidealized real systems. For the materials technologist who is hunting for models and facts in the sciences it is perhaps instructive to recall two recent examples; the existence of icosahedral phases in intermetallic aluminum compounds and high-temperature superconductivity in mixed yttrium barium copper oxides.

Since the work of Haüy two centuries ago, every crystallographer knows that crystal lattices can never have a fivefold symmetry element. When recently Shechtman[15] actually synthesized them he got into trouble with his colleagues in physics but he also started a revolution in crystallography, which was by then as stationary as classical Latin.

The second example is also from solid state physics. For many decades top physicists did advanced work on superconductivity, but in spite of their sophisticated theories they were unable to make materials having a critical temperature for superconductivity in excess of 23 K. The recent successes in constructing high-critical temperature superconductors were not the result of theoretical prediction but of laboratory work, done by solid state chemists with the wrong guidelines.

In 1986 an experimental solid state chemist almost doubled and somewhat later more than tripled that critical temperature. This caused a revolution from which theoretical physicists have still not recovered. These examples show that the practical value of age-old habits in a discipline, even if based on very solid and convincing evidence, remains debatable.

## Chemical Physics

Chemical physics is done by physicists attacking chemical problems in the usual reductionist manner and is quite interesting for chemists. The use of symmetry is an example. Group representations as developed in the first half of this century are now virtually indispensable for chemical spectroscopy. The main source of inspiration for chemists in chemical physics at present is statistical mechanics as a basis for chemical kinetics.[16] Not many practitioners and materials technologists who need chemistry use it but exciting developments in statistical mechanics most likely will affect the matter technologist soon. Homogeneous nucleation, an important subject in materials synthesis, is still in a sorry state and is likely to benefit from these developments.

## Physical Chemistry

Physical chemistry is done by chemists who do chemistry as if it were physics. Generalities are established and simplicities introduced, and matter-specific behavior is not considered to be of primary interest. Physical chemistry is understandably popular and materials chemists hunting for useful techniques will notice that it has a lot to offer but they might find that not all of it is as advertised. At best its methods are valuable but remain semiquantitative.

Similar to inorganic chemistry, physical chemistry is not one single discipline but a collection of widely differing subjects including electrochemistry, colloid chemistry, spectroscopies (not the equipment but the chemical part), magnetic resonance, thermal analysis, kinetics, quantum chemistry, and thermodynamics.[14] In the recent past such subjects dominated academic chemistry: textbooks that appeared from the 1950s to the 1990s presented the subject in a top-down manner starting with the current model. These textbooks were spectroscopy-ridden, and tried the sort of approach that had made such a success of physics. Some of the work was impracticable but many of the experimental techniques that were developed by the physical chemists in these subjects are in general use now. This textbook has many of them. One which has been very successful in the inorganic chemistry of materials is colloid chemistry,[17–19] which is the basis of ceramic processing.

## Solid State Chemistry

Somehow the chemistry of the solid state has split off from the chemistry of the elements as organic chemistry did in the 19th century. Solid state chemistry is a fairly new discipline, which is concerned with atomic defects in solids and their action in chemical processes. As materials are atomic solids, solid state chemistry dominates materials chemistry.[20] Solid state chemistry is not yet generally found in curricula of academic or technological chemistry, so Chapter 5 provides an introduction to the subject. Its contribution to materials chemistry is shown in Chapters 6, 8, 9, and 10. Solid electrolytes of Nernst-type sensors and solid oxide fuel cells are ionic conductors owing to the behavior of ionic defects.

## Biology

The impact of materials technology on biological systems is through the development of materials that can be used in prostheses. When a prosthesis is put in place, the surface of the materials it is made of should, at the very least, not cause inflammation. The surface chemistry of many biomaterials is controlled with biocompatible coatings on substrate materials chosen for their attractive bulk properties such as lightness, strength, and toughness. Diamond coatings made with plasma-enhanced chemical vapor deposition are very suitable for imparting biocompatibility to strong ceramic composites and light metals.

### *Organic Chemistry*

Most synthetic reactions in chemical technology use solid inorganic reactants as heterogeneous catalysts. These are metals, oxides, or composites having carefully tailored, chemically active surfaces. Catalyst fabrication is a large market for inorganic chemists skilled in synthesis as long as vast quantities of simple organic chemicals are produced and sold. It is expected that biotechnology will eventually take over a considerable part of the market for organic chemicals, as organisms have excellent organic-synthetic capabilities. This may lower the demand for inorganic catalysts in the future but so far products from catalyzed reactions and those made biologically are quite different.

The contribution of organic chemistry to inorganic materials chemistry is in ligands of coordination compounds, functional groups in inorganic polymers, and ligands in volatile organometallic compounds. Organometallic and coordination compounds are used as precursors (also called reactants or synthons) in synthesis of inorganic materials.

### *Chemical Technology*

Standard chemical technology is in large measure applied continuum physics of gases and liquids. The chemical engineer controls transport phenomena in processes for making large quantities of bulk chemicals for as low a price as possible. Not much chemistry is involved. For the materials chemist who scales up fabrication processes and uses processes that operate in transport-limited regimes, unit operations from chemical engineering may provide useful methods. Such techniques for materials can be found in treatises about materials production and mining.[1,21–23] A new constraint in chemical technology is the cost of the environmental load. The fabrication techniques used in industry gradually change under societal pressure. Thus electrogalvanic processes for making inorganic coatings are being phased out and replaced by cleaner methods.

### *Inorganic Chemistry*

Inorganic chemistry has several distinct parts: coordination chemistry, main group chemistry, borane chemistry, structural chemistry, cluster chemistry, and several other specialties. All of these subjects lead a life independent of the others,[24] but bonding discussions figure largely throughout the discipline. It might be thought that a thorough training in academic inorganic chemistry is the best chemical basis for materials science; i.e., one should study the elements if one wants to use them. However, inorganic chemistry is a science, an academic discipline. Its objective is therefore to provide a feeling of understanding, irrespective of any pragmatic spin-offs. The methods used and the results obtained in academic chemistry are not automatically of immediate use for the materials technologist. The parts that are have been collected here. However, it must be remembered that what is extensively studied in science but without practical value now may be applicable later and vice versa. Concepts in the sciences are subject to fashion!

### Obtaining the Information One Is After

The inorganic chemistry of materials is not an established subject but a collection of diverse activities. Its practitioners have to build their subject by collecting information from many sources. In doing so they have to take account of the characteristics of these sources. Many have been mentioned in the main text. Some others follow.

*Not Publishing Negative Results*

There is a general tendency not to publish unwelcome (called negative) results of experiments. As a result, many individuals duplicate the same fruitless experiments and of course, in turn, do not publish their efforts because of negative results. What publications leave out can, however, be as relevant as what they do include: "negative" results provide information. The information available in the patent literature is also intentionally restricted because one of the functions of patent description is concealment. Very often the details given are insufficient for repeating the process described. Patents are also interesting for what they do not include.

*Getting Information Fast*

Amassing new information by collecting it from the literature takes some time, but getting it by talking with those who know takes much less. People are prepared to verbalize what they are not ready to publish. By talking with colleagues in other institutes, one can collect information faster than from reports and papers, and discussions are a good way to become aware of negative results if there are any. Conferences are excellent occasions and are worth the expense.

*Use of Tables of Properties*

There are many tables that list properties of components, and the entries are compounds that are identified by their chemical composition. A compound is well characterized by its chemical formula only in the case of molecules, and so such tables of molecules are therefore unambiguous. Solids, however, are different. As the properties of solids depend on the number and nature of their defects and on their morphology, imposed during synthesis, properties of solids depend on the maker and are more variable than those of molecules. Extrinsic features are almost never mentioned in the tables. One therefore is led to expect that solids labeled with a formula (its chemical composition) have certain precise properties. However, tables of properties of solids in which the compounds are identified by their chemical composition and crystal structure alone are incomplete and must be used with caution.

## 1.6. The Use of Models

Models, or in other words, theories in science, and their use and relevance do not constitute a subject that technologists worry about. I mentioned them here only because when consulted, scientists tend to explain their data in terms of the current theory or model. They prefer to deal with generalities rather than matter-specific behavior and they mostly ignore facts that do not fit (called anomalous) within the context of the current model. For an appreciation of the information that can be obtained from academic chemists, one has to realize what it means. In the following the word "model" will be used instead of "theory" because it is less heavy.

Models are necessary not only because they systematize facts but also because they enable one to see those facts in the first place. According to most methodologists, without a theoretical frame there are no facts or at least they cannot be recognized as such. Others do not agree and point out the large contribution of serendipity[25] to developments in science, at least in the complex disciplines such as chemistry. The many facts that have been uncovered by accident were observed outside a theoretical context or at most within a very modest one. A recent book on serendipity[26] lists 36 item groups. Only 4 of them are in physics, while 15 are in chemistry, and 9 are in materials. Methodologists up to now apparently have been strongly influenced by physics.

Apart from allowing us to see and understand facts, models also hide facts that do not fit. As models are approximations, there are inevitably facts that they do not account for.[27] In the sciences in which models overshadow facts, if the latter are anomalous they are not a serious problem and can be disregarded with impunity. If challenged, the answer is usually: "that is not in the model." Unlike scientists, technologists have not merely to understand something, they have to make it, so for them facts have more significance than models. However, models remain helpful if not absolutely necessary and for lack of user-friendly models, technologists often have to accept academic ones unchanged from scientific chemistry.

A recent Japanese report on the development of technology claims that increased attention to basic science would bring a higher return on investment in research and a disproportionately higher innovation rate in technology. However, there does not seem to be much evidence for that. In general, creativity is not easily associated with fundamentalism and myopia is not much less rampant in basic science than in technology. Familarity with chemical facts and experience in dealing with them are more useful for dealing with problems concerning the behavior of materials than mastery of any fashionable fundamental theory that is currently successful in science.

There has always been a tendency to overestimate the value of scientific theory for technology. It is often said that the great advances in technology have a scientific source, that they are made possible by scientific work and developments of scientific models by theorists: electronic chips are said to be unthinkable without a quantum theory of solids. This is putting the cart before the horse. In actual fact we do things first and after that try to understand them through science. Theories are fed by experience and models come after facts, not the other way around (or much less so). We actually had solid state rectifiers in radios before solid state electronics, we also flew before aerodynamics, we rode bicycles before solving the relevant Lagrange equations, and we converted solar energy before understanding photosynthesis. The Scots made locomotives before Carnot analyzed steam engines, and superconductivity was not predicted by Bardeen, Cooper, and Schrieffer, nor was their theory very helpful in increasing the critical temperature. There have been some minor exceptions, unrelated to materials: e.g., elementary particles were found after they were predicted theoretically.

Models are works of fiction, not truths or images of real processes. They are at most dispensable prototruths and acceptable only if they point the way to new facts to be established experimentally.[28] An understandable wish for certainty tends to ignore this, as can be witnessed in heated debates.[29] Scientific models such as quantum chemistry and equilibrium thermodynamics can sometimes "explain" what has been observed, but as tools for prediction in technology such models have as much use as astrology: sometimes they work and sometimes they don't.

Many simplifying models developed in physics and physical chemistry have been very successful. They simplify the system by transforming strongly interacting units into separate weakly interacting ones. Not too complicated systems thereby become simple collections of independent units and are calculable. Examples are:

1. Normal vibrations are constructs that make sense of infrared spectra. Atoms in lattices are strongly bound to each other and their vibrations are strongly coupled to those of their neighbors. The atomic displacements along the three axes cannot be isolated from each other. However, if the displacements are carefully combined into groups to form "normal vibrations" then these are to a good approximation independent of one another.
2. Quasi-electrons are electrons in solids with properties other than those in vacuum. In a condensed phase electrons interact with each other and with the nuclei. In order to deal with them as independent free units their interactions with the environment must be eliminated (or at least substantially reduced). This is done by transforming the interactions into novel particle properties. For example, when the motion of an electron in a solid is impeded by atoms the electron seems to be heavier, and as a result the electron is transformed into a free quasi-electron with a higher mass. The properties have changed under the transformation from those of free electrons in vacuum to those of quasi-electrons in a solid and so have their identities.
3. In a hydrogen atom in free space the nucleus is not stationary and both the proton and the electron classically rotate around a center of gravity. The dynamics of this system can be represented by a stationary proton and an electron moving around it if the mass of the electron is transformed to a value dependent on the mass of the moving nucleus such that the center of gravity is at the proton. The lower "reduced" electron mass permits considering the nucleus as being fixed, which now allows a derivation of the orbital of the transformed electron in the field of the proton with the simplified Schrödinger equation.
4. In ligand field theory, as summarized in Section 2.4, the valence electrons of a metal ion are considered to be occupying orbitals of free metal ions although there is bonding between the metal ion and the ligands in a coordination complex. The interaction of the valence electrons with the ligands is again transformed but now into a "quenching factor," the value of which is determined from measurements. This quenching factor is used to modify the behavior of the valence electrons in the complex while they can still be considered to reside on free metal ions. This simplifies the calculation as it allows one to ignore the rather strong interaction between the valence electrons "on the metal ion" with the ligands.

5. In quantum chemistry, the complete neglect of differential overlap (CNDO) method is an example of eliminating the interactions between atoms by transformation into properties of atoms. It assumes for convenience in the calculation that there is no interaction between the orbitals of neighboring atoms that are bound together. This can be done only if the valence orbitals on the atoms are transformed into so-called "natural" orbitals. This transformation mixes a number of neighboring orbitals into the original atomic orbitals that depends on the strength of the bond (in fact the orbitals overlap). The new set of nonoverlapping atomic orbitals now simplifies calculation.

All models are simplifications but some exaggerate in allowing only one cause for a process. Monocausality has a venerable history. The principle first formulated by Occam in 1360 says, in its most radical form, that if one has found one acceptable reason for something, that suffices and the matter should not be made more complicated by looking for a less simple explanation. This version of Occam's razor is often used as a guide for setting up models without a critical appreciation of what is sufficient. Such models are sometimes less than optimal as there is rarely only one determining factor.

This leads us finally to the question of complexity. It is often difficult to assign one cause or a few causes to complex phenomena. Processes in the realm of physics are simple enough or they are chosen to be simple enough for the reductionist approach. Chemistry is not simple and reductionistic models are not used easily. In chemical processes recognized as being really complex, causality loses its meaning and there are only conditions, as in weather. The rare chemical processes that are so simple that causes can be identified can be considered to have what is known as *stable attractors* in the complex chemical process. In those cases causes can be invented and deterministic models developed, which may be of some use. However, at the boundaries of the condition parameter area for the stable attractor the model stops being useful. Such chemical processes suffer from critical phenomena, hysteresis, nonlinearity, nonreproducibilities, and oscillations in reaction rates and the original causality is lost completely.

As models are simplifications, sticking to a single model has a disadvantage: it restricts the possibilities for finding original solutions because one usually sees only what the model says should be there. Using different and even conflicting models for tackling the same problem helps in finding original solutions in design problems. A model puts one, mentally speaking, into a groove, and lateral thinking requires many intersecting grooves going off in different directions.

Models are used to summarize (or reduce) measurement data and enable a prediction of what can be expected from the system under given conditions. A simulation generally involves letting a computer calculate results that a process would have if it proceeded according to an idealized simple mechanism formulated in a simple mathematical expression. Simulations are alleged to supply the same amount of information as experiments (or even more). Most simulation programs digitize analytical formulas from continuum mathematics, such as partial differential equations. Such programs do not fully use the specific mathematical capabilities of digital machines. Simulations do produce intermediate results in many cases in which such results are not easily found experimentally. Results from simulations are being used increasingly to support models. Simulations are defended by invoking their cost-effectiveness: expensive experiments are claimed not to be necessary if effects can be calculated instead of measured. Others point out that calculations return only what has been put in and can never totally replace experiments. Computers are still slow and software generates much unnecessary information. There is also some contradiction in having ever faster computers (ending up in molecular ones) for simulating atomic and molecular processes: letting the atoms do the job (i.e., doing experiments) would produce results much faster. Simulations might be expected to point the way to novel syntheses, but they have rarely done so in inorganic chemistry.

Simulation of chemical processes is very popular with chemical engineers. It has led to a virtual reality of its own, separate from the world of facts. Many of the papers in chemical engineering are not about experiments but about models, but the jargon is similar and they can only be distinguished because publications that report measured facts have error bars in the graphs. Models have much more authority than hard measurements, not only in virtual reality but also with technologists who have to perform in the real world. In spite of the often voiced claim to the contrary, the actual experiments are essential and the need for them is not precluded by simulations. Doing experiments to check simulations by means of a model is called validation. Remarkably enough, reported validations are always positive: the observed data are invariably consistent with the model. The reason is that discrepancies between model and observations that might occur are not very publishable. Therefore the model that is used is adjusted to get a better fit. Rarely is the model totally replaced by another that is basically different.

Simulations are not restricted to solving differential equations by digitizing them. Any dynamical process can be simulated with simple *cellular automata*, which are algorithms used to simulate complex processes, a method that takes full advantage of the characteristics of the kind of digital data processing that has now become possible. Cellular automata do what continuum models cannot do: they describe observed morphologies successfully, as shown in Chapter 7, so they can also be used to interpret observed microstructures.

## Exercises

1. What are the characteristics of the five types of matter and to what types of materials do they apply?
2. What are the six application areas for the inorganic chemistry of materials?
3. The five different types of matter (ionic, covalent, metallic, hydrogen bonded, and compounds with van der Waals–London intermolecular bonds) are used in various technological fields. Give some of the uses of each of them in crafts such as electronics, optics, catalysis, matter processing, corrosion protection, mechanical engineering (tribology), energy conversion, and communication technology.
4. Choose two elements at random from any group of the periodic table and find their similarities and differences, and describe how they are unique using textbooks such as Greenwood and Earnshaw.[3] Repeat this for the other groups using two elements taken at random from each.
5. Explain the observed periodicities in the periodic table with the electron occupancies of the atomic shells.
6. How does the bond between atoms affect the properties of materials?

7. Name the effect that the type of bond between atoms has on the structure of the compound.
8. Describe the consequences of interatomic bonding for the properties of materials.
9. Describe the consequences of interatomic bonding for the fabrication of materials.
10. How could the method by which materials are synthesized affect their properties?
11. Discuss the role of driving forces and reaction resistances in chemical processes.
12. How are driving forces and chemical resistances for reactions measured?
13. Name three criteria for judging the suitability of scientific models for practical materials chemistry.
14. Which models are effectively used in materials synthesis?
15. Find and describe some models that are popular in academic chemistry but useless for materials chemistry.
16. Describe five other cases of the use of models outside their range of validity.
17. What is the purpose of simulation in applied chemistry?
18. How does the morphology affect the properties of materials?
19. What subjects does inorganic materials chemistry have in common with disciplines such as physics, physical chemistry, and materials science?
20. Sketch the reason for the use of hybrid composites and coatings.
21. Give some examples of molecular compounds that have different properties but the same elemental composition.
22. List a few rules on how to make metastable structures.

## References


1. J. E. Fergusson. *Inorganic Chemistry and the Earth: Chemical Resources, Their Extraction, Use and Environmental Impact.* Pergamon, Oxford (1982).
2. D. Jones. *Nature* **360**, 206 (1996).
3. N. N. Greenwood and A. Earnshaw. *Chemistry of the Elements.* Pergamon, Oxford (1984).
4. C. N. R. Rao (ed). *Chemistry of Advanced Materials.* Blackwell, Oxford (1993).
5. J. J. McKetta (ed.) *Inorganic Chemicals Handbook. Vols. 1 and 2.* Marcel Dekker, New York (1993).
6. Pogge, H. B. (ed.). *Electronic Materials Chemistry.* Marcel Dekker, New York (1995).
7. C. N. R. Rao and J. Gopalakrishnan. *New Directions in Solid State Chemistry.* Cambridge University Press (1986).
8. T. P. Fehlner (ed). *Inorganometallic Chemistry.* Plenum, New York (1992).
9. R. T. Sanderson. *Chemical Periodicity.* Chapman and Hall, London (1961).
10. K. F. Purcell and J. C. Kotz. *Inorganic Chemistry.* Holt Saunders, Philadelphia (1977).
11. J. M. Rouxel, M. Tournoux, and R. Brec. *Soft Chemistry Routes to New Materials,* Chimie Douce. Trans Tech, Aedermannsdorf, Switzerland (1995).
12. G. Wulfsberg. *Principles of Descriptive Inorganic Chemistry.* Brooks/Cole, Monterey (1987).
13. P. Glansdorff and I. Prigogine. *Thermodynamic Theory of Structure, Stability and Fluctuations.* Wiley, Interscience, New York (1977).
14. P. W. Atkins. *Physical Chemistry.* Oxford University Press, Oxford (1982).
15. D. Shechtman and C. I. Lang. Quasiperiodic materials: Discovery and recent developments. *Mater. Res. Soc. Bull.* November 1997, p. 46.
16. N. G. van Kampen. *Stochastic Processes in Physics and Chemistry.* North Holland, Amsterdam (1981).
17. B. I. Lee and E. J. A. Pope. *Chemical Processing of Ceramics.* Marcel Dekker, New York (1994).
18. C. J. Brinker and G. W. Scherer. *Sol-Gel Science: The Physics and Chemistry of Sol-Gel Processing.* Academic Press, Boston (1990).
19. J. T. G. Overbeek. How colloid stability affects the behavior of suspensions. *J. Mater. Educ.* **7**, 393 (1985).
20. L. Smart and E. Moore. *Solid State Chemistry: An Introduction.* Chapman and Hall, London (1995).
21. P. A. Vesilind and A. E. Rimer. *Unit Operations in Resource Recovery.* Prentice-Hall, Englewood Cliffs, N.J. (1981).

22. W. Buechner, R. Schliebs, G. Winter, and K. H. Buechel. *Industrial Inorganic Chemistry*. VCH, Weinheim (1989).
23. T. A. Ring. *Fundamentals of Ceramic Powder Processing and Synthesis*. Academic, San Diego (1996).
24. J. C. Bailar, H. J. Emeleus, R. Nyholm, and A. F. Trotman-Dickinson. *Comprehensive Inorganic Chemistry*. Pergamon, Oxford (1973).
25. Pek van Andel. Serendipity: its origin, history, domains, traditions, patterns, appearances and programmability. In: M. P. C. Weijnen, A. A. H. Drinkenburg (eds), *Precision Process Technology. Perspectives for Pollution Prevention*. Kluwer, Dordrecht (1993).
26. R. M. Roberts. *Serendipity: Accidental Discoveries in Science*. Wiley, New York (1989).
27. J. J. Zuckerman. The coming renaissance of descriptive inorganic chemistry. *J. Chem. Ed.* **63**, 829 (1986).
28. N. Cartwright. *How the Laws of Physics Lie*. Clarendon, Oxford (1983).
29. N. Oreskes, K. Shrader-Frechette, and K. Belitz. Verification, validation, and confirmation of numerical models in the earth sciences, *Science* **263**, 641 (1994).

*Chapter 2*

# THE CHEMICAL BOND

*Measurements, then, are not the only place to look for a failure of the Schrödinger equation. Any successful preparation will do.*

*NANCY CARTWRIGHT, 1983*

## 2.1. Introduction

This chapter presents several empirical models for describing the different types of bonding. All of them assign bonding to the behavior of the valence electrons and all of them are consistent with the basic assumptions of quantum mechanics of electrons. Several of the models account for the whole of the bonding energy and others only for part of it or for the structure. There are also models for selected classes of compounds. Finally a draft for a generally valid empirical model is given that includes the others.

The bond between atoms is described by the curve that traces the way the energy changes as the atomic coordinates change. The potential of the interatomic attraction is the energy as a function of the normal vibration coordinate strain, which is usually taken to be the distance between the bound atoms (Figure 2.1). Many variants exist. A Morse curve is representative for the potential between covalently bonded atoms. If the bond is between ions the interatomic potential is well described by a Lennard-Jones potential with a Coulomb term.

The shapes of the interatomic potential curves are approximations chosen for mathematical convenience. Such potential functions are generally used in discussions on a variety of properties of molecules and lattices: optical absorption and luminescence, laser action, infrared spectroscopy, melting, thermal expansion coefficients, surface chemistry, shock wave processes, compressibility, hardness, physisorption and chemisorption rates, electrostriction, and piezoelectricity. The lattice energies and the vibration frequencies of ionic solids are well accounted for by such potentials. On heating, the atoms acquire a higher vibrational energy and an increasing vibrational amplitude until their amplitude is 10–15% of the interatomic distance, at which point the solid melts.

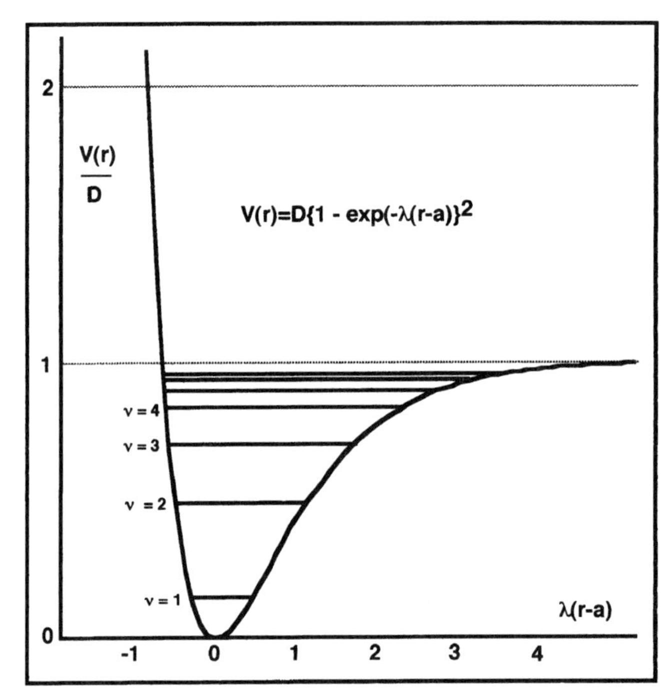


**Figure 2.1.** A graph of the Morse potential (in units of $D$), representative of a potential between two atoms that are linked by a covalent bond. The parameters of the potential are $\lambda$ and $D$. The interatomic equilibrium distance between the atoms equals $a$ and the bond energy is $D$. The lower stretch vibration levels are drawn in the curve as horizontal lines and labeled with vibration quantum numbers $v$.

The melting point of a solid and its response to shock waves depends on the form and depth of the interatomic potential. Its compressibility is simply related to the Born repulsion parameter. The mechanical hardness is a function of the bulk modulus (inverse compressibility), which in turn correlates strongly with the volume density of the chemical bonding energy, i.e., the bonding energy per unit volume: the smaller the atoms and the stronger the atomic bond, the harder the solid. In the box on p. 34, a simple relation is derived between the bulk modulus and the Born repulsion parameter.

To be technologically useful, bonding models must reliably predict the existence, structure, and properties of the compounds (molecular or solid) of particular elements using atomic characteristics. They should be so simple that neither supercomputers nor dedicated specialists are required for their use. None of the existing models given below meets these requirements but each of them is successful for a selected class of compounds or for certain aspects. Thus the molecular orbital (MO) model is good for most of the features noted but needs computers and is weak in properties. The Linnett model[1] predicts structure well in a simple way but is qualitative. The Miedema model[2] is restricted to metals and ignores structure and many properties, the Pearson scheme[3] is best for molecules, and the Johnson model[4] is for metals only and is also qualitative. However, all of them are useful in one sphere or another, and they can be combined in one model that meets all our requirements and describes the bonds in metals, covalent solids, and molecules. This composite model could be used to estimate stabilities and properties of compounds.

Two models based on charge transfer between atoms bonded to each other are good at predicting the presence of certain compounds: the Miedema model for intermetallic solids and Pearson's electronegativity model for molecules. Both models are empirical and limited to a restricted class of compounds; neither provides the entire bonding energy but allows calculation of extra bonding energy in excess of the corresponding homopolar bonding energies that are considered known.

The enthalpy of formation of intermetallic compounds, which is an important quantity for determining presence or stability, is well predicted by the Miedema model (Section 2.6). This model regards the metal atoms as the smallest entities that retain the properties of the macroscopic metal, a return to the ancient atomic model of Democritos. Atoms have the same surface tension, work function, and other electromechanical properties as the parent bulk metal inself. The extra enthalpy of the intermetallic bond is the result of charge transfer and is a simple function of experimentally determined parameters that are characteristic for the elements involved in the bonding. Using this model, one can quickly predict the existence of compounds of two given atoms with surprisingly high reliability.

The electrical, optical, and magnetic properties of metals can be well described using Johnson's interstitial electron model. The metal atoms on the lattice are considered to be partially ionized, and the valence electrons that are stripped off are then localized in the tetrahedral or octahedral interstitials of the stacked metal cations (Section 2.9). According to this model metals can be regarded as salts of metal cations and interstitial electrons as anions. These interstitial "anions" are highly mobile, and a metal is a good conductor. In the interstitial electron formulation metals are analogous to electrides, which are ionic compounds of complexed cations and free electrons as anions. Conversely, electrides can be considered metals with very large complex cations.

In the physics literature models that assume delocalized valence electrons are more common than those based on localized electrons such as described here. Examples of such models for solids are band models based on molecular orbitals (the so-called tight-binding method) or models that describe the electrons in metals as being in an almost flat potential (the nearly-free-electron model). While physicists traditionally think in terms of delocalized electrons, chemists feel more at ease with localized electrons, and almost all the bonding models discussed here are based on localized valence electrons. The dilemma between stationary eigenstates and molecular structure[5,6] is less problematic in these models than in those for symmetry-adapted delocalized electrons.

Molecular structures and lattice structures are sometimes rationalized using the the valence bond method and the Lewis model[7] with the octet rule. In these cases it is necessary to invoke concepts such as resonance or mesomerism. Such models are less convenient for technologists but are nevertheless summarized because they are so often used in a qualitative way to rationalize chemical bonding. The molecular orbital model is invaluable in assigning spectra.

**Interatomic Repulsion and Compressibility**

The potential curve that describes atomic attraction and repulsion has a form that depends on the kind of solid lattice that the atoms form. The lattice energy $U$ of ionic compounds is well described by the expression

$$U = C + B + W$$

where $C$ is the Coulomb attraction:

$$C = -\frac{MNZ_1Z_2e^2}{4\pi\varepsilon_0 r_e}$$

with $M$ the Madelung constant, $N$ the Avogadro number, $e$ the charge of an electron, $Z_i$ the number of unit charges on ion $i$, and $r_e$ the equilibrium distance between the ions; $B$ represents the Born repulsion. One model has $B = B_0 \exp(-r_e/r)$ and in another $B = B_0/r_e^n$ with $B_0$, $r$, and $n$ characteristic constants for the atoms; $W$ represents the van der Waals–London interaction, of which the largest term is attractive:

$$W = -C/r^6$$

The compressibility $K$ of an ionically bound solid depends on the parameter $n$ in the repulsion potential $B_0/r^n$, which can be seen as follows: The lattice energy $L$ as a function of the interatomic distance $r$ is

$$L(r) = \frac{NZ^2e^2M}{4\pi\varepsilon_0 r}\left(1 - \frac{1}{n}\right)$$

where $N$ and $M$ are as defined above and $Ze$ is the ion charge in units of $e$. Then

$$K = -\frac{1}{V}\frac{dV}{dp}$$

## 2.2. Valence Electrons on Atoms and Interaction with Light

This section describes a model for the behavior of valence electrons on a spherical, symmetric, free atom. The subject is of some use for the inorganic chemist. It adds a basic regularity to the periodic table and it is necessary for the assignment of spectroscopic absorption and emission bands of atoms and molecules. It is also the basis for discussions of electrons in chemical bonds.

### 2.2.1. Electrons in Atomic Shells

As discussed in many introductory chemistry textbooks,[8,9] electrons on atoms behave according to the rules of quantum mechanics. Their states are described by products of one-electron wavefunctions known as orbitals. One does not have to refer to atomic orbitals for a discussion of the chemical bond[10] but they are essential in systematic descriptions of atomic spectra. A qualitative understanding of optical depends on the value of $n$. Now,

$$dU = T\,dS - p\,dV$$

As

$$\left(\frac{dU}{dV}\right) = -p = -\left(\frac{dL}{dV}\right) \qquad (T = 0\,\mathrm{K})$$

it follows that

$$\frac{d^2L}{dV^2} = -\frac{1}{KV}$$

The molar volume is proportional to $r^3$: $V = ar^3N$. However, the lattice energy is given in the variable $r$ and not in the volume $V$. This can be expressed as

$$\frac{dL}{dV} = \frac{dr}{dV}\frac{dL}{dr} = \frac{1}{3ar^2}\frac{dL}{dr}$$

$$\frac{d^2L}{dV^2} = \frac{1}{3ar^2}\left(\frac{1}{3ar^2}\frac{d^2L}{dr^2} - \frac{2}{3ar^3}\frac{dL}{dr}\right)$$

and this gives the expression for the repulsion exponent in terms of observables:

$$n = 1 + \frac{36\pi V\varepsilon_0 r_0^4}{KNMZ_+Z_-e^2}$$

which holds for ionic compounds.

properties is important for the materials designer who is interested in color or involved in displays, light transducers, or information-transfer devices. Therefore, some attention is given here to a few key aspects of the interaction of atoms with light.

Electrons on atoms are described by an orbital scheme. Briefly, atoms have electrons in shells around the nucleus and each electron in an atom has a unique set of four quantum numbers. No two sets are the same. Electron orbitals are solutions of the Schrödinger equation for one electron in the field of a central charge, the (screened) nucleus. The electronic state of an atom that has more than one valence electron is a combination of products of such orbitals with the right symmetry. Mathematical details are not important here but three points should be noted because they are significant for the color of materials:

1. Orbitals have levels $\varepsilon_i$, which represent the energy necessary to remove the electron from the orbital to infinity at rest ($\varepsilon_i$ is equal to the $i$th ionization potential, which is the energy needed to remove an electron from an orbital at energy level $\varepsilon_i$).

2. The energies of the atomic states depend not only on the energies of the orbitals occupied by electrons but also on the interaction between the electrons on the atom. The interaction term shows incidentally why in the periodic table (Figure 1.2) there are irregularities in the orbital occupation numbers of the atoms.
3. Optical transitions occur between stationary states involving many electrons and cannot in general be described by the excitation of single electrons that change orbitals. The selection rules determine the intensities of the optical transitions and these rules are a simple consequence of the symmetries of the initial and final states.

Strongly bound electrons in low-lying atomic shells are called core electrons. Valence electrons are the least strongly bound electrons in an atom and are the determinants of the chemical behavior and most properties. The core electrons together with the nucleus create the fixed spherically symmetrical potential for the valence electrons.

The single-electron wavefunctions in a spherical Coulomb potential are the well-known hydrogen orbitals (*ns*, *np*, *nd*, *nf*) having energy levels for the hydrogen atom as shown in Figure 2.2. The optical spectrum of the hydrogen atom orbitals can be derived from these levels together with the selection rules. Atomic states of atoms that have more than one electron are described by the products of orbitals or, better than that, by Slater determinants (Slater determinants have the right permutation symmetry). The lowest-lying orbitals are occupied by electrons (two per orbital). A configuration of the occupation of available orbitals by electrons corresponds to one or more states. As long as the potential is spherically symmetrical, the orbitals have the same angular dependence as the orbitals of pure hydrogen (*s*, *p*, *d*, *f*) but a different radial component because the spherical potential for the electrons has no $1/r$ dependence as hydrogen has. Pseudopotentials in solids are good approximations of the potentials for valence electrons, which allow easier calculation than the sum of the Coulomb potential of the nucleus and the averaged

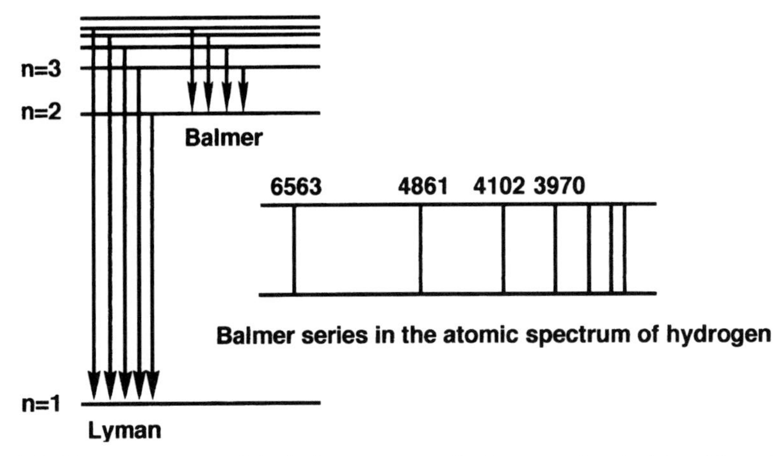


**Figure 2.2.** The energy levels of the hydrogen atom with the allowed optical transitions. The atomic emission spectrum has several series, one of which is shown.

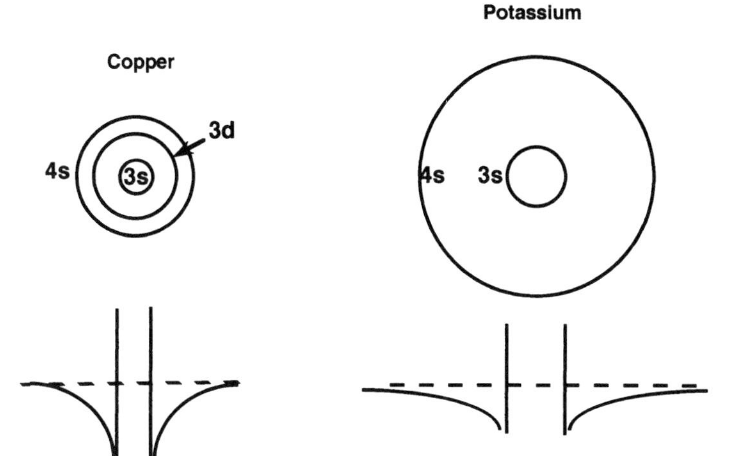


**Figure 2.3.** The pseudopotentials of two metals. The pseudopotential for copper is compact and deeper than the potential for potassium, which is shallow and extended. The ionic radii are 128 and 235 pm, respectively, and the ionization energies are 7.7. and 4.3 eV. The core electrons together with the nucleus form the potential for the valence electrons.

core electrons of the atoms. Two pseudopotentials for the valence electrons in copper and potassium are sketched in Figure 2.3. The copper pseudopotential is compact and deep, while that of potassium is extended and shallow, and these shapes affect the calculated valence electron distribution.

Potentials in atoms are complicated as are the resulting wavefunctions of the valence electrons. However, these orbitals have some features in common with those of electrons in some simple potentials. The orbitals in a cubic flat-box potential (the potential is considered to be constant over the entire volume of the box) turn out to be sine functions if the Schrödinger equation is solved, like the eigenfunctions of a vibrating string. Each different potential imposes a different form on the wavefunctions, which are similar in shape to the sine functions of the flat-bottomed box with nodal planes that increase in number with the quantum number $n$.

The wavefunctions of particles at different potentials or in differently shaped boxes differ. The orbitals of a particle in a planar circular box are Bessel functions, while the orbitals in a spherically symmetrical box have spherical harmonics that are characterized by three quantum numbers. The well-known hydrogen orbitals, which have an angular part and a radial part with quantum numbers $n$, $l$, and $m_l$, are examples of wavefunctions of electrons in the spherically symmetric Coulomb potential of the proton (considered immobile).

The arrangement of the elements in the periodic table follows the occupation with electrons of the lowest-lying orbitals. The periodicity of the orbitals, expressed in their quantum numbers $n$ and $l$, is related to the periodicity of the properties of the atoms. Shells with a higher $n$ allow more subshells with quantum number $l$, which can have a value up to $n - 1$ and each orbital can only be occupied by two electrons. The higher the nuclear charge of the atom, the more electrons in the shells. When the nuclear charge is increased shells gradually fill with electrons from the lowest orbitals up. This is known as the *Aufbau* principle. At the beginning of the first row of the transition metals and in the middle of the row the sequence of

### The Schrödinger Equation for a Particle in a One-Dimensional Box

The Schrödinger equation for a particle in a one-dimensional box is $H\psi = E\psi$, where $H = -\hbar^2/(2m)\nabla^2 + V(x)$ with $V(x)$ the potential. The wavefunction of the particle is $\psi(x)$. This can be rewritten as

$$d^2\psi/dx^2 + \frac{2m}{\hbar^2}(E - V)\psi = 0$$

which has the solutions $\psi_k(x) = \sin kx$ for special values of $E$: $E_k = \hbar^2k^2/2m$. The parameter $k$ depends on the boundary conditions. It is a continuum for free particles and has discrete values for bound particles.

For a particle in a box of length $L$ and a flat bottom, the potential in the box is $V(x) = 0$ and if the walls of the box are hard, $\psi(0) = \psi(L) = 0$. From these boundary conditions it follows that $kL = n\pi$ and $n = 1, 2, 3, \ldots$. There are $n - 1$ nodes in the wavefunctions $\psi_n(x)$ of a bound particle in such a box.

The energy $E_n = n^2h^2/8mL^2$ of the particle in state $\psi_v$ depends strongly on $L$ of the box that contains it. Although electrons in atoms are not in flat-bottomed boxes this discussion provides a useful metaphor by showing the effect of the available space on the kinetic energy ($E_k$ is purely kinetic in this case) of valence electrons.

### The Energy of States in the Term Scheme

Different distributions of electrons over orbitals correspond to different terms. The orbital energies $\varepsilon_i$ correspond to the ionization energies of electrons in orbital $i$. The energy $E$ of one of the many-electron states of an electron configuration is not merely the sum of the energies $\varepsilon_i$ of orbitals that are occupied by electrons; there is an additional interaction term $G$:

$$E = 2\sum \varepsilon_i - G \quad \text{where} \quad G = \sum\sum(2J_{ij} - K_{ij}),$$

with $J_{ij}$ and $K_{ij}$ being the Coulomb and exchange integral, respectively. To a good approximation the $\varepsilon_i$ values are equal to the ionization energies of the electrons in the orbitals $i$. The value of $G$ changes with orbital occupation and the energy of a state does not depend only on $\varepsilon_i$ but on $G$ as well.

electron occupation deviates from the normal. The energy of many-electron state is not merely the sum of the $\varepsilon_i$ values of the orbitals that are occupied by electrons. It also includes the interaction $G$ between the electrons, the strength of which depends on the orbital: orbitals that have a significant overlap (each partly occupies the same space as the other) have a strong interaction. Although the $3d$-orbitals have a lower energy than the $4s$-orbitals (Figure 2.4), in potassium the $4s$-orbital is the one occupied rather than the lower-lying $3d$. As explained below, the reason for this anomaly is the contribution of the interaction $G$ of the core electrons with the $4s$ and $3d$ valence electrons to the ground state energy.[11] The contribution to the total energy of the state can be larger than that of the orbital energies.

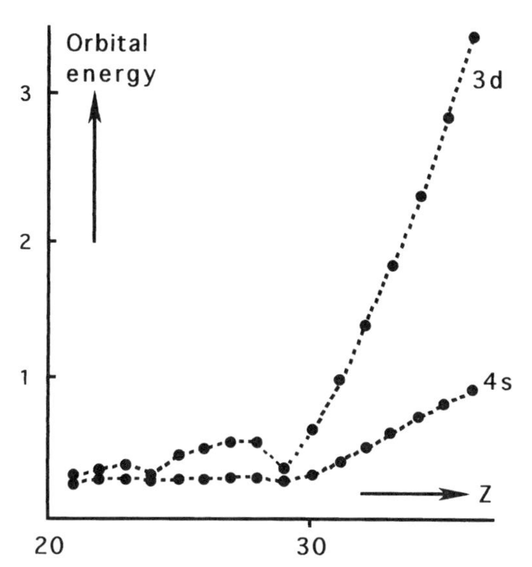


**Figure 2.4.** The energies (in atomic units) of the 3*d*- and 4*s*-orbitals of the elements of the first transition series.[11] The 3*d*-orbital has in all cases a higher energy than the 4*s*-orbital. Reprinted with permission from the *Journal of Chemical Education*, Vol. 55, 1978, pp. 2–6. Copyright © 1978, Division of Chemical Education, Inc.

## 2.2.2. Interaction with Light

Adsorption and emission of light by atoms corresponds to transitions between ground and excited states. There are two types of energy schematics: orbital levels and term levels (which indicate many-electron states). Orbital energy levels represent one-electron states, labeled by lower case letters *s*, *p*, *d*, *f*, etc. Orbital schemes show orbital energies and indicate the electron occupation of the orbitals using arrows for the spin quantum number. Term levels on the other hand are indicated with capital letters (*S*, *P*, *D*, *F*, *G*, etc in the free atom) and give the energy levels of the many-electron states of the atom or ion, the ground state, and the excited states. Term level schemes do not provide information regarding electron occupation of orbitals but are necessary for identifying spectral transitions. Orbital energy levels are unsuitable for assigning bands in spectra.

The states or terms for atoms that have spherical symmetry are labeled with the capital $\Gamma$ with superscripts for the spin and subscripts for the angular momentum: ${}^{2S+1}\Gamma_J$. There is a great deal of information in this symbol: $S$ is the total spin quantum number of the state and $2S + 1$ is the spin multiplicity ($S = 1$ is a triplet state). $\Gamma$ is a symbol that gives the total orbital quantum number $L$ ($\Gamma$ is a capital letter like $S$, $P$, $D$, etc, for $L = 1, 2, 3, \ldots$) and an optional subscript $J$ gives the total orbital momentum of the state ($J = L + S, L + S - 1, L + S - 2, \ldots$). The degree of degeneracy or the number of orbitals with the same energy for a free-atom state that has the orbital quantum number $J$ is $2J + 1$. All of this summarizes some of the spectroscopic jargon that an inorganic chemist may encounter in practice.

Each electron orbital occupation (an electron *configuration*) may correspond to several terms with different energies. The opposite may also occur, i.e., each many-electron state may be a mixture of different electron configurations. Although

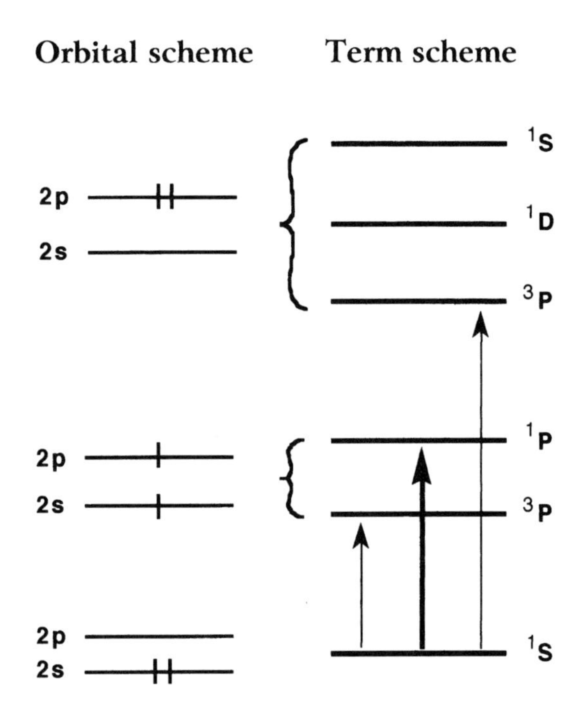


**Figure 2.5.** The schematic spectrum of the states of the free gaseous beryllium atom showing on the left the orbital scheme with electron occupation and on the right half the term levels. One electron configuration can correspond to several terms. The allowed optical transition is indicated with a bold arrow and the "forbidden" transitions that correspond with weak absorption in the optical spectrum are indicated with thin arrows between the ground state and the excited states.

all the terms that belong to one electron configuration have the same electron distribution over atomic orbitals and therefore the same sum of orbital energies, the interactions between the electrons differ in these states (they have different spin distributions), which means that the terms have different energies. Figure 2.5 illustrates the point by showing schematically the lowest-lying terms of the spherically symmetrical beryllium atom with two valence electrons. The configuration $(2s)^2$ has one term $^1S$. The valence electron orbital occupation $(2s)(2p)$ has one singlet and one triplet term with different energies. The electron configuration $(2p)^2$ corresponds to three different terms. The optical transitions between the terms indicated with arrows assign the optical spectrum. The bold arrow indicates an *allowed* transition according to the selection rules of optical transitions and its band is therefore a strong one. The fine arrows indicate transitions that are *forbidden* by the selection rules and so the corresponding bands are weak.

An example of how the electron interaction affects the occupation of the orbitals by valence electrons in free atoms is seen at certain points in the periodic table. When the nuclear charge and the number of electrons is increased, succeeding orbitals are occupied in the order of their energies, but sometimes if the orbital energies are close together the expected filling order is changed. The energy of a 4*s*-orbital is higher than that of a 3*d*-orbital yet the potassium atom has the configuration $\{Ar\}(4s)^1$ and not $\{Ar\}(3d)^1$. The $(4s)^1$ configuration has one state, $^2S$, which has a lower energy

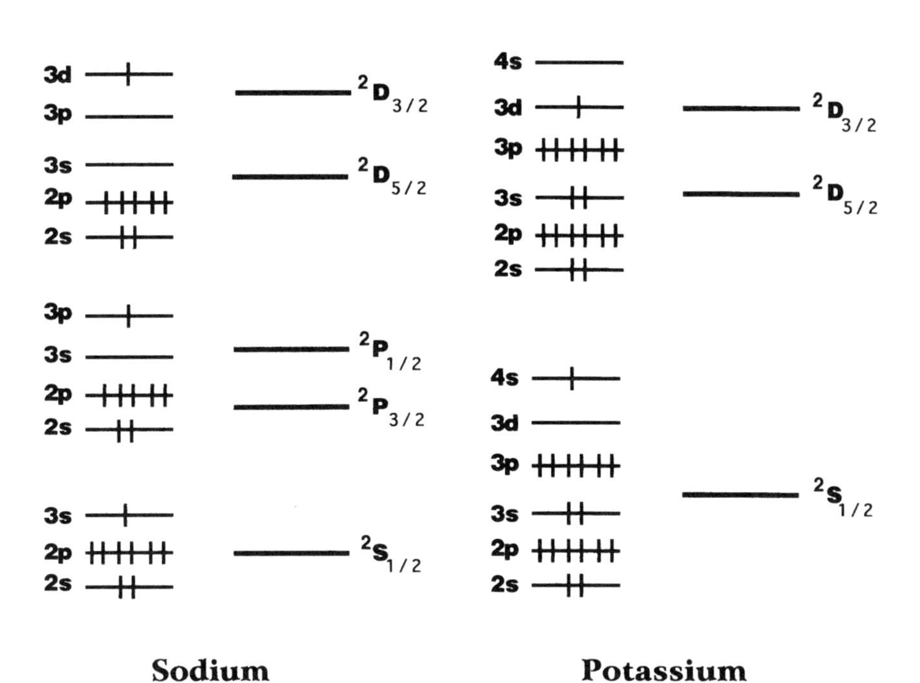


**Figure 2.6.** The relation between the orbital and the term schemes of the free sodium and potassium atoms (in vacuum). Although in potassium the energy of the $3d$-orbital lies below the $4s$ energy, the $4s$-orbital is occupied by the single valence electron and the $3d$-orbital remains empty in the ground state.[11]

than the two states $^2D_{3/2}$ and $^2D_{5/2}$ of the $(3d)^1$ configuration. The repulsion contribution $G$ is larger for the $4s$-orbital than for the $3d$-orbital; hence the energy of the $^2S$ state is lower than that of the $D$-states. In the case of sodium the states have the same sequence as the orbitals and the sequence in the orbital energies determines the sequence in the state energies (Figure 2.6). The repulsion does not change that order. In the case of cations the energy difference between the $3d$- and $4s$-orbitals is greater than in atoms and the orbital energies determine the order in the term energies. Thus the ground state of $Ti^{2+}$ has a configuration $\{Ar\}(3d)^2$ instead of $\{Ar\}(4s)^2$.

The above makes it clear why it can be misleading to use orbital energies instead of term energies when discussing spectra. The term energy of a many-electron state is not merely the sum of the one-electron energies but includes electron–electron interactions. Term schemes rather than orbital schemes should be used to assign atomic spectra.

The basics of atomic spectroscopy as summarized above are strictly valid only for atoms or ions diluted in the gas phase, where the symmetry is spherical. The labels $s$, $p$, and $d$ indicate that. In solids or in molecules the atoms have a lower symmetry and the valence electrons are involved in bonds, which means they are no longer exclusively on the atom under consideration but also partly on its neighbors. Thus it might be thought that atomic spectroscopy does not apply to bound atoms. Yet a similar discussion can be used in connection with spectra of atoms in lattices if their valence electrons are located mainly on those atoms or ions. Adjusting parameters are introduced to account for the interaction with neighbors and such parameters allow one to continue using the formalism for free atoms. The ligands of the ions in the complex or their neighboring atoms in the lattice change the atomic spectra by broadening spectral lines and by shifting or splitting them as will be discussed in more detail in Section 2.3.

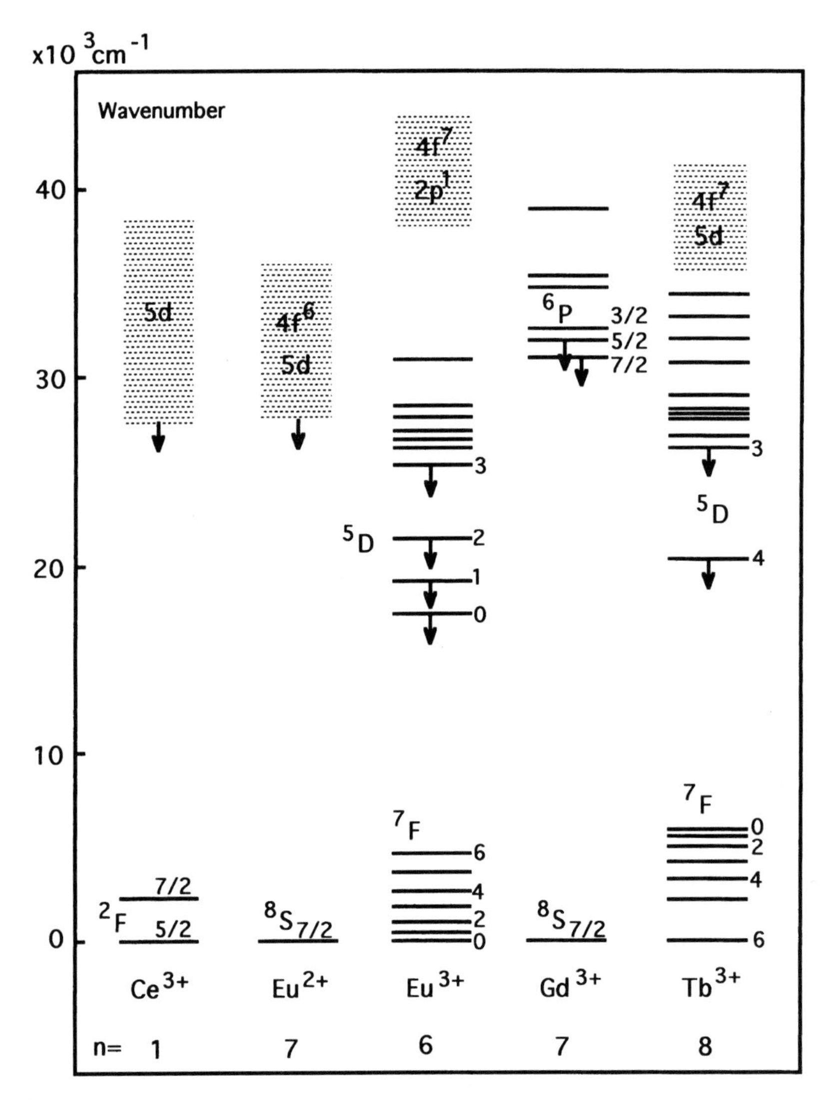


**Figure 2.7.** Term scheme of a few rare-earth ions in oxidic lattices. Spectra are atomic-like if they involve $f$-levels only. Arrows denote levels from which luminescence has been observed. The number of $f$-electrons is $n$. Shaded parts indicate groups of levels corresponding to states involving metal $5d$ or ligand $2p$ electrons. Adapted from J. Blasse. *Progr. Solid State Chem.* **18**, 79 (1988).

An atomic, molecular, or ionic species that determines the color of a substance is called a chromophore. If the spectrum of the chromophore has absorption bands with visible wavelengths the substance is colored because it transmits only those light frequencies that are not absorbed. These absorption bands have frequencies that depend on the strength of the bond of the chromophore with the surrounding atoms in the lattice. Hence the color of the substance depends on the environment of the chromophore.

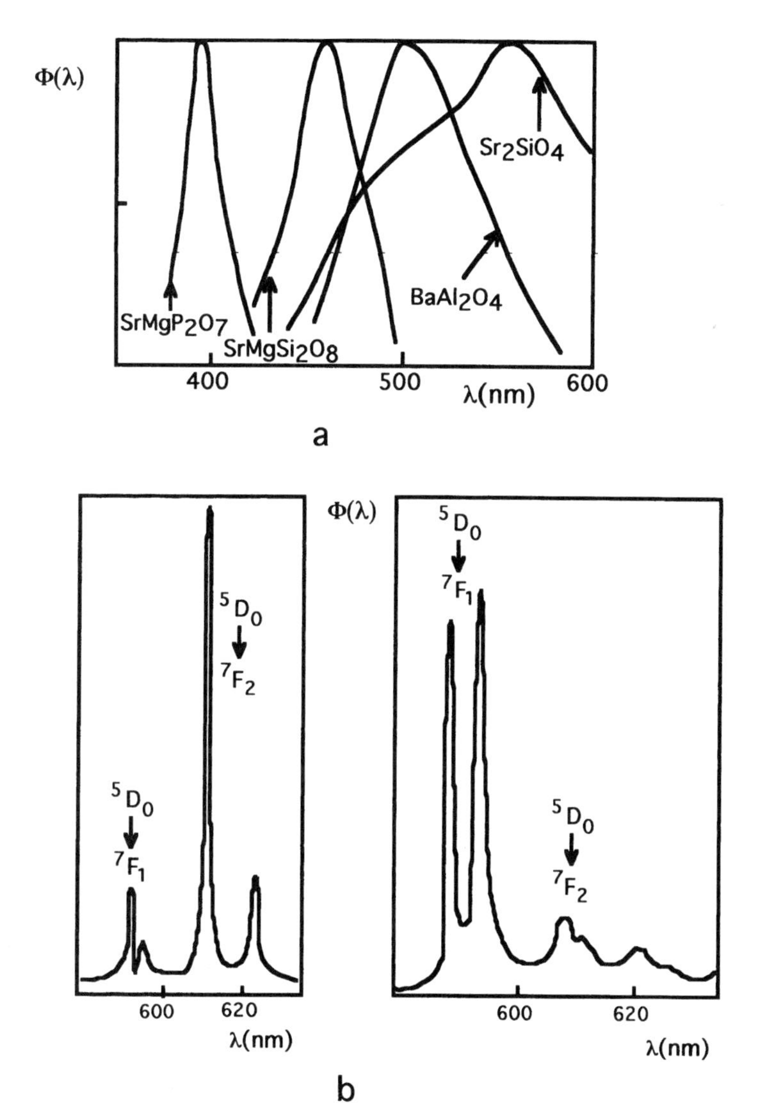


**Figure 2.8.** Luminescence spectra of divalent and trivalent europium ions: (a) Luminescence spectra of $Eu^{2+}$ doped in different lattices. Emission bands in the visible on excitation with ultraviolet light are wide because *5d* electrons are involved; (b) $Eu^{3+}$ doped in $NaGdO_2$ (left) and in $NaLuO_2$ (right), both lattices with an NaCl structure. The first without an inversion center fluoresces with red light, and the second with different selection rules emits orange light.

One example of the use of atomic spectra in luminescence of rare-earth ions in oxidic lattices is the assignment shown in Figure 2.7. The emission spectrum that belongs to a transition from an excited state that corresponds with an electron in a *d*-orbital has broad lines (such as those in Figure 2.8a). In this case the lattice strongly affects the color of the emission because electrons in *d*-orbitals feel the lattice: *d*-orbitals are involved in bonding. On the other hand, fluorescence from *nf* configurations such as $Eu^{3+}$ has sharp lines that look like those in spectra of free gaseous atoms. These sharp lines cannot be shifted by choosing another host lattice for doping with $Eu^{3+}$ because, unlike *d*-electrons, *f*-electrons are buried in the core and insensitive to the surrounding lattice. However, in those cases it is still possible to change the color of the fluorescence by altering the lattice around the chromophore, not because the absorption bands shift but because the selection rules for the transitions depend on the lattice symmetry. For example, the presence or absence of a center of symmetry on the site of the chromophore may strongly affect the intensity of the lines that are present, which indicates that the symmetry of the site determines the color although the lines are not shifted. The red color of the fluorescence becomes orange if the fluorescing $Eu^{3+}$ ion is in a site with an inversion center. The use of luminescence in a ruby laser is shown in Section 2.4.

In summary, one can say that two parameters determine the absorption spectrum:

1. The frequency $\nu$ of a line or band in the spectrum is proportional to the energy difference $\Delta E$ between the ground state and the excited state of the atom or ion ($\Delta E = h\nu$), and not to a difference in orbital energies.
2. The intensity of an absorption band or the oscillator strength is determined by the ease of the transition expressed by the selection rule. The selection rule is determined by the spin of the states that are involved in the optical transition and by the lattice around the chromophore.

Both factors can be calculated given the ion and the symmetry. Atomic spectra can be reliably predicted using atomic orbitals or one-electron states as a basis.

## 2.3. Orbitals: Molecular Orbital and Valence Bond Models

The bonding schemes in the rest of this book rarely rely on orbital methods, and orbitals are reviewed in this section only because they are always used in qualitative discussions of bonding in the chemical literature. This being the case, some familiarity with the relevant terminology is convenient for one to be comfortable with the literature and be able to translate arguments based on orbitals to arguments based on localized valence electrons, as are used in the remainder of this book. Orbital models are discussed extensively in most chemistry texts.[12–15]

If models are works of fiction,[16] the molecular orbital (MO) model is an epic novel.[9] The MO model is almost the only model for the chemical bond that is generally used in academic physics and chemistry for a good reason: it describes certain features of covalent bonding quite well if sufficient computer capability is available and if the structure of the molecule is known. In simple molecules the structure can be derived from the model by calculating the energy of many different atomic arrangements and choosing the configuration with the lowest energy. Usually optical spectra are assigned with the results from MO model calculations. Success in accounting for spectroscopic features is not unexpected for the MO model as it originated in spectroscopy.

Another model based on atomic orbitals is the valence bond (VB) model. In their most simple form both approximations accentuate different aspects of the behavior of valence electrons, but they are equivalent when extended with configuration interaction and resonance, respectively. Both are based on atomic orbitals and allow one to estimate the bonding energies, not just the excess energies as in the Miedema and the Pearson models. Orbital models are not very illuminating in describing bonding in practice because as long as they are used in a qualitative way (the structure is assumed *a priori*) virtually any property can be rationalized, even those that are not in the system. If used quantitatively the laborious results (energy values in tables) are not much help in predicting properties of materials in practice.

These models can describe, e.g., the bonds in silane and methane. The VB description has bonding by the eight valence electrons between four directed $sp^3$-hybrids on silicon and $1s$-orbitals on the four hydrogen atoms. Each of the four bonds in the molecule has two electrons. According to this model all eight valence electrons are equivalent. The MO description has eight molecular orbitals constructed from the silicon $3s$- and $3p$-orbitals and the four hydrogen $1s$-orbitals. The molecular orbitals are symmetry-adapted and as in the case of VB the molecular geometry is presupposed. In the MO model the eight valence electrons are in two (low-lying) orbitals that have different energies.

Both models predict that the valence electrons have a different behavior in ultraviolet photoelectron spectroscopy (UPS). In UPS one "sees" in the spectrum the different occupied orbitals. Monochromatic ultraviolet light is absorbed by the molecule; part of the energy of the quantum is used by the photoelectron to escape from the molecule and the rest is the kinetic energy of the photoelectron. A photoelectron spectrum is a plot of the number of electrons that have a particular kinetic energy against the value of that kinetic energy. To a first approximation, the peaks in such a spectrum indicate the energies of the atomic shells that are occupied by electrons. A UPS spectrum of methane shows that the valence electrons have two orbital energies, which confirms the MO description. However, the tetrahedral form of the molecules suggests equivalence of all the valence electrons in the four bonds with the electron pairs in agreement with the VB description, while the MO model predicts two electron levels. Apparently the relationship between spectroscopy and structure is not so simple that hand-waving arguments explain it.

The VB model given by Heitler and London[17] was the first description of covalent bonding that followed the formulation of quantum mechanics. If $\phi_a(1)$ and $\phi_b(2)$ are the one-electron wavefunctions or orbitals localized on neighboring atoms labeled $a$ and $b$ (e.g., the $1s$-orbitals on the two hydrogen atoms labeled $a$ and $b$ in molecular hydrogen) and each is occupied by one electron (the numbers 1 and 2 represent the coordinates of the two electrons and stand for $x_1$, $y_1$, $z_1$, $x_2$, $y_2$, $z_2$), then the wavefunction describing the binding electrons in this model (ignoring spin) are

$$\Psi_{VB}(1,2) = [\phi_a(1)\phi_b(2) + \phi_a(2)\phi_b(1)]/\sqrt{2}$$

The factor $1/\sqrt{2}$ normalizes the VB wavefunction in order to get the right total probabilities for finding electrons anywhere in space when they are described by this function.

The energy $\int \Psi_{VB} \mathscr{H} \Psi_{VB} dV$ of the two electrons having the wavefunction $\Psi_{VB}$ ($\mathscr{H}$ is the energy operator, the Hamiltonian) is lower in $\Psi_{VB}$ than the energy of the two electrons separately in the orbitals $\phi_a$ and $\phi_b$. Thus this wavefunction $\Psi_{VB}$ means chemical bonding between atoms $a$ and $b$. One wavefunction $\Psi(1,2)$ like that may not be sufficient to describe the ground state of the system sufficiently well, and a mixture of wavefunctions may help. In certain cases resonance occurs (the ground state is then a combination of VB wavefunctions), which lowers the energy by mixing different VB structures, ionic ones among others, into the original single wavefunction that describes the two electrons in the bond.

The VB model for simple molecules can also be used in solids but then there are so many resonating boundary structures that the description needs statistical methods and the model is not very user-friendly. The VB model has the advantage over the MO description that it includes some electron correlation, which means that the electrons feel each other's presence and that therefore their motion and position are interrelated. When one electron is on one atom, the other electron is on the other atom. This is different in the simple MO model, where the valence electrons are completely independent and not correlated: both may be on the same atom simultaneously.

The MO model considers a valence orbital as a single-electron wavefunction of an entire molecule. For a molecule, which is a polynuclear "atom," a molecular orbital is what an atomic orbital (of type *ns*, *np*, *nd*) is for a single atom. The MO model is generally used for covalent bonding in molecules and solids. It uses products of linear combinations of atomic orbitals (LCAOs) of the atoms that are bonded neighbors. The atomic orbitals that it is based on are always centered on the atoms in the molecule or solid. The simple forms of the MO model neglect electron correlation: an electron in it feels the other electrons around it as a smeared-out stationary average charge cloud, not as a collection of instantaneous moving charges. The atomic orbitals ($s$, $p$, $d$, $f$) have energies that increase with the number of angular nodal planes. Similarly, the molecular orbitals (of type $\sigma$, $\pi$, $\delta$) have an increasing number of nodal planes through the interatomic axis. The more nodal planes there are, the higher the energy of the orbital.

In the LCAO approximation a molecular orbital is a linear combination of the atomic orbitals centered on the atoms bonded. For a two-atom molecule the many molecular orbitals occupied by electron $i$ have the form of a linear combination $\psi_{MO}(i)$ of atomic orbitals $\phi_m$ centered on the bonded atoms ($m = a$ or $b$):

$$\psi_{MO}(i) = c_a\phi_a(i) + c_b\phi_b(i)$$

in which $c_m$ are coefficients that give the contribution of the different atomic orbitals $\phi_m(i)$ in the molecular orbital $\psi_{MO}(i)$ that has electron $i$. Bonding (energy lowering) occurs when the bonding molecular orbital (energetically the lowest) is occupied by two electrons (the maximum number allowed). The number of molecular orbitals formed depends on the number of atomic orbitals that combine. If the bonding molecular orbital has the form $p\phi_a + q\phi_b$, ($p$ and $q$ are the orbital coefficients), then the corresponding (orthogonal) antibonding combination (empty on bonding) is $q\phi_a - p\phi_b$. Two atomic orbitals can form two molecular orbitals, one bonding ($\psi$) and one antibonding ($\psi^*$).

Ligand field theory, described in the next section, can be seen as an example of MO theory describing the bonding between the central transition metal ion and the surrounding ligands in ionic lattices and complexes. The highest occupied atomic orbitals in the transition metal ions are 3$d$-orbitals that all have the same energy in spherical symmetry. In octahedral complexes the $e_g$-orbitals $d(x^2 - y^2)$ and $d(z^2)$ of the metal center are the main part of the antibonding $\sigma^*$ molecular orbitals of the complex. These orbitals are no longer purely metal atomic orbitals but have some contribution of occupied ligand orbitals. The metal $t_{2g}$-orbitals $d_{xz}$, $d_{yz}$, and $d_{xy}$ have a nonbonding ($\pi$-type) character and retain their metal character since they do not "mix" with $\sigma$-ligand orbitals for reasons of symmetry (the labels $e_g$ and $t_{2g}$ are group-theoretical jargon). Thus the initially degenerate $d$-orbital levels are split into two groups. The "metal" orbitals contain the valence electrons. The bonding $\sigma$-electrons are mainly localized on the ligands and the bonding between the metal cation and the ligand electron donor atom is a coordinative acid–base associate. The size of the ligand field splitting (the energy difference between the $e_g$- and the $t_{2g}$-orbitals) depends on the strength of the bond. The bond energy is the result of a lower energy of the ligand orbitals that are occupied by the two bonding electrons and this decrease is as large as the increase in the energies of the empty antibonding metal $e_g$-orbitals (that have $\sigma$ character). The increased energy of these empty orbitals on bond formation indicates an increasing ligand field splitting of the orbitals of the central ion of the complex. The position of the metals and ligands in the spectrochemical series (see Section 2.4) can be qualitatively understood in this model as can the magnetic and optical properties of transition metal complexes.

The VB model describes correlated valence electrons while in the MO model the valence electrons are independent. This becomes clear if the two-electron wavefunctions $\Psi(1,2)$ of the bond description in the two models are compared. $|\Psi(1,2)|^2$ represents the probability per unit volume of finding the two electrons somewhere in space with coordinates $x_1$, $y_1$, $z_1$ and $x_2$, $y_2$, $z_2$. For the VB model the electron pair wavefunction is (apart from the normalization factor):

$$\Psi_{\mathrm{VB}}(1,2) = \phi_a(1)\phi_b(2) + \phi_a(2)\phi_b(1)$$

For the MO model the electron pair bond function of a homopolar two-atom molecule (the orbital coefficients $c_a$ and $c_b$ are $\pm 1$ for this case) that has two electrons in the bonding molecular orbital is

$$\begin{aligned}\Psi_{\mathrm{MO}}(1,2) &= [\phi_a(1) + \phi_b(1)][\phi_a(2) + \phi_b(2)] \\ &= \phi_a(1)\phi_b(2) + \phi_a(2)\phi_b(1) + \phi_a(1)\phi_a(2) + \phi_b(1)\phi_b(2)\end{aligned}$$

The first half of this expression is identical to $\Psi_{\mathrm{VB}}$. However, $\Psi_{\mathrm{MO}}$ has in addition two terms that mean that the electrons in $\Psi_{\mathrm{MO}}$ have a considerable probability (0.5) of being on either of the two atoms ($a$ or $b$) simultaneously. Those two ion configurations $a^-b^+$ and $a^+b^-$ contribute strongly to $\Psi_{MO}$ but are absent in $\Psi_{VB}$. In the VB approximation the two valence electrons always sit on different atoms and their spins are paired or antiparallel. Spins are also paired in the MO model.

The value of the resonance parameter $\beta = H_{ab} = \int \phi_a \mathscr{H} \phi_b dV$ is an indication of which of these two models gives the better description of the bonding in the molecule $ab$. If the one-electron part in the matrix element $H_{ab}$ of the Hamiltonian $\mathscr{H}$ is much greater than the electron interaction between the two electrons on a single atom the MO model is a good description of the situation. In that case it does not matter much whether the two valence electrons are on the same or different atoms. If the ionic configurations have a high energy because of strong electron–electron interactions on a single atom then these configurations do not contribute much to the description and the VB model (which does not include any ionic configurations) is better than the MO model. The VB description can be improved by adding some ionic configuration: $\Psi(1,2) = \Psi_{VB} + \lambda\ \Psi$ (ion configuration), where $\lambda$ is a number that gives the amount of the contribution of the ionic configuration to the wavefunction. For $\lambda = 0$ the pure VB model obtains; if $\lambda = 1$ the function becomes the MO description (Figure 2.9). The value of $\lambda$ that corresponds with the lowest energy of the system depends on the particular atoms and is usually neither 0 nor 1 but something between. If that minimum is close to the value $\lambda = 0$, then the VB model is better than the MO model, while if $\lambda$ is close to 1 the MO description is the better one. One chooses the model that works best. Titanium oxide is a metallic conductor and $\lambda$ for the bond between the metal and the oxygen atom in this case is higher than in nickel oxide (a semiconductor). The bonding in metallic conductors such as titanium oxide is better

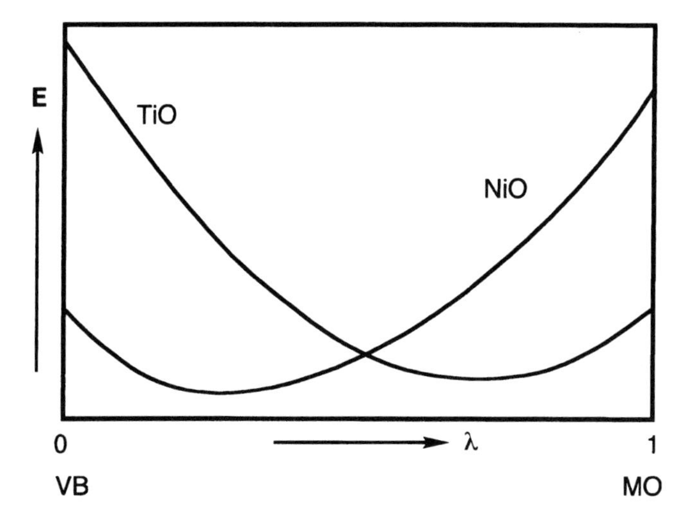


**Figure 2.9.** The qualitative behavior of the ground state energy of titanium oxide and nickel oxide calculated with a VB model that includes some ionic contribution (amount $\lambda$). A pure valence bond wavefunction has $\lambda = 0$ and a molecular orbital state has $\lambda = 1$. For titanium oxide the MO model predicts an energy closer to the real value than a VB calculation. For nickel oxide the valence bond model is more realistic.

described by an MO wavefunction and in insulators such as nickel oxide by localized electrons in VB functions.

The LCAO–MO model is the most popular one in the description of covalent bonding in atomic lattices of metals, semiconductors, and insulators.[9,13] As in the case of the MO model for molecules, the atomic orbitals on the atoms in a solid can be combined into molecular orbitals by linear combination. As many molecular orbitals can be made out of atomic orbitals as there are atomic orbitals for them. In solids that number is very high and the many molecular orbitals made from one atomic orbital on each atom form continuous bands. The number of nodal planes in the molecular orbitals increases with their energy.

The name given to the MO model for solids is the tight-binding method. The MOs in solids are labeled with their wavenumber $k$, which is a good quantum number if the crystalline solid has sufficient translation symmetry (very small particles do not, nor does amorphous matter). The molecular orbitals have orbital energies spread over a range between that of the most bonding molecular orbital (the one that has the lowest energy of the all the constructed molecular orbitals) to that of the most antibonding molecular orbital (the molecular orbital with the highest energy). Loosely speaking the orbitals with low energies are bonding, those with high energies antibonding, and those with energies in the middle between them nonbonding. The most bonding molecular orbital has no nodal plane between the atoms, the most antibonding molecular orbital has as many nodal planes as there are atomic orbitals. Every available atomic orbital forms its own band. Figure 2.10 shows qualitatively the form of the molecular orbitals and their energy and how the wavenumber $k$ increases with increasing orbital energy. The bandwidth, which is the energy difference between the most bonding and the most antibonding orbital in the band, is of the order of $12\beta$, where $\beta$ is the resonance integral $H_{ab}$. The weaker the bonding between the atoms (using that particular orbital), the narrower the band created by that orbital. If atoms have a strong interaction there is an atomic orbital that forms a so-called wide band of molecular orbitals, which means that the bonding orbitals in that band have a rather low energy and its antibonding orbitals high energy. An electron in a wide band moves easily and is lighter than a free electron without interaction with a lattice, while electrons in a narrow band are heavy and are difficult to displace.

The molecular orbitals in a band of an atomic solid form a continuum because there are so many orbitals in a limited energy interval. In this model if a sufficiently wide band is partially filled with electrons (up to the Fermi level $E_F$, which is the highest level of the occupied orbitals), the compound is a metallic conductor. In terms of the band model, a compound is a semiconductor or an insulator if the separate bands do not overlap on the energy scale and are all either completely filled with electrons or completely empty. Solids that have empty bands, e.g., the lowest empty band or conduction band in a semiconductor, do not conduct electrons because there are no electrons in the band. Bands fully occupied by electrons (the valence band in semiconductors is the highest energy band that is completely filled) do not conduct electricity either because the electrons that are there cannot move. In the band model there must be empty orbitals available with approximately the same energy if the electrons are to be mobile. In full bands separated from empty bands by bandgaps, the electrons cannot find orbitals near by (with similar energy) to go into as they are all occupied. Insulators have large bandgaps between the

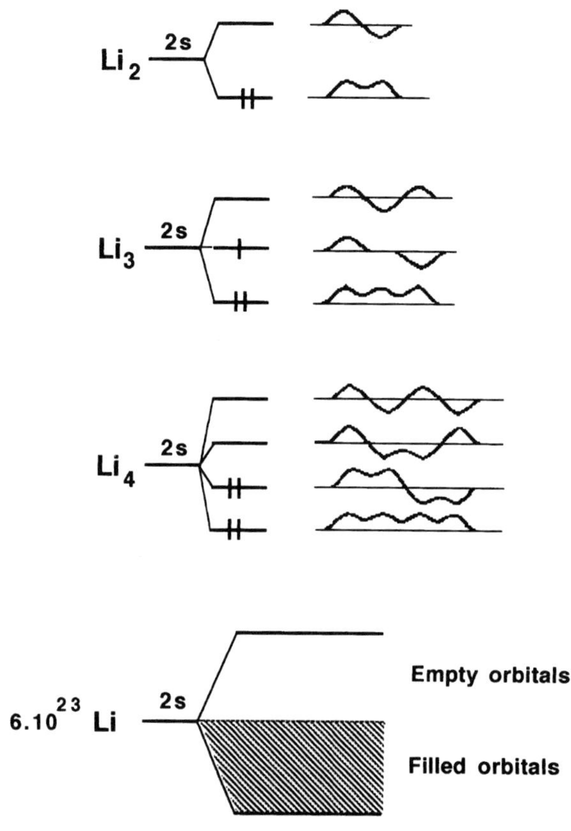


**Figure 2.10.** Molecular orbitals in chains of lithium ions of increasing length. Many atoms bonded together form bands of molecular orbitals made up from atomic orbitals of the same symmetry. In this case the lithium 2*s*-band is sketched. Half of this band is filled with the valence electrons (two per orbital).

highest occupied and lowest unoccupied bands. Semiconductors (insulators at low temperatures) have moderate bandgaps.

Electron conduction in an insulator or semiconductor needs at least some electrons in an empty conduction band or a few electron holes in an otherwise fully occupied valence band. If there is a large bandgap between the highest occupied level (the top of the valence band) and the lowest unoccupied orbital (the bottom of the conduction band), the compound is an insulator. As noted above there are compounds such as nickel oxide that could reasonably be expected to have a partially filled valence band in the MO picture and would therefore be metals but are not. As noted above, for nickel oxide the simple MO description is less appropriate than the simple VB model because of the strong interaction between two electrons on the nickel atom compared to the stabilization that is due to delocalization (the ion configurations require too much energy and contribute little to the ground state). This generally holds in the case of a narrow band (small $\beta$): If $6\beta < I - A$ ($I$ is the ionization energy of the metal ion and $A$ the electron affinity of the nonmetal ion), the compound is an insulator, and electron transport is due to the electrons hopping from atom to atom (a thermally activated process). This is unlike the case in metallic conductors, where electron transport is a result of the excitation of electrons just below $E_F$ to empty levels just above $E_F$. If a compound has partially filled bands or if the filled and empty bands partially overlap on the energy scale, the compound is a metallic conductor.

In binary semiconductors having a sphalerite structure (the structure of zinc blende), the valence band in the tight-binding MO model is a combination of $sp^3$-hybrid orbitals, linear combinations of the *s* and three *p* orbitals on the metal atom that are directed toward the neighboring nonmetal atoms and form bonding combinations with atomic orbitals (*s* and *p*) on them. Similarly the empty conduction band is then built up of the antibonding combinations of the $sp^3$-hybrids on the metal atoms directed away from the bonded neighbor toward the interstices of the lattice.

If the difference in the electronegativities of the two bonded atoms becomes large enough, the charge transfer becomes so high that ions are formed. The ionic description (in Section 2.5) is then more appropriate.

Orbital models have the following disadvantages in applications to bonding problems,[10] so they are not used for bond descriptions in this volume:

1. They are reductionistic models, hence particularly suitable for scientific descriptions. They involve approximations that often do not hold in practice. Many facts remain paradoxical because such facts do not fit in the orbital bonding model.
2. For using models such as these quantitatively, laborious computer calculations are necessary.
3. The building blocks of the model (multicenter integrals) are nonobservables.
4. The properties are not easily derived from the calculated wavefunctions and that does not favor their use by the materials designer.

## 2.4. The Coordinative Bond in Complexes

Coordination compounds are complex molecules that consist of a central metal ion (usually only one) surrounded by ligands (ions, molecules), which supply the electron pair for the bond to the central ion. The coordinative bond between the metal ion and the ligand is also called an acid–base bond, where the metal ion is the Lewis acid and the electron pair donating the ligand is called the Lewis base. The coordinative bond can have both ionic and covalent character. The covalent view is described by the angular overlap model,[18] based on metal *d*-orbitals and ligand orbitals localized on the donor atom, which is directly bonded to the metal. The coordinative bond is used to explain and control color in complexes and also to estimate the rates of charge transfer in redox reactions and ligand exchange. The chemistry of complexes is described in Chapter 3.

Although coordination complexes are molecules, units can be identified in atomic solids that behave optically and electronically like fixed molecular complexes although there are no molecules in atomic solids. There are strong interactions among all the atoms in the lattice. The optical spectrum of a chromium dopant in oxide lattices, e.g., in a ruby laser, is like that of chromium coordination complexes in oxidic solvents.

The structures of typical ligands were shown earlier in Figure 1.9 and will also be shown later in Figures 3.1 and 3.2. Complexes include some metal halides, hydrates, amines, amides and imides, such as $Ti(NR_2)_4$ (R is an alkyl group), oxides, $H_3B{:}NR_3$ (a borane-amine adduct), $Co(MNT)_2$ (MNT = maleonitrile dithiolate), Cupc (pc = phthalocyanine), $Mo(CO)_6$, cluster carbonyls, and metal acetylacetonate derivatives.

In terms of the orbital model the coordinative bond means the following. Ligand and metal orbitals are mixed to form occupied bonding and empty antibonding combinations. The occupation lowers the energy of the electron donor pair on the ligand on bonding with the metal. The bonding orbitals are mainly ligand orbitals and the participating metal orbitals are antibonding.

The central ions in complexes are often transition metal ions that have partially filled *d*-orbitals. The symmetry of the complex determines which of the metal *d*-orbitals are involved in bonding, i.e., are antibonding. The orbital theory was given in Section 2.3. The ligand field splitting strongly affects the optical spectrum and the magnetic behavior of the metal ion and is also a clear example of a structure–property relation. This influence is expressed in the spectrochemical series. Different ligands L for the same metal ion or different metal ions M for the same ligand can be arranged in a sequence of increasing ligand field splitting in the complex. The color of complexes can be tuned by changing the splitting through changing M or L.

The spectrochemical series arranges ligands and metal ions according to increasing ligand field strength or degree of orbital splitting. For ligands around the same metal ion the series is:

$$I^- < Br\text{-} < Cl^- < {}^*SCN^- < F^- < OH\text{-} < \text{acetate} < \text{oxalate} < H_2O < {}^*NCS\text{-}$$
$$< \text{pyridine} < NH_3 < \text{ethylene diamine} < \text{bipyridine} < {}^*NO_2\text{-} < CN \approx CO$$

The spectrochemical series for metal ions and the same surrounding ligand is:

$$(Mn^{2+} < Ni^{2+} < Co^{2+} < Fe^{2+} < V^{2+}) \ll (Fe^{3+} < Cr^{3+} < V^{3+} < Co^{3+})$$
$$< Mn^{4+} < (Mo^{3+} < Rh^{3+} < Ru^{3+}) < Pd^{4+} < Ir^{3+} < Re^{4+} < Pt^{4+}$$

If the metal ions have valence electrons in the antibonding $e_g$-orbitals they make a negative contribution to the bond energy. The ionic radius then is larger than if these orbitals are empty and the bonding between the metal and the ligand is weaker. The distribution of electrons over the available orbitals also affects stability and rates of ligand and electron exchange in reactions of complexes, as will be reviewed in Chapter 3.

Molecular orbital schemes for many complexes are given in most introductory inorganic chemistry texts. Assigning bands in the optical spectrum requires states with many electrons. As these are constructed with orbitals the crystal field that splits orbital levels also splits many-electron state energies of ions surrounded by ligands or ions in a lattice. Figure 2.11 shows the term level scheme of free transition metal ions in a vacuum having $d^2$–$d^5$-configurations. These levels split in a ligand field and the nature and size of the splitting depend on the symmetry and strength of the field and on the symmetry type of the level.

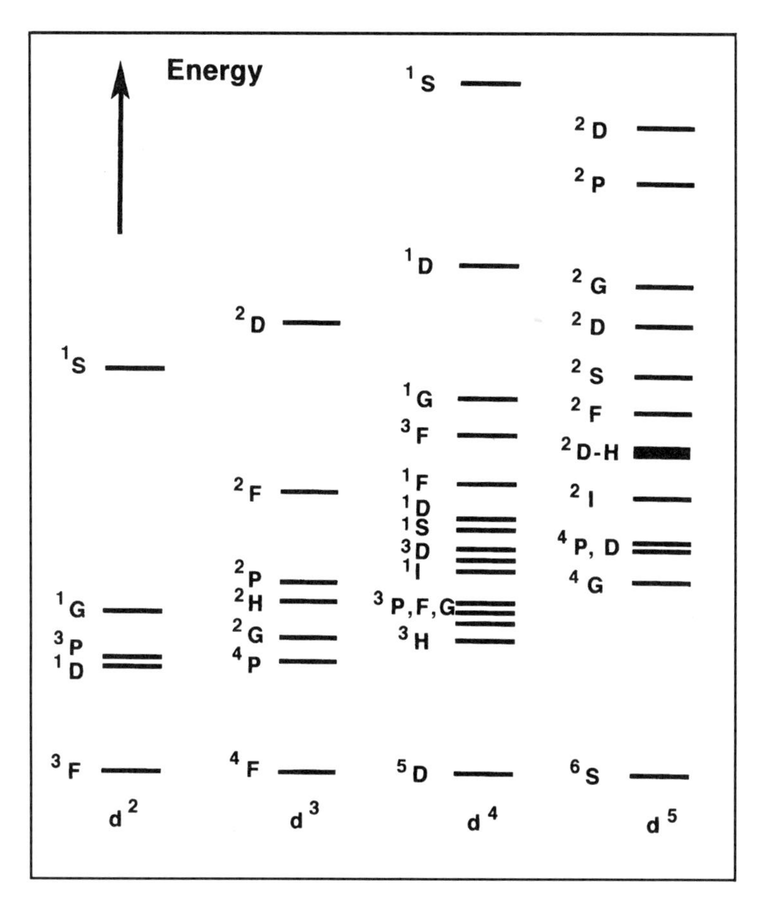


**Figure 2.11.** Term level scheme of free ions (in a dilute gas) having electron configurations with two to five valence electrons in *d*-orbitals.

The Tanabe–Sugano diagrams show how the lowest more-electron ionic levels for the $d^n$- configurations split if the environment is no longer spherically symmetric but has cubic symmetry (octahedral or tetrahedral), such as a transition metal ion has in a complex molecule or in a lattice site. Figure 2.12 shows two Tanabe–Sugano diagrams for octahedra. These diagrams show the influence of the strength of the crystal field, indicated with the parameter $Dq$, on the term level splitting of the central ion in units of $B$, which is a particular electron repulsion energy. The $x$-axis is the energy of the ground state, and term energies are expressed with respect to the ground level. A kink in the energy curves means a change of ground level, usually a change between high-spin and low-spin ground states with increasing ligand field strength.

Tanabe–Sugano diagrams are convenient for assigning optical spectra of transition metal ions in complexes or in lattices and for deriving the strength of the ligand field from those spectra. This can be done as follows: A vertical line is shifted in the appropriate diagram until the crossing with the term lines coincide with the

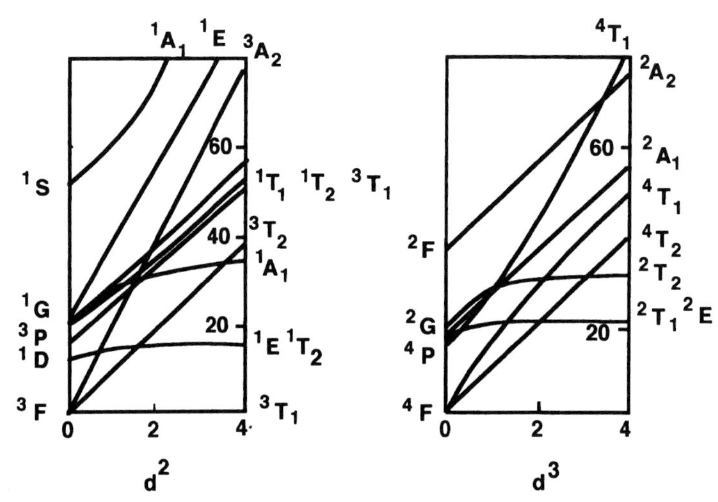


**Figure 2.12.** Tanabe–Sugano diagrams for transition metal ions in cubic six-coordinated complexes (octahedra) with two and three *d*-electrons. Vertically the energy is plotted against the ligand field strength parameter *Dq* (splitting energy of the *d*-orbitals), both energies in units of *B*, a quantity that depends somewhat on covalency. These diagrams can be used to assign transitions in ligand field spectra of complexes and to derive the strength of the ligand field from the position of the bands.

band positions in the observed spectrum. The $x$-value of the vertical line then gives the ligand field, and the energies of the crossing points are approximations of the band energies in the absorption spectrum. Figure 2.13 gives the absorption spectrum of a $d^3$ ion having water molecules as octahedral ligands. This figure also shows the visible light spectrum of trivalent chromium in ruby and emerald. These stones have the same dopant but different host lattices and different colors, as the ligand field of the oxidic lattices on the chromium sites are different. The oxide ligands of the chromium ions are bound to different partners in the three lattices. In this case of $d^3$-states, one of the two prominent bands in the spectrum gives the ligand field parameter $Dq$ directly.

Another example of the optical behavior of a chromophore in the ligand field of a lattice is the pulsed solid state ionic laser. The schematics of the lowest-lying levels that take part in the laser action are given in Figure 2.14. A powerful flash of white light excites the chromium atoms by absorption in an allowed band to a $^4T_2$-state. Usually the excited atoms rapidly return to the ground level and in so doing emit light (they fluoresce). In this case the rate of the radiation-less transition to a different excited state, an $^2E$-state, is much higher than the rate of fluorescence. The transition to the ground state of the chromium ions $^4A_2 \leftarrow {}^2E$ is then spin-forbidden (the spin states of the two levels differ) and the excited chromium ions remain for a comparatively long time in their spin-doublet excited states. Emitting light by returning to the ground state in a forbidden transition is known as phosphorescence, which is a much slower process than fluorescence. The result of the flash is excitation of most of the chromium ions into these low-spin states. The electronic "temperature" according to the Boltzman distribution of energies is now negative because more ions are in electronically excited states than in their ground states. An accidental photon from a phosphorescing ion that fits exactly between the

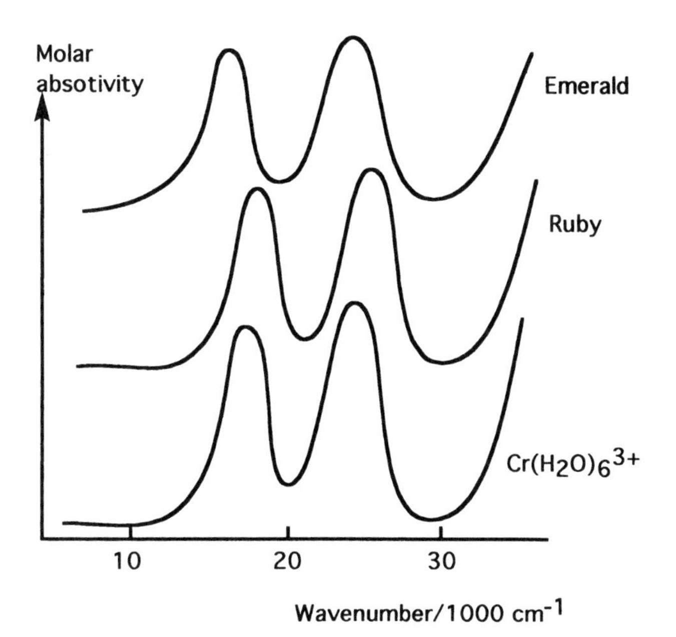


**Figure 2.13.** The spectrum of trivalent chromium in two oxidic lattices and in a chromic salt solution in water. The two absorption bands in the visible part of the spectrum correspond with the two lowest spin-allowed transitions shown in Figure 2.12; the transition with the longest wavelength gives the ligand field splitting directly.

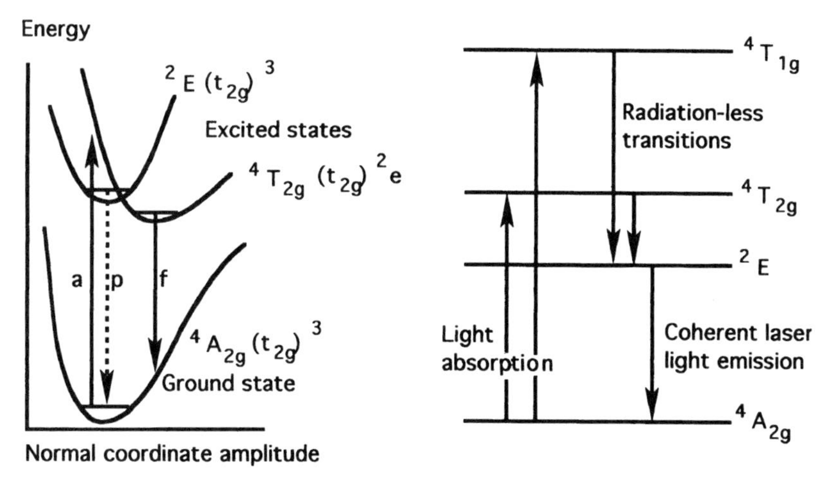


**Figure 2.14.** The potential energy diagram of the lower states of trivalent chromium. At left the state energies vs. the position of the nuclei of the metal and the ligands. Arrows denote optical transitions: *a* means absorption, *f* is fluorescence, *p* is phosphorescence (the dashed line is spin-forbidden). Laser function can also be represented in a diagram as shown on the right.

two levels triggers the forbidden transition in excited atoms. The emitted photon of these atoms is in phase with the exciting photon and amplifies it in a coherent manner. This amplified emitted radiation, if kept long enough in the lattice, causes an avalanche of emitting chromium ions and emission of a strong pulse of coherent light having a wavelength corresponding to an energy difference $^4A_2 - {}^2E$.

The effect of the ligand field in this description follows from the Tanabe–Sugano diagram for the $d^3$ case combined with the crossing of the interatomic potential curves. In the left-hand side of Figure 2.14 the energy of the involved states is given in their dependence on the interatomic distance as discussed in the introduction to this chapter. The involved states are the $^4A_{2g}$-ground state, the excited $^4T_2$-state (to which absorption transition is allowed), and the $^2E$-state for collecting excitations to use in the laser action. The minimum of the $^2E$-state is in this case vertically above the minimum of the ground state $^4A_{2g}$, both states having the same interatomic equilibrium distance because both states have the same electron occupation configuration $(t_{2g})^3$. This means that the emission wavelength is not affected by fluctuations in the interatomic distance and the transition is sharp. This is also seen in the horizontal $Dq$-independent $^2E$ line in the Tanabe–Sugano diagram. The position of the first optical absorption band of the chromium ion depends much more strongly on the size of the ligand field. The so-called "intersystem crossing" means flipping of the spins from $S = \frac{3}{2}$ to $S = \frac{1}{2}$ in the radiationless transition from the excited $^4T_2$-state to the slightly lower excited $^2E$ state. The optical transitions between more-electron states are vertical: during electronic transitions that are very fast the nuclei have no time to move. This so-called Born–Oppenheimer approximation holds well in optical transitions.

This discussion of the coordination bond can be summed up by reiterating that the bond in coordination complexes is an acid–base bond or a sharing of an electron pair between a donor (Lewis base) and an acceptor (Lewis acid). The ligands (bases) surrounding the metal ions (the acids) cause a splitting of the valence electron levels on the metal ions. These valence electrons do not contribute a great deal to the strength of the bond between the metal center and the peripheral ligands but strongly affect the optical and magnetic properties as well as the kinetics of the reactions. This model of bonding is used extensively in discussions of properties but not for descriptions of bonding enthalpies.

## 2.5. Bonding in Ionic Compounds

It has been known for some time that the bond in ionic compounds is reasonably well accounted for by the Coulomb interaction between the ions, which are conceptualized as charged hard spheres. Positive and negative ions attract each other and they stack in such an array as to minimize the lattice energy. The resulting lattice structure depends on the size and charge of the ions, and its energy is calculated using the Coulomb interaction, the electron affinity of the atom that forms the anion, and the ionization energy of the atom that loses one or more electrons and forms the cation. This scheme is known as Born–Haber cycle, which is fully described in undergraduate texts.[19] The ionic model is a first approximation: ions are not really hard spheres, they are somewhat polarizable, and they are subject to some covalent bonding as well. The model cannot predict the observed structure solely from geometrical considerations, nor can it discriminate among different possible structures because the most stable of the possible structures is often determined by small energy effects that are not accounted for in the simple ionic model.

Ion bonding is found in compounds made up of atoms that have very different electronegativities. In bond formation there is so much charge transferred from the atom with the lowest electronegativity to the atom with the highest that the Coulomb attraction between the ions dominates the bonding energy. The valence electrons are then located at the ions. Examples of binary compounds having a large ionic contribution to the bond are LiF, NaCl, CsAu, $CaC_2$, and $SbCl_5$.

Ion conductors are usually compounds in which the atoms are bonded together with an ionic bond. Textbooks on ion conduction do not spend much time on bonding and for ionically conducting solids the classical electrostatic Coulomb interaction suffices. However, there also exist covalent layered lattices, which admit intercalation of ions between the covalent layers, and these lattices can conduct electrons as well as interstitial ions. Good ion conductors have ions that in their bond with the other atoms in the lattice have no strong preference for one particular coordination number, but balance between two values. This lowers the activation barrier of the ion path through the lattice and results in a higher ionic conduction.

The Coulomb attraction between ions with different charges is not strongly directed as covalent bonds are (the valence shells in ions are fully occupied) and so the size of the ions and their charges (see Figure 1.4) determine the stacking possibilities of the atoms in ionics and the structures of those compounds. Ionic radii are derived from measured interatomic distances in salts. The Pauling rules,[20] which take ions to be hard charged spheres, derive optimal coordination numbers from the size of the cations and the anions in the compound. If the ratio $q$ of radii $r$ ($q = r_{\text{cation}}/r_{\text{anion}}$) is between 0.225 and 0.414, a tetrahedral coordination is expected for sterical reasons: with this $q$-value the cation has enough space around it to accommodate four anions without them touching one another. The zinc blende and Würtzite structures are lattices with this atomic coordination. If $0.414 < q < 0.732$, the most favorable ionic stacking means six anions around the cation in an octahedral arrangement. The cubic halite (NaCl) structure is an example. For a bigger cation with $0.732 < q < 1.0$, the eight ligands are arranged on the corners of a cube around the cation according to the Pauling rules. The lower limit in each range means that below that value of $q$ the cations start to rattle in the cage of ligands, which touch each other; in this case a lower coordination number is better for optimal stacking. The Pauling rules sometimes lead to the wrong prediction because the hard-sphere model is too restrictive. Table 2.1 lists the right and the wrong predictions for structures based on the hard-sphere model.

In mixed salts the ionic sizes determine the mutual solubilities and the possible structures.[21] Figure 2.15 shows the relation between the size of the ions and the formation of mixed oxides. The larger the difference in the ion radius of divalent cations, the greater the number of possible compounds according to the phase diagram. Good mixing and wide existence ranges in phase diagrams of oxidic alloys require similar radii. An analogous rule for intermetallic compounds will be discussed in Section 2.6.

**Table 2.1. The Observed Structures of Ionic Compounds, the Ratio of the Ion Radii, and the Prediction According to the Pauling Rules[a]**

| Compound | Ratio $r_+/r_-$ | Match |
|---|---|---|
| CsCl structures | | |
| CsCl | 0.93 | + |
| CsCN | 0.98 | + |
| $NH_4Cl$ | 0.90 | + |
| $NH_4CN$ | 0.85 | + |
| NaCl structures | | |
| LiF | 0.76 | − |
| NaF | 0.97 | − |
| KF | 0.78 | − |
| RbF | 0.72 | + |
| CsF | 0.66 | + |
| NaCl | 0.69 | + |
| NaI | 0.65 | + |
| NaH | 0.65 | + |
| KI | 0.74 | − |
| KOH | 0.81 | − |
| KCN | 0.86 | − |
| MgO | 0.68 | + |
| MgSe | 0.49 | + |
| CaO | 0.90 | − |
| AgCl | 0.77 | − |
| AgBr | 0.71 | + |
| BaO | 0.85 | − |
| MnO | 0.71 | + |
| NiO | 0.66 | + |
| PbS | 0.78 | + |
| TiN | 0.47 | + |
| CrN | 0.44 | + |
| Würtzite structures | | |
| BeO | 0.47 | − |
| AlN | 0.39 | + |
| GaN | 0.44 | − |
| InN | 0.55 | − |
| ZNO | 0.70 | − |
| $NH_4F$ | 0.79 | − |
| Sphalerite structures | | |
| BeTe | 0.29 | + |
| AlP | 0.32 | + |
| GaAs | 0.30 | + |
| CuCl | 0.54 | − |
| ZnSe | 0.48 | − |
| CdTe | 0.53 | − |
| HgS | 0.86 | − |
| AgI | 0.63 | − |
| Fluorite structures | | |
| $CaF_2$ | 0.96 | + |
| $ZrO_2$ | 0.68 | − |
| $HfO_2$ | 0.67 | − |
| $CeO_2$ | 0.80 | + |
| $ThO_2$ | 0.86 | + |
| NiAs structures | | |
| NiAs | 0.32 | − |
| TiS | 0.59 | + |
| NiTe | 0.40 | − |
| CoSb | 0.29 | − |
| MnSb | 0.31 | − |

[a]The plus sign indicates that the compound obeys the rule and the minus that the observed structure does not.

While the size ratio of the cations determines the structural possibilities, the ionic charge determines the bonding strength and the melting temperatures. Figure 2.16 shows the phase diagrams of two ionic systems having the same ion size ratio and different ionic charges. The number of mixed compounds does not differ in the two phase diagrams but the lattice energies do and so do the melting points, which are much higher for the more strongly bound atoms.

## 2.6. The Miedema Model for Intermetallics

The Miedema model[2,22] describes bonding in intermetallic, ordered compounds of metallic elements. In addition to the metallic elements the model also includes the elements hydrogen, boron, carbon, nitrogen, silicon, and phosphorus in their metallic

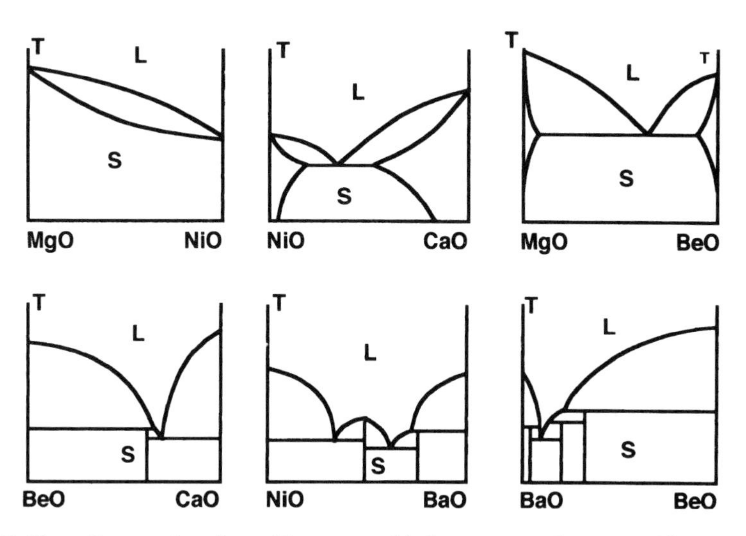


**Figure 2.15.** Phase diagrams for a few oxide systems with the same crystal structure. The more the cation sizes of the two oxides differ, the more compounds are possible in the system.

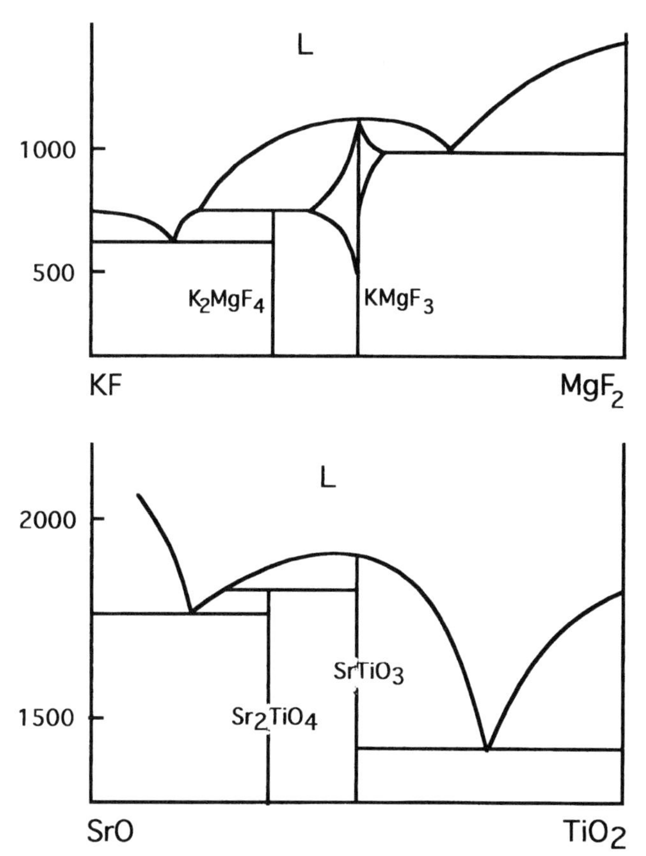


**Figure 2.16.** Phase diagrams of two systems with similar ionic radii and similar compound formation. Note the higher melting points in the oxidic system.

states. One can easily calculate the excess formation enthalpy of heteropolar compounds from the parameters of the elements and these values predict the possible occurrence of those compounds. The atomic parameters for each element are related experimentally to its ionization energy, surface tension, compressibility, and molar volume. Whether two given metallic elements can be dissolved in each other or can form compounds can be predicted with a high reliability from the atomic parameters.

The model explains why in alloys the smaller atoms diffuse and the larger do not, or how traces of certain metals poison heterogeneous catalysts. The equilibrium pressure of hydrogen in ternary hydrides can be controlled by choosing the metallic components because their Miedema parameters are known. The formation energy of vacancies in intermetallic compounds can be estimated using the parameters of this model, which predicts whether compounds of type $M_mN_n$ or $MM'N_n$ exist in which M and M′ are metallic atoms and N a metal or a nonmetal such as hydrogen, boron, carbon, silicon, nitrogen, phosphorus, or arsenic. Examples of compounds made up of a metal and a nonmetal (sometimes called ceramic compounds) are ReH, $MoN_2$, $Au_2B$, and $ZrC_2$. Nonmetallic elements can be incorporated into this scheme if these elements in compounds with metals behave as if they are metals. Metal nitrides and hydrides for instance have a maximum stoichiometry that depends on temperature and pressure in a way well predicted by this model. However, mixing of ternary nitrides is only possible when the corresponding metals have been alloyed. The model works in the case of nonmetallic elements only after an extra transformation enthalpy makes them metallic.

The Miedema model has a hightly original starting point and although developed by physicists, it encountered strong resistance from other physicists when it was first suggested. Atoms are regarded as pieces of metal having the same electronegativity and electron density at the atomic surfaces (or their surface tension) as the bulk metals. These two values are atomic parameters, which have been determined for each element by fitting expected enthalpies from the observed values of a great number of compounds. The excess bonding enthalpy of a heteropolar compound with respect to the bonding enthalpy in the parent metallic elements is the result of electron transfer between atoms having different work functions or electronegativities. The model is only valid for metallic atoms and ignores structure.

In the Miedema model (also called the macroscopic atom model) every atom has two parameters that determine the excess bonding enthalpy of binary compounds: the work function $\phi^*$ and the electron density $n_{WS}$ on the surface of the Wigner–Seitz cell (the surface of the atom). On contact between the atoms, an amount of electric charge flows from the most electropositive to the most electronegative atom. The amount of charge is determined by the difference in $\phi^*$ and the transfer means a lowering of the energy or a negative contribution to the reaction enthalpy $\Delta H$. This transfer does not continue indefinitely: the charge density on the boundary between the atoms must be made the same, which requires energy or a positive contribution to $\Delta H$ of mixing or bonding. The excess enthalpy as a result of the charge transfer in a binary intermetallic compound MN becomes

$$\Delta H = f(c)[-P(\Delta\phi^*)^2 + Q(\Delta n_{WS}^{1/3})^2 - R]$$

In this expression $f(c)$ is a simple function of the atomic fraction $c$ of N in the compound MN. If the atoms M and N are not different in size, $f(c) = c(1 - c)$.

$P$, $Q$, and $R$ are empirical group parameters, which have been obtained by fitting this expression to known $\Delta H$ values of a large number of binary compounds. $\Delta\phi^*$ and $\Delta n_{WS}$ are the differences of the corresponding parameter values of the participating atoms M and N. The expression $\phi^*$ is the scale of work function, which strongly correlates with existing scales of electronegativity. Some values are slightly adjusted to get a better fit. The $n_{WS}$ values that are obtained from the fit correlate strongly with $(KV_m)^{-1/2}$, where $K$ is the compressibility and $V_m$ the molar volume.

For many intermetallic combinations only two group parameters, $P$ and $Q$, are needed. For elements that have $s$- and $p$-type valence electrons the extra term $R$ is required. From a diagram of $\phi^*$ vs. $n_{WS}$ for metallic elements the thermodynamic stability of binaries can be predicted with the known values of $P$, $Q$, and $R$ for the group.

There is excellent correlation between measured values of the excess bonding enthalpy and those calculated from the two atomic parameters determined by a fitting procedure. The existence predictions of this model are highly reliable. Table 2.2 gives the parameter values of the atoms in this model. The values of $P$ and $Q$ are

**Table 2.2. Values of the Atomic Parameters in the Miedema Model**

| Atom | $\Phi^*$ | $n_{WS}^{1/3}$ | Atom | $\Phi^*$ | $n_{WS}^{1/3}$ |
|---|---|---|---|---|---|
| H | 5.2 | 1.5 | Mo | 4.65 | 1.77 |
| Li | 2.85 | 0.98 | Tc | 5.30 | 1.81 |
| Be | 5.05 | 1.67 | Ru | 5.40 | 1.83 |
| B | 5.3 | 1.75 | Rh | 5.40 | 1.76 |
| C | 6.24 | 1.77 | Pd | 5.45 | 1.67 |
| N | 6.68 | 1.65 | Ag | 4.35 | 1.36 |
| Na | 2.70 | 0.82 | Cd | 4.05 | 1.24 |
| Mg | 3.45 | 1.17 | In | 3.90 | 1.17 |
| Al | 4.20 | 1.39 | Sn | 4.15 | 1.24 |
| Si | 4.70 | 1.5 | Sb | 4.40 | 1.26 |
| P | 5.55 | 1.65 | Cs | 1.95 | 0.55 |
| K | 2.25 | 0.65 | Ba | 2.32 | 0.81 |
| Ca | 2.55 | 0.91 | Ln | 2.5–3.25 | 0.88–1.34 |
| Sc | 3.25 | 1.27 | Hf | 3.60 | 1.45 |
| Ti | 3.80 | 1.52 | Ta | 4.05 | 1.63 |
| V | 4.25 | 1.64 | W | 4.80 | 1.81 |
| Cr | 4.65 | 1.73 | Re | 5.20 | 1.85 |
| Mn | 4.45 | 1.61 | Os | 5.40 | 1.85 |
| Fe | 4.93 | 1.77 | Ir | 5.55 | 1.83 |
| Co | 5.10 | 1.75 | Pt | 5.65 | 1.78 |
| Ni | 5.20 | 1.75 | Au | 5.15 | 1.57 |
| Cu | 4.45 | 1.47 | Hg | 4.20 | 1.24 |
| Zn | 4.10 | 1.32 | Tl | 3.90 | 1.12 |
| Ga | 4.10 | 1.31 | Pb | 4.10 | 1.15 |
| Ge | 4.55 | 1.37 | Bi | 4.15 | 1.16 |
| As | 4.80 | 1.44 | | | |
| Rb | 2.10 | 0.60 | Transformation energies (kJ/mol) | | |
| Sr | 2.40 | 0.84 | H | 100 | |
| Y | 3.20 | 1.21 | B | 30 | |
| Zr | 3.45 | 1.41 | C | 180 | |
| Nb | 4.05 | 1.64 | Si | 34 | |
| | | | Ge | 25 | |
| | | | N | 310 | |
| | | | P | 17 | |

not fixed for all metallic elements but differ for different groups of metals. The parameter $R$ is necessary if one of the partners in the bond is a metal atom having occupied $p$-orbitals, in which case the boundary between the areas of positive and negative $\Delta H$ is a parabola instead of straight lines. Figure 2.17 shows graphs with $\phi^*$ and $n_{WS}$ values of the elements. These graphs are used to determine elements that tend to react with each other and those that do not. They also give the polarity of the bond between the elements in the compound. Lines corresponding to certain $P/Q$ values divide the parameter space into four quadrants. The lower and upper quadrants contain the metals with which the metal at the crossing point has a negative $\Delta H$ (and tends to react with). As examples, the elements magnesium, hydrogen, and gold are on the crossing points of those lines in the graphs. Magnesium reacts with tin but not with titanium; hydrogen reacts with zirconium but not with iron, and gold has a higher affinity for niobium than for tin.

The Miedema model also shows how the stoichiometry of compounds of metals with nonmetals depends on the atomic parameters. The thermodynamics of binary mixtures of metals for example is discussed in Section 10.2. The Miedema parameters can be used to determine whether the $\Delta H/x$ diagram of the system is convex or concave from the sign and size of $\Delta H$. In order to apply this to a system of a metal and a nonmetal the latter first has to be converted to a metallic element through a transformation enthalpy. Figure 2.18 shows how the transformation enthalpy affects the calculation of the mixing or reaction enthalpy. The horizontal line in the diagram runs between the standard enthalpy of the metal to that of the nonmetallic element. A system having negative $\Delta H$ (a curve with a minimum) runs from the standard metal value to the metal form of the nonmetal, the transformed value of the

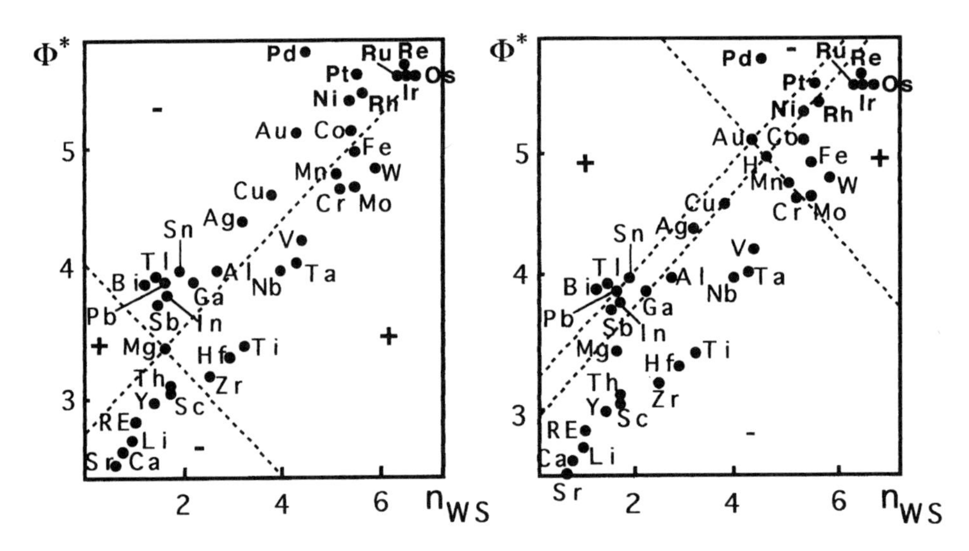


**Figure 2.17.** Several elements arranged according to their parameter values in the Miedema model. The sign of the reaction enthalpy is indicated in the quadrant bounded by the $P/Q$ lines of the group of metals. The stable binary (metallic) compounds can be identified at a glance. The metallic form of hydrogen is also placed with the metals. Hydrogen has a negative charge in lithium hydride and a positive charge in the stable compound with palladium.

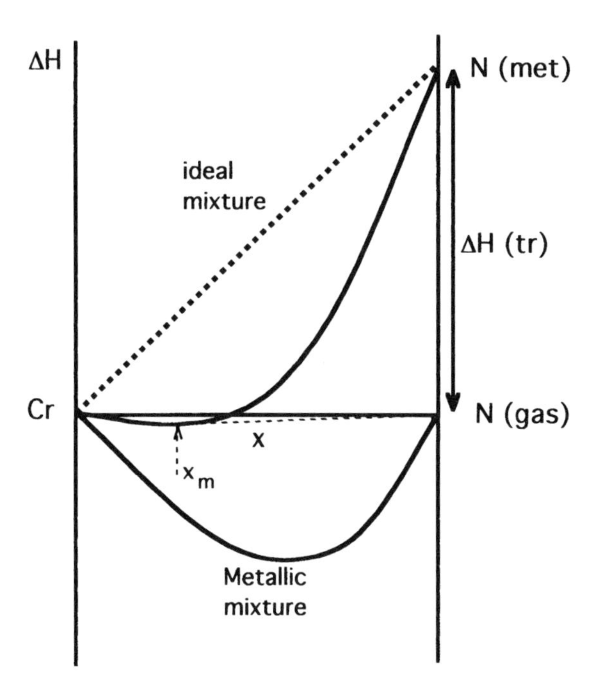


**Figure 2.18.** The $\Delta H/x$ plot for chromium nitride. This couple would have a negative formation enthalpy if both atoms were metallic (lower curve). The transformation enthalpy $\Delta H$(tr) raises the $\Delta H$ value for nitrogen and the $\Delta H/x$ curve is lifted for the real case (upper curve). The maximum stoichiometry of the chromium nitride (in equilibrium with nitrogen gas) is the tangent point of a line to the $\Delta H = 0$ point. For metals with a $\Delta H/x$ curve above the $x$-axis there is no stable nitride.

nonmetal having a higher "standard" enthalpy. The only stoichiometries that are possible (are stable with respect to the elements in their standard form) are those for which the minimum is below the horizontal "ideal" line. The maximum stoichiometry for the nonmetal in the compound is the point of contact of a tangent to a curve drawn from the standard value of the nonmetal. An element that has a high transformation enthalpy (such as nitrogen) requires a large mixing or reaction enthalpy to have stable compounds.

## 2.7. The Pearson Model: Electronegativity Equalization

The Pearson model[3] considers bonding between atoms as a decrease in energy that results from electron transfer from a donor to an acceptor. The model is based on two atomic parameters: electronegativity and chemical hardness. Like the Miedema model, it does not account for the entire bond energy but only for that part of the bonding energy of heteropolar compounds that is the result of charge transfer from the atom having a low electronegativity to one more electronegative. As in the Miedema model, such an electron transfer corresponds with an energy decrease and that indicates chemical bonding. The changing quantity of charge on the atoms during bond formation alters the original electronegativities[23]:

the high electronegativity of the oxidant (electron acceptor), gradually decreases and the electronegativity of the reductant (electron donor) increases by the loss of charge.

The transfer continues until the electronegativities of the bonded atoms are equal. The total amount of charge that has passed between the atoms on bonding depends on the chemical hardness of the two atoms, the second atomic parameter in this model, which expresses the rate at which the atomic electronegativities change with the quantity of charge added or removed. We can picture it as though the atoms behave as communicating vessels: the liquid is electronic charge, the level of the liquid is the electronegativity, and the diameter of the vessels is the chemical hardness. This parametrized empirical model for the bonding of atoms in molecules extends the old concept of electronegativity[24] with a quantitative form of chemical hardness.[3] This chemical hardness was originally postulated as a property of Lewis acids and bases that related the stability constants of their adducts to the qualitative idea of the hardness or softness of the separate partners. Table 2.3 lists the original qualitative categories of acids and bases according to their hardness and softness. In acid–base complexes soft bases prefer to bond to soft acids and hard bases bond well to hard acids. Soft does not bond well to hard. Originally, the quality of softness was thought to be electric polarizability.

The idea of electronegativity has always interested chemists[25,26] and there have been many scales published, which can be categorized into one- and two-parameter models. The two-parameter models assume equalization of electronegativities, i.e., charge is transferred between the atoms until they have the same electronegativity.[2,3,23,27] The single-parameter models use other devices (including hybridization, orbital energies, and group properties) to compensate for the lack of the second parameter.

According to Pauling, who first conceived of the idea of electronegativity, the bonding energy $E_{AB}$ in a heteropolar bond is the geometric average of the homopolar binding energies $E_A$ and $E_B$ (the bond energies of atoms bonded to their own type) to which a contribution is added that is proportional to the difference in elec-

**Table 2.3. Hard and Soft Acids and Bases**

| Characterization | Acids | Bases |
|---|---|---|
| Hard | Cations in groups 1, 2, 3, 4, 5, and 13; Rare-earth metal cations; $VO^{2+}$, $MoO^{2+}$, $WO^{4+}$, $BCl_3$, $AlCl_3$, $Si^{4+}$, $Sn^{4+}$, $As^{3+}$, $CO_2$, $SO_3$, $N^{3+}$, $Cl^{3+}$, HCl, $(CH_3)_2Sn^{2+}$, $AlH_3$ | $NH_3$, $RNH_2$, $H_2O$, $OH^-$, $O^{2-}$, ROH, $R_2O$, $F^-$, $Cl^-$, $CO_3^{2-}$, $NO_3^-$, $SO_4^{2-}$, $PO_4^{3-}$, $CH_3COO^-$ |
| Borderline | $Fe^{2+}$, $Co^{2+}$, $Ni^{2+}$, $Cu^{2+}$, $Zn^{2+}$, $Rh^{3+}$, $Ir^{3+}$, $Ru^{3+}$, $Os^{2+}$, $B(CH_3)_3$, $GaH_3$, $Sn^{2+}$, $Pb^{2+}$, $Sb^{3+}$, $Bi^{3+}$, $NO^+$, $SO_2$ | $N_3^-$, $N_2$, $NO_2^-$, $SO_3^{2-}$, $Br^-$, pyridine |
| Soft | $Cu^+$, $Ag^+$, $Au^+$, $Cd^{2+}$, $Hg^+$, $Hg^{2+}$, $BH_3$, $Ga(CH_3)_3$, $GaCl_3$, $Tl^+$, $Te^{4+}$, N, $I_2$, halogen and metal atoms; | $H^-$, CO, RNC, SCN, $R_3P$, $RO_3P$, $R_3As$, $R_2S$, $RS^-$, RSH, $S_2O_3^{2-}$, $I^-$, $CN^-$ |

tronegativity of the elements bonded:

$$E_{AB} = \sqrt{E_{AA} \cdot E_{BB}} + (\chi_B - \chi_A)^2$$

or more generally,

$$-\Delta H_f(A_pB_q) = na(\chi_B - \chi_A)^2$$

The second expression gives the excess bond enthalpy for a binary heteropolar compound $A_pB_q$ with $n = pZ_A = -qZ_B$ ($Z$ is the valency) and $a$ is a proportionality constant that is 1 if electron volts are the energy units or 96.5 if the energy is expressed in kilojoules/mole. These electronegativities $\chi_A$ and $\chi_B$ have a dimension that is the square root of the energy. Figure 2.19 shows the electronegativities of the elements according to the Pauling scale.

Mullikan has given another definition of the electronegativity of an atom: $\chi(M) = \frac{1}{2}(I + A)$, in which $I$ is its ionization energy and $A$ its electron affinity. In this definition $\chi(M)$ has the dimension of energy expressed in electron volts. Figures 1.4 and 1.5 on page 7 show the experimental values of $I$ and $A$ of the elements.[19,28]

There are several other scales of electronegativity. The Allred–Rochov scale is

$$\chi(AR) = 0.36\, Z_{eff}/r + 0.74$$

in which $Z_{eff}$ is the effective nuclear charge seen by the valence electrons and is

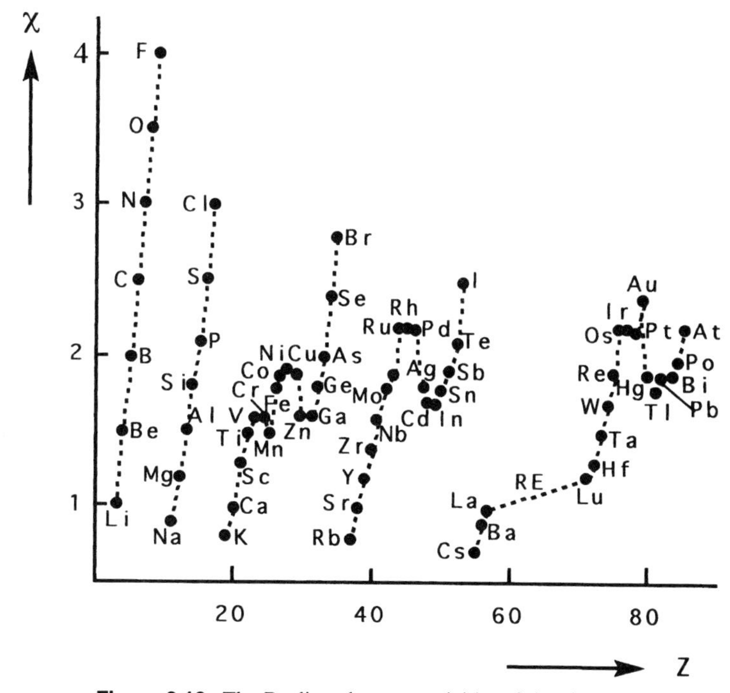


**Figure 2.19.** The Pauling electronegativities of the elements.

derived from the Slater rules, and $r$ is the covalent atomic radius of the atom expressed in angstroms. This definition takes into account the effect of atomic size, which is an advantage, but it also relies on a theoretical concept ($Z_{\text{eff}}$) that is not simply related to an observable.

Sanderson[23] calls his electronegativity stability ratio SR. The electronegativity in this scale is $\chi(\text{SR}) = n/(4.19\,r^3)$, in which $n$ is the number of valence electrons and $r$ the covalent atomic radius. It also expresses the influence that atomic volume has on electronegativity. Sanderson was the first to point out the electronegativity equalization principle. Although the $\chi$-values are different in the various scales as is clear from Table 2.4, all the electronegativity scales correlate strongly with one another.

The two-parameter models of Miedema (Section 2.6) and Pearson (this section) are both based on the idea of electronegativity and its equalization on bond formation but each interprets the positive contribution to $\Delta H$ in a different way: in the Miedema model it means equalizing the electron density at the surface of the atom, while in the Pearson model it is called chemical hardness and also expresses resistance against charge transfer.

In the Pearson model the electronegativity $\chi$ is the first derivative of the energy $E$ with respect to the amount of electronic charge on the atom; the chemical hardness $\eta$ is the second derivative, both at constant internuclear distance and nuclear charge:

$$\chi = -\left(\frac{\partial E}{\partial n}\right) \qquad \text{and} \qquad \eta = \frac{1}{2}\left(\frac{\partial^2 E}{\partial n^2}\right)$$

There is a concept in solid state physics that is similar to electronegativity in chemistry: the thermodynamic potential of the electrons or the Fermi level in solids is equivalent to electronegativity in atoms, except that the sign convention is different.

**Table 2.4. Some Electronegativity Values in Different Scales Compared with Pauling's Values**

| Atom | Pauling | Mulliken | Allred–Rochov | Sanderson |
|---|---|---|---|---|
| H | 2.2 | 2.21 | 2.20 | 2.31 |
| Li | 0.98 | 0.84 | 0.97 | 0.86 |
| Be | 1.57 | 1.40 | 1.47 | 1.61 |
| B | 2.04 | 1.93 | 2.01 | 1.88 |
| C | 2.55 | 2.48 | 2.50 | 2.47 |
| N | 3.04 | 2.28 | 3.07 | 2.93 |
| O | 3.44 | 3.04 | 3.50 | 3.46 |
| F | 4.00 | 3.90 | 4.10 | 3.92 |
| Na | 0.93 | 0.74 | 1.01 | 0.85 |
| Si | 1.90 | 2.25 | 1.74 | 1.74 |
| Cl | 3.16 | 2.95 | 2.83 | 3.28 |
| Ge | 2.01 | 2.50 | 2.02 | 2.31 |
| Br | 2.96 | 2.62 | 2.74 | 2.96 |
| Sn | 1.96 | 2.44 | 1.72 | 2.02 |
| I | 2.54 | 2.52 | 2.21 | 2.50 |

In partial accord with the definition for electronegativity as given by Mullikan, an approximation to the Pearson $\chi$ is the average energy of the frontier orbitals (highest occupied and lowest empty). In this definition the hardness $\eta$ is half the distance between the highest occupied and lowest empty atomic orbital. In semiconductors $\eta$ would be half the bandgap. The approximations are: $\chi = \frac{1}{2}(I + A)$ and $\eta = \frac{1}{2}(I - A)$.

The quantitative concepts in the Pearson model coincide with previous intuitive ideas on electronegativity and hardness:

1. Cations that have a large charge are hard because there is a large energy difference between the highest occupied and lowest unoccupied orbital.
2. Polarizable molecules or ions are usually soft because they have low-lying empty orbitals.
3. If the electron affinity $A$ is low ($A = 0$), the ionization energy dominates the electronegativity and high ionization energy means a high hardness.
4. Metal particles are extremely soft ($I = A$).

In compounds, all the atoms bonded to each other end up having the same electronegativity owing to charge transfer. Figure 2.20 shows how the energy of an atom depends on its charge. Using the definitions in the Pearson model, one can calculate the amount of charge transferred in the bond and the contribution to the bond energy. From

$$E_A = E_A^0 + \mu_A^0 \Delta n_A + \eta_A (\Delta n_A)^2$$

$$E_B = E_B^0 + \mu_B^0 \Delta n_B + \eta_B (\Delta n_B)^2$$

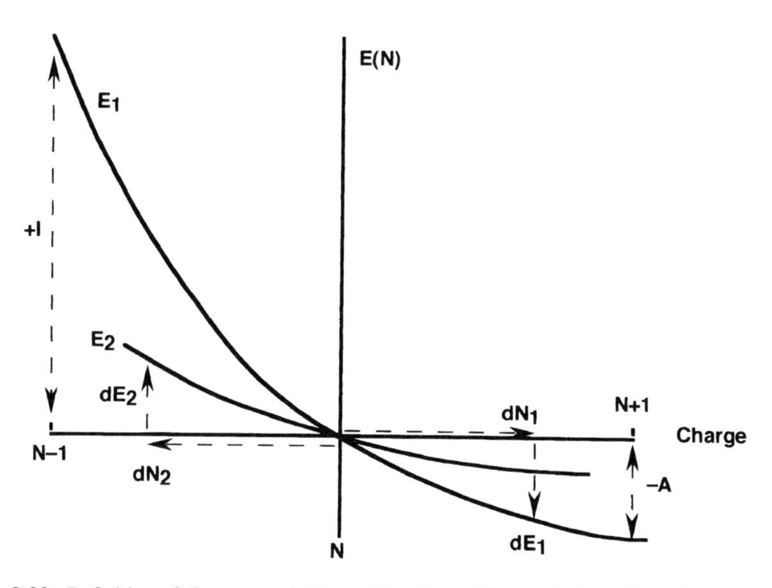


**Figure 2.20.** Definition of electronegativities and hardness. The graph shows how the atomic energies $E_i$ depend on the atomic charge.[27] The neutral atom has charge $N$. Adapted from W. J. Mortier, *Electronegativity Equalization* (1990) with kind permission from Kluwer Academic Publishers.

and

$$\Delta n_A = -\Delta n_B = \Delta n$$

it follows that

$$\left(\frac{\partial E_A}{\partial n_A}\right) = \mu_A = \mu_A^0 + 2\eta_A \Delta n = \mu_A^0 + 2\eta_A \Delta n$$

Thus an amount of charge $\Delta n$ is transferred on bond formation:

$$\Delta n = \frac{\mu_B^0 - \mu_A^0}{2(\eta_A + \eta_B)} = \frac{\chi_A - \chi_B}{2(\eta_A + \eta_B)}$$

The contribution to the bonding energy of that charge transfer is

$$\Delta E = (E_A + E_B) - (E_A^0 + E_B^0) = \tfrac{1}{2}(\mu_A^0 - \mu_B^0)\Delta n$$

or

$$\Delta E = -\frac{(\chi_A - \chi_B)^2}{4(\eta_A + \eta_B)}$$

This contribution to the bond energy is large for soft partners and small if both bound atoms are hard, which is consistent with the old rule that soft bases combine well with soft acids and have stable adducts. The notion that hard bases prefer to bond to hard acids does not follow from this model of covalency. That empirical rule has an electrostatic source; there is no covalency in bonds between hard atoms.

The Pearson model may be extended to include the Madelung potential: ions feel the charges in their environment. The Coulomb contribution affects the electronegativity of an atom by the charges $q$ that surround it[27] at distances $R$:

$$\chi = -\left(\frac{\partial E}{\partial n}\right) + q\left(\frac{\partial^2 E}{\partial n^2}\right) + \sum_f \frac{q_f}{R_f}$$

In the phlogiston model (Section 2.10) the surrounding charges are the result of the charge transfer and the final Coulomb term is shown to compensate for the chemical hardness.

## 2.8. Linnett's Localized Electron Model for Molecules

This qualitative model for the chemical bond describes molecular structure with localized valence electrons. Valence electrons repel each other strongly when they are close to each other in the same shell on one atom and have the same spin. The Linnett model for the bond is based on the recognition that electron correlation exists, which means that the position or the path of a valence electron strongly depends on the positions of the other valence electrons.

This is not the same as in orbital models. The correlation energy is in the first approximation ignored in orbital models where the average position of the other electrons is used to estimate the potential of each individual valence electron. The effect of correlation is introduced in orbital models as a refinement in a quantum-mechanical calculation by mixing excited states in with the ground state using the electron interaction. Linnett's model accounts for the electron correlation from the start.[29] The localized valence electrons in this model are not independent but feel each other's presence in the first approximation, and subsequent corrections become unnecessary. The model, which is qualitative and not parameterized, predicts structures but not bonding strengths, and is described here because it gives a useful overview of covalent bonding and molecular structure. It could also be generalized and parameterized as suggested below.

The Linnett model assumes that localized valence electrons form bonds for the usual reason: lowering of energy. Linnett electrons behave as follows:

1. The valence electrons remain on atoms, repel each other in the atomic shells, and position themselves as far as possible from one another in their shells.
2. Electrons have spin: two electrons having opposite spin repel each other much less than electrons with parallel spins, which behavior follows directly from the Pauli principle in quantum mechanics (Section 2.10).
3. Positioned valence electrons can be shared between two neighboring atoms if some additional space becomes available for them in a neighboring shell; sharing of electrons constitutes chemical bonding.
4. The arrangement of the valence electrons on an atom in a shell determines the direction of the covalent bond with the neighboring atoms; therefore the arrangement of the valence electrons determines the structure of the molecule or lattice.

The first rule means that two valence electrons on an atom sit diametrically opposite each other on either side of the nucleus; three form a triangle around the nucleus; four arrange themselves at the corners of a tetrahedron; five form either a square pyramid or a trigonal bipyramid; six are on the corners of an octahedron; seven form a pentagonal bipyramid; and eight a square antiprisma, a cube, or two interlaced tetrahedra.

According to the periodic table shells offer space for 2, 8, 8, 18, 18, and 32 electrons with increasing main quantum number. A valence shell of a small atom can contain only a small number of electrons; larger atoms can have more electrons in their outer shells (18 in the fourth or fifth period or even 32 in the sixth period) in accordance with the *aufbau* principle of the periodic table.

Usually valence electrons having the same spin are set apart in a separate polyhedron. The first period of elements from lithium to neon has two tetrahedra in the $n = 2$ shell, one for each spin. For a particular atom (given its neighbors) the two spin polyhedra are positioned with respect to each other for the lowest energy of the atom. Figures 2.21 and 2.22 show the arrangement of the valence electrons in their shells around several ions and atoms in molecules. The triplet state of $O_2$ is as simple to describe in this model as the excited singlet $O_2$ and the structures of the molecules HF, $OH^-$, and $NO_2$ (no resonance structures are required in this model), $N_2O_4$, and $O_3$, which are shown, follow easily from the rules. Unlike the valence shell electron pair repulsion (VSEPR) model,[30] the Linnett model does not postulate the necessity of an electron pair for bonding but allows for the occurrence of other bond orders having one, two, three, or more electrons.[1,31]

Nonbonding valence electrons even prefer not to arrange themselves in electron pairs as in the Lewis model unless forced by the situation. Trisilylamine, $(SiH_3)_3N$

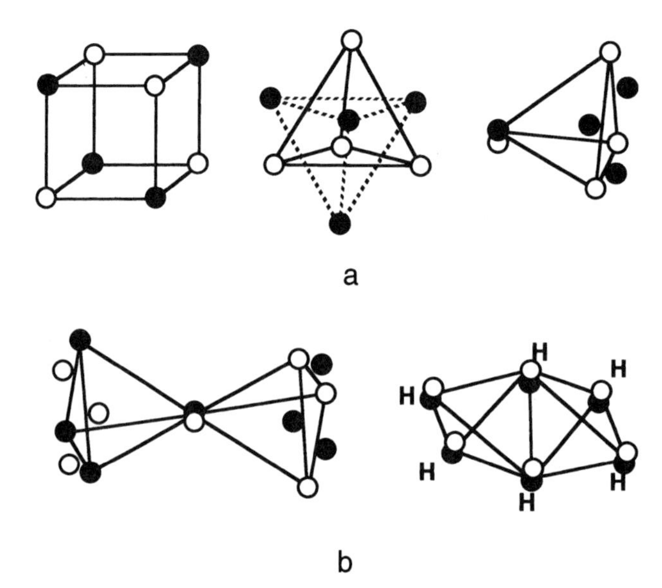


**Figure 2.21.** (a) Some double-quartet structures of eight electrons around one or two nuclei. White and black circles indicate electrons with different spin. At left and in the middle two views of the same arrangement of eight valence electrons in the oxide or halide anions. At right the electron structure of hydrogen chloride or the hydroxyl anion. The hydrogen atom is in the apex of the tetrahedron where the two paired electrons are, and the other nucleus is at the center of gravity of the tetrahedron. The other six electrons do not form spin pairs. (b) At left a single bonded molecule like fluorine or chlorine; it has a bond of two spin-paired electrons, the other electrons are not paired. At right diborane: the electrons are located on the six hydrogen atoms. In both cases the two nuclei (F, Cl, or B) are located in the center of the two tetrahedra.[31] Reprinted with permission from the *Journal of Chemical Education*, Vol. 44, 1967, pp. 206–212; copyright © 1967, Division of Chemical Education, Inc.

is trigonal planar and has two nonbonding valence electrons above and below the plane of the molecule (Figure 2.23). The nitrogen atom in trimethylamine has less space for the valence electrons than silicon, and so the organic amine is not planar but pyramidal. Bonding electrons do occasionally form electron pairs as in the case of borazine ($B_3N_3H_6$) or in benzene. The CO ligand in transition metal carbonyls is bound to the metal ion (which is large enough to accommodate 18 electrons in its shell) by three or sometimes four electrons. Figure 2.23 also shows that if the valence shell of an atom has enough space (has *d*-orbitals in orbital terminology) then it can have more valence electrons. In phosphazenes, the five valence electrons on the phosphorus atom are arranged on the corners of a square pyramid.

The Linnett model not only easily predicts molecular structure, it can derive properties as well. One example is the sign of the magnetic coupling between cations having unpaired electrons in oxide lattices, for which the explanation in the orbital model is conceptually much less easy than in the Linnett model. The coupling between the magnetic moments of the metal ions separated from each other by diamagnetic anions is called superexchange. The sign of the magnetic superexchange interaction $J$ determines whether the lattice is antiferromagnetic ($J < 0$), ferromagnetic ($J > 0$), or ferrimagnetic ($J_1 < 0, J_2 > 0$). The size of $J$ determines the Curie or Neèl temperatures of the compound. The oxide ion has a closed shell of eight valence electrons, arranged in two interlaced spin quartets. A neighboring magnetic cation

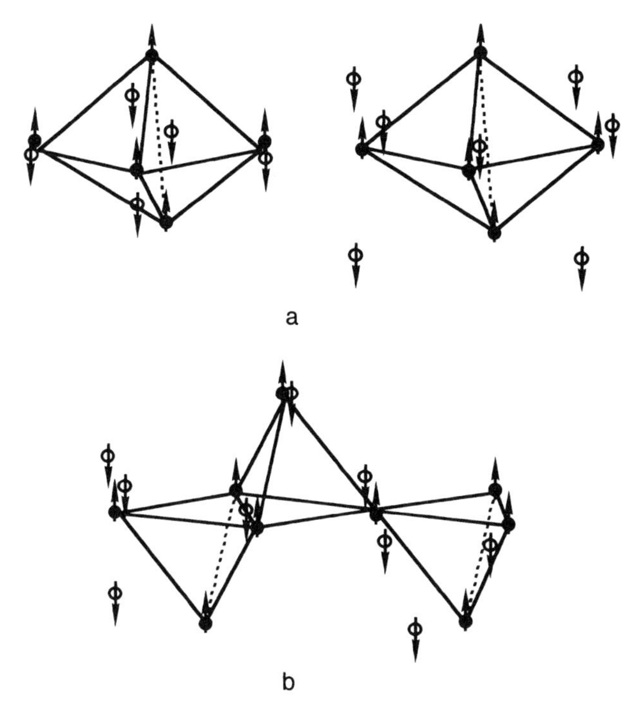


**Figure 2.22.** (a) The double-quartet electrons structures of some biatomic molecules. At left a threefold bond as in nitrogen. The bond in oxygen is shown on the right. The two oxygen nuclei share four electrons and net spin is 1; no electron pairs are formed. (b) The electronic structure of the ozone molecule according to the Linnett model. The oxygen cores in the center of the tetrahedra are bonded to each other with three-electron bonds. There are no electron pairs in the bonds, only the central oxygen atom has one. Electron spin is indicated by arrows.[31] Reprinted with permission from the *Journal of Chemical Education*, Vol. 44, 1967, pp. 206–212; copyright © 1967, Division of Chemical Education, Inc.

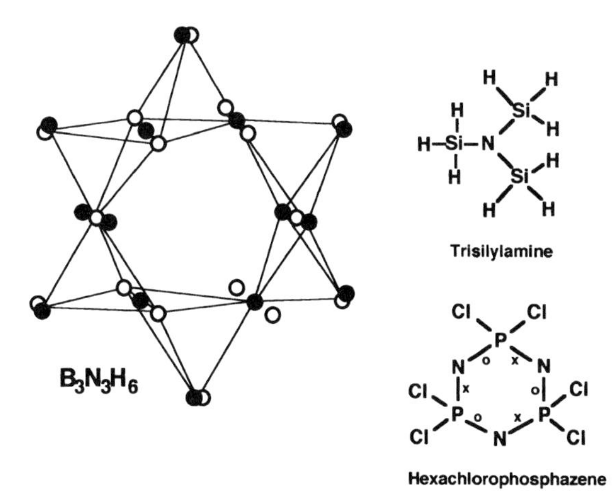


**Figure 2.23.** Arrangement of the valence electrons in a few molecules according to the Linnett model. The boron and nitrogen atoms in the centers of the tetrahedra form a ring; the hydrogen atoms are at the sites of the electron pairs. Electrons of both spins are indicated with filled and open circles.[31] Reprinted with permission from the *Journal of Chemical Education*, Vol. 44, 1967, pp. 206–212; copyright © 1967, Division of Chemical Education, Inc.

$K_1$ orients one spin quartet on the oxide anion because of its spin state. It allows one electron of the four in one spin quartet on the oxide to leak away somewhat on to itself. This electron sharing lowers the energy of the system and is in fact a chemical bond. One of the spin quartets on the oxygen anion (that has a closed shell) is fixed by the position of cation $K_1$ and therefore the other spin quartet is also preferably oriented with its electrons in the middle of the triangles formed by the four electrons of the first quartet. The parallel spins in the other quartet (see Figure 2.21a) now orient the magnetic moment of another neighboring cation $K_2$ antiparallel to the spin of the first cation in the same way: bonding or energy lowering is possible only to an antiparallel cation spin state. If the angle $K_1 - O - K_2$ equals 180° one has antiferromagnetic coupling). An angle $K_1 - O - K_2$ of 90° indicates that the magnetic coupling between the cations is weakly ferromagnetic by a slight trigonal flattening of the two spin quartets (as a result of the electron transfer to the cation). This model also shows how the sign of the coupling depends on the total spin of the cations.

The Linnett model is not parameterized as the Pearson and Miedema models are and bond enthalpies cannot be calculated from atomic data. However, it is strong on molecular structures and provides a simple (but qualitative) explanation for them without the need for much resonance and configuration interaction.

## 2.9. Johnson's Interstitial Electron Model for Metals

There is a third type of strong chemical bond between the atoms in metals, although one school of thought insists that there is no such thing as a metallic bond that is not covalent. Physicists describe the metallic bond in terms of orbitals or bands made up of atomic orbitals localized on the atoms as was summarized in Section 2.3. In that view the so-called metal bond is nothing but a covalent bond. However, in metals the valence electrons are not fixed on the atoms as they are in covalent or ionic compounds, and so they need not necessarily be described in terms of orbitals localized on the atoms but might also be in the interstitial sites. Metals have more interstitial tetrahedral or octahedral holes than valence electron pairs and are somewhat electropositive. This combination makes the solid that has such metal bonds an electron conductor and a metal. Thus metal bonding implies valence electrons in interstices, and it can only occur in atomic solids, being rare if not altogether absent in molecules.

The description of metal bonding by interstitial electrons was given in 1972.[4] In the interstitial electron model, atoms in metals shed some of their valence electrons, which occupy interstitial holes in the metal ion lattice. Not all of the valence electrons are in the interstices. The more electrons that are removed from the atoms, the more electronegative the cations become and the less they tend to donate further electrons to the interstices. Thus some valence electrons stay behind on the metal ions. How many depends on the atom.

A bcc lattice has two octahedral interstices and three pseudotetrahedral interstices per atom. The ccp and hcp lattices have one octahedral and two tetrahedral interstitial sites per atom. In metals the interstitial octahedral and tetrahedral sites between the positive metal ions are not all fully occupied; some of them are only partially filled with valence electrons and some are even empty. Conduction is possible in metals because the interstitial electrons can move through the lattice by occupying neighboring empty or only partially occupied interstices.

Johnson's interstitial electron model provides a qualitative description of the conductivity and magnetism of intermetallic compounds. The valence electrons that remain on transition metal atoms in a metallic compound are in *d*-shells. Their energy levels split under the ligand field of the surrounding interstitial charges (and neighboring cations), which behave like ligands. The valence electrons that remain on the atoms can cause the metal to become magnetic if their total spin is not zero: in that case the atoms have a net magnetic moment. The octahedral interstices are separated from the tetrahedral ones by eight trigonal boundaries of three metal ions each, while the octahedral sites are separated from one another by an edge between two metal ions. Tetrahedral interstitial sites also share an edge between them. The strength of the repulsion between the electrons depends on the distances between the interstices. The distribution of the free valence electrons over the available interstices is such that the electrons have the least interaction among them and are well screened from each other by the cations.

In a metal with an fcc unit cell, the four tetrahedral interstitial sites will be first occupied by the free valence electrons. That arrangement would mean that these electrons have less repulsion energy than if they also occupied the octahedral sites. The mutual orientation of the interstitial cavities of two lattices that occur frequently is shown in Figure 2.24. The metals do not all have the same number of interstitial electrons. The number of valence electrons that could leave the atoms to go to interstitial sites depends on the energy that it costs (ionization and repulsion) and the energy that is gained by transfer (lower kinetic energy). The maximum number of interstitial valence electrons is also limited in a metal by the number of available interstitials: a valence state higher than five is not possible for the atoms in a metallic conductor. If six electrons are placed in the three interstices (there are three per atom in this lattice), there are no empty vacancies left for metallic conduction and the substance would be an insulator if it existed. In actual fact a valence state of six would require too much ionization energy, which would not be compensated for by the lower kinetic energy of the interstitial electrons.

Copper has an fcc lattice. The unit cell is described in the original model by its electronic structure $4Cu^{+}(d^{10})$, $2e_{oct}$, $2e_{tet}$. Half of the octahedral interstitial sites and a quarter of the tetrahedral sites are filled with free electrons. The copper lattice has $Cu^{+}$ cations.

Aluminum (also fcc) has the electronic configuration $4Al^{3+}$, $2^2e_{tet_1}$, $4e_{tet_2}$, $4e_{oct}$, which means that, according to this model, there are two electron pairs in the first type of tetrahedral interstitial hole ($tet_1$), four other electrons in each of the four other tetrahedral interstitial holes of $tet_2$ of the cell, and four electrons in the four octahedral interstices.

The bcc structures of iron, molybdenum, and tungsten are not close-packed as are those of fcc and hcp, and more interstitial sites are available for the free valence electrons. In the bcc lattice the octahedral holes are in the middle of the faces and in the center of the edges of the unit cell. Four tetrahedral sites form a circle in the eight faces of the unit cell. A second group of four tetrahedral sites forms a circle around all twelve edges of a cubic bcc unit cell. A maximum of two free electrons per face and per edge can be accommodated in the octahedral sites and two electrons

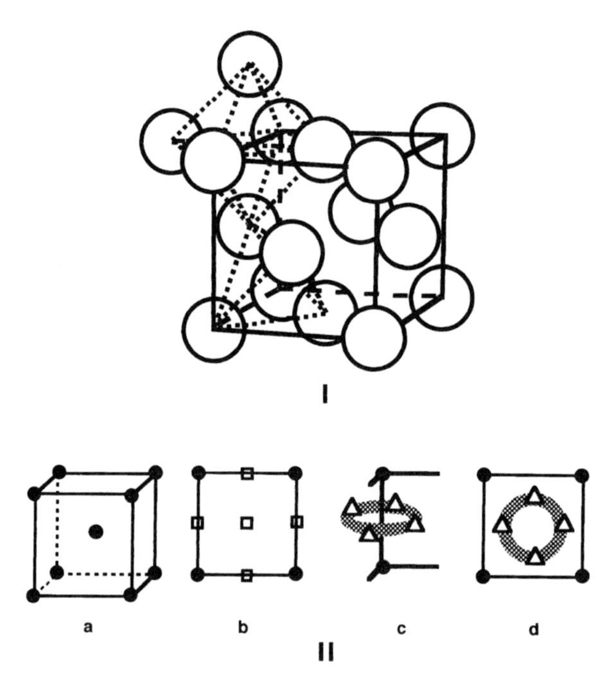


**Figure 2.24.** The interstitial spaces in fcc and bcc lattices: (I) The cubic fcc lattice with an octahedral and a tetrahedral interstitial site indicated. In the interstitial electron model the valence electrons in fcc metals occupy these sites. (II) The cubic bcc lattice a and the types of interstitial sites. The octahedral sites in one of the cube faces are shown as squares in b. The tetrahedral sites (triangles) are closely connected and form rings in the side planes and around the cube ribs as shown in c and d.

per face in the tetrahedral sites. The tetrahedral interstices form a connected net through the lattice through which the electrons can easily be displaced.

The electron configuration for sodium (bcc) is $2Na^+$, $e_{tet_1}$, $e_{tet_2}$. Chromium is a metal that donates four of its six valence electrons to half of the available tetrahedral interstices and keeps two on the atom. It has the electron structure $Cr^{4+}(d)^2$, $4e_{tet}$. Magnesium with the electron configuration $2Mg^{2+}$, $2e_{oct}$, $2e_{tet}$ is an example of a metal that has an hcp structure. All octahedral interstitial sites and half of the tetrahedral sites each have one free valence electron.

The way in which the interstitial electrons act as ligands to split the *d*-orbitals of the bcc metals chromium, tungsten, and iron is illustrated in Figure 2.25, which shows that the tungsten and chromium ions are spin-paired and the iron ions have a triplet ground state in the ligand field of the interstitial electrons and the other cations.

Table 2.5 lists the distances between the interstices in metal lattices. These distances determine the order in which the available interstitial sites are occupied by the free metallic valence electrons. The closer any two interstices are, the stronger the repulsion between the electrons in them.

Johnson's interstitial electron model has not been used by others, the dominant model for metallic bonding being the band model. Recently the interstitial view was revived because new *ab initio* calculations (calculated approximations without

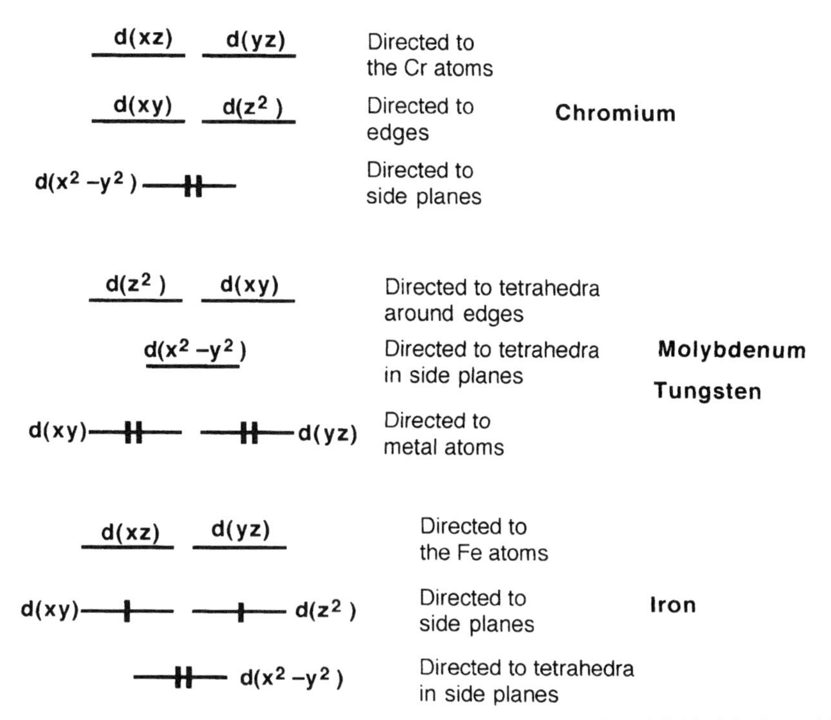


**Figure 2.25.** Splitting of the valence electron levels on metal atoms by the ligand field of the interstitial electrons. The magnetic moments on the iron atoms are ferromagnetically coupled.

empirical parameters) and magic number measurements of lithium clusters have partially supported it (Section 2.10). These recent calculations also corrected the model: the interstitial valence electrons prefer the tetrahedral interstitial sites and avoid the octahedral interstices owing to energy considerations. A limitation of the Johnson model is that it does not derive the number of interstitial electrons from the electronegativities of the atoms and the interstices. In this model for every particular case the number of free electrons is prescribed to be consistent with the observed magnetism.

**Table 2.5. Distances between Interstices in Metal Lattices**

| Distance | ccp | bcc | hcp |
|---|---|---|---|
| Unit cell edge | 1.414 | 1.155 | 1 |
| tet-tet | 1 | 0.82 | 1 |
| tet-tet | 0.71 | 0.41, 0.58 | 0.41 |
| oct-oct | 1 | 0.41, 0.58 | 0.82 |
| oct-tet | 0.61 | — | 0.61 |
| metal-oct | 0.71 | 0.58, 0.81 | 0.71 |
| metal-tet | 0.61 | 0.645 | 0.61 |

## 2.10. The Phlogiston Model

All the models described in this chapter were developed independently and each of them approaches what is basically the subject of the chemical bond from a different direction. Taken together these models provide a wide view of two types of chemical bonds. Their different aspects can be combined into one coherent bonding model, made up here to show the relationships between them, and called the phlogiston model. The phlogiston model is a quantitative generalization of Linnett's localized electron repulsion model that has been parameterized like the Pearson model. It treats interstitial electrons for metals on the same footing as valence electrons on atoms, and for metals it extends Johnson's model by letting the electronegativity determine the number of detached valence electrons.

All of these models are connected through the Ruedenberg principle,[32] which states that the main part of the energy of the covalent bond between atoms is the lower kinetic energy of the valence electrons as they get more space. This is clear from the energy expression of a particle localized in a one-dimensional box that has a length $L$ (derived in Section 2.2):

$$E(n) = \frac{h^2 n^2}{8mL^2}$$

The exact shape or analytical expression of the potential is not essential for the argument; it is sufficient to see it as a potential well. For the simple well considered here, the larger $L$ becomes in a potential of the same depth, the stronger the particle delocalizes (the electron is then distributed over a larger available space) and the lower its kinetic energy (and therefore the total energy). Such a situation obtains for two atoms that approach each other with available space in their wells. When the potential wells are sufficiently close to each other the valence electron in one of them is able to occupy both atomic wells instead of the original single one. The delocalized electron now has more room available, its kinetic energy drops with respect to the situation before bonding, and the lower energy means bonding. Valence electrons that are involved in bond formation must be able to occupy the extra space offered by the atom that is the bonding partner. If the distance between the two atoms is too large, there is a barrier obstructing the exchange of electrons, the kinetic energy is not lowered by delocalization, and there is no bonding.

This model explains why two electrons on two atoms (for example hydrogen and lithium) get more space when the atoms are close enough to each other. Two electrons that have antiparallel spins hardly feel each other in the first approximation while electrons with parallel spins strongly repel. The metaphor that describes valence electron density as a liquid that fills the available space on atoms is convenient to describe the covalent bond, as suggested in Figures 2.26 and 2.27. This behavior of bonding electrons is also implicit in the Miedema and Pearson models. The valence electrons that are not directly involved in the bond are arranged on the atoms according to the simple prescription of Linnett, and this accounts for the directions between the covalent bonds.

In the bond between electropositive atoms (metals) some of the valence electrons leave the atoms and are distributed over the interstitial holes such that the total energy is minimal. Valence electrons that remain behind on the atoms because

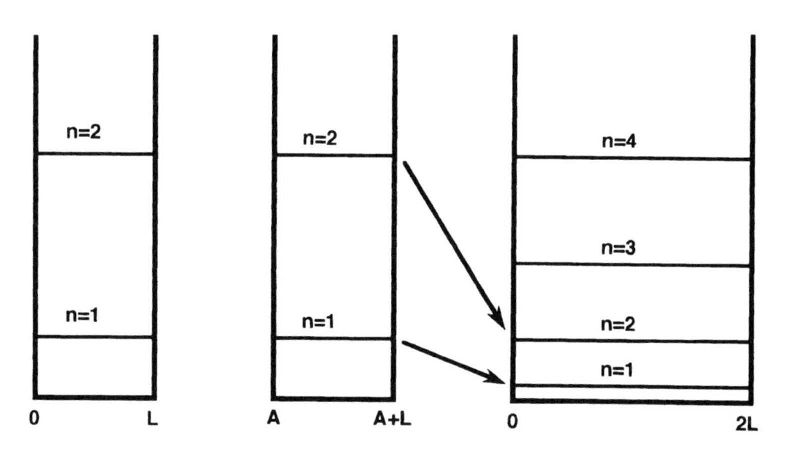


**Figure 2.26.** Chemical bonding represented as lowering of the kinetic energy when the space that is available for the valence electrons increases.[32] To illustrate the point, the potential is simplified to a flat-bottomed one-dimensional box, which makes the electron energy entirely kinetic and inversely proportional to the length $L$. When the two potential wells combine the energy levels for the electrons drop. One electron from each of the two boxes can be paired in the lower $n = 1$ level of the combination box. Reprinted with permission from the *Journal of Chemical Education, Vol. 65*, 1988, p. 581; copyright © 1988, Division of Chemical Education, Inc.

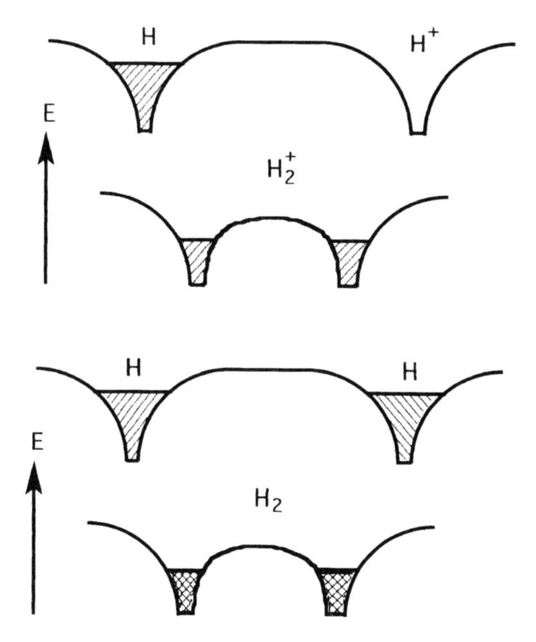


**Figure 2.27.** Graph of the potential wells of hydrogen atoms filled with valence electrons considered as an incompressible fluid[32]. Bonding between two atoms is the result of delocalization of the electron density fluid: the vessel for the fluid becomes bigger and the level drops. Similarly, when an electron with spin *a* (right-slanted) has more space available its average kinetic energy drops. Similarly for spin *b* (left-slanted). The two fluids with different spins can occupy the same space and ignore each other as a first approximation. Reprinted with permission from the *Journal of Chemical Education, Vol. 65*, 1988, p. 581; copyright © 1988, Division of Chemical Education, Inc.

the electronegativity of the cations has become too high again are arranged according to the Linnett formula, or, if more convenient, the crystal field prescriptions.

The interstitials between the stacked electropositive atoms in metals behave as a novel type of atom as they can accommodate electrons. The interstitials are large enough for that and have a positive potential because of the positive ions surrounding them. They have a certain electronegativity and a chemical hardness, which makes them a bonding partner for the lattice atoms. The nature and properties of this novel type of atom depend strongly on the original lattice atom. As the interstitial electron model predicts: more interstitials than free valence electron pairs means that not every interstitial is doubly occupied and the solid is an electron conductor. The behavior of itinerant electrons is analogous to that of ions in solid electrolytes: ionic conduction also needs vacancies.

In molecules, the electronegativity of an atom has a slight dependence on the electronic charges on the neighboring atoms. This Madelung correction on the purely atomic electronegativities is what gives the interstitials in metals their electronegativity. In a later refinement of the Pearson model the electronegativity of an atom is taken to depend on what groups or other atoms are bonded to it. This is only necessary if the field of surrounding charges is ignored. In the phlogiston model, a carbon atom bound to oxygen does not need another electronegativity than when it is bound to other carbon atoms because the Madelung term is explicitly taken into account.[27]

The Pauli principle accounts for crystal and molecular structures and for the fact that covalent bonding is directed in space. Its quantum-mechanical basis is explained with a simple orbital approach: the wavefunction $\Psi(1,2)$ for two electrons that have the same spin (and together form a triplet state) is described by the Slater determinant $|ab|$ in which $a$ and $b$ indicate the atomic orbitals on atoms $a$ and $b$. This notation is shorthand for

$$\Psi(1,2) = a(1)\alpha(1)b(2)\alpha(2) \; - a(2)\alpha(2)b(1)\alpha(1) \; = [a(1)b(2) \; - a(2)b(1)]\alpha(1)\alpha(2) \; = |ab|$$

(the normalization constant has been omitted for convenience). This expression gives the wavefunction of one of the three components of the triplet wavefunction for the two electrons, the component with the total spin quantum number $M_S = +1$. In this expression, $i$ in $a(i)$ indicates the set of spatial coordinates $\{x_1, y_1, z_1\}$ of electron $i$ in orbital $a$, and $\alpha(i)$ is the spin part of the electron $i$ with spin quantum number $m_s = +\frac{1}{2}$. The part between brackets (the spatial part) shows that electrons that have the same spin avoid each other. The wavefunction becomes very small if the electrons approach each other: $x_1, y_1, z_1$ becomes then almost equal to $x_2, y_2, z_2$ and the term in brackets becomes zero. The probability of having them close together is thus seen to vanish.

This traffic-light function of the Pauli principle regulates electron motion and is the source of the correlation of the valence electrons in the molecule. Electrons with the same spin arrange themselves as far as possible from each other and that tendency is ultimately the origin of molecular structure. Four unpaired electrons with the same spin are arranged in a tetrahedron in the atomic shell keeping as much as possible out of one another's way. Linnett's electron repulsion model for molecular structures is based on electron correlation from the outset. Hybridization of the $s$- , $p$- , and $d$-orbitals is not the origin of chemical structure,[10] although it is often presented that way.

The recent quantum-mechanical *ab initio* calculations on lithium clusters referred to earlier have suggested that the valence electrons in these clusters are arranged in interstitial holes and not on the lithium atoms as the LCAO–MO models always assume.[33,34] The structure of small clusters is the result of the tendency of electrons to occupy the interstitials in such an arrangement that the repulsion between them is minimal and that they are best screened by the lithium cations (Figure 2.28). Clusters having 8, 14, and 20 atoms are particularly stable for that reason (Figure 2.29). It has been shown experimentally with mass spectrometry that clusters prefer such magic numbers. Clusters that have 13 atoms or more have fivefold symmetry axes which point to imminent quasi-crystal formations. The form of bond in which electrons are found not on but rather between atoms has been called a "glue bond."[35]

Boron is the only element in the periodic table that ought to be a metal but, owing to the fact that the electronegativity of boron atoms happens to be a bit too high with respect to that of the interstitial spaces between them, it does not conduct electrons easily. The electrons on boron prefer to stay mainly on the atoms in the modifications of the element. Although boron is not a metal, the valence electron density in the lattice indicates a rudimentary tendency to form a metallic bond. They have a nonnegligible density between the atoms in the interior of the polyhedra and in the middle of the interatomic triangles of the chains in elementary boron and in perborides such as $B_4C$ and $B_{13}Si_2$.

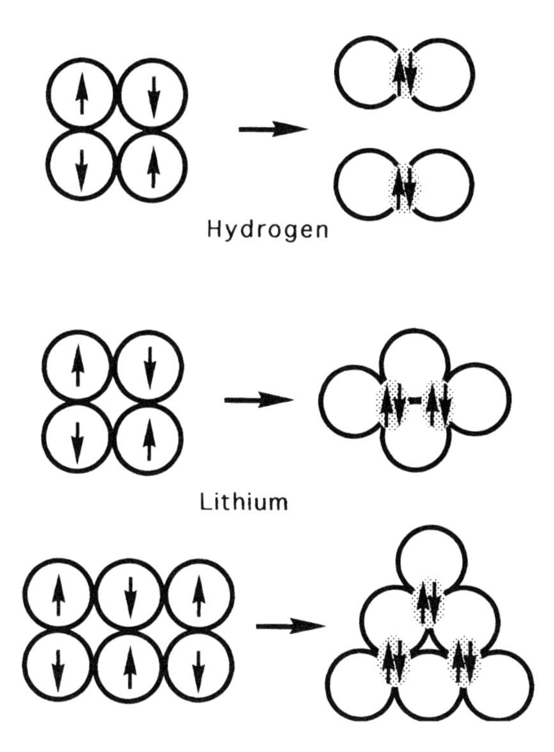


**Figure 2.28.** The difference between covalent bonding and metallic bonding. Four hydrogen atoms break up into two molecules. Four and six lithium atoms form planar clusters with the valence electrons in the triangular interstitial spaces as far as possible from each other. In lithium the bonds are metallic.

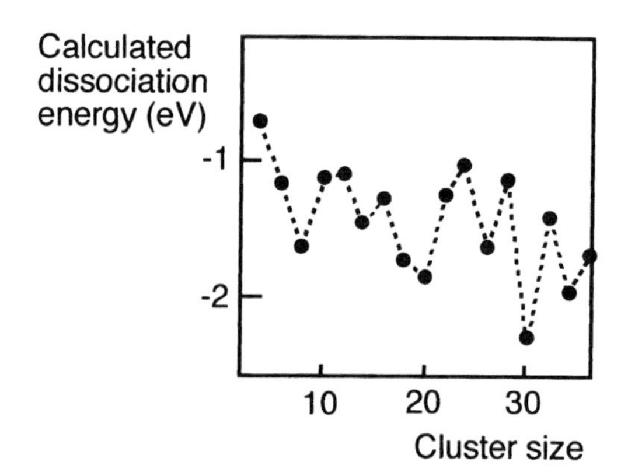


**Figure 2.29.** The bonding energy of lithium clusters calculated with a basis of interstitial atomic orbitals. Clusters of 8, 20, and 30 atoms are relatively stable. This size preference has been observed in mass spectroscopy.

The name phlogiston comes from a formulation developed by Stahl nearly 300 years ago, in which he called the combustible part of matter phlogiston. Lavoisier replaced this concept by the idea that oxidation is not removing phlogiston but rather bonding to oxygen. The old phlogiston is similar to bound valence electrons (as a liquid): oxidation was assumed by Stahl to be removing phlogiston while oxidation is now seen as removing valence electrons.

Bound electrons merit another name to distinguish them from free electrons because their particle properties (mass, charge, spin) are so different. They are subject to interaction with the surrounding nuclei, which can make them heavier or lighter than unbound free electrons in vacuum. Superconductivity, e.g., is one kind of interaction of electrons with the lattice.

The name phlogiston was chosen for a bound electron because the alternative proposed, gluon,[35] is already in use by elementary particle physicists for a different particle and also because the word phlogiston already has the form (with the suffix -on) that is usual for the names of elementary particles in physics. Elementary particles such as electrons, photons, tachyons, mesons, fermions, neutrons, leptons, and so on are examples of that syntax. Phlogistons then are localized renormalized chemical electrons. They behave quite like a liquid,[32] and that electron liquid could be called phlogiston.

Stahl's original idea is now seen to have been partly correct, but as it was not quantitative, it was useless compared to Lavoisier's ideas on oxidation. The phlogiston model fell rightly into oblivion, but recent thinking regarding electronegativity has perhaps partly reinstated it.

The bonding models described in this chapter can be seen as forms of the same localized electron model, called here the phlogiston model, and, as we have seen above, these different forms have been elaborated on by different authors for restricted classes of compounds. The phlogiston model for the chemical bond is based on three physical effects, one classical and two quantum mechanical. The Coulomb attraction between charges that form as a result of charge transfer owing to the difference in electronegativity between the bonded atoms is the classical contribution to the heteropolar bond. The two quantum-mechanical effects are lowering of the kinetic energy of the valence electrons on bonding and the Pauli principle, which prescribes the arrangement of valence electrons and thereby accounts for structure.

The formalism of the phlogiston model, if it existed, would follow closely that of the Pearson model, with differences being in the extension with interstitials, the Madelung potential, and a second electronegativity parameter to include the homopolar bond.

The energy of an atom $i$ ($i$ = A or B) is a simple function of its electronic charge $n$ in excess of the charge present in the neutral atom:

$$E_i(n) = E_i^0 + \left(\frac{\partial E}{\partial n}\right)_{n=0} n + \frac{1}{2}\left(\frac{\partial^2 E}{\partial n^2}\right)_{n=0} n^2 - \sum_k \frac{Ne^2}{4\pi e_0 R_k} n \cdot n_k$$

where $n_k$ stands for the charges on the neighbors and the $R_k$ are the distances to the bonding partners. The fourth term on the right-hand side is the Madelung term, which is due to charges in the neighborhood that are the result of the charge transfer. This term is zero in covalent compounds without charge transfer (when the atoms have equal electronegativities).

By definition:

$$\left(\frac{\partial E_i}{\partial n}\right)_{n=0} = \mu_i^0 = -\chi_i^0 \qquad \text{and} \qquad \left(\frac{\partial^2 E_i}{\partial n^2}\right)_{n=0} = 2\eta_i^0$$

in which $\mu^0$ is the chemical potential of the electrons, $\chi_i^0$ is the electronegativity, and $\eta_i^0$ is the chemical hardness of atom $i$ (all of them for neutral atoms, superscript $n = 0$). The energy of the atom that has some extra electronic charge $n$ is then to second-order for a biatomic compound:

$$E_i(n) = E_i^0 - \chi_i^0 n + \eta_i^0 n^2 - M' \frac{Ne^2}{4\pi\varepsilon_0 R_0} n^2$$

Here $M'$ is the Madelung-like constant for the atom $i$ in the lattice (or in a molecule) under consideration. The Coulomb term lowers the initial chemical hardness $\eta^0$ of the atom.

The binding energy of a biatomic molecule would be

$$\Delta E = (\chi_2^0 - \chi_1^0)n + (\eta_1^0 + \eta_2^0)n^2 - \frac{2Ne^2}{4\pi\varepsilon_0 R_0} n^2$$

The actual electronegativity $\chi_i$ after a certain charge transfer can be derived from the equation for $E_i(n)$ by taking the first derivative to $n$:

$$\chi_i(n) = -\frac{\partial E_i(n)}{\partial n} = \chi_i^0 - 2\eta_i^0 n + M' \frac{Ne^2}{4\pi\varepsilon_0 R_0} n$$

For a binary compound AB with a charge transfer of $n$ electrons B → A, the electronegativity of each atom becomes

$$\chi_A(n) = \chi_A^0 - 2\eta_A^0 n + 2M' \frac{Ne^2}{4\pi\varepsilon_0 R_0} n$$

$$\chi_B(n) = \chi_B^0 + 2\eta_B^0 n - 2M' \frac{Ne^2}{4\pi\varepsilon_0 R_0} n$$

The charge transfer from B to A stops when the two electronegativities have become equal (the image here as was noted before being the liquid level in two communicating vessels). As shown below the final quantity of charge that is transferred on bonding also depends on the chemical hardness of the atoms. The Coulomb term has been taken as equal for both atoms. It then follows that

$$\chi_B - \chi_A = (\chi_B^0 - \chi_A^0) + 2(\eta_A^0 + \eta_B^0)n - 2M\frac{Ne^2}{4\pi\varepsilon_0 R_0}n = 0$$

Here $M = 2M'$, the summation of the terms for both atoms. The upper index 0 for the electronegativity and the chemical hardness indicates the values for the neutral atoms.

The amount of charge transferred can be derived from this expression:

$$n = \frac{\frac{1}{2}(\chi_A^0 - \chi_B^0)}{(\eta_A^0 + \eta_B^0) - M(Ne^2/4\pi\varepsilon_0 R_0)}$$

The last term in the denominator is exactly the lattice energy per mole of the binary ionic compound AB as if a charge of one electron was transferred from atom B to atom A.

The bond energy $E_{\text{bond}} = E_A - E_A^0 + E_A - E_B^0$ is then

$$E_{\text{bond}} = \frac{\frac{1}{4}(\chi_A^0 - \chi_B^0)^2}{(\eta_A^0 + \eta_B^0) - M(Ne^2/4\pi\varepsilon_0 R_0)}$$

The phlogiston parameters $\chi^0$ and $\eta^0$ of all the elements could be derived from the ionization energy and the electron affinity as in the Pearson model, but they also can be determined by fitting the theoretical function of the bond energy with experimentally observed values as in the Miedema model. Given the bond length $R_0$ the parameter $\chi$ and $\eta$ for every element can be derived with a least-squares fit. The existence of any compound, its bond energy, and polarity can be simply estimated with these parameters.

If the formalism were to stop here, it would only be possible to obtain a bonding energy that is due to charge transfer, as in the Pearson or Miedema models. It would not cover homopolar compounds because the electronegativities would be equal ($n = 0$) and there would be no net charge transfer, which would mean that there is no energy lowering or bonding. Yet as shown by the Linnett model, in homopolar bonds the valence electrons get more space, there is equal electron transfer in both directions, and there is bonding through lowering of the kinetic energy. In order to have homopolar bonds in the model as well, each atom is given two electronegativities $\gamma$ and $\chi$ for the donor and the acceptor levels ($\chi < \gamma$ for the same atom). In one shell they correspond to the sublevels of each spin type. Both parameters are important for atoms with equal or almost equal electronegativities and have mutual sharing of electronic charge. In some cases only one or the other of them is relevant, depending on the relative electronegativities of the bonded atoms. To a first approximation the hardness in the donor and acceptor levels can be taken as equal. Thus every atom has three phlogiston parameters, $\chi$, $\gamma$, and $\eta$, and their values can be established once for all the elements by fitting the energy expression to experimentally determined bonding strengths, as in the Miedema model. Part of such a model development was done by Pearson and Mortier in a different scheme.

The basis common to the Pearson, Miedema, Linnett, and Johnson models that have been described in Sections 2.6 to 2.9 can be summarized as follows:

1. Molecules and atomic solids consist of atoms that are bound to each other by localized valence electrons. Certain arrangements of the atoms and of the valence electrons in the atomic shells increase the available free space for certain electrons, and they are no longer restricted to one atom but can delocalize over several. This lowers their kinetic energy and there is bonding.
2. The electronegativity and the chemical hardness are two element-specific parameters that determine how much electron charge is transferred from an atom to the bonded partner and how large the stabilization energy is.
3. If the electronegativity of the atoms is low enough, interstitial holes behave as virtual atoms that have their own electronegativity and hardness and can accept valence electrons. In the phlogiston model, as in the Johnson model, metals are substances that have a cation lattice and interstitial valence electrons.
4. Structure is the result of the arrangement of the valence electrons on the atoms according to the Pauli principle applied to localized electrons in shells. Some of these electrons can be shared by atoms positioned in the right direction.

The phlogiston model can be used to derive properties:

1. All the examples given by the authors of the different models discussed above are covered by the phlogiston generalization of their models.
2. The chemisorption of hydrogen on a metal surface[36] develops via an intermediate configuration that is counterintuitive in other models but that can be easily understood through a localized electron model (see Chapter 6).
3. Metals that have a compressibility of the same order of magnitude as covalent solids are much softer mechanically and more plastic than ceramics because the bonding electrons are in interstices and are much more easily displaced under stress than in ionic or covalent compounds. Owing to the mobility of the valence electrons, defects and dislocations have a different electronic behavior in metals than in rigid semiconductors. The barrier against plastic deformation is approximately proportional to the bandgap (which naturally follows from the model).[37]

There are numerous applications of the model. All phenomena interpreted by orbital models can be viewed in terms of the localized phlogiston model. Thus there are the direct and indirect bandgaps in semiconductors, which express optical behavior. In tight-binding models optical transitions are seen as the transfer of electrons from the valence band to the conduction band. The optical excitation in an indirect bandgap semiconductor is accompanied by phonon effects in the lattice. Translation symmetry is essential for the band model and is assumed throughout by users of the model, even in amorphous solids, where it is absent. In the phlogiston model for simple metals the electron transfer is always spatial displacement of an electron from a metal to an empty interstitial hole nearby. A transition in a direct bandgap semiconductor then means direct displacement to an available neighboring empty interstitial site while an indirect transition means that a lattice fluctuation (phonon-mediated) is necessary to make interstitial room for the electron.

The prime function of a bonding model is not to be a means for understanding which compounds exist and which do not but rather to provide a framework that predicts the existence of novel compounds and novel properties correctly. How fertile the localized electron concept is in this context is left as an advanced exercise.

## Exercises

1. Describe the force–atomic distance relation in several bonding schemes. Give an example of a calculation of the parameters from observables.
2. Explain the three types of strong atomic bonding in terms of the localized electron model.
3. Use the interatomic bonding potentials for a qualitative discussion of properties such as thermal expansion coefficients, compressibility, optical behavior, and chemisorption rate.
4. Explain the role of atomic states with many electrons and their energies in atomic spectra.
5. Give the reasons for color changes in luminescence of the rare-earth ions in oxides.
6. Summarize the molecular orbital and valence bond schemes for chemical bonding between two atoms.
7. Take an example of an inorganic compound from anywhere in this book and rationalize its existence with an orbital model. Using the same model, prove that it is unstable.
8. Discuss the contribution of localization and correlation in the MO and VB schemes; give examples of compounds that can be best described with either of the two orbital models.
9. Describe electron conduction in insulating solids with bands of tight-binding orbitals.
10. Determine the ligand field splitting of the *d*-orbitals in the $Cr^{3+}$ ion in a coordination shell of six $H_2O$ molecules in an octahedron from the spectrum in Figure 2.14 (using Figure 2.3 if necessary). What color could be expected for a solution of $Cr^{3+}$ in liquid ammonia?
11. Determine the ligand field parameters for $Cr^{3+}$ in ruby, emerald, and alexandrite using the Tanabe–Sugano diagrams. What color would you expect a chromium-doped boron oxide glass to have? Why?
12. Explain the bonding between metal ions and ligands in complexes in terms of crystal-field theory and MO theory.
13. In the ruby laser the chromium ion has two low-lying excited states and transitions to them with proper selection rules. Explain the role of these levels in the working ruby laser. Is it possible to change the color of the laser light by changing the strength of the crystal field of the chromium ions in the laser by choosing another host lattice? Does an emerald laser emit green light?
14. Explain the position of the ligands and the metal ions in the spectrochemical series.
15. Derive the Pauling rule for the relative ion-size dependence of the structure of salts.
16. Give two criteria for the occurrence of mixed oxides with predominant ionic bonding.
17. Given any two transition metals, use their Miedema parameters to derive the existence of binary intermetallic compounds.
18. Use the Miedema parameters to derive the existence of borides, carbides, and nitrides of transition metals.
19. Use the Miedema parameters to derive the amount of surface enrichment in solid solutions of metals.
20. Define electronegativity and chemical hardness of atoms and give approximations of them using the optical properties of the atoms.
21. Use the electronegativity and chemical hardness of atoms or groups to estimate the excess bonding energy owing to a heteropolar bond between them.

22. Show that the Madelung potential affects the effective chemical hardness.
23. Explain the structure of some complexes and covalent inorganic molecules in terms of Linnett's localized electron model.
24. Discuss the electric and magnetic properties of metals in terms of the interstitial electron model.
25. Find the Hume–Rothery criterion (15% rule) of mutual solubility of elementary metals and discuss it in terms of the Johnson interstitial electron model.
26. In metallic or ionic lattices there exist nets of linked boron and phosphorus atoms, their oxides, and silicon oxide units. Why are there no nets of carbon and nitrogen atoms in metallic carbides and nitrides as there are in borides?
27. Why has it so far been impossible to make $C_3N_4$? What is wrong there?
28. What is the ground state of a $d^3$ ion in an octahedral crystal field (Figure 2.13)?

## References


1. J. W. Linnett. *The Electronic Structure of Molecules. A New Approach.* Methuen, London (1964).
2. A. R. Miedema, P. F. de Chatel, F. R. de Boer. Cohesion in alloys: Fundamentals of a semi-empirical theory. *Physica* **100B**, 1 (1980).
3. R. G. Pearson. *Chemical Hardness.* Wiley-VCH, Weinheim (1997).
4. O. Johnson. An interstitial-electron model for the structure of metals and alloys *I-VI. Bull. Chem. Soc. Japan* **45**, 1599, 1607 (1972); **46**, 1919, 1923, 1929, 1935 (1973).
5. R. G. Wooley. Must a molecule have shape? *J. Am. Chem. Soc.* **100**, 1073 (1978).
6. S. J. Weininger. The molecular structure conundrum: can classical chemistry be reduced to quantum chemistry? *J. Chem. Educ.* **61**, 939 (1984).
7. S. S. Zumdahl. *Chemistry.* Houghton Mifflin, Boston, N.Y. (1997).
8. K. F. Purcell and J. C. Kotz. *Inorganic Chemistry.* Holt Saunders, Philadelphia (1977).
9. J. K. Burdett. *Chemical Bonding in Solids.* Oxford University Press, New York (1995).
10. J. F. Ogilvie. The nature of the chemical bond—1990: there are no such things as orbitals. *J. Chem. Educ.* **67**, 280 (1990).
11. F. J. Pilar. 4*s* is always above 3*d*, or how to tell the orbitals from the wavefunctions. *J. Chem. Educ.* **55**, 2 (1978).
12. T. A. Allbright, J. K. Burdett, and M. H. Whangbo. *Orbital Interactions in Chemistry.* Wiley, New York (1985).
13. P. A. Cox. *The Electronic Structure and Chemistry of Solids.* Oxford University Press, Oxford (1981).
14. M. Gerloch. *Orbitals, Terms, and States.* Wiley, Chichester (1986).
15. R. Hoffmann. *Solids and Surfaces: A Chemists View of Bonding in Extended Structures.* VCH, Weinheim (1988).
16. N. Cartwright. *How the Laws of Physics Lie.* Clarendon, Oxford (1983).
17. R. McWeeny and B. T. Sutcliffe. *Methods of Molecular Quantum Mechanics.* Academic, London (1969).
18. M. Gerloch and R. C. Slade. *Ligand Field Parameters.* Cambridge University Press, Cambridge (1973).
19. W. W. Porterfield. *Inorganic Chemistry: A United Approach.* Academic, San Diego (1993).
20. L. C. Nathan. Prediction of crystal structure based on radius ratio: how reliable are they? *J. Chem. Educ.* **62**, 215 (1985).
21. O. Muller and R. Roy. *The Major Ternary Structure Families.* Springer, Berlin (1974).
22. F. R. de Boer, R. Boom, W. C. M. Mattens, A. R. Miedema, and A. K. Niessen. *Cohesion in Metals. Transition Metal Alloys.* North-Holland, Amsterdam (1989).
23. R. T. Sanderson. *Polar Covalence.* Academic, New York (1983).
24. L. Pauling. *The Nature of the Chemical Bond.* Cornell University Press, Ithaca (1960).
25. K. D. Sen and C. K. Jørgensen (eds). *Electronegativity.* Springer, Berlin (1987).
26. Y. R. Luo and S. W. Benson. The covalent potential: A simple and useful measure of the valence-state electronegativity for correlating molecular energetics. *Acc. Chem. Res.* **25**, 375 (1992).
27. W. J. Mortier. Electronegativity equalization, solid state chemistry, and molecular interactions. In: J. B. Moffat (ed). *Theoretical Aspects of Heterogeneous Catalysis.* Van Nostrand-Reinhold, New York (1990), Ch. 4.

28. E. C. N. Chen and W. E. Wentworth. The experimental values of atomic electron affinities. *J. Chem. Educ.* **67**, 486 (1975).
29. J. W. Linnett. A modification of the Lewis-Langmuir quartet rule. *J. Am. Chem. Soc.* **83**, 2643 (1961).
30. R. J. Gillespie. The valence state electron pair repulsion (VSEPR) theory of directed valency. *J. Chem. Ed.* **40**, 295 (1963).
31. W. F. Luder. The electron repulsion theory of the chemical bond. I: New models of atomic structure, and II: An alternative to resonance hybrids. *J. Chem. Educ.* **44**, 206, 269 (1967).
32. S. Nordholm, Delocalization: The key concept of covalent bonding. *J. Chem. Educ.* **65**, 581 (1988).
33. M. H. McAdon and W. A. Goddard III. New concepts of metallic bonding based on valence-bond ideas. *Phys. Rev. Lett.* **55**, 2563 (1985).
34. B. Silvi and A. Savin. Classification of chemical bonds based on topological analysis of electron localization functions. *Nature* **371**, 683 (1994).
35. O. Sugino and H. Kamimura. Localized-orbital Hartree–Fock description of alkali-metal clusters. *Phys. Rev. Lett.* **65**, 2696 (1990).
36. P. J. Feibelman and J. Harris. Surmounting the barriers. *Nature* **372**, 135 (1994).
37. J. J. Gilman. Why covalent ceramics are hard. *Mater. Res. Soc. Sympos. Proc.* **327**, 135 (1994).

# Chapter 3

# INORGANIC MOLECULES

*Is one molecule of snow— white?*

STEPHEN THEMERSON

## 3.1. Introduction to Inorganic Molecules

This chapter deals with the use of molecules in inorganic materials chemistry. Materials scientists are interested primarily in solid state and structural chemistry, and not much concerned with molecules as almost all inorganic materials, with the exception of liquid crystals, are atomic solids, which do not consist of molecules. However, materials chemists have to be aware of the behavior of molecules in order to be able to use them to fabricate materials and adapt their surfaces. Whatever is known about the reactivity of particular molecules, e.g., on ligand exchange or on electron transfer, can be applied to solid surfaces or polymers that are modified by immobilized molecules.

One then asks the question, which parts of molecular chemistry are needed? Molecular inorganic chemistry is a large subject, and only the few parts directly applicable to synthesis and surface modification are summarized briefly here. The small molecules that are of most interest in materials chemistry are coordination and metal–organic compounds. Others of importance for materials are inorganic polymers, and these large molecules are not only used as materials themselves but can be precursors for advanced ceramics.

Three aspects of coordination chemistry (described in Section 3.2) are essential for synthesis, for immobilization, and for the operation of molecular devices:

1. Ligand substitution, its reaction rates, and the molecular structure of the reaction products: This touches on aspects of stability versus activity, which are also basic for other fields in materials chemistry. The selection provided below will include substitution kinetics and mechanisms in complexes.
2. Electron transfer rates in redox processes of complexes: Controlling the rates of redox reactions of surface-bound molecules is necessary for the design of active materials in chemical devices.
3. Spectroscopy and magnetism: The optical and magnetic behavior of metal complexes can be tuned if their chemistry is understood.

Metal organics and organometallics are complexes, which consist of a central metal atom bound to organic ligands arranged around it. If a noncarbon atom (oxygen, nitrogen, sulfur, phosphorus) of an organic ligand is directly attached to the metal atom, the compound is called a metal organic. If the metal atom is directly bound to a carbon atom of an organic ligand, the compound is an organometallic. Organometallic compounds are useful reagents in synthesis because they are volatile and can easily react at modest temperatures to form refractory materials (most refractory ceramics have metal atoms). Organometallic precursors are metal atoms equipped with organic "wings."

Polymers and clusters are very large molecules with mechanical characteristics that are related to molecular interactions: tenacity in glue and toughness in plastics, e.g., are explained by intermolecular attraction, rather than by strong interatomic bonding within the polymer chains. In terms of their properties polymers fall between molecules and solids. They are made from molecular precursors but unlike small molecules actually have a microstructure and, if they are large enough, perhaps also a surface. Inorganic polymers are more expensive than organic polymers and for that reason are less popular for daily use. They also tend to be less fragile upon exposure to ultraviolet light, heat, or oxidizing agents than organic molecules, but their chemical inertness has not yet resulted in their being used extensively for consumer products. They will become more important when there is more of a demand for durability in consumer products. An example of a versatile inorganic polymer is polyphosphazene, which is described in Section 3.5. Polymers are potentially useful in nanotechnology as replacements for the shrinking solid-state electronic circuits, which are now still being made with conventional lithographic methods.[1]

There are four types of chemical bonds, in molecular inorganic chemistry, one more than in organic chemistry. Ionic bonding, including dipolar coordination bonding, is seen in inorganic coordination compounds. Covalent bonding occurs within the ligands of complexes (if those ligands are molecules) and of course between all bonded atoms having similar electronegativities. Hydrogen bonding is rare in small molecules but is seen, e.g., in boranes. Finally there is the weak van der Waals type of bonding between the molecules. Metallic bonding is specific for metals and does not occur in molecules although it is sometimes said to exist in mixed-valence compounds such as partly oxidized $(—SN—)_x$ or thiazanes. Incipient metallic bonding is seen in borane molecules and small atomic clusters of metallic elements as was discussed in Chapter 2.

Optical properties are light absorption and light emission (luminescence). Absorption of light by molecules is either allowed (intense colors) or forbidden (weak colors). In spectroscopic terms, there are, respectively, strong charge-transfer absorptions and comparatively weak crystal-field transitions. In a charge-transfer absorption electrons are excited from ligands to the central metal atom, while in crystal-field transitions the excited electrons remain on the metal ion. The optical transition in a charge-transfer band is analogous to a redox process in chemistry: the ligand is photo-oxidized and the metal ion photoreduced in such an excitation. Similarly in chemical charge-transfer or redox reactions an electron passes from the reductant to the oxidant, which may be a molecule, a small particle, or a solid surface. Charge transfer is the basis of molecular electronics and of solid state transducers, devices that interconvert forms of energy. Rates and potentials of electron transfer from and to complexes can be controlled with the right ligands and metals.

Molecular properties are not very different in the different phases of molecular compounds: gases, liquids, solutions, and solids all have the same molecule. As the interactions between molecules, even in solids, are weak, the properties of the chromophores in all phases are similar. However, even if interactions are strong, localized chromophores can often be identified, as shown in Section 2.2.

Inorganic molecules are applied as follows:

1. Interfaces are tailored with molecules. Heterogeneous catalysts can be made from molecules that are known to be homogeneous catalysts by immobilizing them on the surfaces of solids. Surfaces can be made hydrophobic with halogenized silane derivatives. Molecules can act as antenna dyes in novel types of solar cells or in nanoelectronic materials for optical devices. Surface chemistry is molecular chemistry.
2. When they are fixed to inorganic polymer chains they can impart the desired properties, such as color, redox potential, and conductivity, to these polymers. Novel materials for electronic devices are composed of molecules.
3. Molecular compounds are precursors in the synthesis of inorganic solids, e.g., in gas-phase synthesis of coatings.
4. In nonlinear optics, suitable molecules or complex ions are used to transform light. Asymmetric molecules are used in inorganic liquid crystals.
5. Glazes are colored with dissolved transition metal ions when they are complexed with the oxygen atoms of the glass as ligands.

Inorganic molecules, especially the covalent ones, are synthesized using methods that are similar to those used in organic chemistry. Often a vacuum is required or an inert ambient in scaled-up Schlenk tubes. In such tubes all the usual chemical manipulations, such as filtration, distillation, and crystallization under nitrogen or argon atmospheres or under vacuum are possible. The wide range of elements available requires a wide range of methods for processing inorganic materials.

Process temperatures in inorganic synthesis can be very high or very low and require appropriate solvents. Well-known nonaqueous solvents for complex chemistry include $POCl_3$, $CS_2$, $CCl_4$ liquid (condensed) $NH_3$, molten ionic salts, and the organic solvents DMSO and acetonitrile. Water has solvating properties that vary strongly with temperature and at the very high temperatures and pressures reached in autoclaves dissolves silica so well that single crystals of quartz are made hydrothermally on an industrial scale. The use of solvents in synthesis is discussed in Chapter 8.

Clusters are large compact molecules that are, like polymers, intermediate between molecules and grains of solid matter. Metal clusters are model compounds for the active part of heterogeneous catalysts. Such catalysts consist of very small metal particles on the surface of ceramic carriers. To be highly accessible to gases and to be efficient in the reactions, the carriers must have a high specific surface area, i.e., a high surface area per unit of weight. The unexpected behavior of valence electrons in clusters of metallic atoms gives these particles characteristics that differ from those of their parent bulk metallic solids. The absence of translation symmetry in nanosize semiconductor grains induces their remarkable electro-optic behavior.

This is known as the quantum size effect. Small silicon clusters that are the result of an etching treatment of crystalline bulk silicon have a "direct" bandgap (which means loosely speaking that there are allowed optical transitions of electrons from valence to conduction bands). Unlike bulk semiconductors, which have a fixed bandgap that depends only on the nature and crystal structure of the crystalline solid, the luminescence and absorption wavelengths and the virtual bandgap of nanoparticles of semiconductors depend on the grain size. This is one of the many examples of the influence of morphology on the properties of solids. Clusters are made by evaporating the solid in an inert gas at low vapor pressures, so that the mean free path of the atoms is small with respect to the size of the reactor vessel. In that gas atmosphere the evaporated solid condenses to form an aerosol.

The purpose of the present chapter is to introduce the most important aspects of molecular inorganic chemistry for properties and synthesis: electron and ligand exchange in coordination compounds, the reactions of covalent inorganic compounds including organometallic molecules, and the existence of inorganic polymers that are useful materials for a wide range of applications.

## 3.2. Complexes and Their Chemistry

The chemistry of coordination compounds—the chemistry of metal ions in solution[2,3]—is perhaps the largest part of inorganic chemistry as it is taught in university curricula. Examples of significant complexes are $K_3Fe(CN)_6$, $Pt(NH_3)_2Cl_2$, Cu-Phthalocyanine, and $Ru(bipy)(SCN)_3$. Figure 3.1 shows some organic and inorganic ligands and indicates the way in which they are bonded to the metal ion. The complexes have a square planar, pentagonal bipyramidal, or octahedral arrangement of coordinated ligand atoms around the metal ion.

First a few definitions. Complex molecules are composed of Lewis acids (electron pair acceptors, most often metal ions) and Lewis bases (electron pair donors, ligands). Ligands are attached to the metal ion in a complex through an electron pair bond. Polynuclear complexes have more than one metal ion per molecule, and in these compounds the metal ions are linked either directly or by bridging ligands. Ligands may themselves be molecules, often having an electric dipole moment, such as $NH_3$, $H_2O$, or pyridine. Ligands can also be negative ions such as $Cl^-$ or the acetylacetonate anion. Donor atoms (atoms of the ligands that are attached directly to the central metal ion) have a free electron donor pair involved in the coordinative bond. The donor atoms D shown in Figure 3.1 are chalcogenide (O, S, Se) or pnictide (N, P, As) atoms. If the *d*-electrons on the metal have their spins parallel as much as possible, the complex is called high-spin; if *d*-electrons are spin-paired as much as possible, the complex is low-spin.

Ligands are distinguished according to the orbital type of their bond with the metal ion, being $\sigma$- or $\pi$-type, depending on the symmetry of the electron distribution in the bond: the $\pi$-bond has a nodal plane through the bond axis, and the $\sigma$-bond does not (Section 2.3). There are also $\mu$-ligands, which bridge two metal ions in a polynuclear complex. If several donor atoms of one ligand molecule are bound simultaneously to the same metal ion, the ligand is said to be a polydentate ligand or a chelate. Some well-known organic ligands are shown in Figure 3.2. The organic ligands also contain chalcogenide or pnictide donor atoms. Metal complexes with

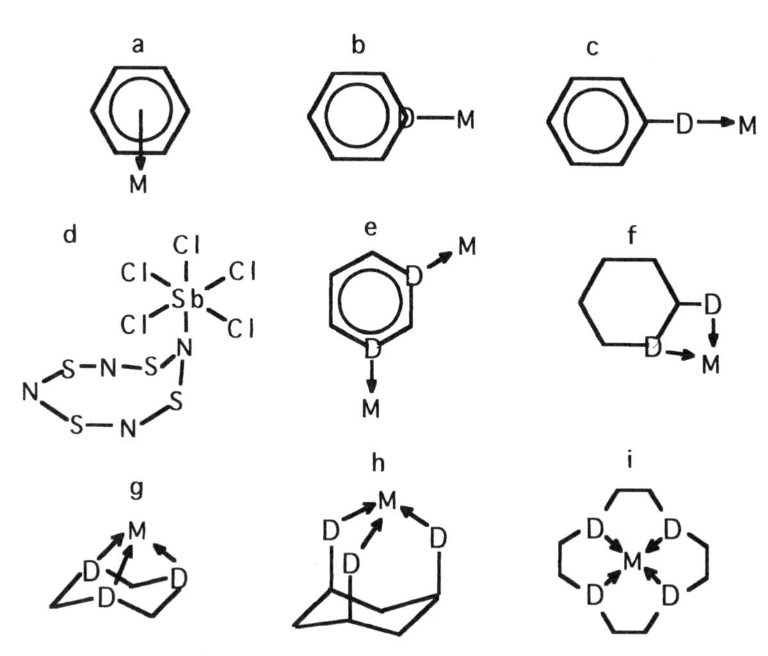


**Figure 3.1.** Types of ligands and the manner in which they are bound to metal ions in complexes. The sixfold rings are (P—N), (B—N), or (S—N) oligomers or aromatic organic molecules. Ligand (a) is $\pi$-bonded to the metal. Monodentate ligands are (b), (c) and (d). Polydentate ligands (e–i) can act as bridging ligands (e).

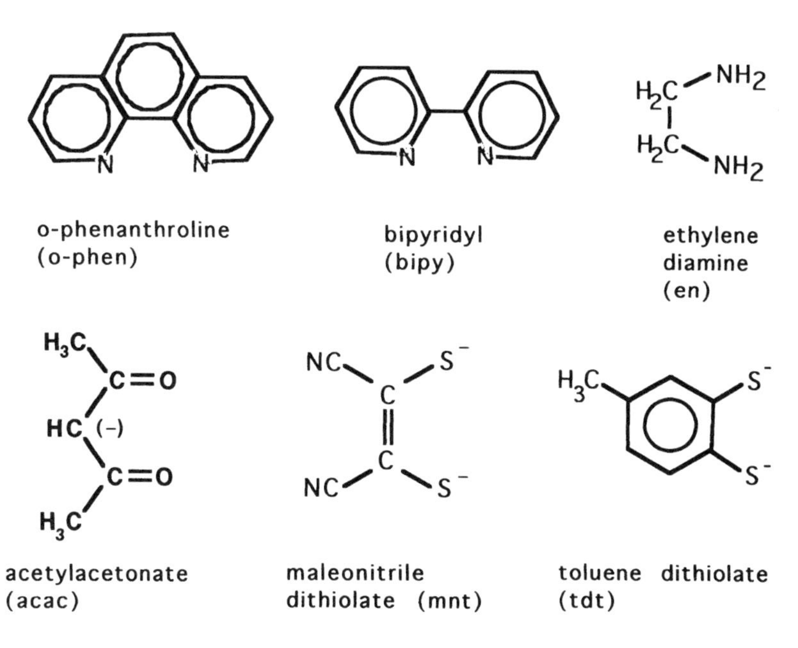


**Figure 3.2.** A few chelating ligands that form stable coordination compounds.

organic ligands are considered to be inorganic compounds although they have organic ligands, because the valence electrons in the complex are mainly on the metal atoms. The coordination number of the central metal ion in a complex (the number of coordinating ligand atoms around the metal and directly bound to it) is not directly related to the conventional valency of the bonding partners but is a function of the size and charge of the ion and ligands. The number of ligands around a metal ion is usually four, six, or eight, forming a tetrahedral, square planar, octahedral, or cubic arrangement. Figure 3.3 shows a complex that has simple ligands and a high coordination number.

Molecular ligands are usually first synthesized and then bound to the metal ions in solution by simply mixing them (see below). Macrocyclic polydentate ligands are sometimes assembled around the metal ion from simple molecules as in the case of nickel phthalocyanine. Such a synthesis is called a "template reaction." Transition metals in complexes can have more than one oxidation or valence state (Table 3.1). The chemical hardness of the ligands and metal atoms determine which of the metal's valence states is the most stable. Soft ligands are used to stablilize low-valence states of metals. A high formal valency of the metal ion does not necessarily imply that the charge on that ion is highly positive. The formal valency of cation and ligand functions merely as a bookkeeping device: it counts valence electrons. In the permanganate ion ($MnO_4^-$) ion, the valency of the manganese ion is formally 7 and that of the four oxide ligands 2 each, but there is much less positive charge ($\approx +1$) on the manganese ion than the valency of manganese ($+7$) indicates.

### 3.2.1. Ligand Exchange

Ligands in dissolved complexes can be replaced with other ligands that are present in that solution, and this ligand exchange is used in synthesis of complexes. The coordination bond can be strong or weak. The bond strength determines the compound's thermodynamic stability (a concept that relates to the energy of the initial and final states) and the size of the stability constant. These temperature-dependent numbers reflect the driving force for the exchange reaction. For the equilibrium:

$$ML_5^{2+} + L \leftrightharpoons ML_6^{2+}$$

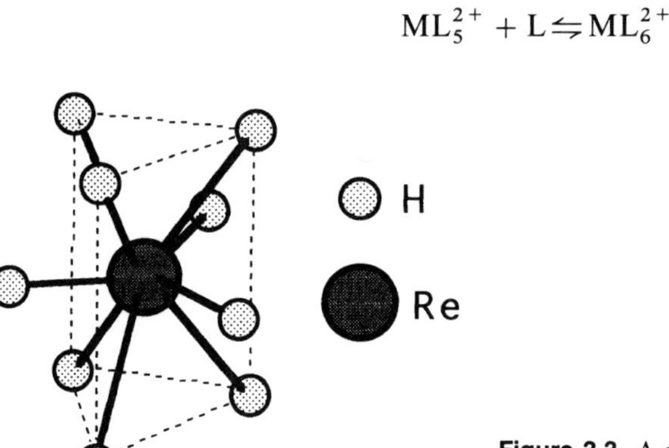


**Figure 3.3.** A nine-coordinate rhenium hydride anion $(ReH_9)^{2-}$.

**Table 3.1. Formal Oxidation States of the Metals in Transition Metal Complexes (x denotes existence, and xs indicates the most stable valence states)**

| | Hard ligands: Fluoride ions, Oxide ions | | | | | | | | Soft ligands: Carbon monoxide, Other metal atoms, Organic molecules | | | |
|---|---|---|---|---|---|---|---|---|---|---|---|---|
| | +8 | +7 | +6 | +5 | +4 | +3 | +2 | +1 | 0 | −1 | −2 | −3 |
| Ti | | | | | xs | x | x | | x | x | | |
| V | | | | x | xs | x | x | x | x | x | | |
| Cr | | | x | x | x | xs | x | x | x | x | x | |
| Mn | | x | x | x | x | x | xs | x | x | x | x | x |
| Fe | | | x | | x | xs | x | x | x | | x | |
| Co | | | | x | x | x | xs | x | x | x | | |
| Ni | | | | | x | x | xs | x | x | x | | |
| Cu | | | | | | x | xs | x | | | | |
| Zr | | | | | xs | x | x | x | x | | | |
| Nb | | | | xs | x | | x | x | | x | | |
| Mo | | | xs | x | x | x | x | x | x | | x | |
| Tc | | x | xs | x | x | x | x | x | x | x | | |
| Ru | x | x | x | x | x | xs | x | x | x | | x | |
| Rh | | | x | | x | xs | x | x | x | x | | |
| Pd | | | | | x | | xs | | | | | |
| Ag | | | | | | x | x | xs | | | | |
| Hf | | | | | xs | x | | x | | | | |
| Ta | | | | xs | x | | x | x | | x | | |
| W | | | xs | x | x | x | x | x | x | | x | |
| Re | | xs | x | x | x | x | x | x | x | x | | |
| Os | x | x | x | x | xs | x | x | | x | | | |
| Ir | | | x | x | xs | x | | x | x | x | | |
| Pt | | | x | x | xs | | x | | x | | | |
| Au | | | | | | xs | x | x | | | | |

the stability constant $K$ is defined as

$$K = \frac{[ML_6^{2+}]}{[ML_5^{2+}][L]}$$

In this expression $K = \exp(-\Delta G_f/RT)$, where $\Delta G_f$ is the free energy change of the association reaction between the ligand L and the rest of the complex ($ML_5^{2+}$).

The complex can be stable or unstable as well as inert or labile. The degree of inertness is a measure of whether the rate of ligand exchange in solution is slow or rapid, not of whether the complex is thermodynamically stable or not. While the stability constants are the equilibrium constants $K$ for certain equilibria, the reaction rate constants $k$ indicates the ease with which the complex reacts, i.e., whether it is labile or inert. For example, the $Ni(CN)^-$ ion has a strong bond between the nickel cation and the ligand cyanide anion and thus has a high stability constant. Nevertheless it is labile because the cyanide ligand can be rapidly exchanged with other bases present in an aqueous solution. $Cr(H_2O)_3^{3+}$ is an inert compound in which the ligand molecules are slow to exchange with water molecules, while $Cr(H_2O)_3^{2+}$ has an exchange rate coefficient that is $10^{16}$ times as large. The relation between the stability and rate constants is $K = k_1/k_2$, in which $k_1$ is the reaction rate constant belonging to the formation reaction and $k_2$ the rate constant for dissociation back. A certain $K$ (a certain stability) can be the ratio of small or large $k$'s and a stable (or an unstable) complex can be labile or inert for ligand exchange.

Figure 3.4 illustrates how the replacement of $H_2O$ by $NH_3$ in an aqueous solution depends on the concentration of $NH_3$ in a labile copper complex. The concentration curves of the complexes in the graph indicate which mixed-ligand complexes of copper exist in the solution and in what concentrations.

As a complex has several association–dissociation equilibria, it has several stability constants. The stepwise stability constant $K_n$ is the equilibrium constant for the reaction

$$ML_{n-1} + L \leftrightharpoons ML_n$$

The so-called instability constant that is sometimes mentioned corresponds to the dissociation equilibrium, which is the same equilibrium in the reverse direction, the two sides in this equation being interchanged.

Subtotal dissociation equilibrium constants $\beta_m$ are sometimes used instead of $K$-values. The $\beta_m$-values are equilibrium constants of the reaction

$$M + mL \leftrightharpoons ML_m$$

Hence the two types of equilibrium constants are related as:

$$\beta_m = K_m K_{m-1} K_{m-2} \cdots K_1$$

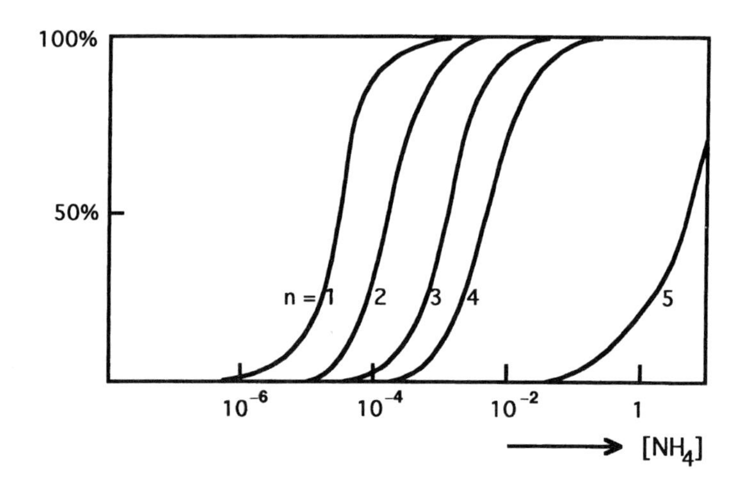


**Figure 3.4.** Solvated water molecules exchanged for ammonium molecules in an aqueous solution of cupric ions. The graph shows how the concentrations of the different complexes depend on the concentration of the ammonium ions. Vertically the percentage of copper present in the solution bound as $[Cu(NH_3)_n(H_2O)_{6-n}]$ is given for $n = 1$ to 5.

**Table 3.2. Values of the Partial Stability Constants of Nickel Complexes of Polydentate Amines[a]**

| | $Ni(NH_3)_6^{2+}$ | $Ni(en)_3^{2+}$ | $Ni(dien)_2^{2+}$ | $Ni(trien)aq_2^{2+}$ |
|---|---|---|---|---|
| $\beta_1$ | $5 \times 10^2$ | $5 \times 10^7$ | $6 \times 10^{10}$ | $2 \times 10^{14}$ |
| $\beta_2$ | $6 \times 10^4$ | $1 \times 10^{14}$ | $8 \times 10^{18}$ | |
| $\beta_3$ | $3 \times 10^6$ | $4 \times 10^{18}$ | | |
| $\beta_4$ | $3 \times 10^7$ | | | |
| $\beta_5$ | $1 \times 10^8$ | | | |
| $\beta_6$ | $1 \times 10^8$ | | | |

[a]Abbreviations: en = ethylene diamine; dien = diethylene triamine; trien = triethylene tetramine.

The stability of complexes depends on the entropy and enthalpy change on reaction. Complexes of polydentate ligands are more stable than those with a corresponding number of equally "strong" monodentate ligands owing to changes in the entropy. If one atom of the chelate is bound to the metal, the chelate's other donor atom follows more easily than a different ligand would. This is the chelate effect: the thermodynamic stability of a complex having a polydentate ligand is greater than the stability of a complex having the same coordination number but monodentate ligands of the same coordinating atoms. Table 3.2 gives the $\beta_m$-values of polydentate (mono-, bi-, tri-, and tetradentate) nickel complexes, which show the chelate effect clearly.

Another factor that determines the stability of transition metal complexes is the crystal-field stabilization energy, which affects the enthalpy of the reaction with ligands. In the ligand-field model this is the extra energy that the valence electrons in the *d*-orbitals of the central transition metal ion get or lose on reaction with the ligand. It strongly depends on how many valence electrons are in the *d*-shell of the metal ion. This contribution is shown in Figure 3.5, which gives the $\beta_m$-values for

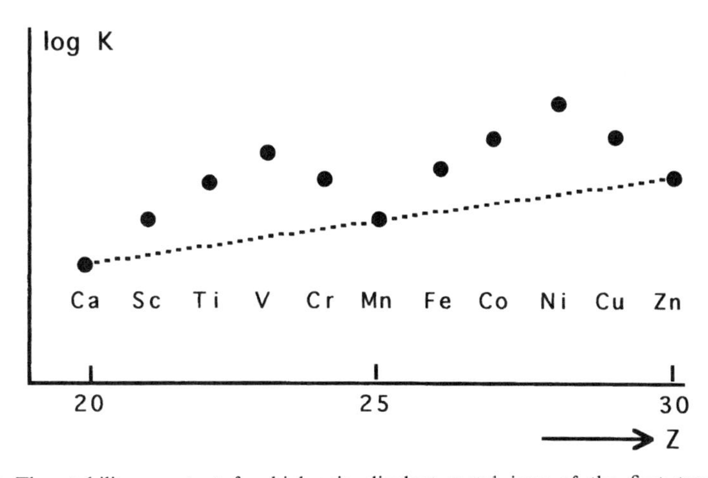


**Figure 3.5.** The stability constant for high-spin divalent metal ions of the first transition series coordinated by six monodentate ligands. The dashed line gives the expected increase in $K$ without the contribution from the nonbonded valence electrons on the metal ion. The characteristic double hump is the result of the crystal-field effect.

**Table 3.3. Stability Constants of the Metal–Hydroxide Bonds of Several Metal Ions**[a]

| Metal ion $M^{n+}$ | Radius (pm) | Charge/radius | $K\{MOH^{(n-1)+}\}$ |
|---|---|---|---|
| $Li^{+}$ | 60 | 1.7 | 2 |
| $Ca^{2+}$ | 99 | 2.0 | $3 \times 10$ |
| $Ni^{2+}$ | 69 | 2.9 | $3 \times 10^{3}$ |
| $Y^{3+}$ | 93 | 3.2 | $1 \times 10^{7}$ |
| $Th^{4+}$ | 102 | 4.0 | $1 \times 10^{10}$ |
| $Al^{3+}$ | 50 | 6.0 | $1 \times 10^{9}$ |
| $Be^{2+}$ | 31 | 6.5 | $1 \times 10^{7}$ |

[a]These values depend on the cation charge and radius.

divalent high-spin complexes of first-period transition metals. The double hump in log $K$ vs. the metal is attributed to the electron occupation of the metal shells: antibonding electrons weaken the bond. Another feature that affects the stability of an associate is the formal electrostatic field strength on the valence electrons in an ion shell, as can be seen in Table 3.3. If the formal field strength increases too much, $K$ decreases again because the other ligands lower the apparent field strength again. In other words the electrostatic approach breaks down.

Thus the stabilities of different complexes can be very different. The reaction rates for ligand exchange can also be considerably different, as is shown in Figure 3.6. The reaction rate constant $k$ for a ligand exchange reaction

$$ML_n + L' \rightarrow ML_{n-1}L' + L$$

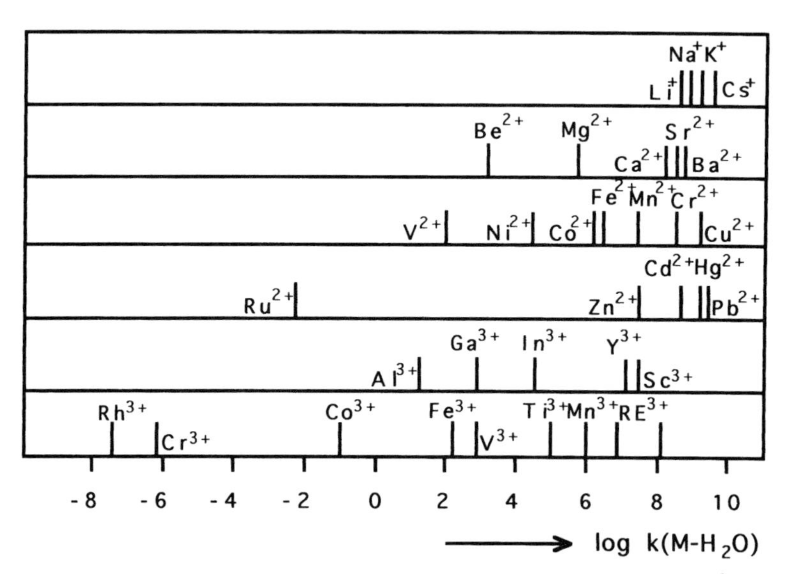


**Figure 3.6.** The rate of ligand exchange in hydrated metal ions in aqueous solution.[2] Rate coefficients range over 15 orders of magnitude, depending on the metal ion. Adapted with permission from A. E. Martell. *Coordination Chemistry. Vol. 2.* Copyright (1978) American Chemical Society.

determines the reaction rate $r$ in this equation according to the expression

$$r = k[ML_n][L']$$

Small highly charged cations are more inert than large low-valency cations. The occupation of the orbitals by the valence electrons (crystal-field stabilization) affects not only the stability but also the ligand exchange rate (Table 3.4) in different ways.

The *trans* effect is a kinetic effect that describes the directing influence of the ligands in square planar complexes on the increased ligand exchange rate at the coordinating site at the *trans* position with respect to that ligand. The series of decreasing *trans*-directing strengths for ligands is

$$CN^- \approx CO \approx NO > PR_3 \approx H^- \approx SC(NH_2)_2 > CH_3^- > C_6H_5 > SCN^- > NO_2^- \\ > I^- > Br^- > Cl^- \text{ pyridine} \approx NH_3 > OH^- > H_2O$$

which happens also to be the sequence of increasing chemical hardness of the ligands: hard ligands have no *trans* effect. The *trans* effect is linked to covalence and is used in synthesis to attach a certain combination of ligands in a particular arrangement to a metal ion in a square planar complex. Figure 3.7 shows the synthesis path for *cis* and *trans* dichlorodiamine platinum(II). The *trans* influence is a thermodynamic effect in which a ligand affects the bonding strength of the *trans* ligand, and should be distinguished from the *trans* effect.

An interesting class of complexes that have sulfur-containing chelates (see Figure 3.2) includes the dithiolenes and dithiolates. These are stable compounds that can have several electron occupations or valencies of the metal ion and maintain one molecular structure. The dithiolate complexes are stable in a state with a charge from $-2$ to 0, which implies a valence state of the metal ion in the complex ranging from

**Table 3.4. Degree of Inertness and Lability of Octahedral Transition Metal Complexes for Dissociative Ligand Exchange**[a]

| Number of $d$-electrons | Square pyramidal $C_{4v}$ | Trigonal bipyramidal $D_{3h}$ |
|---|---|---|
| 0 | Inert | Inert |
| 1 | Inert | Inert |
| 2 | Inert | Inert |
| 3 | Inert | Inert |
| 4 low-spin | Inert | Inert |
| 4 high-spin | Labile | Labile |
| 5 low-spin | Inert | Inert |
| 5 high-spin | Labile | Labile |
| 6 low-spin | Inert | Inert |
| 6 high-spin | Labile | Labile |
| 7 low-spin | Labile | Inert |
| 7 high-spin | Labile | Labile |
| 8 | Labile | Inert |
| 9 | Labile | Labile |
| 10 | Labile | Labile |

[a]Depends on the electron configuration of the central metal ion and also on the configuration of the intermediate five-coordinate transition complex.

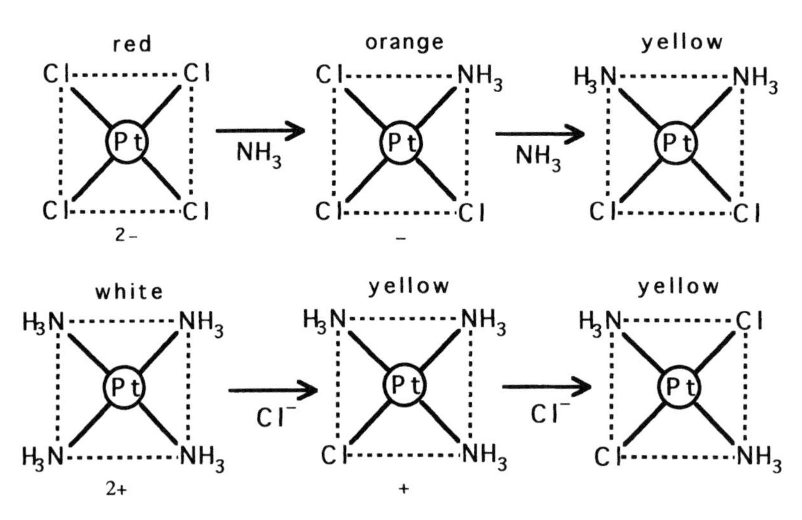


**Figure 3.7.** The *trans* effect in the synthesis of two isomers of a mixed platinum complex.

+4 to +2. Such complexes may therefore be suitable groups in "molecular electronics" and can be used as catalytic dyes for redox reactions. Colored inorganic compounds of this type are optically and chemically stronger than many organic dyes. The redox potentials of complexes can be coarsely tuned by choosing the metal ion and fine-tuned by small changes in the ligand substituents (Table 3.5).

### 3.2.2. Redox Reactions

Reactions accompanied by electron transfer can be fast or slow, depending on the ease of access of the threshold states (like a changed geometry) that are intermediate between the initial and final states of the reacting molecule. A redox reaction can strongly accelerate a ligand exchange reaction by a slight temporary reduction or oxidation of an inert metal ion to a labile valence state. The chromium cation $Cr^{3+}$, which is kinetically sluggish for ligand exchange, can be temporarily reduced to $Cr^{2+}$, which is very labile (see Figure 3.6). If the rate coefficient for the reduction reaction is large and the reduced form labile, then the ligand exchange is

**Table 3.5. Oxidation Potentials in Several Cobalt and Iron Complexes**[a]

| Reaction | Oxidation potential |
|---|---|
| $Fe(H_2O)_6^{2+} \rightarrow Fe(H_2O)_6^{3+} + e^-$ | −0.8 |
| $Fe(CN)_6^{4-} \rightarrow Fe(CN)_6^{3-} + e^-$ | −0.4 |
| $Fe(EDTA)^{2-} \rightarrow Fe(EDTA)^{6-} + e^-$ | +0.1 |
| $Co(H_2O)_6^{2+} \rightarrow Co(H_2O)_6^{2+} + e^-$ | −1.8 |
| $Co(NH_3)_6^{2+} \rightarrow Co(NH_3)_6^{3+} + e^-$ | −0.1 |

[a]These values depend strongly on the nature of the metal ion and the ligand.

accelerated (catalyzed) by impurities that react by reducing chromium and making it labile for a period long enough to exchange the ligands.

The kinetics of redox reactions between complexes is very significant for many types of molecular transducers. Two mechanisms are usually distinguished: double-ligand (outer-sphere) and single-ligand (inner-sphere) electron transfer. Electrons cross, respectively, two ligands between the metal centers or only one bridging ligand when they change metal ions during collision in the redox process. The two mechanisms have different kinetics. An example of an outer-sphere redox reaction is sketched in Figure 3.8. The reacting species in the redox reaction that occurs simultaneously with a ligand exchange reaction are rather inert in this case, while the products of the reaction are labile. An inner-sphere redox reaction has a transition state where one of the reacting complexes has ejected a ligand and forms an associate with the other complex; there is one bridging ligand between the metal ions. Table 3.6 lists several representative reaction rate coefficients for the two types of electron exchange reactions.

The Marcus law of redox reactions is one of the few rules that relate the rate of the reaction to its driving force:

$$k^2 = k_1 k_2 K f$$

where $k$ is the rate coefficient of the electron transfer reaction, $K$ is the equilibrium constant for that reaction, $k_1$ and $k_2$ are the reaction rate coefficients for electron transfer between the oxidized and reduced form of the reductant and oxidant molecules, respectively, and $f$ is the factor that contains the collision frequency. Within certain restricted groups of similar reactions the activation energies may be found to change monotonically with the reaction energies. How that can occur in those particular cases is shown qualitatively in Figure 3.9, which illustrates the energetics behind the Marcus law for double-ligand electron transfer reactions. The curves of the energy of the system of metal complexes during electron transfer

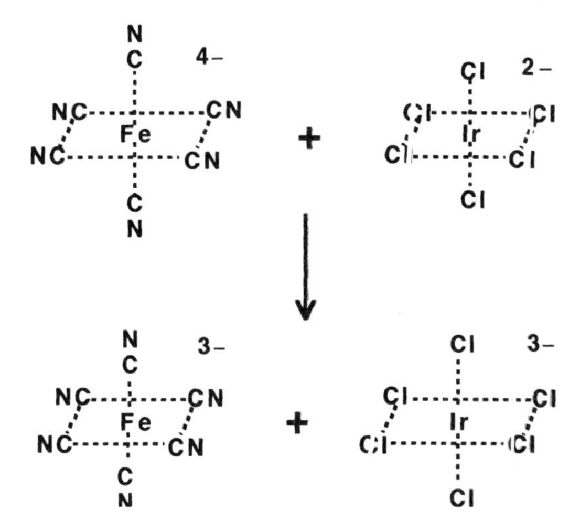


**Figure 3.8.** An example of a double-ligand or outer-sphere redox process. An electron is transferred from the reductant to the oxidant metal over two ligand anions, $CN^-$ and $Cl^-$. The rate constant is $4 \times 10^{-1}\,M^{-1}s^{-1}$.

**Table 3.6. Some Second-Order Rate Constants for Inner-Sphere and Outer-Sphere Redox Reactions in Aqueous Solution**

| Type of reaction | Oxidant | Reductant | Rate constant $M^{-1}s^{-1}$ |
|---|---|---|---|
| IS | $Co(NH_3)_5Cl^{2+}$ | $Cr(H_2O)_6^{2+}$ | $2.5 \times 10^5$ |
| IS | $Co(NH_3)_5Cl^{2+}$ | $Fe(H_2O)_6^{2+}$ | $1.4 \times 10^{-3}$ |
| IS | $Co(NH_3)_5NCS^{2+}$ | $Cr(H_2O)_6^{2+}$ | $1.9 \times 10^1$ |
| IS | $Co(NH_3)_5NCS^{2+}$ | $Fe(H_2O)_6^{2+}$ | $3.0 \times 10^{-3}$ |
| OS | $Fe(H_2O)_6^{3+}$ | $Fe(H_2O)_6^{2+}$ | 4 |
| OS | $Fe(o\text{-phen})_6^{3+}$ | $Fe(o\text{-phen}_6^{2+}$ | $3.0 \times 10^7$ |
| OS | $Co(H_2O)_6^{3+}$ | $Co(H_2O)_6^{2+}$ | 5 |
| OS | $Co(NH_3)_6^{3+}$ | $Co(NH_3)_6^{2+}$ | $10^{-9}$ |
| OS | $Co(o\text{-phen})_6^{3+}$ | $Co(o\text{-phen})_6^{2+}$ | 11 |

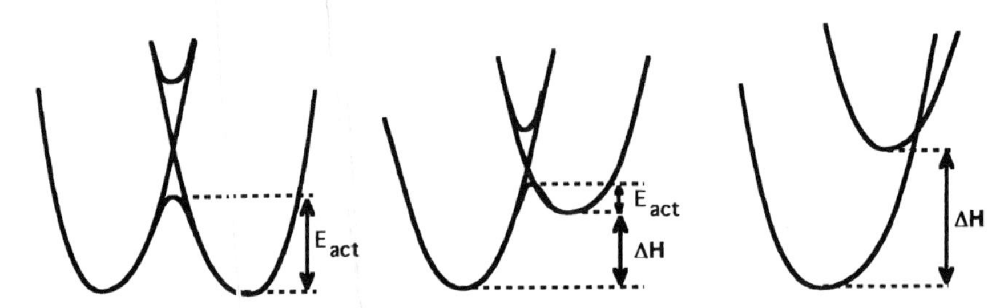


**Figure 3.9.** Use of the potentials for a redox process between two neighboring atoms to explain the relation between the stability and the reactivity of complexes in reactions involving electron transfer. Other things being equal, reaction potentials are related to activation energies, or driving forces to resistances.

against the reaction coordinate show how the driving forces affect the reaction resistances. Within the proper limits, the activation energy is seen to increase with a smaller reaction energy difference.

The foregoing discussion of the chemistry of complexes showed how molecular reactions are influenced by stability and lability. Both can be changed by choosing different metal ions and ligands. The synthesis of metal complexes is not difficult and will not be discussed in detail here. The six methods used most frequently are:

1. SOLVATION OF METAL ATOMS OR IONS:

$$M(A)_n + xL \rightarrow \{ML_x\}A_n$$

In this equation M is the metal ion, A the complexed ligand, which may be an anion (preferably a large one so that it is not an easy complexer itself), and L is the solvent ligand molecule. The reaction can also occur in the gas phase, as in the case of nickel carbonyl.

2. LIGAND EXCHANGE:

$$ML'_y(A)_n + xL'' \rightarrow \{ML''_x\}A_n + yL'$$

The product can precipitate by a low solubility or the ligands L′ can distilled off for a complete reaction. Ethylorthoformiate can be used to bind strongly complexed water (L′) if one wants to replace it with weaker ligands (L″). This reaction can also occur in the gas phase, as in vapors of titanium and boron halides.

3. ANION EXCHANGE WITH SOLVATION:

$$M(A)_n + KA'_m + xL \rightarrow \{ML_x\}A'_m + KA_n$$

When exchanging anions A by anions A′ concomitantly with coordination of solvent molecules L the anions are themselves ligands. An example is the synthesis of $K_3Fe(CN)_6$ in water.

4. FORMATION OF LIGANDS L ON THE METAL ION DURING COORDINATION OF THE PARTS L′ OR L TO THE METAL ION:

$$M(A)_n + xL' \rightarrow \{ML_y\}A_n$$

An example is the template synthesis of phthalocyanines.

5. FORMATION OF THE METAL ION FROM METALLIC SOLIDS DURING COMPLEXATION:

$$M + 2HA + 6L \rightarrow ML_6A_2 + H_2$$

This method is applicable if anhydrous metal salts are not available as precursors.

6. FORMATION OF THE ANION DURING COMPLEXATION:

$$MA_2 + 2KA_m + 6L \rightarrow \{ML_6\}\{KA_{2+m}\}_2$$

This method is suitable if $K$ is a strong acid that is capable of binding the excess anions A.

## 3.3. Molecules with Covalent Bonds

In this section I will describe a small number of inorganic molecules from the large and diverse collection that have covalent bonds between their atoms. Molecules that have covalent bonds are used as precursors for synthesis of advanced solids either from the gas phase or from liquid solution using colloidal intermediates.

In solid state chemistry the binary compounds between the elements of groups 13 and 15 (or 12 and 16) are similar to the elements from the central group 14 or to binaries of those elements (SiC). There are also inorganic molecules made up of elements neighboring carbon, and these are similar to well-known organic molecules, e.g., borazine (also called inorganic benzene), substituted cycloborophosphane, and oligomeric silazanes, shown in Figure 3.10. The inorganic analogy of porphine (porphines are organic tetradentate chelating molecules ubiquitous in living organisms) has boron and sulfur atoms (Figure 3.11). Aluminum bound to nitrogen can

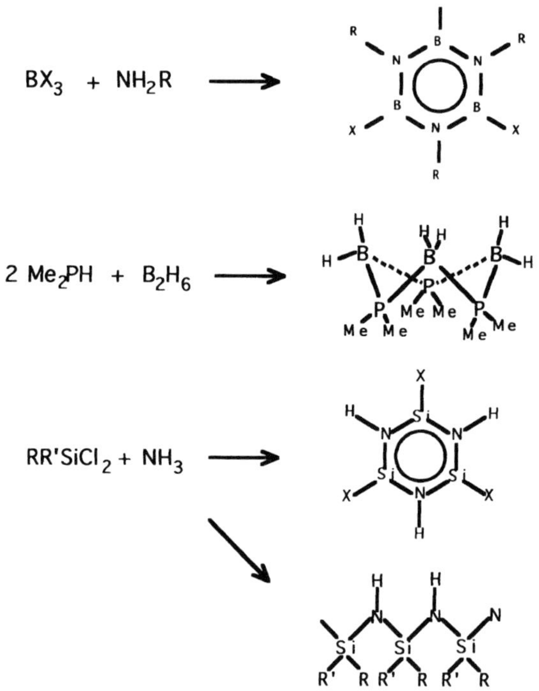

$BX_3 + NH_2R \longrightarrow$

$2\ Me_2PH + B_2H_6 \longrightarrow$

$RR'SiCl_2 + NH_3 \longrightarrow$

**Figure 3.10.** Examples of molecules with covalent bonds.

form extended networks and also cubanelike molecules (Figure 3.12), as carbon does. Organic groups can participate in those networks: aluminum alkanolates have the same remarkable ligand-bridged dimeric structure as diborane and aluminum chloride. Like silicates, phosphates and oxysulfuric acids occur in many different polymeric structures in which tetrahedrons share corners.

Boranes and borane ions are boron–hydrogen compounds that can also act as ligands. Carboranes are boranes in which some boron is replaced by carbon. Borane

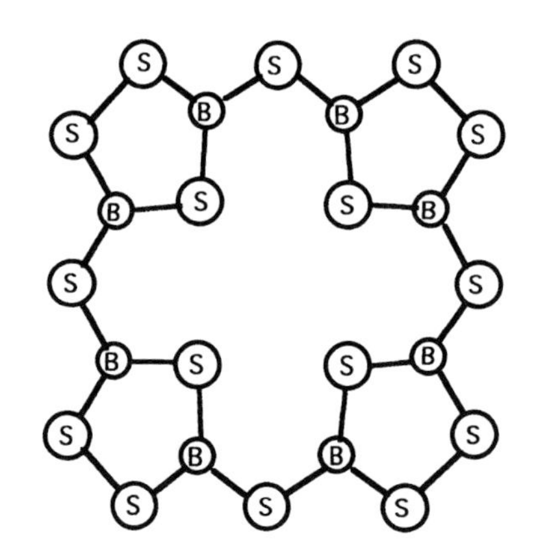

**Figure 3.11.** An inorganic porphine, $B_8S_{16}$.

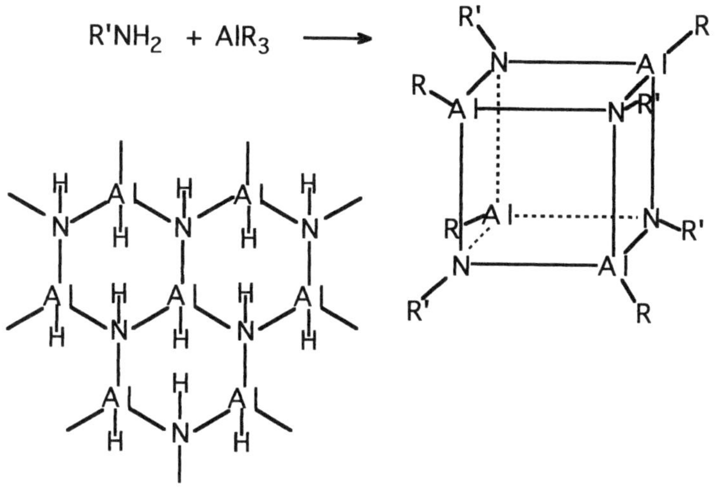

Figure 3.12. Two aluminum–nitrogen cluster compounds.

derivatives constitute a separate branch of molecular inorganic chemistry, which is almost as large and interesting as organic chemistry. Their remarkable structures are characterized by comparatively large coordination numbers for the boron atoms and the occurrence of triangles in the molecular structures. Boron atoms do not supply enough valence electrons for conventional two-electron $\sigma$-bonds as carbon does in organic molecules. Electron-deficient bonds prefer triangles and tetrahedra of atoms that share their valence electrons in the center. Carboranes are thermally much more stable than boranes and boranates. They can also be used as ligands either alone (Figure 3.13) or in combination in one complex with organic groups such as cyclopentadienyl ($C_5H_5^-$).

Metal–organic compounds are used as precursors for the synthesis of inorganic materials, engineering ceramics, and composites. The metal–ligand bonds in these

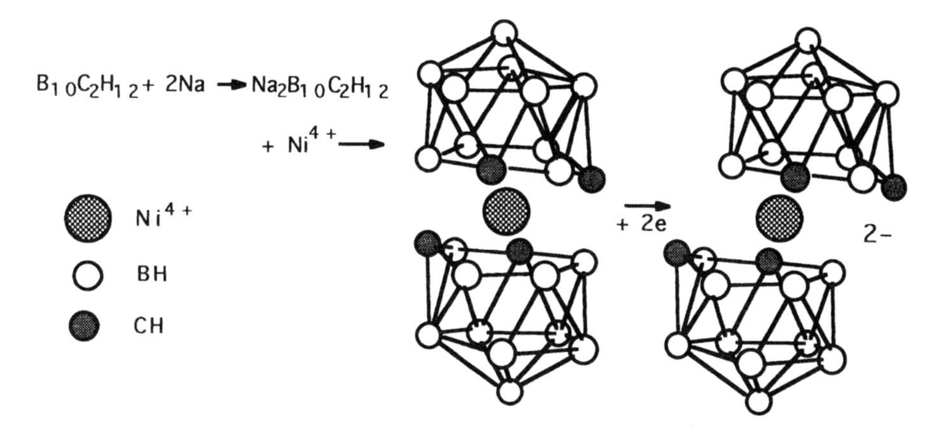


Figure 3.13. *Nido*-carboranes as ligands in a nickel complex that can undergo two reduction reactions.

compounds have some ionic character. In aqueous or alcoholic solution they are reactants for the sol-gel synthesis of ceramic colloids. If they are volatile enough such compounds are used as precursors in gas-phase techniques such as chemical vapor deposition. Examples of such metal–organic compounds are transition metal alkanolates, acetylacetonates (Figure 3.14) and amide-adducts such as the non-pyrophoric alane-dimethyl-ethyl-amine adduct $H_3Al\text{-}NMe_2Et$, which has a melting point of $-12°C$ and a boiling point of 125°C.

The nature of the bond between the metal atoms and the organic groups evidently depends on their electronegativities, size, and hardness. Organic ligands can have a bridging position as shown in Figure 1.12. The volatile covalent organometallics are those of metals to the right of group 2 in the periodic table (transition metals). The bonding energies increase with increasing atomic weight. Organometallic compounds can be made in the gas phase by reacting organic molecules with evaporated metal atoms. Metal is evaporated in a gas of organic molecules and the reaction product is condensed. The mixture of organometallic products is then separated in the usual way.

According to the orbital bonding model, metal alkyls can have $\sigma$- or $\pi$-type bonds or both. In metal cyclopentadienyls the bonds between the metal and the ligands involve the $\pi$-electrons of the ligands. Such compounds have strong colors and are volatile. Some of them such as ferrocene($FeCp_2$) are inert in air but others, especially the paramagnetic ones such as cobaltocene ($CoCp_2$), are very air-sensitive.

The carbon monoxide molecule (CO) is a ligand that coexists well together with organic ligands in one organometallic molecule. In metal carbonyls and organometallic compounds the molecule prefers 16 or, more often, 18 valence electrons, while in coordination compounds (Section 3.2) that number is more variable and depends on the valency of the metal atom. This difference indicates that the bonding in the two classes of compounds is not the same: organometallics have covalent bonds between metal and ligand (organic molecule or carbon monoxide), while the coordination compounds (also many metal organics) have ionic-type bonds. Table 3.7 lists several properties of various metal carbonyls. The ligands in carbonyls can

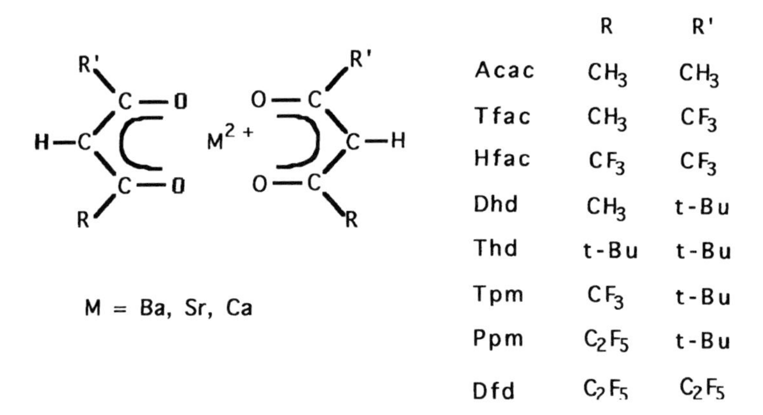


**Figure 3.14.** Metal complexes of acetylacetonate (acac) and derivatives that are used as precursors in gas-phase syntheses of mixed oxides. Examples of additional ligands in these volatile complexes are ethers, dimethyl sulfoxide, or ortho-bipyridine.

**Table 3.7. Properties of Some Transition Metal Carbonyls**

| Compound | Color | Melting point °C | Symmetry | Remarks |
|---|---|---|---|---|
| $V(CO)_6$ | Dark green | 70 | $O_h$ | One unpaired electron |
| $Cr(CO)_6$ | Colorless | 130 | $O_h$ | Sublimes |
| $Mo(CO)_6$ | Colorless | — | $O_h$ | Sublimes |
| $W(CO)_6$ | Colorless | — | $O_h$ | Sublimes |
| $Mn_2(CO)_{10}$ | Yellow | 154 | $D_{4d}$ | |
| $Re_2(CO)_{10}$ | Colorless | 177 | $D_{4d}$ | |
| $Fe(CO)_5$ | Colorless | −20 | $D_{3h}$ | B.P. 103°C |
| $Ru(CO)_5$ | Colorless | −22 | $D_{3h}$ | $Ru_3(CO)_{12}$ with light |
| $Os(CO)_5$ | Colorless | −15 | $D_{3h}$ | Forms $Os_3(CO)_{12}$ |
| $Fe_2(CO)_9$ | Yellow | — | $D_{3h}$ | |
| $Co_2(CO)_8$ | Orange-red | 51 | $C_{2v}$ | |
| $Ni(CO)_4$ | Colorless | −25 | $T_d$ | B.P. 43°C, toxic |

be exchanged, and the exchange rate in very volatile carbonyls such as $Ni(CO)_4$ is three orders of magnitude higher than in group 6 carbonyls.

Similar to metal organics, organometallic compounds have a sufficient vapor pressure and stability at their evaporation temperatures to be useful as reactants in gas-phase synthesis such a chemical vapor deposition. Given the central metal atom, there are some design considerations for increasing the vapor pressure in complexes, which is helped by choosing the appropriate structure and ligands[4]:

1. The complex should be coordinatively saturated for high volatility and the central metal should be screened by the ligands. If a complex is not volatile because it is not coordinatively saturated, adding neutral ligands would help to make them more volatile.
2. The complex should be uncharged and the ligand–metal bonds should be covalent.
3. Ligand groups should not interfere spatially with other ligands.
4. Replacing ligand hydrogen by methyl groups or by ternary alkyl groups lowers volatility in coordinatively saturated complexes and increases volatility in complexes that are coordinatively unsaturated.
5. Replacing ligand hydrogen with fluorine generally increases volatility, for instance, in $\beta$-diketone alkyl groups.

## 3.4. Inorganic Polymers

Very large molecules behave like small pieces of solid. Oxidic colloidal particles are basically strongly cross-linked inorganic polymers. Such compounds can best be made using colloidal methods, such as the sol-gel technique for covalent or ionic metal oxides. These methods will be discussed in Chapters 6 and 8.

Inorganic polymers have backbone chains that do not contain carbon, as organic polymers do, but atoms such as oxygen, nitrogen, phosphorus, boron, silicon, and metals or alternating combinations of them.[5,6] The side groups attached to these backbone chains are often organic. Examples of inorganic polymers are phosphazenes $(=P-N=)_n$, siloxanes $(-Si-O-)_n$, silazanes $(-Si-N-)_n$, borosiloxanes $(=Si-O-B=)_n$, geopolymers $(-Si-O-Al-O^-)_n$, borazanes $(=B-N=)_n$, and borophosphenes $(=B-P=)_n$. Two of these inorganic polymers that have their own market niche, the siloxanes (known as silicones) and the polyphosphazenes, will be described below.

Inorganic polymers often have better mechanical properties (low glass point), optical properties (ultraviolet-transparent), thermal properties (heat and fire resistance), and chemical properties (widely adjustable behavior) than organic polymers but they also are more expensive when processed in the same way as organic polymers. Organic-substituted inorganic polymers are made in two different ways:

1. By substituting bound halide atoms in cyclic oligomers or monomers with the necessary side groups and then polymerizing the substituted oligomers by ring opening or aggregation.
2. By polymerizing the unsubstituted oligomers and then substituting the halogen atoms that are bound to the inorganic polymer backbone by the appropriate organic or inorganic side groups.

### *Polysiloxanes*

Polysiloxanes consist of a $(-Si-O-)_n$ chain having small, mostly organic, side groups attached to the silicon atoms. They are made by hydrolysis and dehydration of organohalosilanes (Figure 3.15). If the synthesis reaction is catalyzed by alkalies, high-molecular-weight linear polymers are formed while acid catalysis yields lower cyclic oligomers. High molecular weights obtain with ring-opening polymerization.

Polydimethylsiloxane $[-Si(CH_3)_2-O-]_n$ or PDMS has a very low glass point ($T_g$) of $-125°C$, which in a polymer is the transition temperature between the brittle glass regime and the rubbery elastomeric region. Its low $T_g$ results from a comparatively large Si—O distance (164 pm) and a large Si—O—Si angle of 180°, which means that like most other inorganic polymers it suffers less from steric obstruction than organic chains that have a C—C distance of 154 pm and a bonding angle of 109.5°. Of the siloxanes, PDMS is the one used most frequently. It has good surface properties because of weak interactions between the methyl groups, a high flexibility thanks to the —Si—O—Si— chain, and a high thermal stability. Such polymers are durable and resistant against temperatures up to 500°C. Silicone oil is used as a heating fluid. In spite of their exemplary stability, silicone wastes are broken down by hydrolysis when they are accidentally released into the environment. Figure 3.16 illustrates the synthesis of polysilanes Above 900°C alkyl-substituted polysilane precursors form silicon carbide ceramics.

### *Polyphosphazenes*

Polyphosphazenes are inorganic polymers with properties and applicability that can be varied within exceptionally wide ranges with the functional substituent groups attached to the phosphorus atoms in their —P=N— chains.[7,8] In general, the structure of a polymer is determined by the precursor monomer, and its properties are more the result of the statistical size distribution than of deliberate

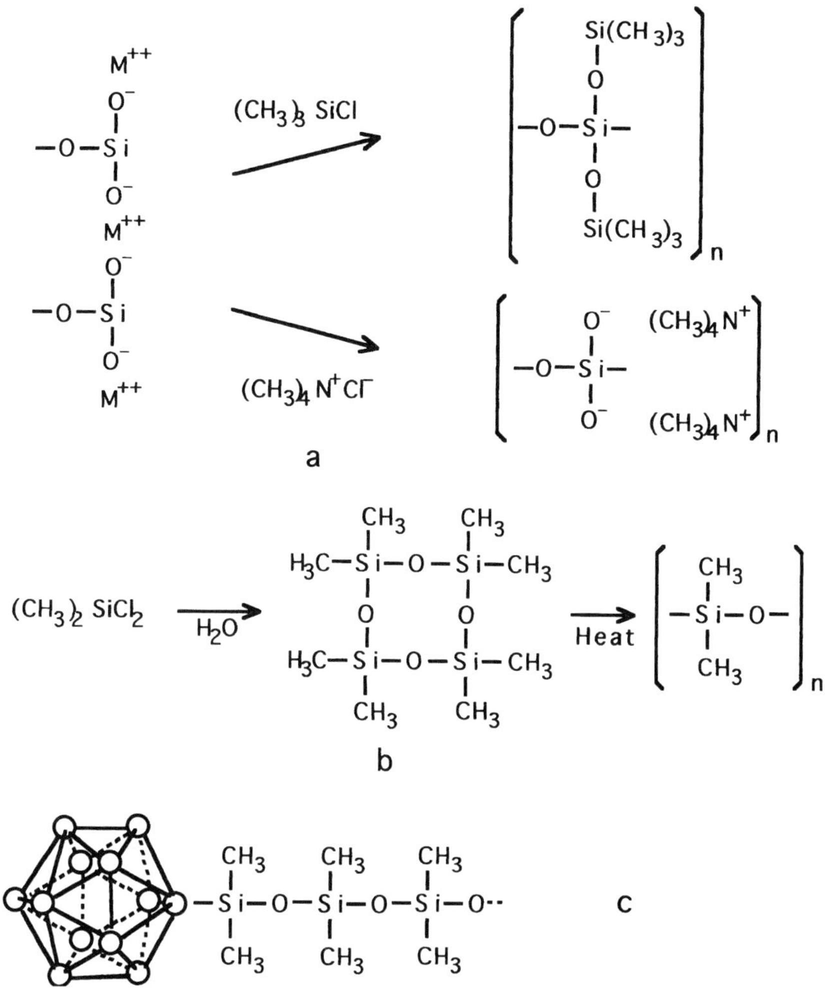

**Figure 3.15.** Some siloxane chemistry: (a) Silicates react with organic silane derivatives to form siloxanes. (b) Siloxanes may also be formed by partial hydrolysis of substituted silanes followed by heat treatment with a basic catalyst. (c) A highly refractory polymer consisting of a carborane-substituted siloxane.

monomer property engineering. However, this is not true of the polyphosphazenes: various monomers yield an unlimited number of structures and different properties can be realized as needed. The synthesis (first polymerization, then substitution) is schematized in Figure 3.17. If an organic substituent is needed that has a carbon atom attached to the phosphorus atom in the central chain, the chlorine atom in the monomer precursor should be replaced by fluorine because Grignard reactants split central chains that have chlorine in their side groups. The second synthetic route in this scheme, first substitution of halogen in the monomers and then polymerizing the substituted monomers, yields partially substituted chlorine only if the chlorinated trimer is the reactant.

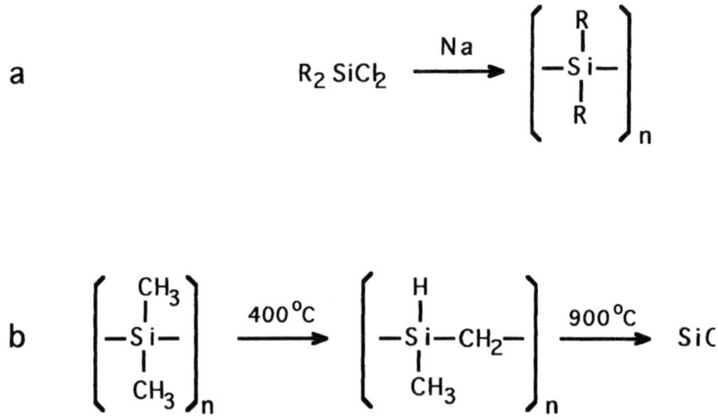

a $R_2SiCl_2 \xrightarrow{Na} [-SiR_2-]_n$

b $[-Si(CH_3)_2-]_n \xrightarrow{400^\circ C} [-SiH(CH_3)-CH_2-]_n \xrightarrow{900^\circ C} SiC$

**Figure 3.16.** Polysilanes are polymers with a silicon chain as a backbone to which organic side groups are attached. Sidegroups R can be aliphatic, cyclic, or unsaturated organic groups: (a) Synthesis of polysilanes. (b) Pyrolysis of polysilanes yields silicon carbide ceramics.

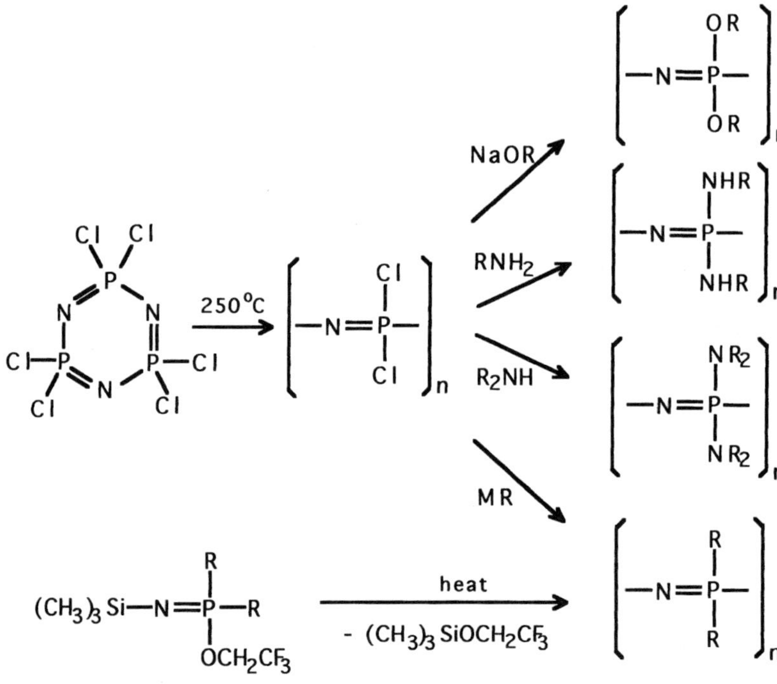

$(NPCl_2)_3 \xrightarrow{250^\circ C} [-N{=}PCl_2-]_n$

$[-N{=}PCl_2-]_n \xrightarrow{NaOR} [-N{=}P(OR)_2-]_n$

$[-N{=}PCl_2-]_n \xrightarrow{RNH_2} [-N{=}P(NHR)_2-]_n$

$[-N{=}PCl_2-]_n \xrightarrow{R_2NH} [-N{=}P(NR_2)_2-]_n$

$[-N{=}PCl_2-]_n \xrightarrow{MR} [-N{=}PR_2-]_n$

$(CH_3)_3Si-N{=}PR_2(OCH_2CF_3) \xrightarrow[-\ (CH_3)_3SiOCH_2CF_3]{heat} [-N{=}PR_2-]_n$

**Figure 3.17.** The synthesis of polyphosphazenes by substituting organic groups for halogen after polymerization by ring opening of chlorine-substituted oligomers. An alternative synthesis uses a more reactive silicon derivative.

Polyphosphazenes have a wide range of applications, depending on the side group R. MEEP (its side group is methoxy ethoxy ethoxy ether: $—R = —OC_2H_5OC_2H_5OCH_3$, hence the name) is a good lithium ion conductor and is eminently suitable as a solid electrolyte in lithium batteries. Inorganic molecules can also be easily immobilized on this polymer, which opens a vast field of application. Examples of some materials with organic groups —R in $\{—PR_2{=}N—\}_n$ are:

- $R = OCH_3$ or $OC_2H_5$, a polymer that has a very low glass point and is a low-temperature elastomer.
- $R = OCH_2CF_3$ or other fluorine-substituted alkoxy group is hydrophobic, chemically stable, has easily formed elastomers, and is very fire-resistant.
- $R = OC_2H_5OCH_3$ is a water-soluble ion conductor, suitable for use as an electrolyte in batteries and as an antifoam agent and film-former (MEEP is a variant).
- $R = NHCH_3$ is stable in aqueous solutions.
- $R = NHC_4H_9$ is insoluble and stable in water.
- R = Procaine is an immobilized anesthetic for implantation.
- R = Pyrrole linked to its neighboring pyrrole groups is a conducting polymer.

Thiazanes are inorganic polymers like polyphosphazenes but with phosphorus atoms in the chain replaced by sulfur. They are electron-conducting covalent polymers that are analogous to solid polysulfur nitride. Electron-conducting polymers such as polythiazanes are too reactive for use in air because they are radicals: a free unpaired valence electron or an electron hole is in chemical language a radical, and radicals happen to be very reactive.

## Exercises

1. What is the use of molecules in materials chemistry?
2. In what ways do molecules help in creating novel properties of solids? Find some examples in the field of sensors and nanoelectronics.
3. Explain the difference in charge transfer in complexes that have $\sigma$ or $\pi$ donor ligands. Which of the two types of organometallic compounds is the more stable.
4. Design some preparation processes to make mixed-ligand square planar complexes using the *trans* effect.
5. What are stability and lability in complexes or other molecules? Name the factors that determine these characteristics.
6. Describe the two mechanisms for redox processes between complexes. How are they related to ligand exchange?
7. Give an example for each of the six main synthesis paths that coordination chemists use.
8. Describe the Marcus law and explain what it means. Is it generally valid? Why?
9. Which molecules mentioned in this chapter are in principle suitable as: (1) catalysts for redox reactions; (2) precursors for synthesizing solids? Give the reaction equations for these processes.
10. Give two methods for making thin films of the following inorganic polymers: (a) MEEP; (b) Dexil; (c) a fire-resistant polymer; (d) an anion-conducting polymer membrane.
11. What type of molecular catalyst would accelerate the ligand exchange in inert complexes?

12. What is the relation between chemical hardness and the *trans* effect (if there is one)? Does that fit with the phlogiston model?
13. Devise a method to incorporate trivalent chromium (which strongly prefers octahedral coordination and is very inert in aqueous solution) in tetrahedral sites in the cation sublattice of a mixed oxide.
14. A ligand field stabilization energy treatment accounts for the inertness of trivalent chromium complexes for ligand exchange. Is the phlogiston model applicable here? If so, how?
15. Give the synthesis of the two possible isomers of $Ni(PR_3)_2Br_2$.
16. Describe the bonding between the metal atom and the organic ligand in all the organometal compounds discussed in this chapter according to the phlogiston model.
17. How can certain metals be selectively etched away from mixtures of fine metal particles?

## References

1. D. Gust. Molecular wires and girders. *Nature* **372**, 133 (1994).
2. A. E. Martell (ed). *Coordination Chemistry, Vol. 2.* ACS Monographs 174. Washington (1978).
3. G. Wilkinson, R. D. Gillard, and J. A. McCleverty (eds). *Comprehensive Coordination Chemistry. The Synthesis, Reactions, Properties and Applications of Coordination Compounds* (7 volumes), Pergamon Press, Oxford (1987).
4. Y. Buslaev (ed). *Electron Structure and High-Temperature Chemistry of Coordination Compounds.* Nova Science, Commack, N.Y. (1996).
5. H. R. Allcock, J. E. Mark, R. West. *Inorganic Polymers.* Prentice Hall, Englewood Cliffs, N.J. (1992).
6. M. Zeldin, K. J. Wynne, and H. R. Allcock (eds). *Inorganic and Organometallic Polymers.* ACS Symposium Series 360, Washington (1988).
7. P. Potin and R. de Jaeger. Polyphosphazenes: synthesis, structures, properties, applications. *Eur. Polym. J.* **27**, 341 (1991).
8. M. Witt and H. W. Roesky. Transition and main group metals in cyclic phosphazanes and phosphazenes. *Chem. Rev.* **94**, 1163 (1994).

*Chapter 4*

# STRUCTURAL SOLID STATE CHEMISTRY

*Deux dangers menaçent le monde: le désordre et l'ordre.*

PAUL VALÉRY

## 4.1. Crystal Chemistry

Solids occur in very diverse crystal structures and a large part of solid state chemistry deals with crystals structures.[1-3] Structural chemistry describes the atomic arrangements in solids and some of the consequences of the various structures for their chemical and physical properties. A small selection is given here to illustrate key aspects that are of interest to the materials chemist. The three primary bonding types are represented in the examples and some properties are shown to be the result of structural features.

Crystal structures get a great deal of attention in materials science because structure has a strong effect on intrinsic properties.[4] Perhaps the most spectacular evidence of the influence of structure on properties is seen in the graphite and diamond modifications of carbon. However, the relationship between structure and properties is not a simple one. Similar structures can have different properties: the group of ionic compounds NaCl, MgO, and NiO have the same halite structure and the group 4 elements (carbon, silicon, germanium, gray tin) also have the same structure, but even within these groups the compounds or elements have very different electrical, thermal, and mechanical properties. Alternatively, compounds with similar properties can have very different structures: GaAs and $Zn_3P_2$ have the same bandgap, SiC and $B_4C$ almost the same hardness, and $Si_3N_4$ and $ZrO_2$ have a similar toughness, but their structures are all very different.

Solid state chemistry itself is not a single self-contained subject but comprises several parts and sometimes the subject is even identified with those parts:

1. Structural chemistry, or the description of lattice arrangements, their energetics, and their physical properties.
2. The chemistry of covalent lattices, the effect of doping on electrical properties, stoichiometries, and solid solubilities.

3. Solid state ionics, the energetics and transport behavior of ions in ionic lattices, solid–gas reactions, corrosion, and diffusion.
4. Catalysis and crystal growth, which are not strictly solid state reactions but rather surface reactions that produce molecules or solids, respectively.

Properties of solids are intrinsic if they are functions of the nature of the atoms and their arrangement in the lattice, or in other words of the chemical formulation and the crystal structure only. It thus makes sense to list the intrinsic properties of solids in tables according to chemical formula. Intrinsic properties are best measured on large perfect single crystals in order to minimize the effect of the boundaries. Sometimes property values are anisotropic. Some extrinsic properties are discussed in Chapter 7. Examples of intrinsic properties are thermal expansion coefficient, dielectric constant, magnetic susceptibility, refraction index, formation enthalpy, the existence of phase transitions, and compressibility. Extrinsic properties, on the other hand, depend on factors such as dissolved impurities, nonstoichiometry, microstructure, and the state of the grain surface. As they are functions of the way the solid is made, they cannot be placed in a one-to-one relation with the chemical formula. Some extrinsic properties are discussed in Chapter 7. Examples of extrinsic properties are hardness, toughness, strength, electrical and thermal conductivity, work function, coercive force, and light scattering.

Some properties show hysteresis, and these are not unique functions of the conditions (pressure, temperature) but depend also on the history or the way the state has been arrived at. Examples of such properties are the degree of magnetic or electric polarization in ferromagnetic or ferroelectric compounds at temperatures below the Curie temperature and phase transition such as melting or solidification. The transition temperature depends on whether it is approached from below or above the Curie temperature. The phase transitions in some ferroelectrics are described in Section 4.5.

Crystal lattices have symmetry elements such as rotation axes, mirror planes, inversion points, and combinations of these. A crystalline lattice has translation symmetry, except for quasi-crystals or icosahedral phases, which have lattices with point symmetry elements only. Glasses are amorphous solids that do not have any symmetry element in their lattices.

The simpler structures with a high symmetry such as sphalerite, calcite, the A15 structure, spinel, garnet, and the silicates are discussed in other introductory texts[2,3] and need not be dealt with here. The examples that are discussed in the following sections have been selected because of their characteristic behavior. Two types have been selected—ceramic compounds and intermetallic compounds:

- Borides of metals and nonmetals: Solid borides involve covalent and metallic bonding or both simultaneously in the same compound.
- Metal carbides and nitrides: These compounds, which are very hard and inert, are usually metallic conductors. They demonstrate that nonmetal atoms can behave in lattices as if they were metallic atoms.
- Perovskites, mixed ternary oxides: These have ionic bonds and exceptional physical properties (such as ferroelectricity) related to their peculiar crystal structure.
- Intermetallic solids: These compounds between metallic elements are metals and may behave somewhat like salts; for instance, they may be brittle. Unlike

**Table 4.1. Crystal Structures Based on Close-Packed Ion Stacking with Ions in Its Available Octahedral and Tetrahedral Interstitials**
[after R. E. Newnham. In: A. M. Alper (ed). *Phase Diagrams, Vol. 5.* Academic, New York (1978)]

| Compound | Close-packed atoms | Layer sequence | Octahedral atoms (occ) | Tetrahedral atoms (occ) |
|---|---|---|---|---|
| $\alpha$-$Al_2O_3$ | 3O | AB | 2Al($\frac{2}{3}$) | — |
| $Mg_2SiO_4$ | 4O | AB | 2Mg($\frac{1}{2}$) | Si($\frac{1}{8}$) |
| $TiO_2$ rutile | 2O | AB | Ti($\frac{1}{2}$) | — |
| ZnS (Wurtzite) | S | AB | — | Zn($\frac{1}{2}$) |
| NiAs | As | AB | Ni(1) | — |
| $Al_2ZnS_4$ | 4S | AB | — | 2Al, Zn($\frac{3}{8}$) |
| $Al_2Se_3$ | 3Se | AB | — | 2Al($\frac{1}{3}$) |
| $CsNiCl_3$ | Cs, 3Cl | AB | Ni($\frac{1}{4}$) | — |
| $BiI_3$ | 3I | AB | Bi($\frac{1}{3}$) | — |
| $SrTiO_3$ | 3O, Sr | ABC | Ti($\frac{1}{4}$) | — |
| $ReO_3$ | 3O | ABC | Re($\frac{1}{4}$) | — |
| $TiO_2$ anatase | 2O | ABC | Ti($\frac{1}{2}$) | — |
| NaCl | Cl | ABC | Na(1) | — |
| $CaF_2$ | Ca | ABC | — | 2F(1) |
| $K_2PtCl_6$ | 6Cl, 2K | ABC | Pt($\frac{1}{8}$) | — |
| $CrCl_3$ | 3Cl | ABC | Cr($\frac{1}{3}$) | — |
| $MgAl_2O_4$ | 4O | ABC | 2Al($\frac{1}{2}$) | Mg($\frac{1}{8}$) |
| ZnS sphalerite | S | ABC | — | Zn($\frac{1}{2}$) |
| $TiO_2$ brookite | 2O | ABAC | Ti($\frac{1}{2}$) | — |
| SiC (4H) | Si | ABAC | — | C($\frac{1}{2}$) |
| $K_2MnF_6$ | 6F, 2K | ABAC | Mn($\frac{1}{8}$) | — |
| $CdI_2$ | 2I | ABAC | Cd($\frac{1}{2}$) | — |
| $BaTiO_3$ | 3O.Ba | ABCACB | Ti($\frac{1}{4}$) | — |

covalent or ionic compounds they have stoichiometries unrelated to the usual atomic valencies. The aluminum–titanum–nickel intermetallics are described as an example of their type.

Table 4.1 lists some structures based on closest dense packing of ions, usually anions, in which the octahedral or tetrahedral interstices are occupied by their partners, usually the cations.

## 4.2. Amorphous Solids and Icosahedral Phases

A solid that has a disordered atomic arrangement, no unit cell, and no symmetry element at all, like a liquid, is amorphous and is called a glass.[5,6] The name glass is also used for those solids that have a crystalline atomic arrangement but are disordered in another way in that they have observables that are not linked by crystal symmetry. Cyanide ions substituted for bromine in crystalline KBr form a dipole glass because the electric dipole moments of the cyanide ions have an orientation without crystalline order. A spin glass is a crystalline, often metallic, alloy that has localized, disorderly oriented electron spins owing to the presence of unpaired electrons on the atoms on the lattice sites. In the discussion on glasses here,

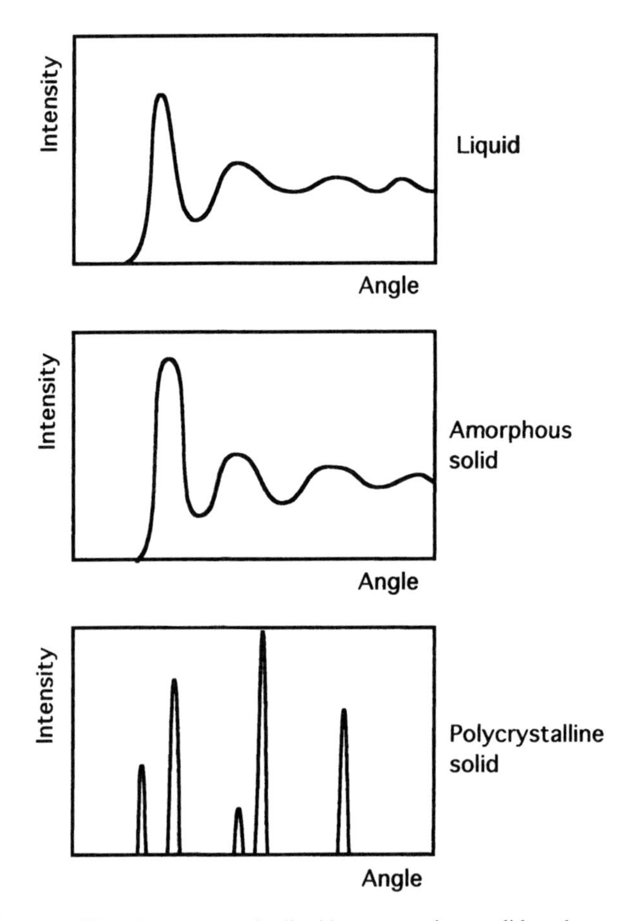


**Figure 4.1.** X-ray diffraction pattern of a liquid, an amorphous solid, and a crystalline solid.

it will be assumed that the glass is amorphous in the structural sense; a glass has no positional symmetry as crystalline solids have.

There are two types of glasses in a structural sense—metallic and covalent. The structural models for both types of glass are made either by random stacking of atoms or by random linking of neighboring atoms. If the atoms are not stacked according to a crystalline lattice and have no directed bonds (i.e., if they have a metallic or an ionic bond) the dense random close-packing model is used, which is a disordered closest packing. The interatomic distances are more or less uniform but the packing is random. Metal glasses are examples. The packing density of random close-packed glasses is slightly lower than in cubic or hexagonal close-packed lattices. For amorphous solids that have directed covalent bonds between the atoms, the structure model is the continuous random network. Neither the bond angles nor the distances between the atoms have precise values as in a crystal but vary around a mean. Silicate, phosphate, and borate glasses belong to this group.

The amorphous state is a nonequilibrium structure in a thermodynamic sense: a crystalline arrangement has a lower energy than a glassy state. If the crystalline lattice has point defects, there is some favorable entropy contribution to the stability.

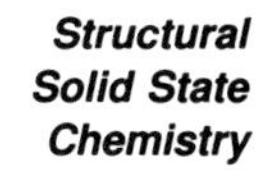


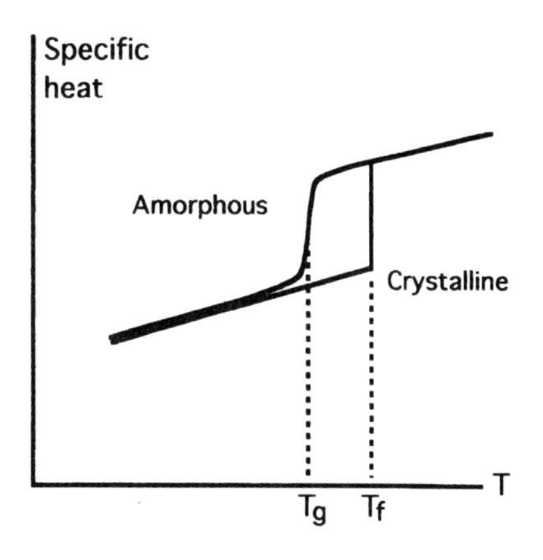


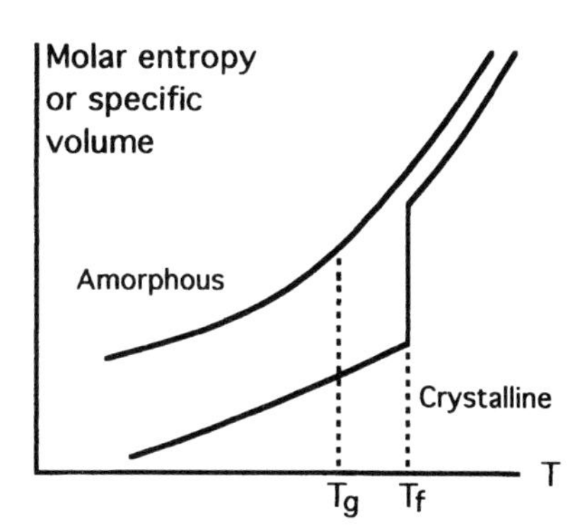


**Figure 4.2.** The glass point $T_g$, the temperature at which an undercooled liquid solidifies.

In ideal crystals the atoms are arranged on the lattice sites that have translation symmetry. Some disorder, such as point defects or stacking faults, does not seriously damage the average long-range order, and an X-ray diffractogram has sharp lines in a pattern that is characteristic of the phase (Figure 4.1). If the number of defects increases and becomes comparable with the number of atoms, the long-range order can disappear and the solid is amorphous. The X-ray pattern then has wide bands instead of sharp lines. The diffractograms of glasses are similar to those of liquids.

While crystalline solids have a precise melting point, glasses have a more or less reversible transition temperature $T_g$ (their glass point) between the solid and the liquid phase. At the glass point several properties besides the mechanical ones change, as they would at the melting point in crystalline solids. Such properties are the thermal expansion coefficient, specific heat, and enthalpy (Figure 4.2). The glass temperatures of some amorphous compounds are listed in Table 4.2, together with applications. Window glass used in buildings and cars is a covalent amorphous phase based on silicates.

When crystalline solids are prepared in an amorphous form they acquire other properties. Glasses are much easier to shape than crystalline inorganic materials such as ceramics. The amorphous modification can be made from the crystalline parent solid by intense, repeated deformations (e.g., in a ball mill), by means of an ion bombardment, by mechanical shock, or by low-temperature vapor deposition. Such

**Table 4.2. Some Amorphous Solids**

| Compound | Nature of solid | Bonding | Application |
|---|---|---|---|
| Silica | $SiO_2$, 10% $GeO_2$ | Covalent | Fiber optics, silica reactors, $T_g = 1450°C$ |
| Nasiglas | $Na_3Zr_{2-x}Yb_xSi_{2-x}P_{1+x}O_{24}$ | Ionic and covalent | Solid electrolyte |
| Chalcogenide glass | $As_2Se_3$ | Covalent | Photoconductor in xerography; $T_g = 470°C$ |
| Metal glass | $Fe_{0.7}P_{0.2}O_{0.1}$<br>TiCu | Metallic | Low-loss transformer cores<br>Hydrogen storage |
| Semiconductors | $Si + Si_{0.9}H_{0.1}$<br>$Te_{0.8}Ge_{0.2}$ | Covalent | Photovoltaic cells<br>Computer memories |

glasses may show remarkable catalytic properties. More often a glass is made from the melt by cooling it to a temperature below the glass point so fast that the solid is not given enough time to crystallize. Cooling the melt down to a glass state need not be done fast in mixtures with low-lying eutectics or in solids that have a high $T_g/T_m$ ratio (the ratio between the glass and melting temperatures). Deep eutectics occur in covalent glasses such as silicates and borates. These systems are in fact difficult to crystallize. Metals, however, crystallize easily and very high cooling rates of the order of $10^6$ °C/s are required to make metallic glasses from the melt.

Another method used to make amorphous solids is thermal amorphization by interdiffusing solid crystalline reactants. In certain cases (if formation of the crystal is kinetically frustrated) an amorphous reaction product is formed from crystalline reactants. This occurs if: (a) the two reactants have a high affinity for each other or, in other words, if the reaction has a high reaction free energy; (b) one of the reactants only is able to diffuse easily in the other at the reaction temperature, which must be low (50–200°C), much lower than the crystallization temperature of the product. Couples of elements from groups 4 (Zr, Hf) and La, B, or H with group 10 (Ni, Pd) elements and Au, Co, and Rh from the periodic table are suitable. The end product of this thermal process is not the crystalline state with the lowest possible thermodynamic potential but an amorphous nonequilibrium modification. The product cannot crystallize because one of the reactants is immobile at the reaction temperature.

### 4.2.1. Glass

The melt of $SiO_2$ or $B_2O_3$ crystallizes slowly if at all and these oxides, which have atomic networks in their crystal structure, are glass formers. The glass point is lowered by alloying the covalent network formers with alkaline metal oxides that shorten the polymer chains. The value of the glass point depends somewhat on the cooling rate. The more slowly the melt is cooled, the lower the observed glass point $T_g$. Table 4.3 lists the common additives to silicate and borate glasses.

Vycor and Pyrex are sodium borosilicate glasses in which use is made of spinodal demixing in the glass phase to provide strong resistance against chemical

**Table 4.3. Components of Oxidic Glass**

| | Coordination number | Bond energy (eV) | | Coordination number | Bond energy (eV) |
|---|---|---|---|---|---|
| Network formers | | | Modifiers | | |
| $SiO_2$ | 4 | 2.1 | $Na_2O$ | 6 | 0.4 |
| $B_2O_3$ | 3 | 2.8 | $K_2O$ | 9 | 0.3 |
| Intermediates | | | $Li_2O$ | 4 | 0.9 |
| $Al_2O_3$ | 4 | 1.3 | $Rb_2O$ | 10 | 0.2 |
| $Al_2O_3$ | 6 | 0.9 | CaO | 8 | 0.5 |
| ZnO | 2 | 0.8 | MgO | 6 | 0.6 |
| ZnO | 4 | 0.4 | SrO | 8 | 0.6 |
| PbO | 2 | 2.0 | | | |
| $PbO_2$ | 6 | 0.7 | | | |

attack. Common sodium borosilicate glass can lose alkali ions to acid solutions and show some surface corrosion. The phase diagram (Figure 4.3) shows the initial composition of the glass mixture and the binodal or demixing curve, which indicates that at 600°C the mixture decomposes to form two phases, a silica-rich phase (which is corrosion-resistant) and a phase that is rich in sodium borate (good solubility in water). The mixture is kept for a while at this temperature to achieve sufficient demixing. In Vycor the comparatively large amount of soluble borate phase is then removed by dissolving it, and the remaining porous silica-rich phase is sintered to a corrosion-resistant dense glass. In this way shaping of the material is possible at much lower temperatures than would have been possible with the final product. As

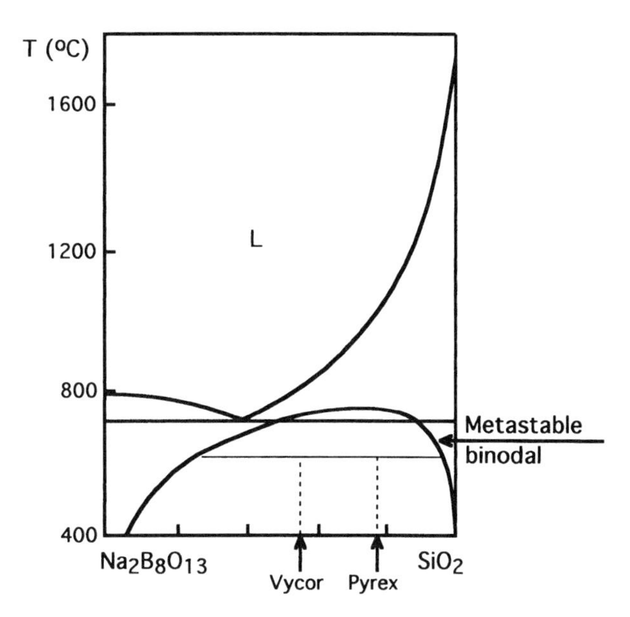


**Figure 4.3.** The borax–silica phase diagram showing the compositions of Vycor and Pyrex glass.

the end product has a concentration of 96% amorphous $SiO_2$, it would require shaping at a temperature slightly above the much higher glass point of fused quartz if the product were made from molten silica. Pyrex has an initial mixture with less sodium borate than Vycor. It is also subjected to a heat treatment for demixing and as in Vycor, the silica-rich phase determines its corrosion resistance. However, in Pyrex the borate phase is never removed from the demixed glass but left behind in the composite, where it is occluded by the silica phase.

The microstructures in composites that are the result of spinodal decomposition differ in a characteristic way from those in composites obtained by nucleation and growth processes. Figures 4.4a and b illustrate the differences between the morphologies that result from each of the two types of formation processes. The boundaries between the phases in products of spinodal processes are not sharp, concentrations vary gradually in the initial stages of demixing, and the concentration difference between the demixed phases continuously increases with processing time. The characteristic time for completion of a demixing reaction is

$$\tau = \lambda^2/8\pi^2 D$$

in which $\lambda$ is the diffusion distance (of the order of 10 nm) and $D$ is the diffusion coefficient. When another phase nucleates and grows into an initially homogeneous solid the boundaries between the two phases are sharp during the entire reaction.

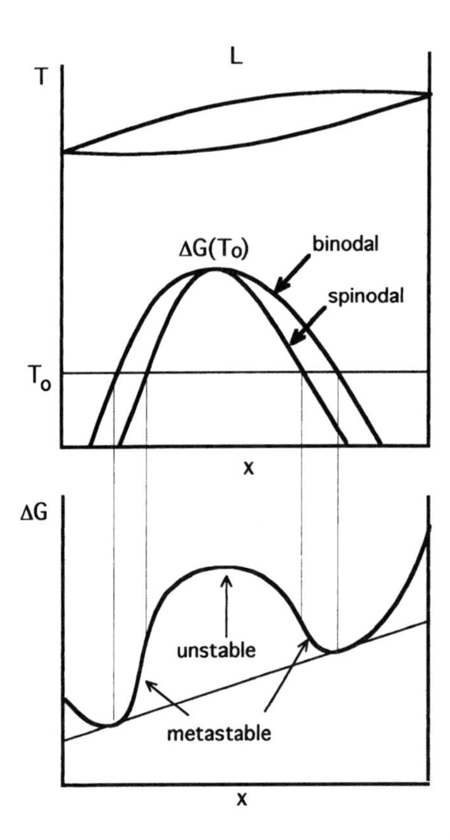


**Figure 4.4a.** The phase diagram and the Gibbs energy plot that correspond with one temperature to indicate the concentrations at which spinodal deposition or nucleation and growth occur. The spinodal area is at concentrations between the inflection points of the Gibbs energy curve. The binodal is the locus of concentrations that represent stable phases in coequilibrium.

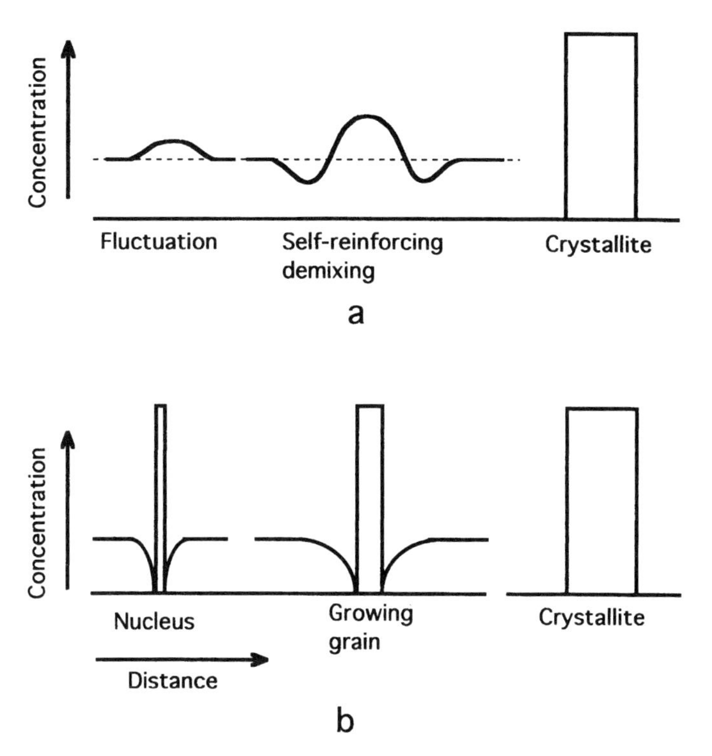


**Figure 4.4b.** The typical morphologies that are the result of spinodal decomposition and binodal growth. (a) Spinodal decomposition: local fluctuations in concentration amplify themselves; boundaries are fuzzy initially. (b) Crystallization: nucleation and growth within the binodal but outside the spinodal area; concentration fluctuations are locally unstable; boundaries are sharp throughout the process.

## *4.2.2. Glass Ceramics*

Shaping glasses is much easier than shaping ceramics and if there is a choice between the two glass processes are preferred. There is no process for making a ceramic (crystalline) product that provides the same ease as spinodal demixing does in shaped glasses. The glass-ceramic process is also a process that profits from the ease of shaping that glass allows. Under the right conditions a formed glass can be converted entirely to a polycrystalline product by precipitation and growth of the stable crystalline phases from the glass with heat treatment. In this way a dense and chemically inert glass ceramic can be formed that is stronger and tougher than the glass it is made from. A suitable system for a family of glass ceramics is $Li_2O/SiO_2$. The phase diagram is shown in Figure 4.5. A melt of 30% $Li_2O$ in $SiO_2$ has an eutectic point at 1030°C and the mixture turns into a glass on solidification. During heat treatment, as shown in Figure 4.6, the crystalline phases $Li_2Si_2O_5$ and $SiO_2$ are formed with the help of nucleation catalysts, such as $TiO_2$ or $P_2O_5$. Sometimes ZnO is added to improve the strength, while adding $Al_2O_3$ helps in the formation of $\beta$-spodumene crystallites (for a low expansion coefficient in the product). From the thermodynamic data of the different phases, one can derive the driving forces for the formation of the phases in the $Li_2O/SiO_2$ system and also the equilibrium compositions in an alloy (Figure 4.5).

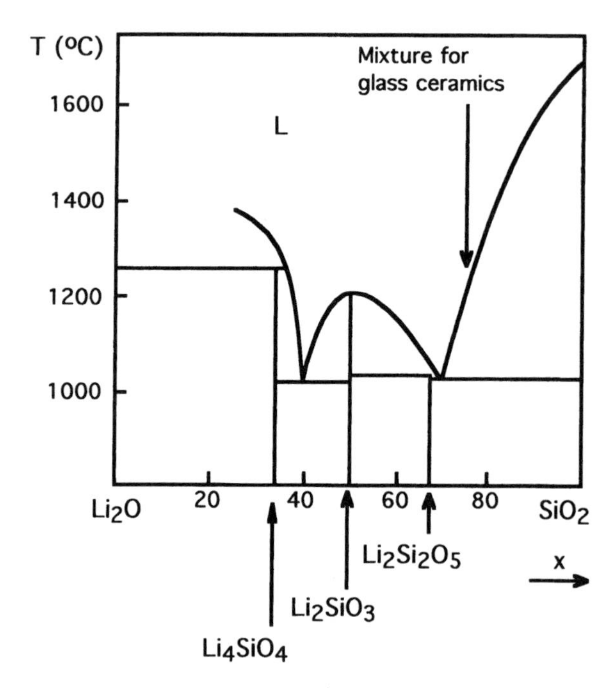


**Figure 4.5.** The phase diagram of the silica–lithia system showing three compounds.

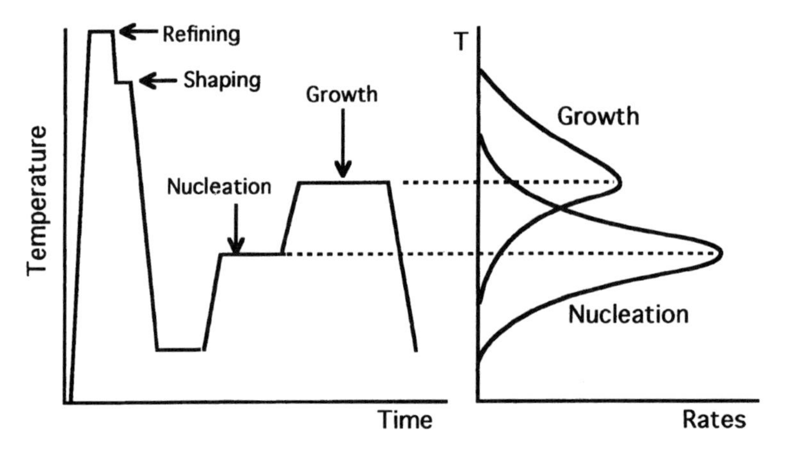


**Figure 4.6.** The thermal process for glass ceramics. After shaping with glass techniques nucleation continues until the proper concentration is reached. The nuclei are then grown at a temperature that does not favor additional nucleation.

### *4.2.3. Quasi-Crystals or Icosahedral Phases*

A decade ago Dan Schechtman synthesized a crystalline solid of manganese and aluminum that according to its X-ray diffractogram has a fivefold symmetry axis.[7] If this material is heated, it first becomes amorphous, but at higher temperatures the

X-ray diffractogram shows the expected crystalline phases for this composition. This is a discovery that can hardly be said to be the result of a well-established theory. After the initial discovery other quasi-crystals or icosahedral phases were found, and most of them contained aluminum. Quasi-crystals have no translation symmetry, but they do have long-range order (necessary for sharp X-ray lines). They are called icosahedral phases (an icosahedron is a regular body with 20 triangular faces and 6 fivefold axes), but solids such as boron (Section 4.3) and the higher borides that have icosahedra in their unit cells are not quasi-crystals. These lattices have translation symmetry and therefore no fivefold lattice symmetry. A quasi-crystal can be constructed from icosahedral units by arranging the units as fractals. Twelve atomic icosahedra are arranged in a super-icosahedron to form a new unit, twelve of which are again arranged in a larger icosahedron, and so on. Professional crystallographers now consider actual quasi-crystals to be three-dimensional projections from higher-dimensional lattices that do have translation symmetry. Icosahedral or quasi-crystalline materials can be applied as low-friction and self-healing coatings and as hydrogen storage materials.

## 4.3. Boron and Borides

The modifications of boron and boron compounds are characterized by nets or chains of boron atoms bound to each other by covalent bonds.[8] The coordination number of the boron atoms varies between 1 and 7. The nets have the form of linked icosahedra, cuboctahedra, octahedra, graphite-like planes, or linear branched boron chains. Two-electron three-center bonds in boron-cornered triangles are the rule in the lattice structure. Boron can only form ions in combination with hydrogen ($BH_4^-$) and oxygen ($BO_3^{3-}$). The borates (boron oxide derivatives) form covalent nets like the silicates. They are often amorphous.

### 4.3.1. Elementary Boron

The electron configuration of the free boron atom is $(1s)^2(2s)^2(2p)^1$. The atomic radius (88–90 pm) is small and the electronegativity (2.0 on Pauling's scale) is too high for a metallic bond between small atoms. Boron does not have enough valence electrons for conventional bonds and so forms nets of triangles. These nets have some interstitial space into which the valence electron pairs can leak. The boron nets lack valence electrons and tend to accept electrons from metallic binding partners for strengthening. Boron-rich compounds $M_xB_y (y > x)$ also have covalently bound boron chains or nets.

Elementary boron has several modifications. Figure 4.7 shows the structure of α-rhombohedral boron. Each boron atom is linked with covalent bonds to six or seven neighboring boron atoms. The slightly distorted icosahedra are in what is approximately a cubic close-packed arrangement. Six of the twelve boron atoms in one icosahedron are bound with a three-center, two-electron bond to two other icosahedra simultaneously, and the other six are bound to one other icosahedron with a conventional covalent two-center, two-electron bond. Six boron atoms contribute one electron each to the common bond between the icosahedra, the six

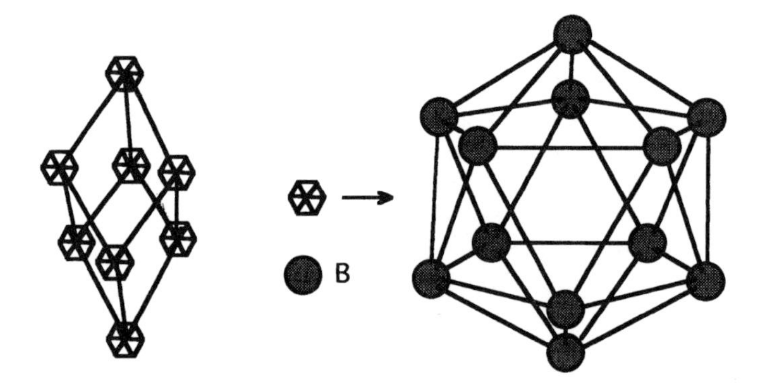


**Figure 4.7.** One of the crystal structures of boron. The boron icosahedra are rhombohedrally stacked and chemically linked to one another with strong boron–boron bonds.

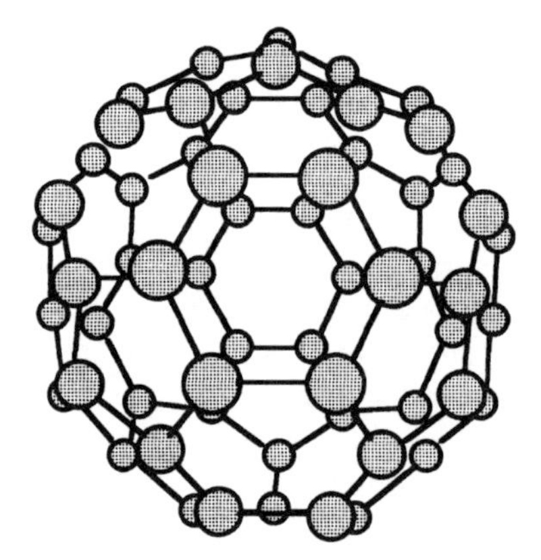

**Figure 4.8.** A basic unit in the structure of $\beta$-rhombohedral borone, an icosahedron with its points chopped off.

other boron atoms each contribute two-thirds of an electron. All together 36 electrons per icosahedron (the 3 × 12 valence electrons of the 12 boron atoms) are involved in $\alpha$-rhombohedral boron bonding.

The most stable of the boron modifications is $\beta$-rhombohedral boron (Figure 4.8), which has a unit cell made out of boron icosahedra with each atom being connected to the top of a pentagonal $B_6$ pyramid. This part of the structure is similar to a buckyball (Buckminster-Fullerene or $C_{60}$), but this boron ball is filled with a boron icosahedron. The unit cell is finished with a single boron atom and two $B_{10}$ units that are draped as three fivefold rings around a central boron atom. The total number of boron atoms per unit cell in this modification is 105.

The third modification is $\alpha$-tetragonal boron, which is based on four icosahedra per unit cell, which are bound to each other either directly or through two single boron atoms. This allotrope has the structure of many binary boron-rich compounds of nonmetals.

### 4.3.2. Metal Borides

In compounds with metals, isolated boron atoms behave as metal atoms and, with an excess of transition metals, boron can form intermetallic compounds and alloys if $n > 2$ in the stoichiometry $M_nB$. Although boron atoms can exist as metallic atoms in metallic alloys they have the tendency to bind to each other if given the opportunity, even in metals. If the boron content increases to values greater than in $M_2B$ covalently bound boron pairs, chains and nets are formed in the metal matrix (Figure 4.9). As in the case of network silicates the stoichiometry of the metal boride is related to the type of boron network in the lattice. A few representative metal borides are listed in Table 4.4.

In metal borides the metal atoms are not stacked as in the corresponding metal lattices (hcc, fcc, or bcc) but form stacks to accommodate the boron nets or chains. In $MB_2$ the metal atoms form face-linked trigonal prisms. The perovskite structure is rare in borides but it does occur in some ternary subborides, e.g., in $LnRh_3B_{1-x}$ in which Ln is a lanthanide.

The metal borides are useful compounds. Some semiconducting borides find application because of their very high thermoelectric power. Like all borides, they are very hard, corrosion-resistant, metallically conducting, good diffusion barriers even at high temperatures, and have a very high melting temperature, and a work

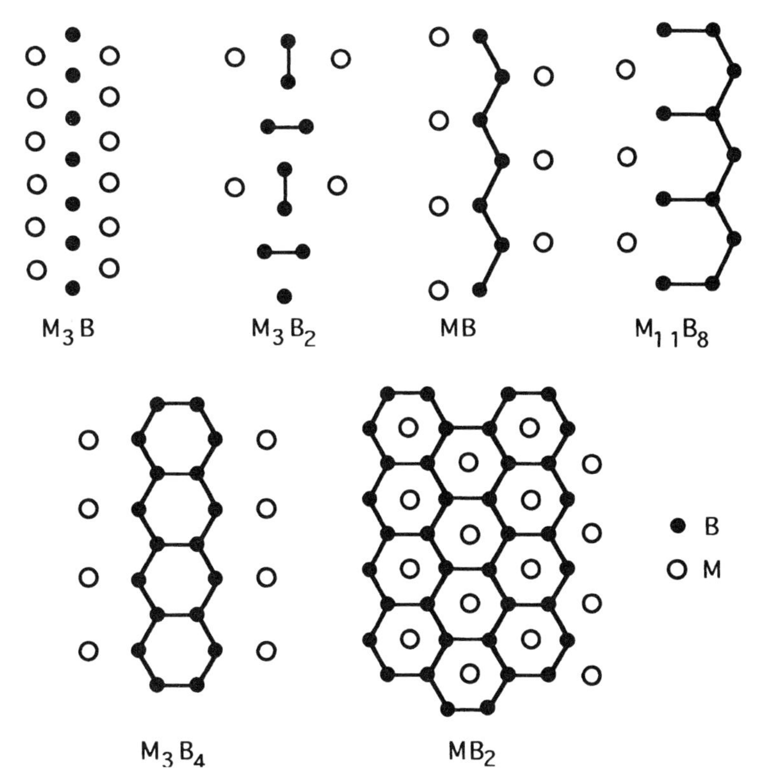


**Figure 4.9.** Schematic of the crystal structure of the lower metal borides. The boron atoms are covalently linked to one another, and the metal atoms are interstitial in the boron frames.

**Table 4.4. Stoichiometries and Structures of Metal Borides**

| Composition | Boron network | Metal arrangement | Observed with: |
|---|---|---|---|
| $M_4B$ | Isolated atoms | Square antiprism | Mn, Cr |
| $M_3B$ | Isolated atoms | Trigonal prism | Later transition metals |
| $M_5B_2$ | Isolated atoms | Trigonal prism | Pd |
| $M_7B_3$ | Isolated atoms | Trigonal prism | Re, Ru, Rh |
| $M_2B$ | Isolated atoms | Square antiprism | Be, transition metals |
| $M_3B_2$ | Pairs | Trigonal prism | V, Nb, Ta |
| $M_4B_3$ | Isolated and chains | Trigonal prism | Ni |
| $M_{11}B_8$ | Branched chains | Trigonal prism | Ru |
| MB | Single chains | Trigonal prism | Transition metals |
| $M_3B_4$ | Double chains | Trigonal prism | V, Nb, Ta, Cr, Mn, Re |
| $MB_2$ | Hexagonal plane | Trigonal prism | Transition metals |
| $M_2B_5$ | Pluckered plane | Trigonal prism | Ti, Mo, W, Ru, Os |
| $MB_4$ | Octahedra and atoms | | RE, Mg, Ca, Mo, W, Mn |
| $MB_6$ | Linked octahedra | | Be–Ba, Sc–La, RE |
| $MB_{12}$ | Cuboctahedra | | Sc, Y, Tb–Lu, Zr |

function that varies strongly with the metal atom. The stoichiometries of the lower borides are not simply related to the valency of their component elements. Which metals can form certain types of borides depends on the fit of the atomic size of the metal in the boron networks.

Monoborides of transition metals contain single boron chains in the matrix of metal atoms. The monoborides of Zr, Hf, and Ta are superconductors that have a critical temperature of 3–4°C. $LuRh_4B_4$ is a superconductor having a transition temperature of 11.7°C and the superconducting transition temperature of $YPd_5B_3C_x$ ($0.3 < x < 0.4$) is 23 K. NiB is used as a catalyst in fuel cells to lower the overvoltage.

The metal diborides, $MB_2$, have two-dimensional networks of boron atoms in a hexagonal crystal symmetry. They are very hard, brittle, metallic conductors, and are chemically inert (especially $TiB_2$ and $ZrB_2$). Transition metal diborides are resistant against carbon at 1000°C and are therefore used to passivate steel in coal liquefaction equipment. The diboride alloys $Zr_{0.13}Mo_{0.87}B_2$ ($T_C = 10$ K) and $Sc_{0.1}Nb_{1.9}B_2$ ($T_c = 6.6$K) are superconductors.

In metal tetraborides $MB_4$ (M = Y, Th, U, or a lanthanide) the boron nets extend into the third dimension. They form octahedral clusters that are linked to each other through $B_2$ pairs. In these tetraborides, the metal atoms are arranged in tunnels parallel to the $c$-axis. The $CrB_4$ structure keeps the two-dimensional boron nets corrugated.

The metal hexaborides $MB_6$ (with M from group 1–3 or a lanthanide) have a cubic lattice that is similar to that of CsCl (Figure 4.10). All six boron atoms in one octahedron are linked to another octahedron. The distance between the boron atoms within the octahedra is slightly longer than the boron–boron bonds that link the various octahedra. This latter distance changes somewhat with the size of the metal atoms. The hexaborides often have a wide existence range. The value of $x$ in $M_xB_6$ can vary between 0.77 and 1 for M = La and between 0.68 and 1 for M = Sm. Most hexaborides are better electrical conductors than the metals themselves and have an

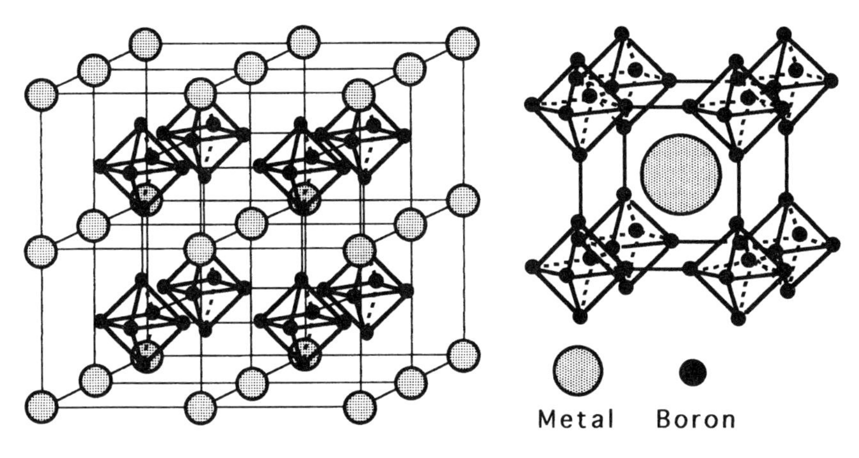


**Figure 4.10.** The crystal structure of cubic metal hexaborides The boron network consists of interlinked octahedra.

exceptionally low work function: 2.22 eV for M = Y and 2.74 eV for M = La. The hexaborides of the alkaline-earth metals (Ca, Sr, and Ba) are semiconductors that have a bandgap ranging from 0.40 to 0.12 eV. Table 4.5 lists the known rare-earth borides.

Europium hexaboride is a semiconductor ($E_g = 0.38$ eV) but becomes a metallic conductor if 1% of the europium atoms is replaced by lanthanum. The boron nets can accommodate some of the valence electrons from the lanthanum. The compound is metallically conducting because of the excess of valence electrons that go into interstitials. The hexaborides have a high melting point and a high strength. They are hard and stiff because of the strong covalent bonds in the lattice.

**Table 4.5. The Rare-Earth Borides with Different Stoichiometries and Their Structures**

| Ion | Radius (pm) | $UB_{12}$ | $AlB_2$ | $YB_{66}$ | $ThB_4$ | $CaB_6$ |
|---|---|---|---|---|---|---|
| $Eu^{2+}$ | 115 | | | | | + |
| $La^{3+}$ | 113 | | | | + | + |
| $Ce^{3+}$ | 108 | | | | + | + |
| $Pr^{3+}$ | 106 | | | | + | + |
| $Nd^{3+}$ | 105 | | | + | + | + |
| $Sm^{3+}$ | 103 | | | + | + | + |
| $Gd^{3+}$ | 100 | | + | + | + | + |
| $Yb^{3+}$ | 99 | + | + | + | + | + |
| $Tb^{3+}$ | 98 | + | + | + | + | |
| $Y^{3+}$ | 97 | + | + | + | + | |
| $Dy^{3+}$ | 96 | + | + | + | + | |
| $Ho^{3+}$ | 95 | + | + | + | + | |
| $Er^{3+}$ | 93 | + | + | + | + | |
| $Tm^{3+}$ | 92 | + | + | + | + | |
| $Yb^{3+}$ | 91 | + | + | + | + | |
| $Lu^{3+}$ | 90 | + | + | + | + | |

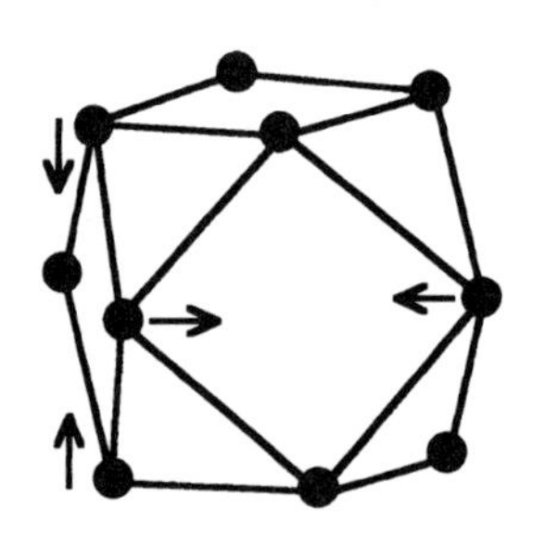


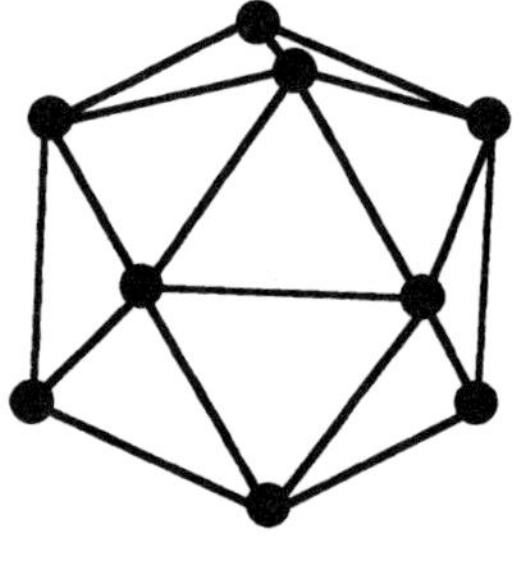


**Figure 4.11.** Structures of the higher borides of metals. These usually contain boron cuboctahedra while nonmetals have icosahedral boron clusters. A slight atomic displacement changes one to the other.

The higher metal borides that have more than six boron atoms per metal atom have structures with boron nets based on icosahedra or cubo-octahedra (Figure 4.11). Cubo-octahedra are regular bodies that have 14 corners. They can be seen as truncated cubes or octahedra with the eight or six points chopped off. Zeolites (aluminosilicates) are the only inorganic compounds that have structural elements with an atomic arrangement similar to these borides. Cubo-octahedra can be converted to icosahedra by bringing the corners of the square sides closer to each other, which means only a slight displacement of the boron atoms.

Dodecaborides $MB_{12}$, in which M = Sc, Zr, Y or one of the late lanthanides, form cubic lattices (NaCl-like) and are based on linked cubo-octahedra. They are good electronic conductors: $ZrB_{12}$ has a specific resistance of 60 $\mu\Omega$ cm and below 6 K it becomes a superconductor. Examples of higher metal borides in which icosahedra occur are $AlB_{10}$, $AlB_{12}$, $NaB_{15}$, and $MgAlB_{14}$. In their structures and properties these compounds are more similar to the higher borides of nonmetals than to those of the metal dodecaborides that have been described above.

### 4.3.3. Compounds of Boron with Nonmetallic Elements

Boron has a great affinity for oxygen and occurs in nature only in boric acid or borates. Borates are composed from clusters of flat trigonal $BO_3$ and tetrahedral $BO_4$ groups. The structural chemistry of borates is as rich and complicated as those of silicates, borides, or boranes. Boron oxide is an essential part of borosilicate glasses such as Pyrex. Boron halides are volatile molecular compounds. They are Lewis acids and react violently with water. The subhalides consist of boron chains or clusters that have terminally bound halogen atoms. They are substitution derivatives of the lower boranes.

Compounds of boron with nitrogen are isoelectronic with carbon, and there are many analogies with organic compounds in molecular boron–nitrogen chemistry. A hexagonal modification of boron nitride (BN) (Figure 4.12) has a graphite-like structure, is very soft as well as heat-resistant, but less oxidizable than graphite, and is a white insulator. Cubic BN has the sphalerite structure and is the hardest material known after diamond. As it is not reactive with iron at

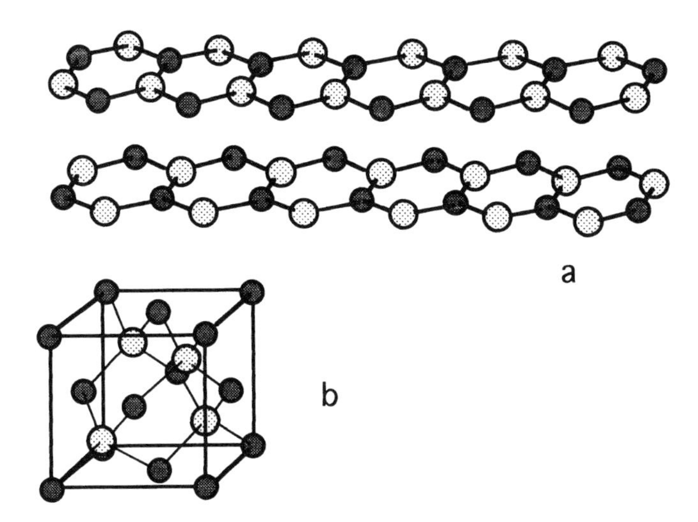


**Figure 4.12.** Two modifications of boron nitride, which is isoelectronic with carbon: (a) layered hexagonal α-BN, and (b) cubic β-BN with a sphalerite structure. Both are electrical insulators, but like carbon they are either very soft or very hard.

high temperatures as diamond is, the very hard cubic BN can be used to cut iron at high feed rates.

Boron phosphide (BP), a very hard refractory semiconductor, has a zinc blende structure as most III–V (13–15) semiconductors, such as GaAs and InP. It has a bandgap of 2 eV and a slightly larger interatomic distance than SiC or AlN.

The boron-rich compounds $N_nB_{12}$ of nonmetals (N = C, Si, P, As, O; $n = 1–3$, but usually 2) are structurally related to both α-rhombohedral and α-tetragonal boron, which are built up of boron icosahedra and intericosahedral boron atoms (Figure 4.13). In the boron carbides $B_{12}C_n$ ($n = 1–3$) the boron atoms in and between the icosahedra are replaced by carbon atoms. In the other compounds in this category the atoms (called N above) are between the polyhedra and not in them.

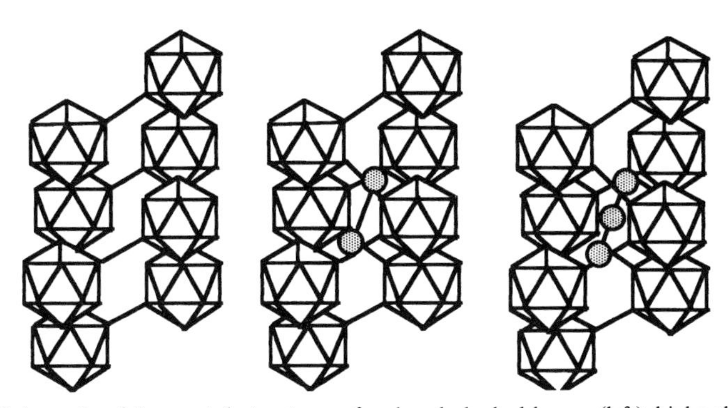

**Figure 4.13.** Schematic of the crystal structures of α-rhombohedral boron (left), higher-boron pnictides (middle), and boron carbide (right). The icosahedra are boron networks, and the single atoms other nonmetals.

They are semiconductors with large bandgaps. $B_{12}O_2$ is extremely hard and $B_{12}Si_2$ is very resistant to corrosion and thermoshock: heating 50 times up to 1100°C and quenching in water is said not to affect even the microstructure of this compound. The phosphorus and arsenic derivatives are refractory semiconductors that can be doped to *p*- or *n*-type materials. They have a comparatively high charge carrier mobility (100 $cm^2/Vs$). In $B_{13}P_2$ and $B_{13}As_2$ every pnictide atom is coordinated tetrahedrally by boron and contributes one electron to the bonds between the boron atoms in the icosahedra.

The borides of this type can be considered as sort of "inverse molecular lattices." In molecular compounds the bonding between the molecules is weaker than between the atoms in the molecule. An external force will strain the intermolecular distances more than the molecular configuration. In these borides, on the other hand, the bonds between the icosahedra are stronger than the ones with them, which explains their remarkable combination of physical properties. Their lattice stiffness and high sound propagation rate are inconsistent with their very low heat conductivity.

In boron nets each boron atom is bound to several (three to six) other boron atoms. Each boron atom has only three valence electrons and can accommodate at most five others in its shell. The icosahedral arrangement of the boron atoms with three-center bonds gives the relatively few valence electrons that are available for bonding as much atomic space to delocalize in as possible. Given the number of electrons per atom and the electronegativity of boron, this arrangement has the lowest kinetic energy for the valence electrons in the system. If extra electrons can be transferred to the electron-deficient networks the bond will be strengthened, as in the case of the metal hexaborides. This happens in the case of the higher borides of metals and nonmetals.

## 4.4. Carbides and Nitrides

Like boron, carbon and nitrogen form ceramic compounds with both metals and nonmetals.[9] These compounds are hard, have very high melting points, and are chemically inert, although less so than borides.

Nitrogen is a gas in its precursors, $N_2$, $NH_3$, $N_2H_4$, but nitrogen atoms can behave as metal atoms when they are bonded in binary intermetallic compounds. Metal nitrides behave like intermetallic compounds; being brittle and good electron conductors and having remarkable stoichiometries.

Carbon in its hexagonal (graphite) or cubic form (diamond) is itself a much used material. Diamond, for instance, is very biocompatible and is therefore used as a coating on implants and bone prostheses. Another form of carbon, diamond-like carbon (DLC), is amorphous and has mechanical and electrical characteristics that are intermediate between diamond and graphite.[10] The degree to which a particular DLC exhibits these properties is closely related to the average coordination number of its carbon atoms. Three-coordinated carbon is soft and has a low bandgap, while four-coordinated carbon is very hard and the bandgap is large. The values between the two extremes (graphite-like and diamond-like) are controlled with the deposition conditions of DLC. As precursors for synthesis of metal carbides organic molecules are most convenient. Some of them can also contain the metal atoms that form the carbides. A combination of graphitic carbon and silicon oxide (pillared graphite) is a very hard and light solid.

Most binary metal carbides and nitrides have the halite structure (NaCl-structure) or an fcc stacking of metal atoms that has carbon or nitrogen atoms in the octahedral interstices. Sometimes the carbide or the nitride can support a sizable concentration of vacancies in its metal or nonmetal sublattices, and, as the phase diagrams indicate, the compound has a large existence range. In $Ti_2C$ and $Ti_2N$ the vacancies (there can be as many vacancies as there are nonmetal atoms) are ordered.[11] The ease of diffusion through these lattices is strongly dependent on the concentration of the vacancies. Figures 4.14 and 4.15 show the phase diagrams of Ti/N and Ti/C and Cr/C and Ta/C, respectively. The different carbides are clearly berthollides and have intermetallic-like stoichiometries ($Ti_2N$, $Cr_7C_3$). There are three chromium carbides and two tantalum carbides, all of them carbon-deficient.

Substoichiometric TaC has the highest melting point (4000°C) of any binary compound, which is higher than the maximum for an element (of all the elements, tungsten has the highest melting point). As many other carbide alloys, TiC and ZrC can form a continuous solid solution above 2000°C.[9]

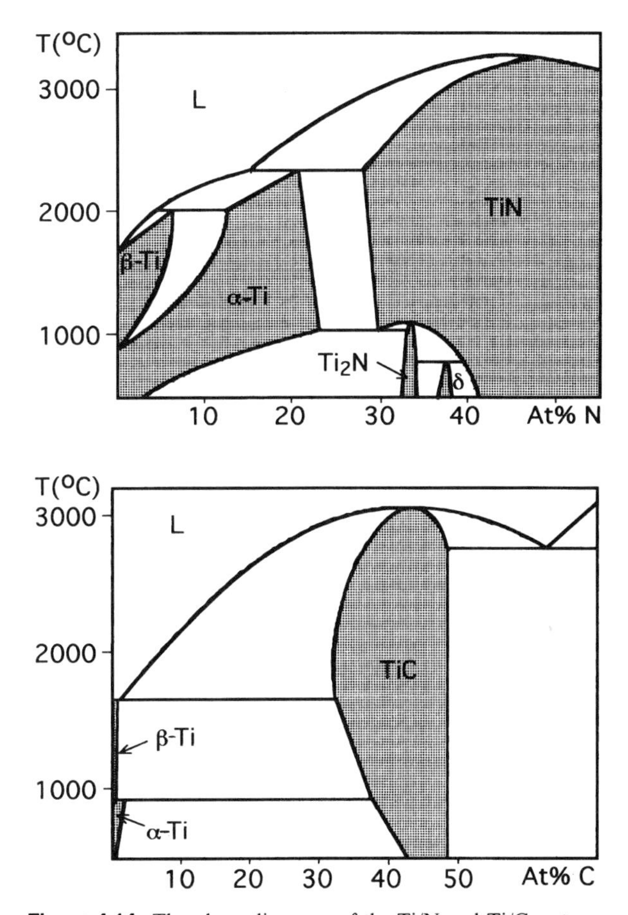


**Figure 4.14.** The phase diagrams of the Ti/N and Ti/C systems.

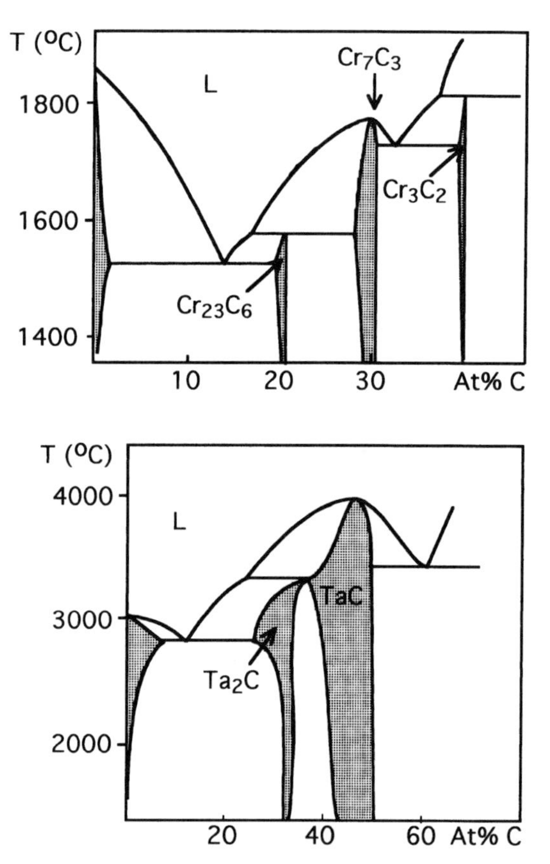


**Figure 4.15.** The phase diagrams of the Cr/C and Ta/C systems.

Silicon carbide (SiC) is a hard semiconductor, and is an inert and inexpensive ceramic used in sandpaper and in grindstones. It is made in bulk on an industrial scale by heating flint with coal tar up to very high temperatures or by means of chemical vapor deposition. SiC fibers are used in technical ceramics as a toughening aid. Like many other III–V semiconductors, it occurs in two modifications, the sphalerite (cubic) and the Würtzite (hexagonal) structures. In both structures carbon and silicon are tetrahedrally surrounded by silicon and carbon. The two structures do not differ much in material properties.

Silicon nitride ($Si_3N_4$) is an inert compound that is useful in technical ceramics because it is chemically inert and can be obtained in a comparatively strong and tough microstructue. It has two modifications with a different habit. Each silicon is bound to four nitrogen neighbors and each nitrogen is linked to three silicon atoms.

Carbon nitride ($C_3N_4$) should exist according to expectations based on pseudopotential calculations. The solid should be approximately as hard as diamond and have the structure of $Si_3N_4$. There have been many attempts to make it but so far its extreme inertness has only permitted the synthesis of a hard glassy form from ion beam deposition that comes close to the expected stoichiometry.

The early transition metal nitrides of titanium, zirconium, hafnium, tantalum, and tungsten as well as titanium and tantalum carbide are effective diffusion barriers.[11] In Section 10.4 on defect chemistry, the diffusion rates in several carbides are given. As is to be expected, the properties of the binary ceramic compounds can be improved upon by alloying. A very stable compound that is exceptionally resistant to aqueous corrosion is the mixed nitride $Ta_{.72}Si_{.28}N$. It is also a good diffusion barrier at 1100°C and starts to react with copper only at 900°C.

In the case of the nitrides and carbides the structures have some effect on the electric properties. For compounds that have an NaCl structure, the mononitrides and carbides are better metallic conductors than the metals themselves. Those that have zinc blende or Würtzite structures (SiC, AlN, InN) or other ($Si_3N_4$) are useful semiconductors or insulators. Tungsten carbide when sintered with cobalt shows some catalytic activity in fuel cells.[11] This hybrid composite is so hard and tough that it is widely used in drill bits for concrete drilling.

## 4.5. Oxides

Metal oxides constitute the largest class of applied nonmetallic inorganic materials. All of the traditional ceramics (composites composed mainly of alumina and silica) and most of the fine ceramics (perovskites, garnets, zirconia) are oxides. A few that have been found to be invaluable in structural and functional ceramics will be described here.

### 4.5.1. Structural Ceramics

The oldest materials known are traditional ceramics, which have been made for at least 200 centuries and are still in use as structural materials in buildings. They are based on a combination of three oxides, clay (colloidal aluminosilicate), flint ($SiO_2$), and feldspar (alkaline and alkaline-earth aluminosilicates).

Silicates are built up from $SiO_4$ tetrahedra linked into one-, two-, and three-dimensional networks, depending on how many oxygen atoms are shared by two neighboring $SiO_4$ tetrahedra (Figure 4.16). This number affects the stoichiometry of silicates and the number of cations that the lattice needs for charge compensation. Aluminum can partly replace silicon atoms in these chains. As aluminum is trivalent and silicon quadrivalent, the lattice needs extra cations for charge compensation. Such cations are alkaline and alkaline-earth ions. Some minerals have ordered stackings of two-dimensional $SiO_4$ nets alternating with $Al(OH)_3$ layers, which are nets of interlinked $Al(OH)_6$ octahedra. These two nets fit neatly together as can be seen in Figure 4.17. Examples of such minerals are kaolinite [$Al_2Si_2O_5(OH)_4$] and montmorillonite, ($MgO.Al_2O_3.5SiO_2.nH_2O$). The hard orthosilicates forsterite ($Mg_2SiO_4$) and zircon ($ZrSiO_4$) have three-dimensional nets or isolated $SiO_4$ ions. The aluminosilicates cyanite ($Al_2O_3.SiO_2$) and mullite ($3Al_2O_3.2SiO_2$) have a cubic closest-packing of oxide ions with silicon in the tetrahedral holes and aluminum ions in the octahedral interstices of the oxide sublattice.

In traditional ceramics, powder particles are sintered together by heating to form a dense, hard, durable material. The basis for ceramics is clay with a variable composition, which is mixed with silica and feldspar powders, depending on the particular application. All ceramics are composites, composed of several phases.

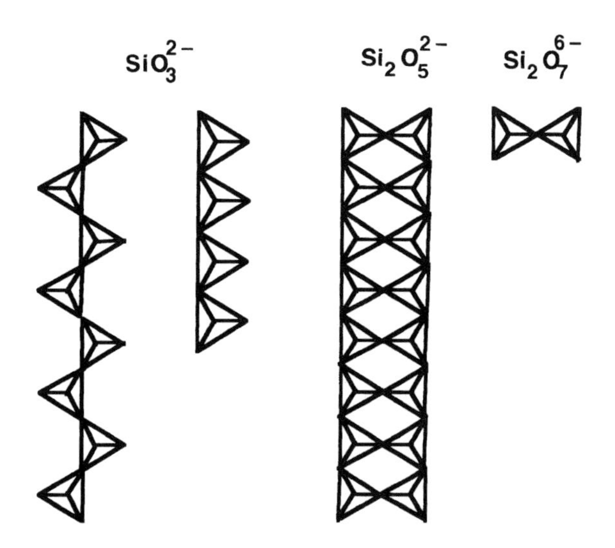


**Figure 4.16.** Chains of linked $SiO_4$ tetrahedra as structural elements in silicates.

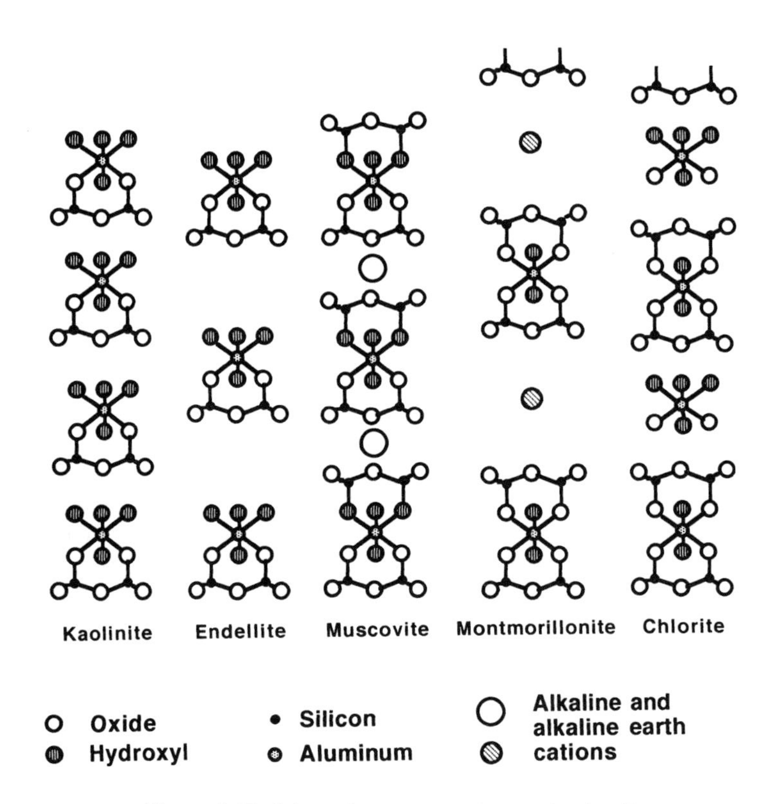


**Figure 4.17.** Schematic structures of some aluminosilicates.

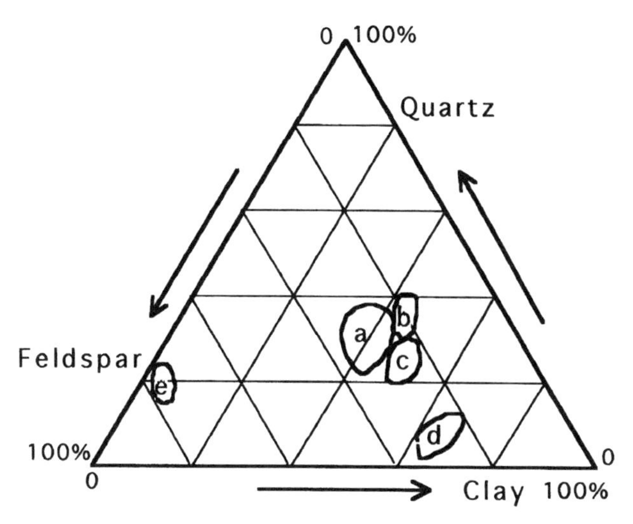


**Figure 4.18.** The composition of different types of triaxial porcelain: (a) soft porcelain, (b) electrical insulators, (c) tableware, (d) chemical porcelain, (e) dental porcelain.

The relative amount of each of the three components in the "green" unsintered shaped precursor product determines the properties and hence the possible applications of the end product. The composition of such "triaxial" ceramics can be seen in Figure 4.18.

The compositions of a few feldspars are:

| | |
|---|---|
| $K_2O.Al_2O_3.6SiO_2$ | Microcline |
| $Na_2O.Al_2O_3.6SiO_2$ | Albite |
| $CaO.Al_2O_3.2SiO_2$ | Anorthite |

Feldspars have melting points that are themselves not much lower than those of $SiO_2$ or $Al_2O_3$ but they form deep eutectics with silica. The function of a feldspar in silicate ceramics is to lower the sintering temperature, and it acts as a flux. Notwithstanding the high melting points and glass transition points of alumina, silica, and the aluminosilicates, these can be sintered at modest temperatures (e.g., in mixtures for enamels and glazes) by adding lead oxide, borates, or alkaline or alkaline-earth oxides. Fluxing additions generate phases that have low melting points.

The two large groups of clays used in ceramics, kaolinite and montmorillonite, are both erosion products of weathered feldspar minerals. For example when the rocky mineral anorthite is eroded by weathering, kaolinite particles are formed:

$$CaO.Al_2O_3.2SiO_2 + 3H_2O + 2CO_2 \Rightarrow Ca(HCO_3)_2 + Al_2Si_2O_5(OH)_4$$

with particles that are platelets with a range in size of 0.1–100 $\mu$m. Kaolinite is $Al_2O_3.2SiO_2.2H_2O$ with some additional oxides as impurities. If calcinated it loses water and is converted to amorphous silica and mullite.

Montmorillonite is a clay group with composition $(Mg,Ca)O.Al_2O_3.5SiO_2$.-$nH_2O$. Bentonite, which contains iron, magnesium, and calcium oxide, belongs to this group. Attapulgite forms thixothropic suspensions as does bentonite but unlike the latter, it can be used in seawater, and so finds applications in offshore exploratory drilling. An attapulgite colloid is used in mixtures with hydrated sodium sulfate for latent heat storage.

The plasticity of clay is the result of the microstructure, particle size, the amount of water bound at the surface of the particles, and the surface charge of the particles. The charge is always negative on the flat sides and when it is positive on the edges of the clay platelets, clay suspensions can form a gel as discussed in Section 6.3. Clay shapes shrink on drying to an extent that depends on the individual clay.

Oxide phase diagrams are used in ceramics to determine the phases that are to be expected on sintering. Figure 4.19 shows the binary *T–x* diagram of the $SiO_2/Al_2O_3$ system. Mullite is the only thermodynamically stable compound in this system. The zeolites are very porous, nonstable aluminosilicates, but they do exist in nature and are found as minerals. Figure 4.20 gives the phase diagram for the system leucite/silica. This diagram clearly shows why feldspars are used as fluxing agents in porcelains.

Pure zirconia has two phase transitions. One is from the low-temperature monocline modification to the tetragonal lattice at 1100°C, which is accompanied with a 5% increase in the size of the lattice cell. The second phase transition occurs at 2370°C from the tetragonal to the high-temperature cubic fluorite structure with a 1% decrease in lattice cell volume. Zirconia when undoped is unsuitable as a ceramic that is made at high sinter temperatures (or is to be used at high temperatures) because of these phase transitions. After sintering a zirconia compact will shatter on cooling owing to stresses that develop during the local increase of specific volume at the phase transition at 1100°C. Doping the zirconia lattice with $Ca^{2+}$, $Y^{3+}$, or $Sc^{3+}$ ions increases the oxygen vacancies and simultaneously stabilizes the cubic high-temperature form down to room temperature as the phase diagram of Figure 4.21 shows for doping with yttria. This characteristic makes

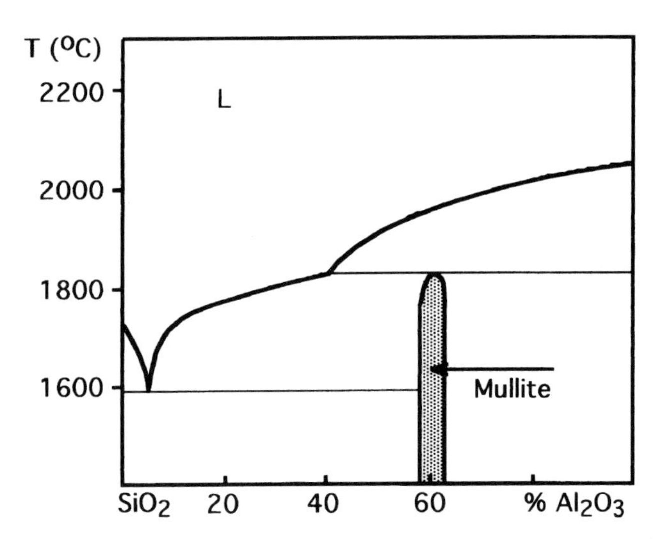


**Figure 4.19.** Phase diagram of the silica/alumina system.

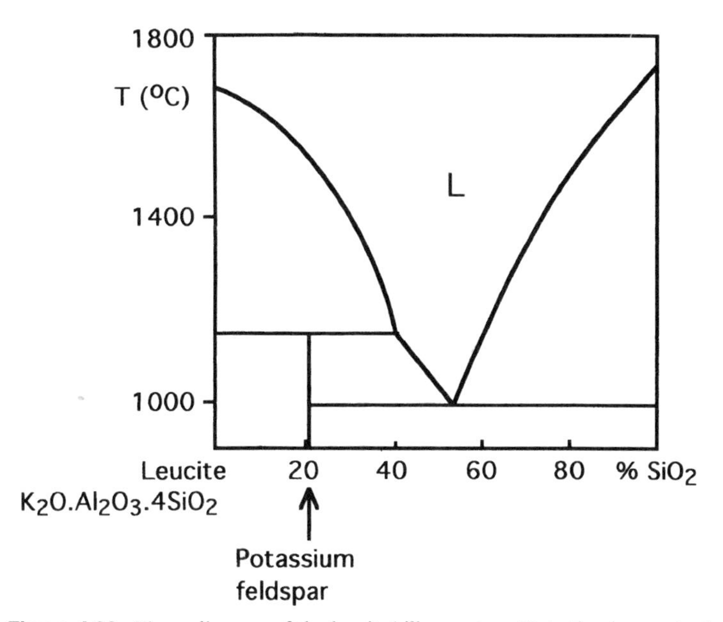


**Figure 4.20.** Phase diagram of the leucite/silica system. Note the deep eutectic.

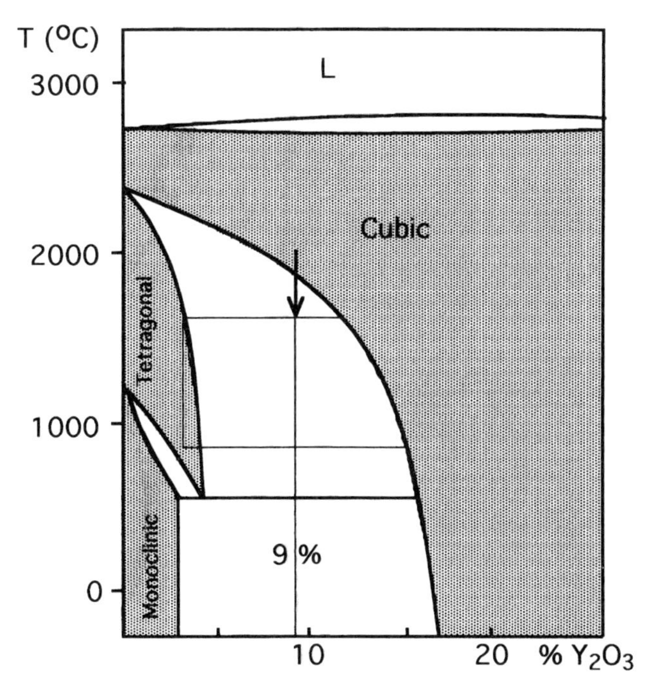


**Figure 4.21.** Part of the zirconia/yttria phase diagram that is of interest for the fabrication of partially stabilized zirconia. In the coexistence region some tetragonal phase is formed on cooling cubic zirconia doped with yttria. At still lower temperatures the tetragonal grains in the cubic matrix cannot convert to the monoclinic modification.

suitably doped zirconia an attractive ceramic for two applications, one structural and one functional: transformation-toughened ceramics and an oxygen-ion-conducting electrolyte for solid oxide fuel cells.

### 4.5.2. Functional Ceramics

The electrical properties of the perovskites are strongly coupled to their crystal structure.[12] Perovskites are mixed oxides with cations that do not fit well together in the cubic anion lattice, some of them being slightly too small for their lattice sites. As a result these mixed compounds display a unique combination of electrical, thermal, and electromechanical properties. Perovskites are the active components of many active materials and transducers.

The crystal structure of perovskites is shown in Figure 4.22. They are ternary compounds $ABO_3$ with A and B cations having, respectively, valencies of 3 and 3, 2 and 4, or 1 and 5. There are also mixed fluorides with this structure in which $O^{2-}$ is replaced by $F^-$ and the cations have corresponding less positive charge. The anion cavity for the larger of the two cations is slightly too large, and the cation rattles in its cage but at high temperatures the structure is still cubic. At a temperature below the Curie temperature (the temperature of the highest solid–solid phase transition) the rattling stops and the symmetry lowers from cubic to tetragonal, orthorhombic, and rhombohedral (Figure 4.23).

The unit cells in the lattice of the lower-symmetry phases that are stable below the Curie temperature have ea permanent electrical dipole moment. In analogy with the magnetic phases, these are called ferroelectric if the moments are coupled in parallel and antiferroelectric if they are antiparallel. When the temperature is decreased to below the Curie temperature the symmetry becomes lower than cubic, the inversion center disappears, and the compound becomes piezoelectric. Figure 4.24 shows the change in properties at the phase transition of $BaTiO_3$. Further down

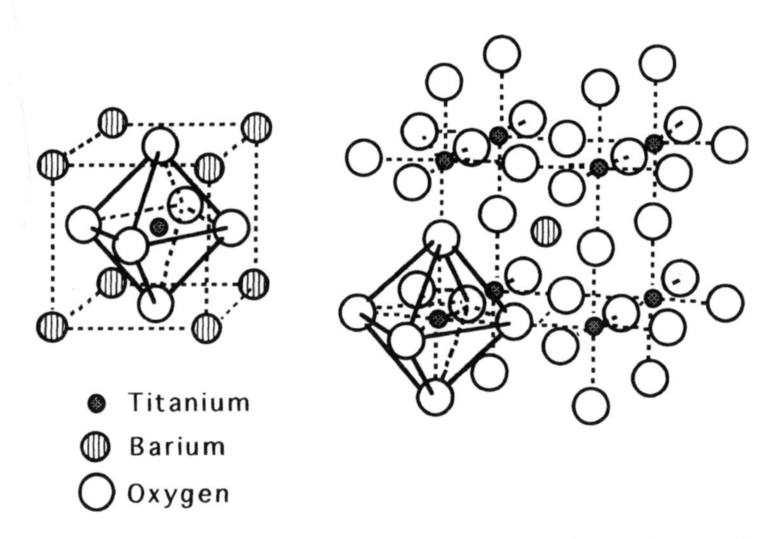


**Figure 4.22.** The crystal structure of $BaTiO_3$ as an example of the cubic perovskites.

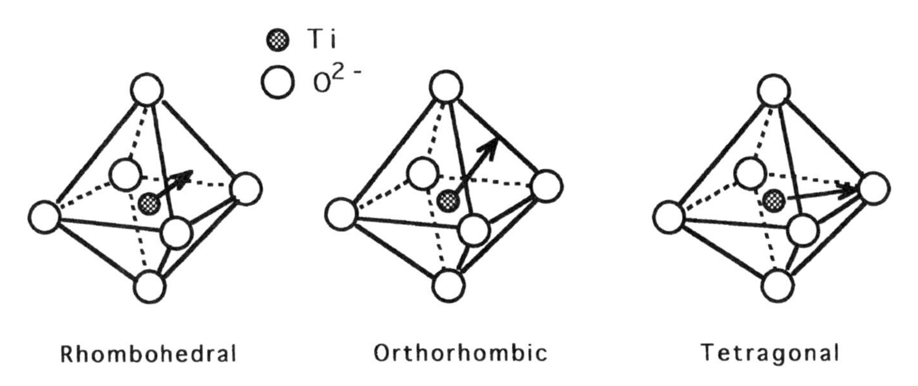


**Figure 4.23.** In the three modifications of $BaTiO_3$ below the Curie temperature, the titanium ion is displaced toward the trigonal plane, the midpoint between two oxide ions, or to a neighboring oxide ion. The resulting electric dipole moment has a corresponding orientation with respect to the crystal lattice. Printed with permission from N. Braithwaite and G. Weaver. *Electronic Materials.* The Open University (1990).

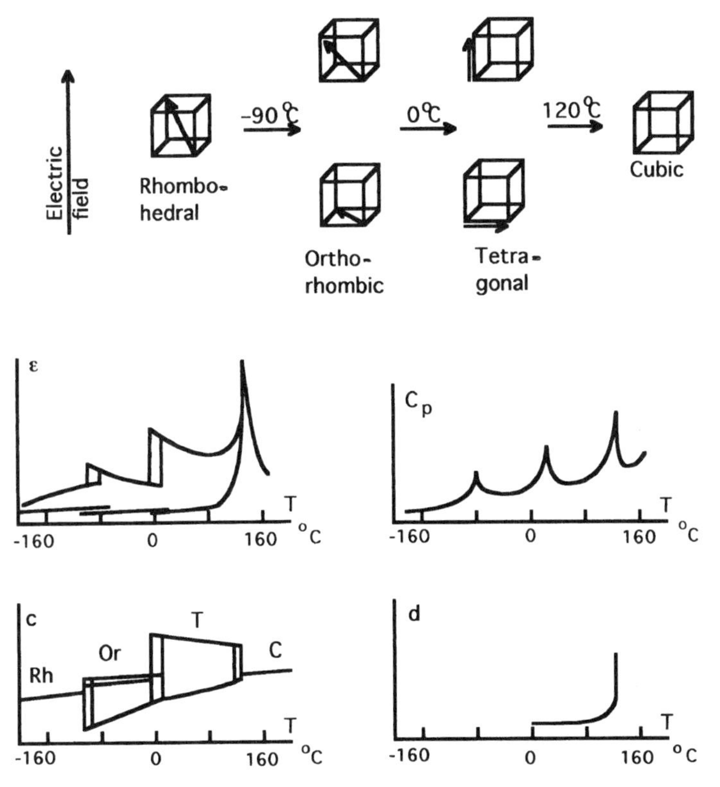


**Figure 4.24.** The change in the dipole moments and the physical properties of $BaTiO_3$ at different temperatures. At the phase transitions between rhombic (Rh), orthorhombic (Or), tetragonal (T), and cubic (C), the dielectric constants (*e*), the cell constants (*c*), the heat capacity $C_p$, and the piezoelectric parameter *d* change discontinuously.

the temperature scale there are other phase transitions to still lower symmetries but the compound remains ferroelectric.

A piezoelectric solid (e.g., quartz) acquires an electrical dipole moment upon mechanical deformation and, conversely, if it is subjected to an electric field **E** it becomes distorted by an amount proportional to the field strength $E$. The dipole moment disappears without the mechanical force. Piezoelectricity is only possible in lattices that do not have an inversion center. Electrostriction is also mechanical distortion in an electric field (strain proportional to $E^2$) but ionic lattices that have a center of symmetry also show this effect. Figure 4.25 is a schematic representation of the source of these effects using the interatomic potential curve. A ferroelectric material is not only piezoelectric but its lattice has a permanent electric dipole moment (below its Curie temperature), which most other piezoelectric materials (such as quartz) do not have.

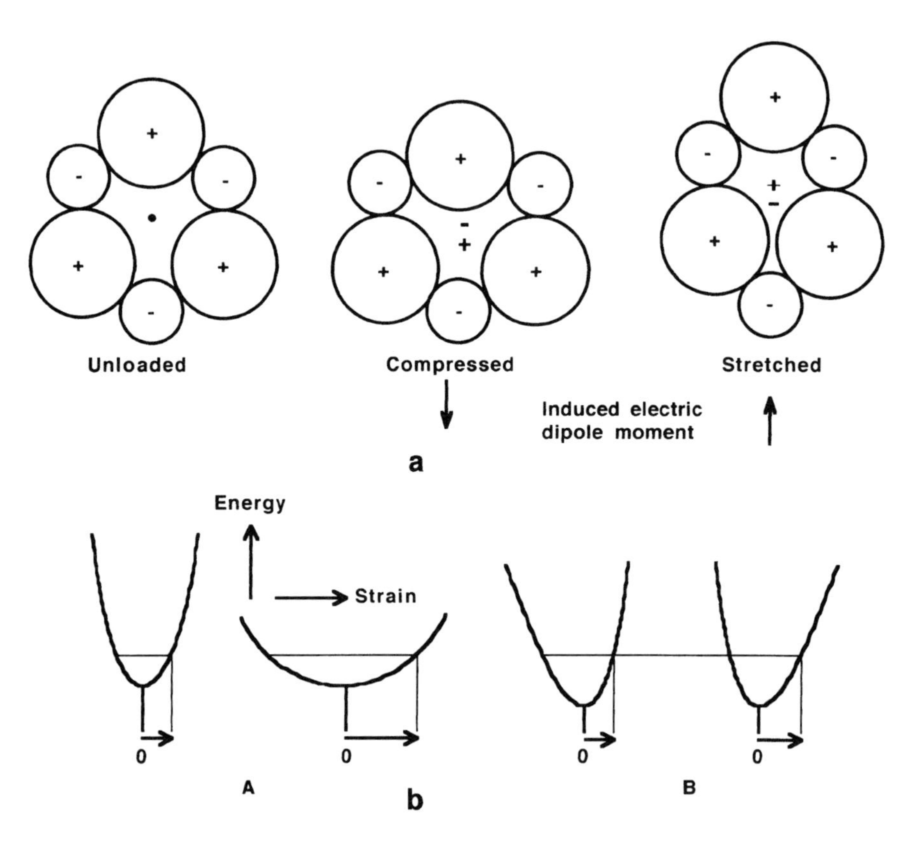


**Figure 4.25.** Electromechanical effects: piezoelectricity and electrostriction. (a) A lattice without an inversion center shows an electric dipole moment when strained. If the atoms are charged the centers of gravity of the charges no longer coincide under stress. In a piezoelectic solid the aligned dipole moments correspond to a surface charge. The strain/charge ratio is given by the piezoelectric parameter $d$ or $e$. (b) The difference between piezoelectric strain (A) and electrostriction (B) explained with interatomic potentials of two neighboring atoms in the lattice. If there is no inversion center (A) and the differently charged atoms are displaced in different interatomic potentials there is piezoelectricity. If the degree of displacement differs in the same interatomic potential (case B with an inversion center) the solid shows electrostriction.[13,14]

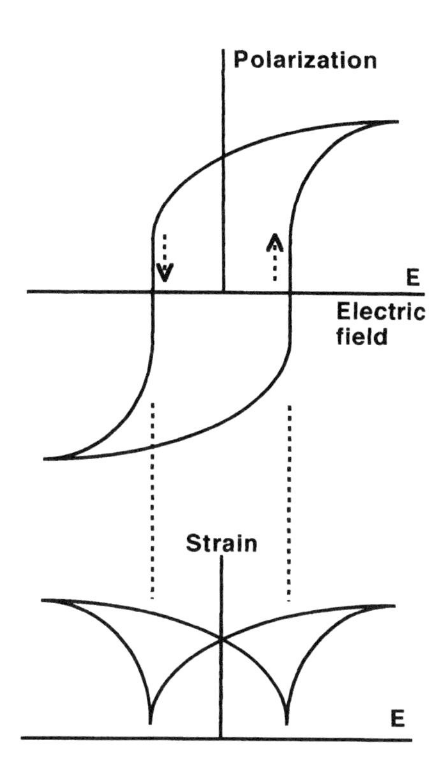


**Figure 4.26.** Hysteresis in the polarization and in the mechanical strain of ferroelectric solids in an external electric field.

As in the case of a ferromagnetic material, the polarization of a ferroelectric solid shows hysteresis: changes in the direction of the electric polarization by an external electric field depend on the history of the system in the field (Figure 4.26). The field also causes relatively large strains, of the order of 1%. This means that ferroelectrics are also strongly piezoelectric. These effects are clearly temperature-dependent (because of the phase transitions) as can be seen in Figure 4.27. Because of hysteresis during changes in polarization, the graph of the stress versus the applied electric field has a butterfly shape.

The relationship between the mechanical stresses and strains and electrical surface charges and polarization is given by the piezoelectric coefficients $d$ and $e$. These parameters are defined in terms of the stress $\sigma$, the strain $\tau$, the elasticity modulus $E$ ($\sigma = E\tau$), the electric field strength $\Phi$ (in $\mathrm{Vm^{-1}}$) and polarization $P$ (dipole moment per unit volume in $\mathrm{C/m^2}$):

$$d = \left(\frac{\partial \tau}{\partial \Phi}\right)_\sigma = \left(\frac{\partial P}{\partial \sigma}\right)_\Phi \qquad e = \left(\frac{\partial P}{\partial \tau}\right)_\Phi = E\left(\frac{\partial P}{\partial \sigma}\right)_\Phi = Ed$$

The piezoelectric generator coefficient $d$ indicates the mechanical strains that are induced by electric potentials and the piezoelectric motorcoefficient $e$ is a measure of the dipole moments that are created by strains. These two parameters are similar in nature and proportional to one another. The electromechanical coupling coefficient $k$ is proportional to both $d$ and $e$, its square giving the efficiency of conversion between mechanical and electrical energy that can be achieved with the ferroelectric

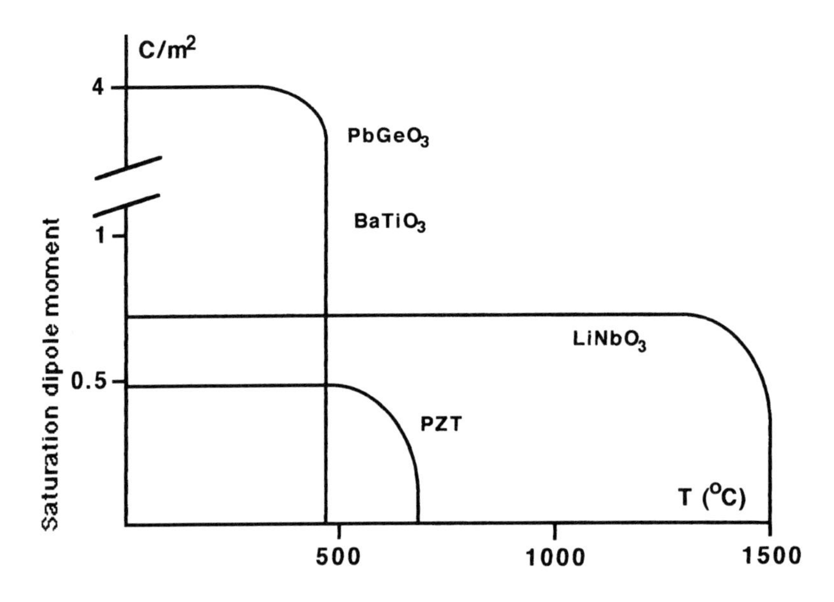


**Figure 4.27.** The temperature dependence of the saturation moment of several ferroelectrics. At the Curie temperatures the ferroelectricity disappears as the lattice becomes cubic.

material:

$$k^2 = E\frac{d^2}{\varepsilon\varepsilon_0} = \frac{e^2}{E\varepsilon\varepsilon_0}$$

Table 4.6 shows the values of these parameters for several solids.

Lead zirconate titanate ($PbTi_xZr_{1-x}O_3$ with $x \approx 0.5$), also abbreviated as PZT, is a ferroelectric mixed perovskite with a Curie temperature that depends on the composition $x$ (Figure 4.28). It also has a very high value of $k$ on the so-called morphotropic boundary (the value of $x$ at the boundary between the tetragonal and rhombohedral structures). It is a material of choice for making sensors and actuators, e.g., in medical echoscopes and sonar equipment.

**Table 4.6. Characteristic Piezoelectric Parameters of Some Single Crystals**

| Material | $\varepsilon$ | $d$(C/N) | $k$(%) | $T_C$(°C) |
|---|---|---|---|---|
| Quartz | 4.5 | 2.3 | 11 | 550 |
| Rochelle salt (X-cut) | 450 | 430 | 78 | 45 |
| Id. (Y-cut) | 9.4 | 27 | 29 | 45 |
| $BaTiO_3$ | 400–1700 | 60–190 | 20–50 | 120–140 |
| PZT | 900–1500 | 80–320 | 23–76 | 350–490 |

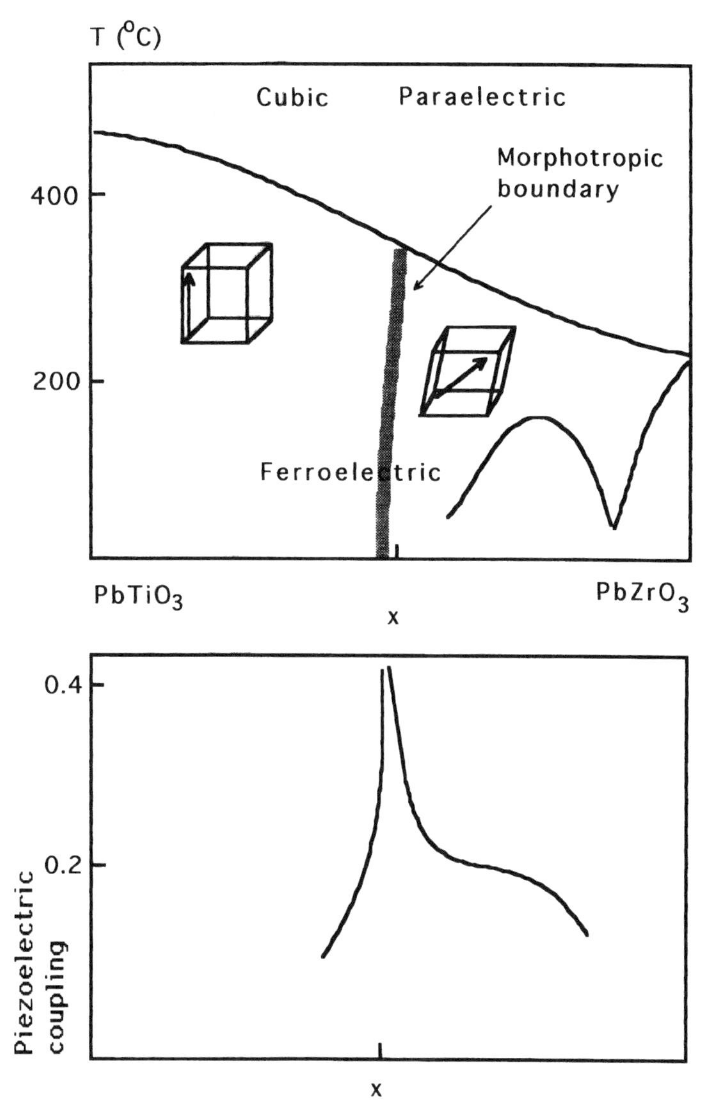


**Figure 4.28.** Phase diagram of the mixture of lead zirconate and lead titanate (PZT). Poling is particularly easy at the morphotropic boundary of the mixed oxide.

Doping PZT with lanthanum ions ($La^{3+}$ replaces $Pb^{2+}$) shifts the morphotropic boundary to higher $x$-values and lowers the Curie temperature of the oxidic alloy. At 5% $La^{3+}$, the Curie temperature can even drop below room temperature if $x$ is high enough. The solid has become cubic and paraelectric and no longer has a permanent electric dipole moment at room temperature but is very polarizable. This material is called PLZT (short for lead lanthanum zirconate titanate) and is used for its nonlinear optical properties, e.g., in doubling the frequency of laser radiation.

Ferroelectricity in perovskites provides a good example of how the crystal structure of a solid affects its electrical properties and uses.[13,14]

## 4.6. Intermetallics

Intermetallics are solid compounds made up of only the metallic elements themselves in the periodic table.[11,15,16] Most of the elements in the table are metals and the intermetallic compounds represent a large group, arguably bigger than the group of organic compounds. Many of the former are not known and are not of central interest in standard molecular inorganic chemistry, but being metals themselves they are increasingly being studied by metallurgists in the neverending search for novel properties. Intermetallics are refractory, are harder than the metals themselves, and like salts are brittle for the same reason: dislocations cannot move because atoms in compounds prefer not to bind to their like. Intermetallic binary compounds between elements from the left- and the right-hand sides of the transition periods in the periodic table sometimes have a stoichiometry that could be expected from their valence, but usually the conventional valence of the metallic elements does not have much to do with the observed stoichiometry of the intermetallic compounds.

Intermetallic compounds are of interest for structural applications (e.g., in gas turbine rotors) but are also exploited for their chemical behavior (corrosion resistance and hydrogen absorption) and physical properties (memory effect).

Some intermetallics are strongly resistant to wet corrosion, while others react very easily with water and rapidly form hydrogen. Examples of corrosion-resistant intermetallics are $Cu_6Sn_5$, CoSn, and NiSn. $CuSn_5$ resembles silver but is more tarnish-resistant and can be used in aggressive electrolytes. CoSn is used as a coating to protect nickel and is a better protector against low-temperature corrosion than nickel–chromium alloys. NiSn is inert in alkalis, in $HNO_3$, and in other acids below pH = 1.2, and does not tarnish in $SO_2$- and $H_2S$-containing atmospheres. It is also used as a conductor on printed circuit boards.[11]

Some intermetallics of magnesium such as $Mg_2Ni$ and $Mg_2Cu$ can make alloys with magnesium metal supercorrosive. These alloys are intimate mixtures of short-circuited anodic and cathodic particles that rapidly corrode in seawater and develop hydrogen. Two pounds of the composite can make a cubic meter of hydrogen in five minutes.

Cerium particles ignite in air but cerium metal is soft and sparks cannot be drawn from it. $CeFe_2$ with small amounts of magnesium, copper, and tin makes an excellent lighter flint material because it is hard and brittle.

Intermetallics such as NiAl, $Ti_3Al$, and $MoSi_2$ are used as matrix materials in composites for special structural applications. If the reinforcing second phase is a dispersion of fibers or particles (SiC, $Al_2O_3$, TiC, or $TiB_2$) the result is a strong, tough, and light composite suitable for aeronautical and space applications.

The case of Al–Ti–Ni is representative for other intermetallics and this combination will be shown here in some detail. Many of the compounds that these elements can form with each other have interesting properties. The two-component phase diagrams are shown in Figure 4.29, and the crystal structures of several compounds are shown in Figure 4.30.[16] These phase diagrams[11] show line compounds ($Al_3Ni$, $TiNi_3$), compounds that have a broad existence region (AlNi, TiAl), and compounds that allow only a small deviation in their stoichiometry ($TiAl_3$, $Ti_2Ni$, $AlNi_3$). $Ni_2AlTi$ is very strong[16] in a two-phase particulate composite with

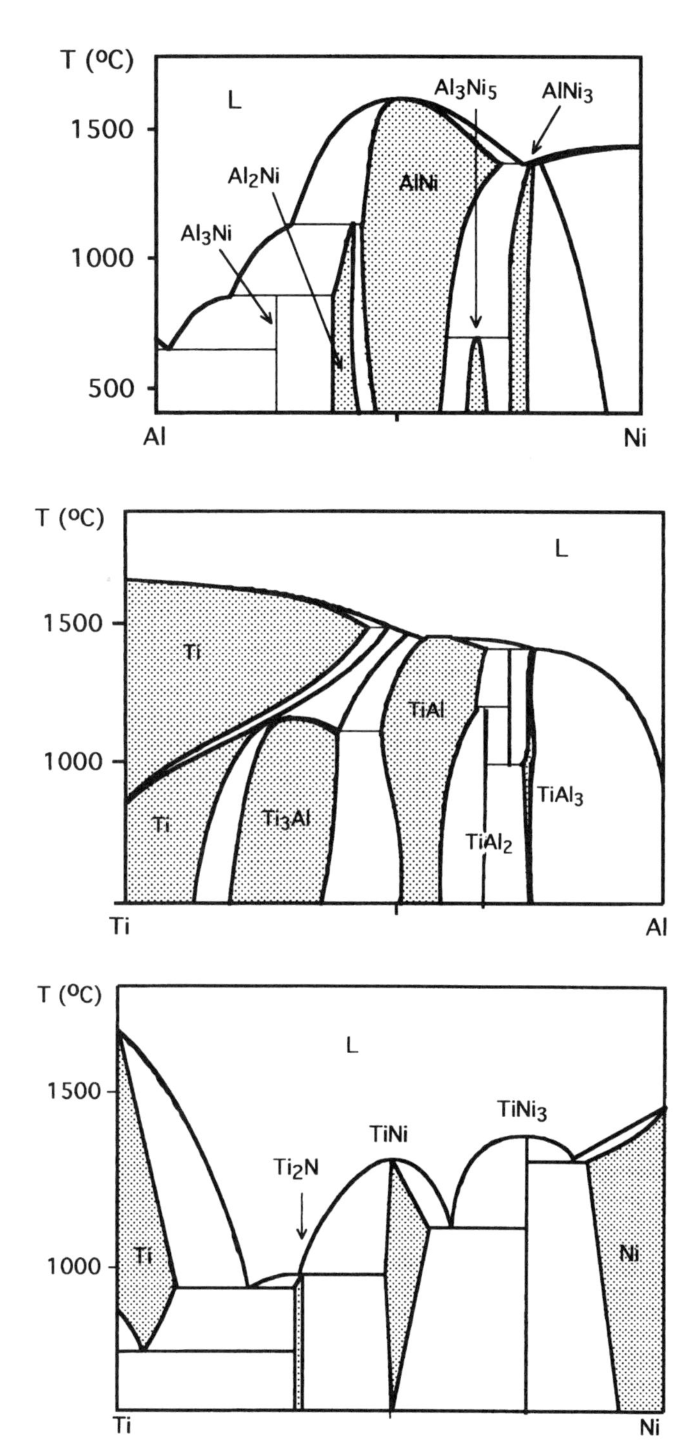


**Figure 4.29.** Binary phase diagrams of intermetallics in the Ni/Ti/Al system: (a) The phase diagram of the Al/Ni system. (b) The phase diagram of the Ti/Al system. (c) The phase diagram of the Ti/Ni system with three intermetallic compounds. From J. H. Westbrook and R. L. Fleischer (eds.). *Intermetallic Compounds.* Copyright John Wiley & Sons (1995). Reproduced with permission.

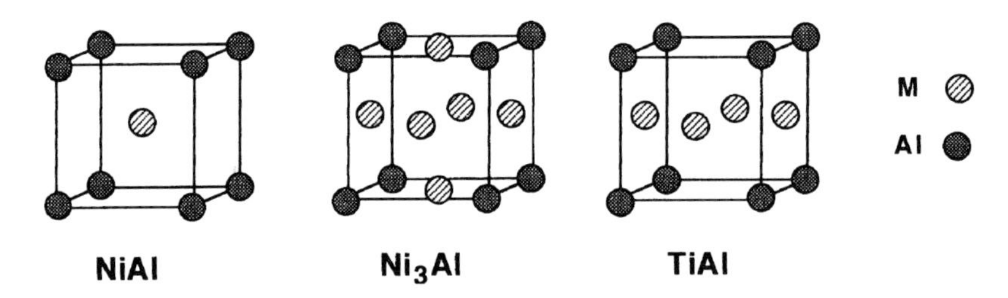


**Figure 4.30.** Some simple crystal structures of intermetallic compounds.

NiAl but somewhat brittle. If $Ni_3Al$ is present in the material as a third phase, the composite becomes less brittle. Because of its lightness, strength, and ductility $Ni_2AlTi$ is now considered as a replacement for superalloys in aeroplane engines.

The corrosion-resistant intermetallic compounds are passive because a protecting surface oxide layer grown by an initial oxidation reaction stops further corrosion if the layer does not support diffusion of oxygen or metal ions. The anticorrosive performance of these compounds varies because the protecting oxide scales that form have different formation rates. TiAl, $TiAl_2$, and $TiAl_3$ develop protecting scales at high temperatures in oxygen which slows further reaction.[11] These compounds have enough aluminum for that. Traces of Nb and Cr help passivation. All Ti/Al compounds are subject to embrittlement by atmospheric oxygen at high temperatures. On oxidation $Ni_3Al$ forms nonadherent scales of alumina and nickel aluminum spinel ($NiAl_2O_4$) but traces of Y, Zr, and Hf improve scale adherence.

NiTi is one of the intermetallic compounds that show shape memory. It has a CsCl structure and undergoes a martensitic phase transition to a sheared rhombohedral modification on cooling to a temperature below a transition temperature of 110°C. If the solid is strained in the low-temperature rhombohedral state it reverts to its original form when it is heated to the cubic modification. The deformation at low temperatures is accompanied by changes in growth of rhombohedral domains, which disappear at high temperatures. The material shows high damping of vibrations and superelasticity and is used in active materials for its thermomechanical properties. The temperature of the martensitic phase transition can be varied by dissolving other metals in it. Pd increases the transition temperature and Fe, Cr, or an excess of Ni lowers it. NiTi is as corrosion-resistant as stainless steel and can be used as an implant material.

Almost all intermetallics are embrittled by hydrogen absorption.[11] The ready reaction of metals with hydrogen to ternary hydrides is used in hydrogen storage devices. Higher hydrogen densities are possible in intermetallic hydrides than in liquid hydrogen. In the compounds hydrogen acts as a metallic element and the hydrides behave as intermetallics. Well-known compounds that show this are $LaNi_5$, which forms $LaNi_5H_{6.5}$ (at a hydrogen pressure of 2 bars) and FeTi that reacts to form $FeTiH_{1.9}$. NiTi and $NiTi_2$ can take up one hydrogen atom per nickel atom. The hydrogen can be removed from the intermetallic hydride by lowering the pressure. The compounds swell on hydrogenation, cyclic hydrogenation breaks up the crystallites, and fine powder forms. Nickel hydride is used as a battery anode.

Raney nickel is a catalyst much used in organic reactions. It is made from AlNi by leaching aluminum from it with an NaOH solution. When aluminum is dissolved

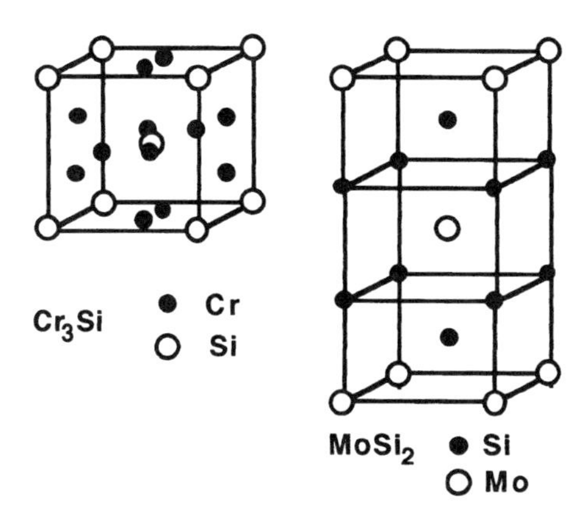


**Figure 4.31.** Crystal structures of some silicides.

from the intermetallic compound, porous nickel is produced with a very high specific surface area.

Synthesis of intermetallics makes use of metallurgical techniques, simply mixing the components in the right proportions at high temperatures followed by the proper heat treatment. $Ti_3Al$ and $Ni_3Al$ can be forged.

Metal silicides are sometimes regarded as intermetallics even if silicon is not strictly speaking a metal. They are quite useful in metallurgy and are applied in planar technology in electronics as metallic conductors to make contact with the silicon semiconductor. Some silicide structures are shown in Figure 4.31.

The semiconductor $CrSi_2$ has a hexagonal structure, while $NbSi_2$ and $MoSi_2$, with the same structure are metals, as is $CoSi_2$, which has an fcc stacking of cobalt atoms with silicon in all the tetrahedral interstices.[11] $NiSi_2$, which has a fluorite structure (the cubic structure of $CaF_2$) can be alloyed with Ni and Al to form blue, yellow, and white materials.[16] FeSi and CoSi can be alloyed together in all concentrations and range from a semiconducting alloy at the iron end to a metallic alloy when the cobalt concentration is highest.

$MgSi_2$ is not a metal but a semiconductor and is light, strong, and somewhat brittle. Its properties make it suitable for use as a cylinder material in internal combustion engines.[16] $MoSi_2$ is a binary compound that is very corrosion-resistant at high temperatures in air (see Section 6.8). It is a metallic conductor and it is used to make super-Kanthal wire, which is used for heating elements in high-temperature ovens.

## 4.7. Intrinsic Properties

The crystal structure, i.e., the way atoms are stacked and the nature of the component atoms together determine a compound's intrinsic properties. In this respect crystalline solids do not differ from molecules and have intrinsic properties that depend only on the nature and the arrangement of their component atoms.[4] Familiarity with the intrinsic properties of solid compounds is invaluable for the materials designer.[17] Intrinsic properties cannot be tailored with the preparation conditions, they have definite values for each compound, and do not depend on the history of the preparation. There are some structural features that affect properties in the same way in different compounds and such general structure–property relationships are discussed briefly below.

### 4.7.1. Electron Conductivity

Almost all metals and intermetallic compounds are electron conductors but the various nonmetallic inorganic materials have very different conductivities; there are very good electron conductors, insulators, and semiconductors among them. Solids are sometimes grouped according to their characteristic properties: e.g., electronic conductivity, where they are insulators, semiconductors, or conductors. Such groupings are related to the type of bonding. Oxides can be insulators ($Al_2O_3, SiO_2$), semiconductors ($SnO_2, Fe_2O_3$, NiO), or metallic conductors (VO, $ReO_3$), and their bonds are covalent, ionic, or metallic. Similarly, nitrides are metallic conductors (TiN), semiconductors (GaN), ionic salts ($NaN_3$), or insulators ($Si_3N_4$, AlN). Carbides are metallic conductors (TaC), semiconductors (SiC), or ionic salts ($CaC_2$). Most borides are also either metallic conductors ($TiB_2$) or semiconductors such as $CaB_6$ or $EuB_6$. While covalent bonds in materials usually mean semiconductivity (SiC, AlN) and ionic solids are insulators for electrons ($Al_2O_3, ZrO_2$), the electronic conductivity of ionic oxides can differ by many orders of magnitude (Figure 4.32). Generally the more ionic the bonding is, the larger the bandgap. The relationship between valence electron behavior and crystal structure of inorganic solids is hidden as usual. $TiS_2$ and $NbS_2$ are metallic conductors and $MoS_2$ is a semiconductor, but all are layered lattices. The differences in the conductivity of these compounds is related to the number of empty interstitials that are gradually occupied with an increasing number of valence electrons.

Intrinsic electron conductivity of semiconducting compounds has a characteristic temperature dependence. The logarithm of the conductivity vs. $1/T$ is a straight line with a slope that is a simple function of the bandgap and the activation energy required for the electron to hop from one site in the lattice to the next. If the bonds in the solid are covalent, the electrons have a high mobility, there are no filled impurity states in the bandgap, and the slope is the bandgap. This only holds for very pure compounds, where the conductivity is really intrinsic. More details can be found in Chapter 10. If, as discussed there, the charge carriers in the solid are not generated thermally but are mainly the result of ionized impurities, the line of the plot is still straight but the slope is the activation energy for displacement of conduction charge carriers. The ease of displacement (mobility) for electrons or electron holes depends on the charges on the ions in the lattice. A higher ionic charge means stronger interactions with a traveling electron or hole. A local electron repels negative atomic charges, attracts positive ones and distorts the lattice at its position. When electrons pass through an ionic lattice they drag lattice distortions (called a phonon cloud) with them, which slows them down. The mobility of charge carriers in polar solids is in general lower than in covalent compounds. A charge carrier together with its surrounding cloud of lattice distortions (sometimes called a dressed electron or electron hole) is a polaron.

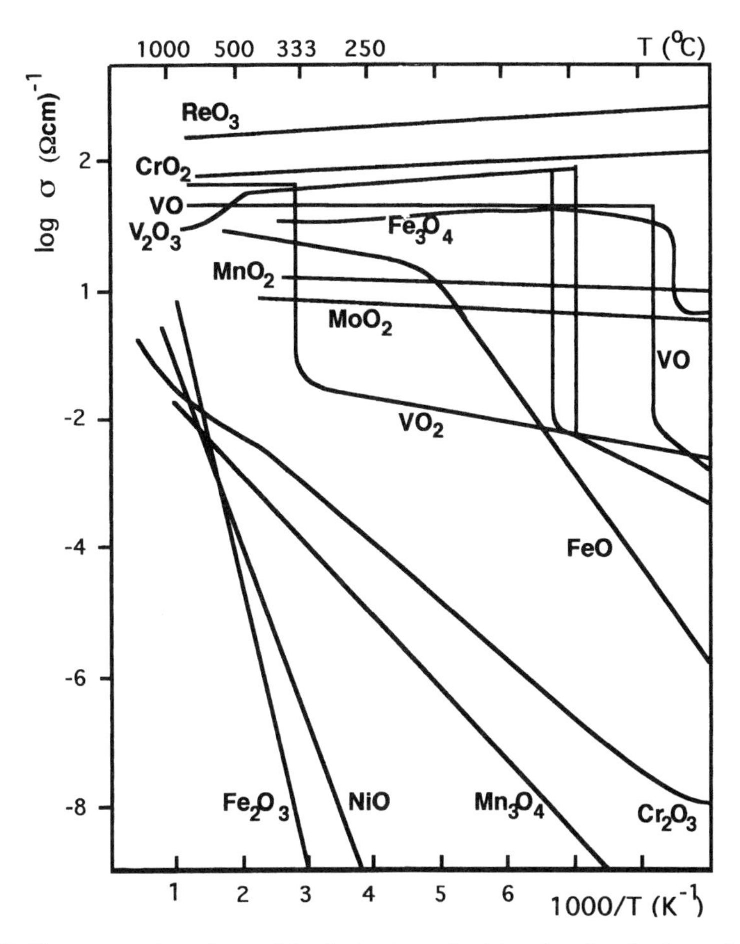


**Figure 4.32.** Temperature dependence of the electronic conductivity of semiconducting and conducting oxides.

The number of charge carriers in an insulating compound or a semiconductor is related to the concentration of the impurities, which can be aliovalent ions or vacancies if the compound is nonstoichiometric. Binary nonstoichiometric compounds (many transition metal oxides are berthollides) can be *n*-type or *p*-type semiconductors (Table 4.7) even if they are not doped with other ions. The excess of electrons or holes is the result of a higher or lower valency of the lattice ions owing to oxidation or reduction that is caused by an excess or a depletion of oxygen ions. Whether a given oxide is *n*-type or *p*-type depends on the valency of the metal ion and not primarily on the structure of the lattice.

Titanium, e.g., cannot have a higher valency than four and it has that valency in $TiO_2$. However, some of its atoms can have a lower valency if some oxygen is lost from the lattice. This results in some loss of energy but there is an increase in entropy. The stoichiometry then determines the temperature-dependent conductivity. Cobalt oxide (CoO), on the other hand, tends to be *p*-type semiconducting because it is easier to form trivalent cobalt by some uptake of oxygen

**Table 4.7. Some Compound Semiconductors with a Type of Electron Conduction Determined by the Stoichiometry**

| | | | | | | | |
|---|---|---|---|---|---|---|---|
| *n*-type conduction | | | | | | | |
| BaO | $TiO_2$ | $V_2O_8$ | $MoO_2$ | $Fe_3O_4$ | $Ag_2S$ | $BaTiO_3$ | $PbCrO_4$ |
| $Cs_2S$ | $Cs_2Se$ | $Nb_2O_5$ | $WO_3$ | $Hg_2S$ | ZnO | CdS | $SnO_2$ |
| *p*-type conduction | | | | | | | |
| $Cr_2O_3$ | $Pr_2O_3$ | MnO | CoO | NiO | SnO | $Cu_2O$ | $Cu_2S$ |
| $Bi_2Te_3$ | $MoO_2$ | $Hg_2O$ | $Sb_2S_3$ | $Ag_2O$ | SnS | CuI | |
| Amphoteric conduction | | | | | | | |
| $Al_2O_3$ | SiC | $Mn_3O_4$ | $Co_3O_4$ | $Ti_2S$ | PbS | PbSe | PbTe |

from the air than it is to form monovalent cobalt through a loss of oxygen. CoO has an excess of electron holes and has *p*-type conductivity, while $TiO_2$ is an *n*-type semiconductor with some extra conduction electrons. In general semiconducting oxides in which the metal ion has its maximum valency have an excess of conduction electrons (and oxygen vacancies) and therefore show *n*-type conductivity, while oxides in which the metal ion can have a higher valency have an excess of electron holes and are *p*-type semiconductors.

Doping a solid can have a big influence on the Fermi level or the chemical potential of the electrons, and solid state electronic devices (diodes, transistors, solar cells) are based on adjusting that level.

Certain binary compounds undergo a Mott transition, e.g., $V_2O_3$, $VO_2$, $V_3O_5$, $V_6O_{13}$, $Ti_2O_3$, $Ti_3O_5$, $Ti_4O_7$, $Ti_5O_9$, $NbO_2$, $Fe_2O_3$, NiS, CrS, and $NiS_2$. A Mott transition is a phase transition in which the solid changes from a semiconductor to a metallic conductor above the transition temperature. When the temperature of the Mott insulator (or semiconductor) is increased, the resulting thermal excitation increases the number of electron–hole pairs in the solid. The excited electron stays near the hole on the atom that is ionized. At a critical temperature $T_m$ these electron–hole pairs have become so numerous that they start to overlap and the electrons are so heavily screened from their positive centers that the attractive force is reduced and they become delocalized, turning the solid into a metallic conductor. Such oxides are used in critical temperature thermistors, resistors that conduct better above the critical temperature.

Many nonmetallic inorganic compounds are superconducting at sufficiently low temperatures. The derivatives of mixed oxides such as $YBa_2Cu_3O_{7-x}$ have the highest critical temperatures. Other superconducting inorganic solids have lower transition temperatures. Of these the niobium compounds have transition temperatures that are comparatively high: NbC (11 K), NbN (16 K), $NbC_{0.3}N_{0.7}$ (18 K), $NbS_2$ (6.2 K), CuS (1.62 K), $CuS_2$ (1.62 K), $TaS_2$ (2.1 K), $Rh_{17}S_{15}$ (5.8 K). Superconducting borides were discussed above.

### 4.7.2. Dielectric Properties



Some features of dielectric properties depend strongly on the structure of the crystal and some are purely atomic. An electric field **E** polarizes materials by inducing an electric dipole moment **p**, which is proportional to the strength of the (internal) field $\mathbf{E}_i$:

$$\mathbf{p} = \alpha \mathbf{E}_i = \varepsilon_0(1 - \varepsilon)\mathbf{E}$$

where $\alpha$ is the polarizability, $\varepsilon$ is the dielectric constant of the material, and $\varepsilon_0$ is the permittivity of the vacuum [$\varepsilon_0 = 8.854\ 10^{-12}$ As/Vm].

The Clausius–Mosotti equation relates the polarizability $\alpha$ to the dielectric constant $\varepsilon$:

$$\alpha = 3\varepsilon_0 \frac{\varepsilon - 1}{\varepsilon + 1}$$

There are four contributions to the polarizability:

$$\alpha = \alpha_e + \alpha_i + \alpha_d + \alpha_s$$

where $\alpha_e$ is the electron polarization, $\alpha_i$ is the ion shift polarization, $\alpha_d$ is the polarization owing to the orientation of the dipoles, e.g., dipoles of defect associates, and $\alpha_s$ is the polarization of the electrode surface by mobile ions.

In an alternating electric field, the four parameters all depend on the frequency of the external field, as shown in Figure 4.33. At the lower frequencies the four polarization components all contribute to $\alpha$. As the frequency is increased, first the space charge polarization $\alpha_s$ disappears because the direction of the field changes too rapidly for the traveling ions to build up a double layer at the electrodes; the precise frequency at which this occurs depends on the mobility of the ions and the concentration of charged point defects. At some higher frequency the oriented dipole moments can no longer follow the alternating field and the polarization drops again because $\alpha_d$ disappears. The dipolar contribution also requires ion or dipole mobility. When the frequency (by now in the infrared range) has become so high that the ions at the lattice points cannot move in phase with the external field the ion contribution to the polarization $\alpha_i$ also disappears. ($\alpha_i$ can be very large in ferroelectric perovskites, which is the reason they are used in ceramic capacitors.) The only polarization that remains at the high, near-optical frequencies is in the electron cloud. The final (electronic) polarizability component is not structure-dependent, being determined primarily by the atomic electron clouds. The dielectric polarizabilities in light atoms are low; in heavy atoms they are higher. Their value is a function of the third power of the atomic radius, which means large atoms have a higher electronic polarizability than small ones. If a low dielectric constant is required, an insulating, non-ion-conducting ceramic with small atoms is chosen (BN).

The optical (infrared and visible) thresholds are described by complex dielectric constants, which accounts for the shape of the polarizability curve at these frequencies.

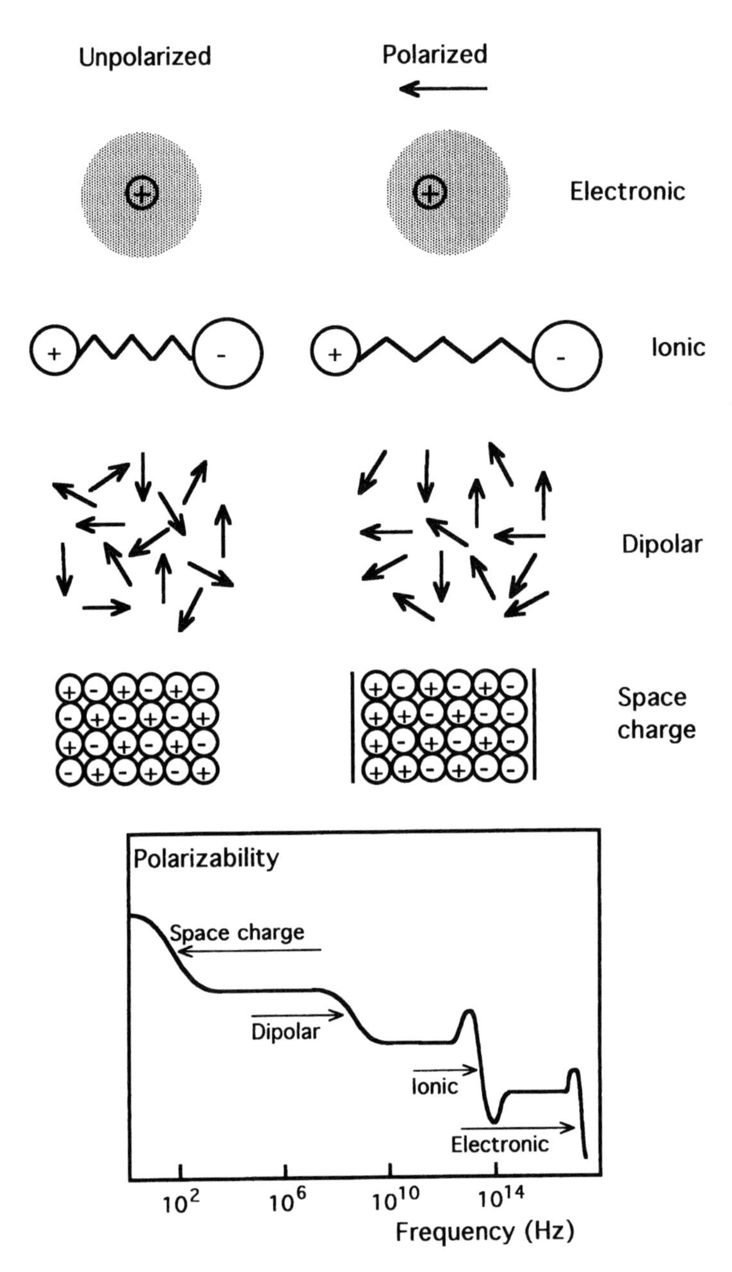


**Figure 4.33.** The four contributions to the polarizability and their frequency dependence.

Piezoelectricity and ferroelectricity are the result of electric dipole effects that are due to ions shifting from their initial positions in the lattice. This effect, which is strongly structure-dependent, was discussed above. Pyroelectric materials have permanent electric dipole moments, their magnitude varying with temperature. In infrared detectors, use is made of the ferroelectric phase transitions in perovskites.

### 4.7.3. Ion Conductivity

The conductivity of ions in ionic lattices is also strongly dependent on the crystal structure. For instance, the calcite structure of zirconia has one empty interstitial space per unit cell, which can accommodate anions in an anti-Frenkel disorder. Another example is Na $\beta$-alumina, which has a structure that consists of spinel blocks alternating with layers that contain the sodium ions. There are many empty interstitial sites in these interblock layers through which the alkali metal ions can move and so this compound is a good conductor of monocharged cations. There are many structures that are good conductors of ions and if such compounds are made in amorphous modifications, the ion conductivity is generally lower. On the other hand, if a certain crystal lattice does not conduct ions well, the amorphous modification of that compound generally shows better ion conductivity because glasses offer pathways in their disordered lattices that are not present in the crystal lattice. Some applications of solid ion conductors are given in Section 10.3.

### 4.7.4. Magnetic Properties

The magnetic susceptibility $\chi$ relates the magnetic polarization **M** to the magnetic displacement **B**. The relation is $\mathbf{M} = \chi\mathbf{B}$. The susceptibility $\chi$ is positive and linearly dependent on the inverse of the temperature in paramagnetic solids and small and negative in diamagnetic compounds. Paramagnetism is the response of noncoupled localized magnetic dipole moments of unpaired electrons in the ions to a magnetic field. Spin-paired electrons do not contribute to paramagnetism, but all fully spin-paired electrons in a solid contribute to its diamagnetism.

If the ions have magnetic dipole moments owing to unpaired spins, the dipole moments of neighboring atoms or ions may be coupled. If the ground state has parallel spins the solid is ferromagnetic. If the spins are antiparallel and the dipole moments in both directions compensate each other fully, the solid is antiferromagnetic. If they do not completely cancel one another out and there is some uncompensated moment left, the solid is ferrimagnetic. Magnetic coupling can be either direct or via an intermediate diamagnetic ion (called superexchange). Magnetic coupling is a strong function of the crystal structure but the relationship is not simple. There are some correlations with distance between the localized magnetic moments and the angle between the bonds of the metal ions with the intermediate anion.

At temperatures above the critical temperature ferromagnetic, antiferromagnetic, and ferrimagnetic compounds become paramagnetic. The critical temperature for ferromagnetic and ferrimagnetic compounds is the Curie temperature and for antiferromagnetic compounds it is the Neèl temperature. Examples of ferrimagnetic oxides with their Curie temperatures are $Fe_3O_4$ (848 K), $\gamma$-$Fe_2O_3$ (858 K), yttrium iron garnet or YIG (560 K), $Mn_{.6}Zn_{.4}Fe_2O_4$ (443 K), and $MgFe_2O_4$ (593 K). The ferrites have a spinel structure. There are not many ferromagnetic oxides, but some examples are $CrO_2$ (400 K), EuO (77 K), and $La_xSr_{1-x}MnO_3$ (210–385 K).

Small parts of a ferromagnetic or ferrimagnetic solid in which the atomic magnetic dipoles are all oriented in the same direction are known as Weiss domains. In a nonmagnetized ferromagnetic (or ferrimagnetic) solid each Weiss domain has its magnetic moment oriented in a different arbitrary direction and the random orientations average out to zero when the solid is not in an external magnetic field. Weiss domains are bounded by Bloch walls. If the solid is now polarized in a magnetic field the domains become oriented, which means that the domains with a favorable direction (parallel with the external field) become larger and those with other polarization directions become smaller. In other words, the Bloch walls move through the solid when the magnetic field is increased until the magnetization directions of all domains are aligned along the magnetic field. There is now one big domain and the solid is magnetically saturated.

If the external magnetic field is then decreased, domains with a reversed polarization direction start to nucleate and grow by displacing their Bloch walls at the expense of domains with the original opposite orientation. When the external magnetic field is zero, there is still a net magnetization left in the solid because the domains are not randomly oriented and their magnetic moments do not compensate each other. The rest magnetization at zero external field is called the magnetic remanence. If the field is now increased in the opposite direction, the small domains with polarization in this direction grow further at the expense of the domains with polarization in the other direction, again by displacement of the Bloch walls. At a certain field strength, called the coercive force, the net magnetization of the solid becomes zero. The coercive force is the counterfield that is able to reduce the total net magnetization of a material to zero. A further increase of the external field finally saturates the magnetic moment of the solid in the reverse direction.

The magnetization vs. magnetic displacement curve is characterized by magnetic remanence, the coercive force, and the magnetization saturation moment. The magnetization curve (which looks like the dielectric polarization-field curve of Figure 2.26) shows some hysteresis. It is large in hard magnets, which have a high coercive force, and small in soft magnets, where the coercive force is small. In hard ferromagnets the displacement of the Bloch walls requires a lot of energy, which makes both the remanence and the coercive force large. In soft ferromagnets the displacement of the domain walls is much easier and the coercive force is thus smaller, which explains why a strong alternating magnetic field heats up a hard magnetic material but not a soft one.

The coercive force and the magnetic hardness are somewhat extrinsic and to some extent can be adjusted by modifying the microstructure: the crystallite size and the grain boundaries, which affect these magnetic parameters, are controlled with the techniques described in Chapters 5 and 8. On the other hand, the Curie temperature and the value of the saturation magnetization are intrinsic properties that depend on the crystal structure and the number of unpaired electrons and their magnetic moments, i.e., on the nature of the elements in the material and the way they are bound.

Information storage in magnetic memories needs intermediate magnetic hardness. If the coercive force is too low (as in soft ferromagnets) there is loss of memory with time because the polarization is low and easily changed. If the polarized material is too hard the information is difficult to replace.

### 4.7.5. Mechanical Properties

Most of the mechanical properties of solids, such as strength and toughness, are not closely related to the crystal structure or to the strength of the bonds between the atoms. As long as the bonds between atoms are strong, all solids would be extremely strong if the flaws that cause local stress concentrations could be completely eliminated. This is more or less possible in monocrystalline whiskers (such whiskers can be elastically strained to 15% of their original length before they fail by exploding when the elastic energy is released) but polycrystalline solids have two- and three-dimensional defects that determine the strength. Yield strengths are related to the surface energy by some (Table 4.8) but the high values for some polycrystalline solids (jades) show that strength is extrinsic. Strength in materials is not a function of the strength of interatomic bonds but is determined by flaws and macrodefects, i.e., it is extrinsic.

Sometimes changes in lattice structure improve the strength: objects made of lithium or sodium glass can be strengthened by dipping them in a $KNO_3$ melt. The potassium ions replace the smaller lithium and sodium ions in the glass by ion exchange and the surface is brought into compression. Hardness is not related only to the chemical formulation as can be concluded from many tables of properties. The diverging values indicate that it is also extrinsic. Yet compressibility and hardness are related to the lattice energy density, which is an intrinsic quantity (Figure 4.34).

Bulk modulus and mechanical hardness of materials (mechanical hardness is not to be confused here with the chemical hardness of atoms) are properties that correlate more with bonding than with structure. The bulk modulus is somewhat structure-dependent: Of two compounds that have the same interatomic spacing the compound that has an NaCl-structure has a higher bulk modulus than a sphalerite-like compound.

There are two relationships between mechanical hardness and the chemical bond:

1. A high density of bond energy, i.e., the amount of bond energy per unit volume favors mechanical hardness and incompressibility.[18] Small, strongly bonded atoms, close together, make for hard materials (diamond, c-BN).

**Table 4.8. The Surface Energy of Some Materials**

| Material | Face | Energy (ergs cm$^2$) | Material | Face | Energy (ergs cm$^2$) |
|---|---|---|---|---|---|
| $Al_2O_3$ | $(10\bar{1}0)$ | 600 | Halite | (100) | 300 |
| $Al_2O_3$ | Polycrystalline | 24 000 | Kaolinite | Polycrystalline | 300 |
| Quartz | $(10\bar{1}0)$ | 1000 | Muscovite | (001) | 4000 |
| Quartz | $(10\bar{1}1)$ | 500 | Periclase | (100) | 4000 |
| Quartz | Polycrystalline | 4000 | Jadeite | Polycrystalline | 120 000 |
| Calcite | $(10\bar{1}1)$ | 900 | Nephrite | Polycrystalline | 230 000 |

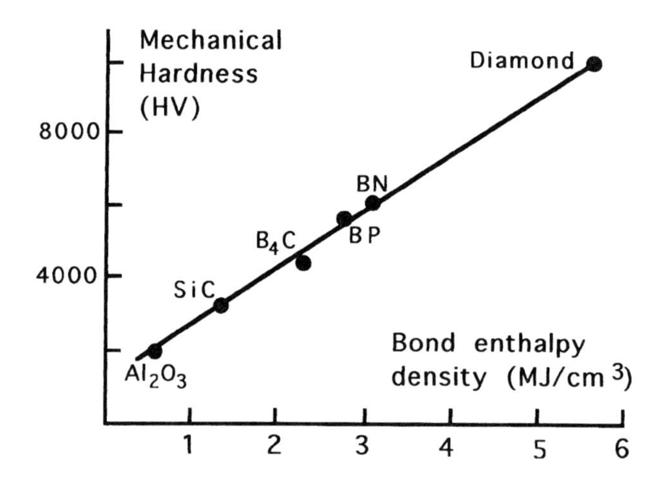


**Figure 4.34.** Correlation between the mechanical hardness of compounds and the density of interatomic bond strength.

2. The mechanical hardness of carbides and nitrides in mixed compounds of group 4 and 5 metals is observed to have a maximum if there are $8\frac{1}{4}$ to $8\frac{1}{2}$ valence electrons per stoichiometric unit (e.g., $Ta_xHf_{1-x}C$).

Mechanical hardness is a complicated function of compressibility, plasticity, and elasticity and is not really an intrinsic property in polycrystalline materials but depends on the microstructure.

Plasticity is an extrinsic property but it is favored by good mobility of dislocations and depends as well on the type of bond in the lattice. With highly mobile valence electrons in interstices, dislocations in metals have a low activation barrier for displacement. Metals therefore often have a high plasticity and deformability as long as the dislocations can travel unhindered by defects and grain boundaries. Strongly directed covalent bonds do not permit dislocations to move easily, nor do ionic lattices, which have a high barrier for shear in the transition configuration due to repelling charges on neighboring atoms. Such compounds are brittle as a consequence of the bond type in the lattice.

## 4.7.6. Optical Properties

Nonmetallic inorganic materials are widely used for optical purposes: lenses, pigments, interference filters, laser hosts, luminescent coatings, displays, solar cells, fiber optics, lamp bulbs, and tubes. For optical applications use is made of the refractory index, light absorption, luminescence, and nonlinear optical behavior of materials. These are intrinsic but may depend on the concentration of impurities. Refraction index and optical absorptivity in insulators are atomic properties and are only indirectly related to the structure, but the structure affects the selection rules and the term splitting in the atomic chromophores. The coordination number determines the intensity and wavelength of absorption and emission bands and also the degeneracy of the valence electron shells, which determines the magnetism.

Spinels have anion lattices with tetrahedral and octahedral sites for transition metal dopants. Garnets have three types of cation lattice sites. Some cation sites are coordinated by eight oxide ions on the corners of a distorted cube around the cation, four-coordinated (tetrahedral), and six-coordinated (octahedral) lattice sites. By substituting the right transition metal or lanthanide ions in these sites the color and optical activity of garnet pigments can be controlled.

The refractory index $n$ depends on the band gap $E_g$ (in eV); for oxides there is an empirical relation:

$$n^2 = 1 + 15/E_g$$

For perpendicular incident light the reflecting power $\rho$ of an interface between two materials that differ in refractory index is

$$\rho = \frac{n_1^2 - n_2^2}{n_1^2 + n_2^2}$$

The greater the difference between refractory indexes of the two materials the better the interface reflects light. Titania particles scatter light well in an organic matrix (paint) and $TiO_2$ is an efficient white pigment because it has a high refractory index ($n = 2.5$) compared to that of the transparent organic matrix. As a pigment the white compound calcium carbonate ($n = 1.6$) has much less covering power. The refractory index is an intrinsic property but in order to function well the titania pigments need a surface treatment (Chapter 8) to prevent the photogenerated electron holes in the oxide from reacting with the organic matrix.

Light absorption by materials in the frequency range from far-infrared to ultraviolet is sketched in Figure 4.35. Infrared absorption is due to lattice vibrations, which are used to characterize compounds because the bands can be assigned to specific interatomic bond vibrations. The electronic transitions in chromophores are in the visible range, as was discussed in Chapter 2. In the ultraviolet range the material becomes opaque at frequencies corresponding to energies higher than the bandgap. In orbital terms the electrons are excited from the valence band to the conduction band at these frequencies. The bandgap of a material determines the lower boundary of its wavelength window for light transmission. The relation is $\lambda_{min} = hc/E_g$ in which $\lambda_{min}$ is the boundary wavelength, $E_g$ is the value of the bandgap interval, $h$ is Planck's constant, and $c$ is the velocity of light. The transmission window is bounded on the lower side by the infrared absorptions that are due to lattice vibrations. Between the upper and lower bounds of the transmission window is a minimum in light absorption. Glasses for fiber optics for information transmission are optimized for a low extinction here, which is expressed in dB/km. The minimum of very pure germanium-doped silica glass is at a wavelength of 1.55 $\mu$m and the light source used for fiber optics is a laser emitting this radiation. Figure 4.36 shows that the transparency is also determined by Rayleigh scattering by inevitable density fluctuations in the glass. A polycrystalline material would be unsuitable for fiber optics because of light scattering at the grain boundaries.

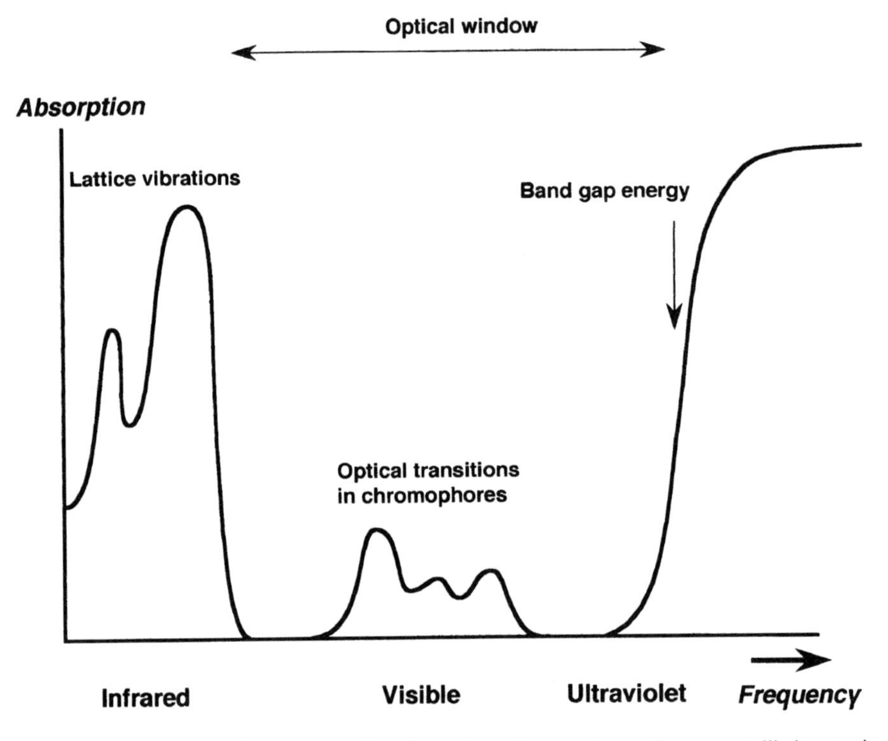


**Figure 4.35.** The schematic representation of an absorption spectrum of a nonmetallic inorganic compound.

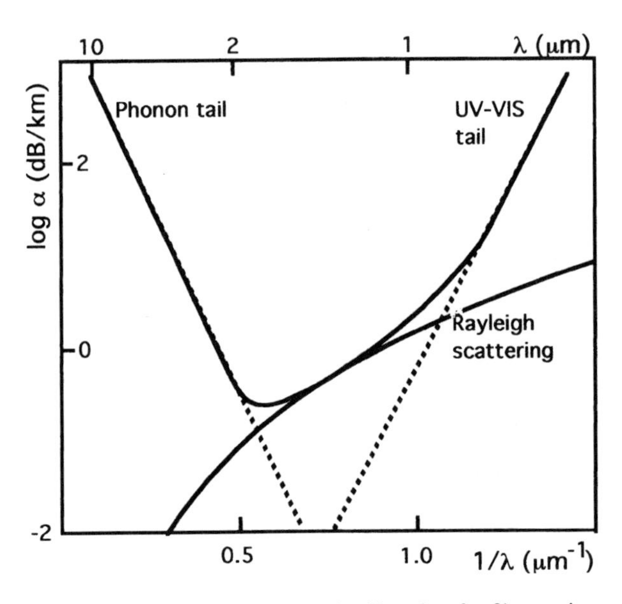


**Figure 4.36.** Transparency gap in silica glass for fiber optics.

### 4.7.7. Chemical Properties

Chemical properties of solid compounds are not intrinsic but extrinsic as the chemistry of solids is for the most part the kind of chemistry that occurs at the surface or at an interface between solid phases. Interface chemistry is the subject of Chapter 6. Bulk chemistry of solids is linked to the presence of defects. An introduction to defect chemistry in crystals is given in Chapter 10. Insofar as bulk chemistry relates to ion and electron transport it has been discussed above under electrical properties.

### 4.7.8. Thermal Properties

Solid compounds react to heat by expanding or with phase transitions such as melting. The thermal properties[19] that are most interesting for the designer are the melting point (or glass point), the solid-phase transition temperature, the thermal expansion coefficient, and the heat conductivity. These last two properties determine the thermal shock resistance.[20]

#### Melting Points

The melting points of covalent inorganic compounds are generally high (Table 4.9 gives a few values). Glass points and sinter temperatures are a fraction of the melting points ($\approx \frac{2}{3} T_m$) and refractory solids can be sintered only at comparatively high temperatures.

#### Phase Transitions

Phase transitions in solid compounds[21] are intrinsic and there are tables that give transition temperatures (Table 4.9) of materials. However, the transition temperature can often be changed by alloying the compound with another one while leaving the other properties more or less intact. As was shown in Section 4.5 the cubic modification of $ZrO_2$ at high temperatures can be stabilized

**Table 4.9. Melting Points of Some Refractory Ceramics**

| Oxides | $T_M$(°C) | Carbides | $T_M$(°C) | Nitrides and borides | $T_M$(°C) |
|---|---|---|---|---|---|
| $MgO$ | 2800 | $B_4C$ | 2450 | $BN$ | 3000 |
| $Al_2O_3$ | 2050 | $SiC$ | 2700 | $AlN$ | 2200 |
| $SiO_2$ | 1780 | $Al_4C_3$ | 2800 | $Si_3N_4$ | 1900 |
| $ZrO_2$ | 2600 | $TiC$ | 3250 | $TiN$ | 2940 |
| $Cr_2O_3$ | 2260 | $ZrC$ | 3500 | $SiB_6$ | 1950 |
| $CeO_2$ | 2730 | $HfC$ | 3900 | $TiB_2$ | 2980 |
| $ThO_2$ | 3300 | $VC$ | 2800 | $ZrB_2$ | 3060 |
| $HfO_2$ | 2789 | $NbC$ | 3500 | $HfB_2$ | 3250 |
| $MgAl_2O_4$ | 2135 | $TaC$ | 3900 | $NbB_2$ | 3000 |
| $3Al_2O_3.2SiO_2$ | 1810 | $MoC$ | 2700 | $TaB_2$ | 3100 |
| $ZrSiO_4$ | 1775 | $WC$ | 2770 | $CrB_2$ | 2760 |

down to room temperature by dissolving oxides such as $Y_2O_3$ in it. Removing phase transitions in passivating layers generally improves protection. For example, the phase transitions of silica and the devitrification of protecting glassy layers that form on silica-containing ceramics ($MoSi_2$) at high temperatures in air cause the breakdown of the diffusion barrier and increased corrosion. If the silica is reacted with another oxide to form a ternary compound that has no phase transitions the corrosion protection under thermal cycling is improved.

### *Specific Heat*

The specific heat of solids is only indirectly related to structure. Its temperature-dependence is described by a universal curve of the specific heat with respect to the reduced temperature $T/\theta_D$, where $\theta_D$ (the Debye temperature) is specific for the compound. It can be shown that the specific heat increases at low temperatures with $T^{d-1}$ in which $d$ is the dimension of the lattice. For linear chains $d = 1$, for planar nets $d = 2$, and for three-dimensional stackings $d = 3$. The parameter $d$ can also be a noninteger in solids that have a special microstructure.

### *Thermal Expansion Coefficient*

The thermal expansion coefficients are related to the crystal structure and to the type and strength of the bond.[22] Many ceramics have low to medium expansion coefficients, as can be seen in Table 4.10. The coefficients may or may not depend strongly on temperature. In many cases the thermal expansion is anisotropic and can have negative values in certain temperature regions.

Thermal expansion is related to bonding through the anharmonicity (deviation from parabolic behavior) of the interatomic potential. Figure 4.37 shows this potential for two extreme cases, short-range covalent bonding and longer-range ionic bonding. On vibrational excitation the mean interatomic distance shifts to higher values, which means expansion of the lattice. Figure 4.37 illustrates

**Table 4.10. Thermal Expansion Coefficients of Several Ceramic Materials**[a]

| Compound | Expansion coefficient ($10^{-6}$ K) | Compound | Expansion coefficient ($10^{-6}$ K) |
|---|---|---|---|
| BeO | 9.0 | $Y_2O_3$ | 9.3 |
| MgO | 13.5 | $SiO_2$ glass | 0.5 |
| $Al_2O_3$ | 8.8 | Na-Ca glass | 9.0 |
| Mullite | 5.3 | Porcelain | 6.0 |
| $MgAl_2O_4$ | 7.6 | $B_4C$ | 4.5 |
| $ThO_2$ | 9.2 | SiC | 4.7 |
| $ZrSiO_4$ | 4.2 | TiC | 7.4 |
| $ZrO_2$ | 10.0 | | |

[a]Values are averages in the temperature range 25–1000°C.

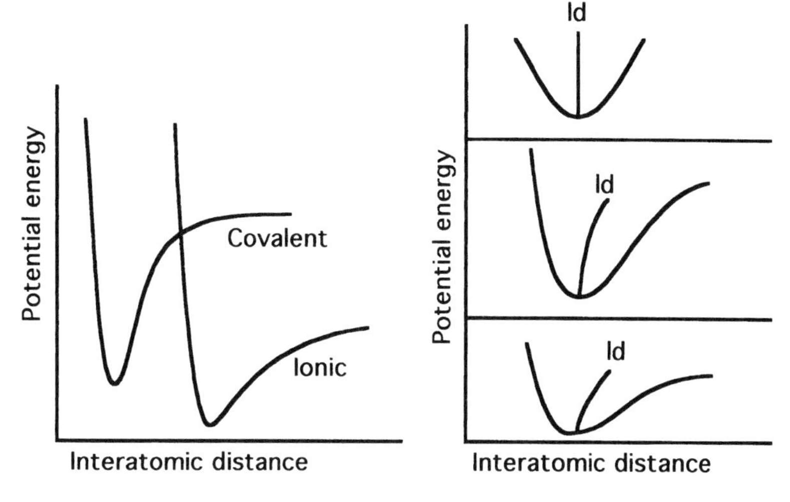


**Figure 4.37.** The interatomic potentials for a covalent lattice and an ionic lattice (left). The locus of interatomic distances (labeled *ld*) indicates that the average equilibrium distance shifts to higher values when the vibration amplitude of the atoms is higher. This effect is larger for ionic potentials than for covalent potentials because the former are more asymmetric.

qualitatively that:

1. A lattice with a purely harmonic interatomic potential does not expand very much.
2. Strong bonding entails low initial thermal expansion because the potential is deep and has a harmonic shape near its bottom.
3. A strong contribution of Coulomb attraction (in salts) makes the potential anharmonic and flatter on the high-distance side of the equilibrium position; this potential means high expansion coefficients.

A low expansion coefficient is beneficial in ceramics that are exposed to intense thermal shocks, as is discussed more fully below.

There are some physical generalities concerning thermal expansion coefficients. One empirical correlation is that $\alpha_v T_m$ is constant for a wide range of cubic and close-packed compounds, where $T_m$ is the melting point and $\alpha_v$ is the volume coefficient of thermal expansion. The Grüneisen equation relates $\alpha_v$ to the compressibility $K_0$, the heat capacity $c_v$, and the molar volume $V$; here $\gamma$ is the Grüneisen constant, a proportionality constant of first order:

$$\alpha_v = \gamma \frac{K_0 c_v}{V}$$

This equation indicates that the thermal expansion coefficient will be as temperature-sensitive as the heat capacity and that very strongly bonded materials have low thermal expansion coefficients. In fact MgO has a thermal expansion coefficient one-fourth that of NaCl.

There is no unique direct correlation between composition and a low expansion coefficient but there are some well-established relationships with structure: Very low thermal expansion coefficients are observed in materials that have: (1) strongly bonded coordination polyhedra linked in three dimensions; (2) an open structure with large interstitial spaces in the lattice to accommodate the thermal energy in transverse vibrations perpendicular to the bond directions. Fused silica, cordierite ($2MgO.2Al_2O_3.5SiO_2$, also called indialite), spodumene ($LiAlSi_2O_6$), and eucryptite ($LiAlSiO_4$) are examples of such open crystal structures. More closely packed lattices, on the other hand, have high expansion coefficients, especially if the interatomic potential curve is asymmetric, as in ionic lattices); (3) ferroelectric or ferromagnetic microdomains, which can absorb thermal energy by changing their dipole order (Invar and $PbMn_xNb_yO_3$).

### Heat Conductivity

Metals have a high heat conductivity owing to the presence of mobile electrons, which transport kinetic energy easily. There are also insulators (diamond, SiC, BeO, AlN) that are good heat conductors. A structural feature that implies a high conductivity in solids is the interatomic potential: the more parabolic its shape, the less heat transfer is hindered.

Heat conductivity of solids is also partially extrinsic because it depends strongly on defects and grain boundaries.

The rate of heat transfer in a thermal gradient is lower than might be expected for transport of vibrational energy: heat transport by conductivity is much slower than sound propagation. The accepted physical model for heat transfer is a process of diffusing phonons (wave packets in the vibrating lattice); heat conduction in solids and in fluids is observed to be diffusive.

Heat is observed to flow down a temperature gradient according to the relationship

$$\frac{dQ}{dT} = \kappa A \frac{dT}{dx}$$

where $\kappa$ is the thermal conductivity in $Wm^{-1}K^{-1}$, and $A$ is the area through which heat $Q$ flows. A constant related to $\kappa$ that is useful under nonstationary conditions is the thermal diffusivity $k_d = \kappa/c\rho$, where $c$ is the specific heat and $\rho$ the density. This parameter (in $m^2s^{-1}$) is a measure of the rate of transfer of thermal disturbance (temperature $T$) through the lattice:

$$\frac{dT}{dt} = k_d \nabla^2 T$$

Thermal conductivities can be determined by measuring the temperature increase of the back side of a plate of the material, which is heated with a laser pulse. As in a molecular gas the thermal conductivity of a phonon gas is

$$\kappa = \tfrac{1}{3} c v \lambda$$

where $c$ is again the specific heat, $v$ the intrinsic phonon speed (speed of sound in the lattice), and $\lambda$ the mean free path of the diffusing phonons.

Traveling phonons may collide with one another and scatter, which would impede their drift rate. At high temperatures there are many phonons; the mean free phonon path and thus the heat conductivity decrease with increasing temperature. In a perfect lattice without scattering flaws and defects, phonons interact and scatter one another only to the extent that the interatomic potential is nonparabolic. An ionic lattice with a highly nonparabolic potential has more resistance for heat conduction than a covalent, strongly bonded solid that has a potential that is more parabolic. The closer the interatomic potential is to parabolic the fewer phonons scatter and the higher are phonon transport rate and heat conductivity. Strong covalent bonds (as in diamond or silicon carbide) favor heat transport. Ionic lattices have asymmetric bonding potentials, a high scattering of phonons, and low heat conductivity.

Apart from phonon–phonon scattering, which is allowed by the anharmonicity of the interatomic potentials, the *umklapp* or flipover process is another effect that decreases the mean free path of the phonons and the conductivity: two short-wavelength waves may interfere to form one long-wavelength wave going back (Figure 4.38). The lattice periodicity contributes to the umklapp process, a reflection of drifting phonons.

At temperatures much lower than the Debye temperature, where there are only long-wavelength phonons and not many of them, conductivity is relatively high and

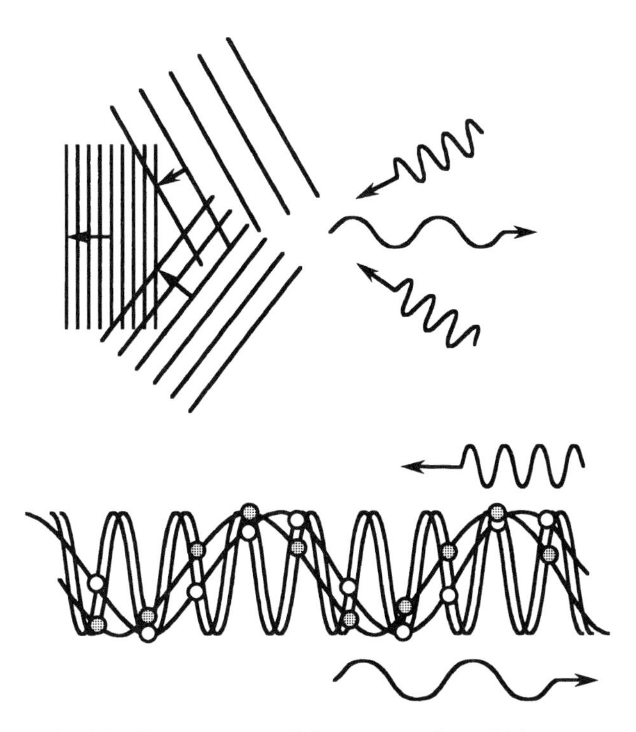

**Figure 4.38.** Schematic of the flipover process of phonon scattering at high temperatures. Two phonons (wave packets) interfere and form a phonon with a shorter wavelength. In a periodic lattice this is the same as a long-wavelength phonon traveling in the opposite direction. The number of short-wavelength phonons in the lattice increases with temperature.

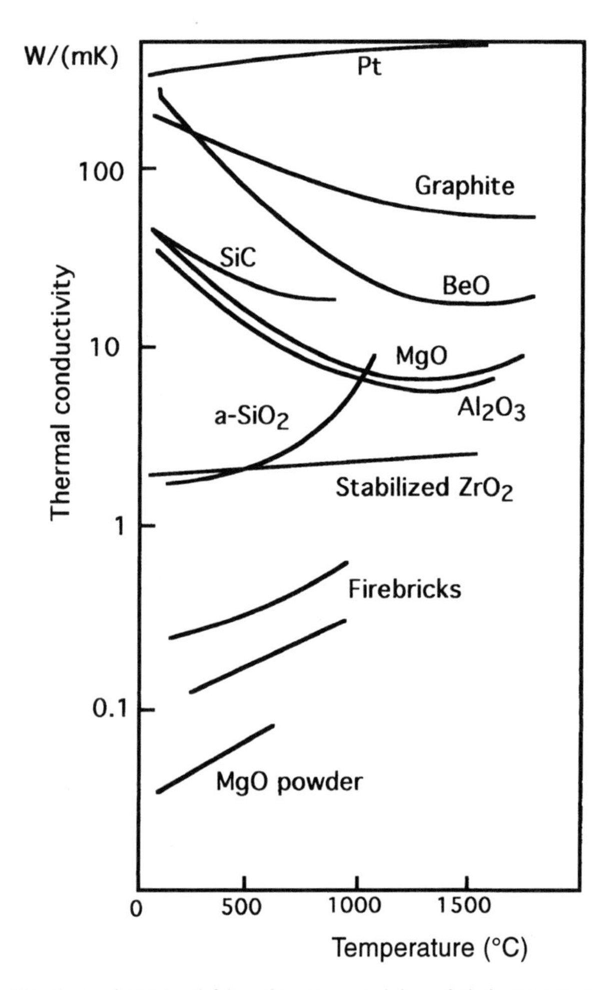


**Figure 4.39.** The thermal conductivities of some materials and their temperature dependence.

is dominated by scattering of the phonons by impurities, point defects, grain boundaries, and interphase boundaries. At low temperatures the mean free path is constant and the temperature dependence of the conductivity is due to changes in the specific heat and the speed of sound. At high temperatures the mean free phonon path decreases because the phonons collide with one another and the thermal conductivity drops. If free conduction electrons are present they may contribute to the increase in thermal conductivity with temperature. Figure 4.39 shows the thermal conductivity of a few refractory ceramics.

### *Thermal Stresses and Thermal Shock*

Temperature differences in solids generally lead to thermal stresses,[20] which may cause cracking and failure if they exceed the critical fracture stress. Thermal shock results from rapid heat transfer at the surface of a solid (quenching or laser beam impact), which may cause failure owing to large local transient thermal stresses.

The factors that affect the ability of a ceramic component to resist degradation of its structural properties by thermal shock are:

1. The geometry of the component, i.e., thickness and radii of curvature, which control the rate of heat transfer to the component.
2. The thermal conductivity of the material, which determines thermal gradients that are set up by the transfer of heat.
3. The thermal expansion characteristics, which determine the level of thermal strain developed under thermal gradients.
4. The elastic properties, which determine the stress developed by the thermal gradients.
5. The fracture strength at the sites where the tensile or shear stresses develop to high levels under thermal gradients.
6. The fracture toughness of the material, which controls the resistance to the propagation of cracks once initiated.
7. The porosity, which controls the resistance to thermal shock damage through reducing the elastic moduli.

The thermal strain $\varepsilon = \alpha \Delta T$ (where $\alpha$ is the coefficient of thermal expansion) leads to a stress $\sigma = E\alpha\Delta T$, where $E$ is Young's modulus. Taking typical values for alumina yields $\Delta T = 55°C$ for thermal stress failure if shrinkage of the alumina bar is blocked.

A hot ceramic component subject to a sudden chilling of its surface by immersion in a cooler medium has its surface layer subjected to a high transient tensile stress. If the stress is sufficiently high, cracks can develop from surface defects. Excessive cooling rates lead to crazing of the surface. If the material is brittle, the cracks may cause complete fragmentation of the component. In a material of moderate toughness, the cracks may be partial and only weaken the material, which remains intact. In a coarse-grained or porous weak material, the cracking may be minor and lead to little degradation. In "inferior" porous materials crack growth may be less catastrophic than in strong dense technical ceramics.

Components tend to be able to survive greater upward thermal shocks than downward ones as cooling subjects the surface to tensile stress and heating to compressive stress. Usually materials, certainly ceramics, can resist much higher compressive stresses than tensile ones. The internal stress is lower than the surface stress. If the thermal expansion coefficient is very anisotropic, as, e.g., in $Al_2TiO_5$, microcracks can form on cooling from synthesis or sintering at high temperatures. The material is not very strong but because of the microcracks is comparatively tough. It is an excellent heat shield.

The severity of a thermal shock depends upon the rate of heating or cooling and is a function of the heat transfer coefficient $h$ between the solid and the surrounding environment, the thermal conductivity $\kappa$, and the characteristic dimension of the material $a$. The Biot modulus $\beta = ah/\kappa$ describes the intensity of the thermal shock. On cooling in air $\beta \approx 1$; in a water quench $\beta \approx 10$. In an infinitely fast quench $\beta = \infty$, which means that the surface is instantaneously brought to the temperature of the environment. In that case the surface tensile stress $\sigma = E\Delta T\alpha/(1 - \nu)$, where $\Delta T$ is the temperature difference between the bulk and the surface and $\nu$ is the Poisson ratio.

**Table 4.11. Thermal Properties and Shock Resistance of Some Simple Ceramics**

| Material | Bend strength (MPa) | Young modulus (GPa) | Expansion coefficient ($10^{-6}$ K) | Thermal conductivity (W/mK) | $R$ ($K$) | $R'$ (kW/m) |
|---|---|---|---|---|---|---|
| Alumina | 500 | 400 | 9 | 8 | 100 | 1 |
| Beryllia | 200 | 400 | 8.5 | 63 | 40 | 2 |
| $Si_3N_4$ | 850 | 310 | 3.2 | 17 | 630 | 11 |
| RBSN | 240 | 220 | 3.2 | 15 | 250 | 4 |
| RBSC | 500 | 410 | 4.3 | 84 | 220 | 20 |
| WC/Co | 1400 | 600 | 4.9 | 86 | 350 | 30 |

Thermal stresses depend on the average temperature because the values of the contributing parameters ($\alpha, E$) change with temperature. Two figures of merit for comparing thermal shock performance are often used: $R = \sigma(1 - \nu)/(E\alpha)$ the failure temperature difference at fast shock; and $R' = \kappa R$, which gives the heat flow limit and is a better criterion at less severe thermal shock. Table 4.11 lists these parameters with the contributing thermal characteristics of several ceramics.

## Exercises

1. Why do materials scientists sharply distinguish intrinsic and extrinsic properties?
2. Give a few characteristic examples of structures with the three types of bonding and discuss the relationship of structure to their intrinsic properties.
3. What are the reasons for the characteristic electron conductivity in different materials?
4. Find some transition metal oxides that have Mott transitions and suggest some applications for those oxides.
5. Describe the thermal behavior of a glass near its glass point.
6. What are the four atomic contributions to the electric polarizability? How can their characteristic frequency be changed in materials?
7. What are the characteristic magnetic parameters of oxidic ceramics and their order of magnitude?
8. Explain from the structure why so many perovskites are ferroelectric. What structural features affect the parameters and how?
9. Describe the thermal and electromechanical behavior of piezoelectric and ferroelectric compounds.
10. Give a few applications of lead zirconate titanate or PZT ($PbZr_xTi_{1-x}O_3$, with $x \simeq 0.5$).
11. Identify the four tetrahedral and the eight octahedral interstitial sites in the fcc lattice. Identify the pseudotetrahedral interstitial sites in the bcc lattice and in the hcp lattice.
12. Give a few characteristic intrinsic properties of $ZrO_2$, $VO_2$, $CrO_2$, and $MoO_2$ that are used in application of these oxides.
13. Compare CdS and SnS. Are they $n$-type or $p$-type semiconductors? Does deviation from their 1:1 stoichiometry affect the type of conductivity?

## References

1. U. Müller. *Anorganische Strukturchemie*. Teubner, Stuttgart (1991).
2. F. S. Galasso. *Structure and Properties of Inorganic Solids*. Pergamon, Oxford (1970).
3. D. M. Adams. *Inorganic Solids*. Wiley, New York (1974).
4. R. E. Newnham. *Structure–Property Relations*. Springer-Verlag, Berlin (1975).
5. G. P. Johari. Introduction to the glassy state in the undergraduate curriculum. *J. Chem. Educ.* **51**, 23 (1974).
6. M. H. Lewis (ed.). *Glasses and Glass Ceramics*. Chapman and Hall, London (1989).
7. D. Shechtman and C. I. Lang. Quasi-periodic materials: Discovery and recent developments. *Materials Research Society Bulletin*, November 1997, p. 46.
8. V. I. Matkovich. *Boron and Refractory Borides*. Springer-Verlag, Berlin (1977).
9. H. Holleck. *Binare und ternare Carbid- und Nitridsysteme der Ubergangsmetalle*. Bornträger, Berlin (1984).
10. J. Robertson. Mechanical properties and structure of diamondlike carbon. *Diamond and Related Materials* **1**, 397 (1992).
11. J. H. Westbrook and R. L. Fleischer (eds.). *Intermetallic Compounds: Principles and Practice*. Wiley, Chichester (1995).
12. N. Braithwaite and G. Weaver (eds.). *Electronic Materials*. Materials in Action Series. The Open University. Butterworths, London (1990).
13. K. Uchino. Applied aspects of piezoelectricity. *Key Engineering Materials* **66/67**, 311 (1992).
14. J. Unsworth. Piezoelectricity and piezoelectric materials. *Key Engineering Materials* **66/67**, 273 (1992).
15. N. Stoloff and V. K. Sikka (eds.). *Physical Metallurgy and Processing of Intermetallic Compounds*. Chapman and Hall, New York (1994).
16. G. Sauthoff. *Intermetallics*. VCH, Weinheim (1995).
17. F. E. Fujita (ed.). *Physics of New Materials*. Springer-Verlag, Berlin (1994).
18. P. W. Atkins. *Physical Chemistry*. Oxford University Press, Oxford (1982).
19. G. Grimvall. *Thermophysical Properties of Materials*. North-Holland, Amsterdam (1986).
20. D. P. H. Hasselman. Thermal stress resistance: Parameters for brittle refactory ceramics: a compendium. *Cer. Bull.* **49**, 1033 (1970).
21. C. N. R. Rao and K. J. Rao. *Phase Transitions in Solids*. McGraw-Hill, New York (1978).
22. R. Roy, D. K. Agrawal, and H. A. McKinstry. Very low thermal expansion coefficient materials. *Ann. Rev. Mater. Sci.* **19**, 59 (1989).

*Chapter 5*

# SOLID STATE REACTIONS

*I thought of another moral, more down to earth and concrete, and I believe that every militant chemist can confirm it: that one must distrust the almost-the-same, . . . the practically identical, the approximate, the or-even, all surrogates, and all patchwork. The differences can be small, but they can lead to radically different consequences, like a railroad's switch points; the chemists' trade consists in good part in being aware of these differences, knowing them close up, and foreseeing their effects. And not only the chemist's trade.*

PRIMO LEVI, *The Periodic Table*

## 5.1. Introduction

The subject of this chapter is the chemistry that can take place in solids, i.e., in a lattice of atoms. In solids, as in all chemistry, a necessary condition for reactions to be possible is sufficient atomic mobility. Atoms in the interior of crystallites or in grain boundaries are fixed. This means that reactions in solids need a comparatively high temperature.[1] Moreover, there is no mobility in lattices without point defects such as vacancies or interstitials. In solids that have perfect lattices the atoms are not mobile and there is no chemistry.[2-5]

An understanding of solid state reactions is necessary for designing microstructures, properties, and processes for making novel solids. Processes that depend on solid state reactions and atomic transport are conversion reactions, ceramics processing, high-temperature corrosion, and ionic device operation.

Most reactions of a solid are heterogeneous, occurring on the interface between the phases where the reaction and the reaction product are located. The microstructures of solids affect reaction rates and conversely, heterogeneous growth reactions can form many kinds of microstructures. Chapter 6 deals with surface chemistry and the chemical consequences of morphology are discussed in Chapter 7. Another typical feature of solid compounds apart from the presence of a surface is that nonstoichiometries often occur in solids and the stoichiometry strongly affects the properties and the reactions.

Diffusion of reactant atoms determines the overall reaction rate and solid state reactions are diffusion-limited and have low rates. What is called the "reaction mechanism" in solid state reactions is not like a reaction mechanism in molecular chemistry. In the latter a mechanism for a reaction describes the path of the atoms in the reactant molecules during the conversion to the product molecules. The mechanisms in solid state chemistry are really diffusion mechanisms combined with atomic balances.

This chapter will first describe the different types of solid state reactions (the sintering reactions in ceramics processing, e.g., are solid state reactions in materials technology[6]) and the typical reaction rates that are observed. Then the mechanisms for these reactions or, more precisely, the diffusion mechanisms and how they are studied will be discussed.

## 5.2. Types of Reactions of Solids

This section gives an overview of the sort of chemistry that is observed in or on solids. Solid state reactions are those chemical reactions in which at least one solid product is formed from at least one solid reactant. This definition excludes processes for synthesis of solids from nonsolid reactants. If solids are made from gases and liquids as in chemical vapor deposition or in sol-gel reactions, it is not strictly speaking solid state chemistry. Defect chemistry is a central subject in solid state chemistry, and the physical chemistry of defects is discussed in Chapter 10.

Phase transitions in solids are also excluded here from the category of solid state reactions if they are not accompanied by chemical reactions or if they do not involve breaking and making chemical bonds. Yet phase transitions can affect the chemistry of solids. The reaction rates in solids strongly increase at temperatures near the temperature of the phase transition because the mobility of the atoms at that point is high. This is called the Hedvall effect and solid state synthesis makes use of it. For example, the measured diffusion rate of lead ions in $PbSiO_3$ is at a maximum at the phase transition at 610°C. The reaction between solid $Fe_2O_3$ and $SiO_2$ has a much higher rate near the phase transition between the $\alpha$- and $\beta$-tridymite phases of $SiO_2$ at 570°C than at more elevated temperatures. Similarly, the reaction between $Fe_2O_3$ and ZnO has a maximum rate just above the Néel temperature of $Fe_2O_3$ at 680°C and zircon ($ZrSiO_4$) is preferably formed at the temperature of one of the phase transitions of $ZrO_2$.

An example of a suppression of a phase transition by doping is nasicon, a sodium ion conductor having the chemical formula $Na_3Zr_2Si_2PO_{12}$. The monoclinic form transforms reversibly to a trigonal structure at 150°C. This transition can be suppressed by partial substitution of $Zr^{4+}$ by $Yb^{3+}$ and simultaneously decreasing the silicate/phosphate ratio to maintain the charge balance. The structure of the alloy $Na_3Zr_{2-x}Yb_xSi_{2-x}P_{1+x}O_{12}$ gradually changes from monoclinic to trigonal as $x$ increases from 0 to 0.6.

Reactions of solids can be grouped into four categories[7]

1. Decomposition reactions of solids in which a gas is produced.
2. Conversion of solids that react with gases.
3. Reactions of solids with a liquid.
4. Reactions between solid reactants to form solid products only.

1. Decomposition of a solid under gas development: Such decomposition reactions are, e.g.,

$$Al_4Si_4O_{10}(OH)_8 \rightarrow Al_4(Si_4O_{10})O_4 + 4H_2O(g) \text{ (kaolinite} \rightarrow \text{metakaolinite)}$$

$$CaCO_3 \rightarrow CaO + CO_2(g)$$

$$BaCO_3 + TiO_2 \rightarrow BaTiO_3 + CO_2(g)$$

$$Cu_2S + 2Cu_2O \rightarrow 6Cu + SO_2(g)$$

2. Reactions of a solid with a gas: Examples are:

$$2Cu + \tfrac{1}{2}O_2(g) \rightarrow Cu_2O$$

$$MoO_3 \rightarrow MoO_2 \rightarrow Mo \text{ (in hydrogen at 750 and 1000°C, respectively)}$$

$$CaCO_3 + SO_2(g) + 2\,H_2O(g) + \tfrac{1}{2}O_2(g) \rightarrow CaSO_4 \cdot 2H_2O + CO_2(g)$$

$$2SiCl_4(g) + 4H_2(g) + Mo \rightarrow MoSi_2 + 8HCl(g)$$

$$ZrMn_2 + 1.8\,H_2(g) \rightarrow ZrMn_2H_{3.6}$$

$$3Si + 2N_2(g) \rightarrow Si_3N_4$$

$$Ti + \tfrac{1}{2}N_2(g) \rightarrow Ti(N) \rightarrow TiN$$

The last two examples are the synthesis of reaction-bonded silicon nitride (RBSN) and reaction-bonded titanium nitride (RBTN). Ti(N) is titanium with dissolved nitrogen.

High-temperature corrosion (of metals) in air also belongs to this group of reactions. The oxidation rate depends on the nature of the product that forms as a more or less protective coating on the surface of the metallic reactant of the corrosion reaction. If the molar volume of the solid oxide formed by the reaction of diffusing oxygen with metal is much lower or higher than that of an equivalent amount of the metallic reactant, the reaction rate is constant in time and does not depend on the thickness of the solid. In these cases the product is porous because the oxide cracks or spalls off the substrate and cannot prevent the oxygen from getting to the underlying metal and reacting there. The Pilling–Bedworth ratio is the ratio of the molar volumes of the oxide and the metal per unit of metal. If this ratio is near 1 growth stresses in the layer are low, a closed oxide cover can grow and further oxidation of the metal can occur only by diffusion of one reagent, metal or oxygen, through the dense solid product layer, which means parabolic oxidation kinetics according to the parabolic scale growth model[8]:

$$\frac{dL}{dt} \propto \frac{1}{L} \quad \text{or integrated } L(t) = \sqrt{kt + C}$$

where $L$ is the thickness of the layer and $k$ the parabolic reaction constant. This model assumes that a simple diffusion of cations, anions, vacancies, electrons, or electron holes determines the reaction rate of the conversion. When growing $Cu_2O$ on copper, the diffusion rate of $Cu^+$ in $Cu_2O$ determines the corrosion rate. In the case of the formation of ZnO on Zn, the rate of growth depends on the concentration and mobility of the zinc interstitials in the oxide.

As any other model, this one on scale formation is an approximation. The growth of thin layers can deviate from the predicted rate. The rate has been observed to be nonlinear, logarithmic, exponential, cubic, or to have a power behavior with an exponent higher than 3. This topic is discussed in the next section. Sometimes the space charges in the layer that is grown are invoked to explain the rates.

3. REACTIONS OF A SOLID WITH A LIQUID: Wet corrosion and galvanic deposition are in this group. The temperature dependence is similar to that in gas-phase reactions. Examples of synthetic reactions are:

(a) $CaCO_3$ (argonite in coral) $\rightarrow Ca_5(PO_4)_3$ (apatite for bone implants), a hydrothermal reaction that converts the solid but keeps the microstructure of the original material intact.
(b) Anodization or electrochemically covering a metal such as aluminum or tantalum with a protecting aluminum oxide scale, which forms by reacting the metal. The thickness of the oxide layer at which the oxidation stops is proportional to the anodizing voltage used.

4. REACTION BETWEEN TWO SOLIDS: There are two types, addition and displacement reactions:

(a) Examples of addition reactions are:

$ZnO + Fe_2O_3 \rightarrow ZnFe_2O_4$ (zinc ferrite, a spinel)

$MgO + SiO_2 \rightarrow MgSiO_3$ (enstatite) or $Mg_2SiO_4$ (forsterite)

Reaction sintering: $Ti + 2B \rightarrow TiB_2$ (for other SHS-reactions see Chapter 8)

(b) Displacement reactions mean the exchange of atoms between solid phases. Examples are:

$ZnO + CuSO_4 \rightarrow ZnSO_4 + CuO$ (doping ZnO with $Li_2O$ increases the rate)

$Ge + 2MoO_3 \rightarrow GeO_2 + 2MoO_2$ (here *n*-type Ge reacts faster than *p*-type)

Reactions between solids require high temperatures for better atomic mobility, but even at high temperatures reactions are sluggish, diffusion-limited, and often incomplete. The solid product of a reaction between solids may still contain unreacted precursors. To continue the reaction, the product has to be milled, pressed, and reacted again at high temperatures. Besides high temperatures for high atomic mobility, reactants with small powder particles that are in good contact with each other favor high rates in solid state reactions. In that case the atoms only have to travel over small distances before they react. The reactivity can therefore be increased by bringing the reactants close together on an atomic scale in a suitable solid precursor by, e.g., making a mixed salt by coprecipitation. Although solid state reactions are sluggish compared to reactions in fluids, they are often used in industry because of their simplicity: they involve nothing more complex than pressing, heating, milling, and repeating these steps a number of times.

If the atomic mobility is good, phase diagrams can be consulted to see whether reactions are in principle feasible and what reaction products can form in combinations of reactants.[9,10] If, according to the phase diagram, more compounds can form, e.g., in a coexistence region, the question arises as to which product will be formed first in the reaction. There is an empirical rule that predicts that in a demixing reaction the compound that lies closest to the lowest eutectic in the phase diagram is the first to form. This rule has been explained by a model based on equilibrium assumptions[11] although the system is far from equilibrium. The rule is supposed to reflect an effect of the driving forces but what it describes is actually a kinetic effect. The atomic mobility in the phase nearest to an eutectic is higher than in the other phases because the composition of the reacting solid in the $T$–$x$ diagram is closer to the liquid state. It has nothing to do with some activation parameter that is used in the chemistry of concentrated liquids to match observations to thermodynamic arguments.

## 5.3. Kinetics of Solid State Reactions

In solution chemistry it is customary to express the reaction rates in terms of concentrations if the reactants and products are dissolved and well mixed and if the reactions are not diffusion-limited. This cannot be done in solids because solids are never "perfectly stirred" reactors. Reaction rates in solids do not have the same or similar relationships to the concentrations as reactions in liquids but they are diffusion-limited. Solid state reactions have other types of reaction "order" than reactions in liquid solution[3,8] and the kinetic rules in solid state chemistry are different from those of reactions between molecules. The reaction rates depend on morphologies, the geometry of the reaction front, the diffusivities of the reacting species, the possibilities of nucleation, and the anisotropy of crystallites. In summary:

1. Reactions occur mainly on surfaces or grain boundaries, the state and form of the surfaces affect the rates, and the kinetics are heterogeneous.
2. Diffusion in solids is less easy than in liquids and the reaction rates in solids are usually limited by the diffusion, which dominates not only the overall formation rates but also the morphology of the solid product of the reaction.

Many solid state reactions have a time-dependent rate as is indicated in Figure 5.1. Three stages can be distinguished in the conversion curve: (AB) formation of nuclei, no growth; (BC) growth of the nuclei, the reaction surface of the growing grains increases in time; and (CD) slowing down of the reaction by a decrease in the available reactant and the smaller area of reaction interface.

The time-dependent relative degree of conversion $\alpha(t)$, the fraction of the initially present reactant that is converted, is used to describe the reaction kinetics. The steepest part of the $\alpha(t)$-curve, BC, which represents the growth of the nuclei, can be fitted to a power law $\alpha(t) = kt^n$. The exponent $n$ suggests a mechanism. If $n > 1$, both ongoing nucleation and growth determine the conversion. If $n = \frac{1}{2}$ there is parabolic growth at the surface of a semi-infinite planar-like reaction interface. The oxidation of metals when oxygen has to diffuse through a growing oxide cover is an example of $\sqrt{t}$-behavior. A value of $n = \frac{3}{4}$ occurs in those cases where the diffusion through grain boundaries is rate-limiting and $0.125 < n < 0.3$ indicates limitation by

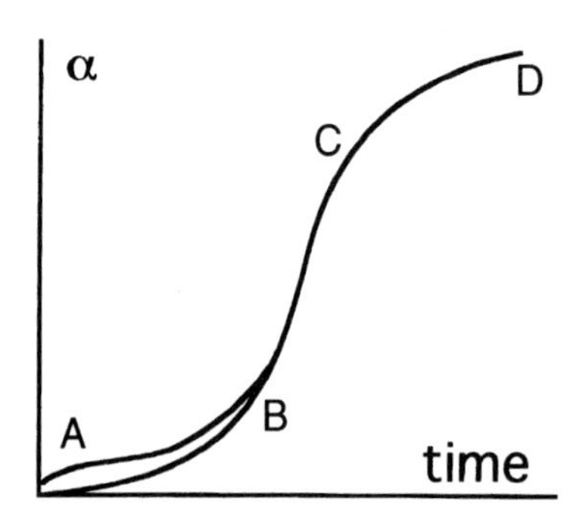


**Figure 5.1.** Extent of two typical solid state reactions as they develop in time: a one-stage reaction and a reaction with an intermediate product.

diffusing electrons. Very high values of $n$ (up to $n = 23$) have been observed, and if $n$ is larger than 6 an exponential dependence such as $\alpha = C\exp(kt)$ or $\alpha = C\exp(kt)^n$ is a more usual choice. $C$ and $k$ are constants.

The conversion can also be described by the Johnson–Mehl–Avrami equation:

$$\alpha(t) = 1 - e^{-(kt)^n}$$

where $k$ is the reaction rate coefficient and $t$ is the time. The exponent $n$ can have different values as can be seen in Table 5.1. This equation does not only describe part BC in the $\alpha$-curve but also part CD. Again, nucleation dominates the reaction in part AB, the induction period. This period can be shortened by increasing the defect concentration by, e.g., letting the solid absorb ionizing radiation, which sometimes increases the rate in this stage. Dislocations also enhance nucleation and a large concentration of them shortens the induction period, domain AB. BC again means grain growth, and a decreasing rate in CD is the result of a diminishing reaction front area (Figure 5.2). Typical solid state reactions are the conversion of gray to metallic tin (Figure 5.3) and the conversion of anatase to rutile. These phase

**Table 5.1. Values of the Avrami Exponent to Be Expected for Cases with a Simple Mechanism**

| Avrami exponent $n$ | Mechanism |
|---|---|
| Reaction-limited | |
| 1 | Homogeneous nucleation at the beginning and continuous heteronucleation on the grain surfaces |
| 2 | Homogeneous nucleation at the beginning and continuous heteronucleation on the grain edges |
| 3 | Homogeneous nucleation initially followed by crystallite growth |
| 4 | Continuous homogeneous nucleation |
| Diffusion-limited | |
| $\frac{1}{2}$ | Scale growth in thickening layers |
| 1 | Porous scale growth |
| 1.5 | Initial homogeneous nucleation followed by crystallite growth |
| 2.5 | Constant homogeneous nucleation with parallel crystallite growth |

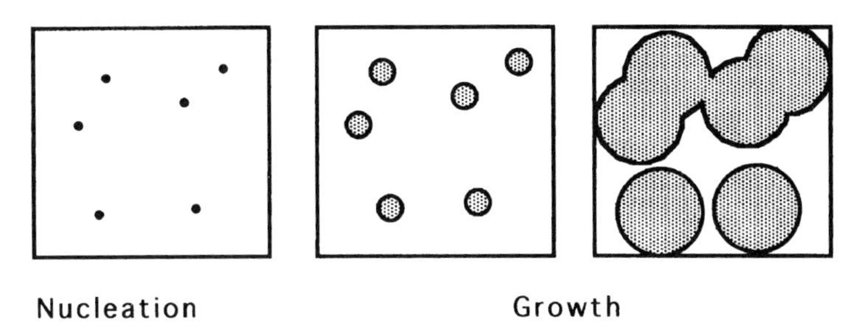


**Figure 5.2.** Nucleation and growth in a solid state reaction. The measure (area) of the interface between the reactant and the product phase increases at first when the interface grows and the reaction rate increases. Near the end of the reaction when the interface diminishes the reaction rate decreases. The reaction occurs at the surface and is proportional to its measure.

transitions may be considered to be chemical reactions because chemical bonds are broken and remade.

The mechanism of a solid state reaction or the transport mechanism of the reacting atoms is measured in a diffusion-couple experiment. An example of the growth of a ternary spinel from separate binary oxides will be sketched below. The discussion of point defects and their behavior and concentration in Section 10.3 provides the physical chemistry background for this discussion of reacting atoms.

Two spatially separated solid particles in contact with one another react at high temperatures to form one or more of the reaction products that are possible according to the phase diagram. The atoms of the two reactants have in general different diffusion rates in the lattice of the solid product. The particular reactant that has the lowest mobility determines the overall formation rate for that product. The spatial arrangement of several possible compounds that are produced as well as the concentrations in them are also determined by the $T-x$ diagram as is shown in Figure 5.4.

Diffusion couples are experimental setups for determining the mechanism for the formation of the solid products. In a diffusion couple two solid reactants are pressed together with an inert net of marker wires made, e.g., of platinum between them. The diffusion couple is kept at the reaction temperature for a while, after which the spatial distribution of the reaction products with respect to the marker wires is

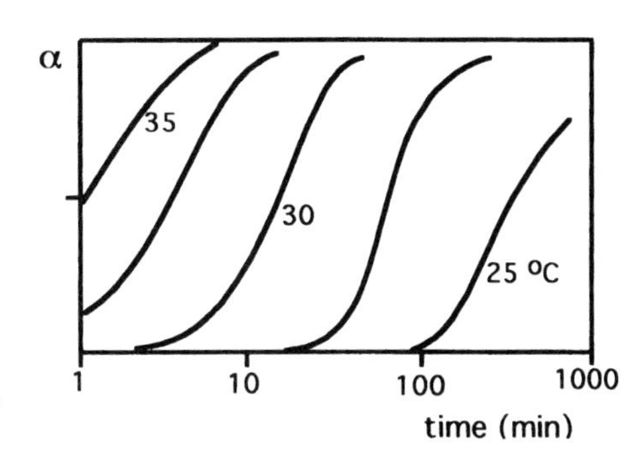


**Figure 5.3.** Degree of conversion of gray tin to white tin at different temperatures (°C).

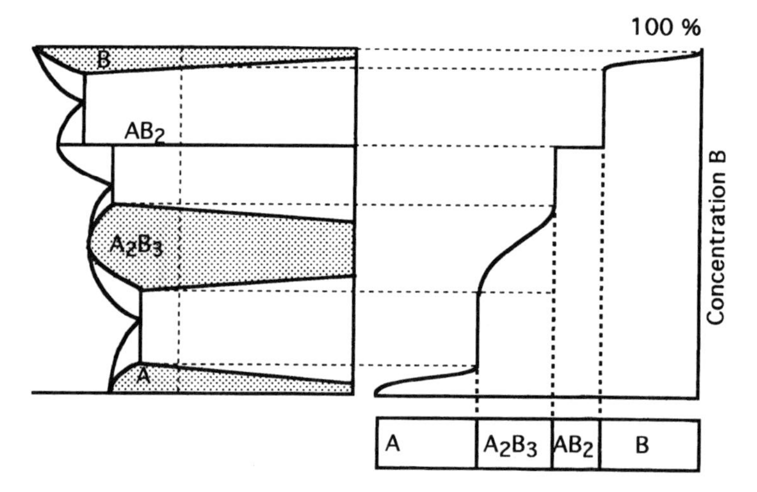


**Figure 5.4.** The concentration distribution in a diffusion couple of a fictitious system A/B that has two compounds, a line compound $AB_2$ and one with considerable nonstoichiometry $A_2B_3$. The reactants have some solubility in each other. If both compounds form during annealing at $T_1$ (dotted line in the phase diagram) a line profile over the diffusion couple shows a distribution of [B] as indicated. In the vertical parts of this graph (in the coexistence regions) the concentration signal oscillates between two values, depending on which of the two crystallites happens to be measured (by SEM).

observed. The place of the marker wires between the products after the reaction depends on the prevailing mechanism.

When two generalized solid oxides, AO and $B_2O_3$, react to the spinel $AB_2O_4$, the atoms move from the solid reactant lattices to the sites of the solid product. The ions A and B can be any main group or transition metal ions. As a specific example these are here taken to be CoO and $Fe_2O_3$, which form only one compound, cobalt ferrite ($CoFe_2O_4$). The marker wires end up somewhere in the spinel area or on the boundary of this phase, depending on how fast the ions of the two reactants are able to diffuse. Some conceivable diffusion mechanisms are in their pure form:

1. BOTH CATIONS, $Co^{2+}$ AND $Fe^{3+}$, ARE ABLE TO DIFFUSE AND THE OXYGEN IONS ARE NOT DISPLACED MUCH IN THE REACTION BUT REMAIN NEAR THEIR INITIAL POSITION:

The trivalent ion arrives through the left interface between the divalent precursor and the reaction product and reacts as follows:

$$4CoO + 2Fe^{3+} \rightarrow \mathbf{1}CoFe_2O_4 + 3Co^{2+}$$

CoO is the left reactant in the diffusion couple and $Fe_2O_3$ is at the right. At the right-hand side of the interface between the trivalent oxide precursor and the solid product, the diffusing divalent ion reacts with the other oxide:

$$4Fe_2O_3 + 3Co^{2+} \rightarrow \mathbf{3}CoFe_2O_4 + 2Fe^{3+}$$

Charge compensation requires that for every two iron ions that diffuse through the spinel from the iron oxide to the cobalt oxide, three cobalt ions leave the cobalt oxide phase and diffuse to the iron oxide side. The bold numbers in these equations indicate how much spinel forms for a certain amount of ion diffusion (six charges) at either side of the markers. According to the equations for this mechanism (both cations can diffuse) the marker wires are expected to end up in the interior spinel layer at a distance that is one-quarter of the total depth of the product layer from the interface with the divalent oxide. This mechanism needs a good contact between the particles.

2. ONLY THE DIVALENT COBALT CATIONS ARE ABLE TO DIFFUSE, THE TRIVALENT IRON CATIONS DO NOT, AND OXYGEN DOES NOT DIFFUSE THROUGH THE LATTICE BUT IS TRANSFERRED AS A GAS:

At the left-hand interface the reaction is:

$$CoO \rightarrow Co^{2+} + 2e' + \tfrac{1}{2}O_2(g)$$

The cobalt cations and the electrons diffuse through the product spinel to the other side. So does oxygen through the gas phase.

At the right-hand interface:

$$Fe_2O_3 + Co^{2+} + 2e' + \tfrac{1}{2}O_2(g) \rightarrow \mathbf{1}CoFe_2O_4$$

The spinel product can only form at the iron oxide side, where it reacts with the entering cobalt ions, the electrons, and gaseous oxygen. These equations show that the marker threads end up on the left-hand boundary between the spinel and the divalent oxide (CoO) after reaction because there is no spinel formation at this interface.

3. THE TRIVALENT CATIONS ARE FIRST REDUCED AND THEN MOVE, THE COBALT CATIONS DO NOT, AND OXYGEN IS AGAIN TRANSPORTED AS A GAS:

Left-hand interface:

$$CoO + 2Fe^{2+} + 4e' + 1\tfrac{1}{2}O_2(g) \rightarrow \mathbf{1}CoFe_2O_4$$

Right-hand interface:

$$Fe_2O_3 \rightarrow 2Fe^{2+} + 1\tfrac{1}{2}O_2(g) + 4e'$$

Generally, highly charged cations do not diffuse as well as cations with a smaller charge. The mechanism proposed here supposes that the trivalent iron ions are temporarily reduced to divalent iron, which is then assumed to be able to diffuse. The cobalt ions in this model are expected to remain stationary. The spinel is now formed only at the left-hand boundary between the spinel and the divalent oxide (CoO); hence the marker wires are found at the right-hand interface between the spinel and the trivalent oxide ($Fe_2O_3$) after the reaction.

4. BOTH IONS ARE ASSUMED TO BE ABLE TO DIFFUSE BUT AS IN THE PREVIOUS CASE, THE TRIVALENT ION IS REDUCED BEFORE TRANSPORT AND OXIDIZED AT THE END WHEN THE SPINEL FORMS. ELECTRONS DO NOT PARTICIPATE IN THE DIFFUSION NOW AND OXYGEN IS TRANSPORTED AGAIN THROUGH THE GAS PHASE:

Left-hand interface:

$$3CoO + 2Fe^{2+} + \tfrac{1}{2}O_2(g) \rightarrow \mathbf{1}CoFe_2O_4 + 2Co^{2+}$$

Right-hand interface:

$$3Fe_2O_3 + 2Co^{2+} \rightarrow \mathbf{2}CoFe_2O_4 + 2Fe^{2+} + \tfrac{1}{2}O_2(g)$$

The bold-faced coefficients in these reactions show that after the reaction, the marker wires are in the interior of the spinel at a point that is one-third of the distance from the CoO interface.

In actual practice the markers in the spinel phase are found somewhere between the extreme values that correspond to these idealized cases, and several mechanisms can be operative simultaneously. Experiments with diffusion couples of ferrite systems indicated that mechanisms 1 and 4 both actually occur to some degree. In a vacuum, there cannot be oxygen transport through the gas phase, and any oxygen that does form is removed.

For solid state diffusion of ions there must be good contact among the grains of the solid. If there is reaction with volatile intermediates such contact among the particles of the reactant powders is not necessary. Thus zinc ferrite is formed from zinc oxide and ferric oxide:

$$ZnO + Fe_2O_3 \rightarrow ZnFe_2O_4$$

This reaction occurs in stages. Zinc oxide evaporates at the surface of the solid and is transported as zinc vapor and oxygen to the $Fe_2O_3$ crystallites. There the zinc condenses on the ferrite layer, diffuses through it, and reacts with oxygen and $Fe_2O_3$ to the ferrite spinel. The rate-determining step is the diffusion of zinc and oxygen through the ferrite layer. As the zinc ferrite layer becomes thicker the initially linear growth eventually becomes parabolic.

## 5.4. Measuring Solid State Reaction Kinetics

The diffusion-couple experiment is one way to measure the kinetics of solid state reactions. The rate of a reaction in the solid state or the rate of a phase transition that involves bond rearrangement can also be measured in another way: by measuring a property change in a homogeneous solid isothermally or at a gradually increasing temperature. The increase in the degree of conversion of the reaction $\alpha(t)$ is measured in time and this gives information on the reaction kinetics and ultimately on the mechanism. The observed curve of $\alpha(t)$ vs. $t$ could in principle be simulated with a function that is calculated from a reaction model and the parameters in this model can be determined by fitting the function to the measured data. Even if the shape of this curve is not known in detail, the activation energy can still be determined from a few measurements at a constant degree of conversion $\alpha$ at different temperatures.[12]

The degree of conversion of a reaction or phase transformation can in principle be followed by measuring the change of a characteristic property such as hardness, length, magnetism, or the electric resistance with time at the reaction temperature. The degree of conversion $\alpha$ depends on the measured property values as follows:

$$\alpha = \frac{p_t - p_0}{p_e - p_0}$$

Here $p_t$ is the property at time $t$, $p_0$ is the initial value of the property at $t = 0$ (when only reactant is present), and $p_e$ is the value of the property of the end product, i.e., when all the reactants have been converted.

The degree of conversion $\alpha$ is a function $\alpha(\beta)$ of a state variable $\beta$, which depends on time. For an isothermal reaction, $\beta$ is taken to be proportional to the time: $\beta = kt$, with $k$ the reaction rate constant, which is assumed to be time-independent because the temperature is. For a nonisothermal conversion: $\beta = \int k d\tau$ (the integral between $\tau = 0 \rightarrow \tau = t$), as $k$ is exponentially dependent on the temperature $T$, which is time-dependent.

In the case of isothermal transformations $k$ can be determined as follows: The solid is kept at a constant elevated temperature and the reaction is monitored by measuring the property $p$ and its change over time while the reaction proceeds. For each of two temperatures, $T_1$ and $T_2$, the two times, $t_1$ and $t_2$ are established at which $\alpha$ has attained two chosen values, $\alpha_1$ and $\alpha_2$. Then,

$$k(t_2 - t_1) = \beta_2 - \beta_1$$

and

$$k = k_0 e^{-E/RT}$$

Taking logarithms of the first equation and substituting the second, we find that

$$\ln(t_2 - t_1) = E/RT - \ln k_0 + \ln(\beta_2 - \beta_1)$$

Now, $\alpha = \alpha(\beta)$ and because $\alpha_2 - \alpha_1$ is chosen constant, $\beta_2 - \beta_1$ does not depend on $T$ either. If the value of $\ln(t_2 - t_1)$ is plotted against $T^{-1}$ for two or more different temperatures, the slope gives the activation energy $E$ even if the functional dependence $\alpha(\beta)$ is not known in detail.

It is not necessary to measure the rates of the solid state conversion reaction at a constant temperature as described above in order to determine $k$. A solid conversion can be studied by observing the reaction rate in a gradually rising temperature. The reaction rate coefficient can be derived from the conversion vs. temperature curve. Experimentally this is easier than measuring it isothermally. The method is similar to the one described for the isothermal case but the equation for $\ln(t_2 - t_1)$ is slightly more involved and contains several heating rates.

In summary:

1. The activation energy can be derived from a few measurements at isothermal or nonisothermal transformation processes without the need to establish a detailed kinetic model.
2. From the axis cutoffs $k_0$ can be determined if a kinetic model is assumed, in which case $\beta_2 - \beta_1$ or $\beta(\phi)$ has to be known ($\phi$ is the rate of heat flow).

## 5.5. The Chemistry of Ceramics and Sintering

Ceramics are composites that are for the most part solid compounds with at least one nonmetallic element, i.e., an element from the part of the periodic table on the upper-right side of the diagonal Be–Po. Although there are some powderless processes for ceramics, these materials are usually made by sintering particles together at high temperatures.[13–16] If the particles are sintered at temperatures below the melting points of all the phases present, the process is called dry sintering; if there is a liquid phase present it is called liquid-phase sintering. During sintering the particles react and are linked together, so the process can be considered a form of chemistry between particles,[17,18] but it also involves a considerable amount of solid state chemistry. Ceramics are materials that are made *after* the product has been shaped. In this aspect, processing ceramics is different than processing steel or plastics, which materials are made before they are shaped.

Sintering of ceramic powder particles during firing is the single most important step during powder processing. It is a complex process in which several subprocesses occur simultaneously or sequentially. Which particular one of these subprocesses affects the sintering most strongly depends upon the material and the process conditions. Each material has its own specific sintering behavior.

Sintering is the result of mass transport by diffusion owing to a driving force, which can be a concentration gradient. In addition to atomic mass transport viscous flow may contribute to densification, which is an important objective of sintering. Many materials properties are determined by the physical and chemical changes during sintering at high temperatures. Powder preparation and compaction processes also have a significant effect on the microstructure and the properties of the sintered ceramic product.[19–23]

A ceramic object is first shaped from powder particles loosely stuck together, the so-called green form. The porous green form is subsequently baked to sinter the particles into a dense object. During sintering of the green form the free surface area of the powder particles decreases substantially. Moreover free surface area is replaced by grain boundaries. In physical sintering (without chemical reactions), the driving force for mass transport during pore-filling is the decrease in free surface area.

There are three different types of sintering processes (Table 5.2):

1. Dry sintering: only solid phases are present at the sinter temperature.
2. Liquid-phase sintering: small amounts of a liquid phase are present during sintering.
3. Reactive sintering: particles react with each other or with the ambient to new product phases. The green form is composed of particles of one or more reactants. Thus synthesis is combined with sintering in a single process step.

### *5.5.1. Physical Solid State Sintering*

At a temperature that is roughly one-half to two-thirds of the melting temperature (in K) the crystallite faces show a disorder, which may be called "two-dimensional melting." This disorder enables rapid surface diffusion. The powder remains dry but the surface becomes slippery. Conventionally, three stages are distinguished in the sintering process (Figure 5.5) in each of which a different mechanism dominates: neck formation, pore shrinkage, and grain growth. Similar stages can be seen in liquid-phase sintering.

1. First stage: On further heating of the green form after burnout of any organic additives, two things happen to the powder particles when the mobility of the surface atoms has become high enough: the initially rough surface of the particles is smoothed and neck formation occurs. Powder particles that touch each other in the green form are strongly attached to each other and "necks" form between them.

**Table 5.2. Sintering Mechanisms**

| Process phase | Material transport mechanism | Driving energy |
|---|---|---|
| Vapor phase | Evaporation–condensation | Differences in vapor pressure |
| Solid state | Diffusion | Differences in free energy or chemical potential |
| Reactive liquid | Viscous flow, solution-precipitation | Capillary pressure, surface tension |

Mass transport is by surface diffusion and possibly through the gas phase by evaporation and condensation. The mobile surface atoms tend to be trapped at the necks between the particles because the average atomic coordination number is slightly higher for an atom on a concave surface than on a convex surface and atoms will be slightly better bound on the former. Driving forces like these are most important in the first stages of the sintering process and gradually decrease when the concave areas at the necks become flatter.

However, the average particle radius is not the only or even the most important driving force for mass transport. The nature of the atoms, the particle roughness, face indexes, and defect properties of the surfaces and grain boundaries contribute to the sinter rates to an extent that depends on the particular compound. These factors are less easily quantified than particle radii and are therefore often ignored in the literature on sintering.

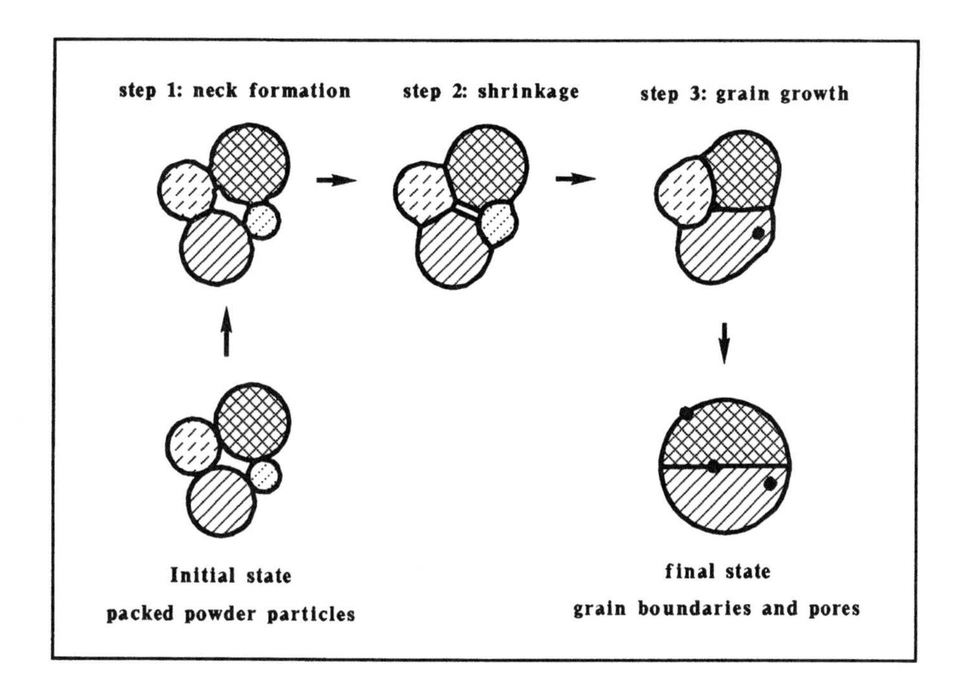


**Figure 5.5.** The three stages in dry sintering.

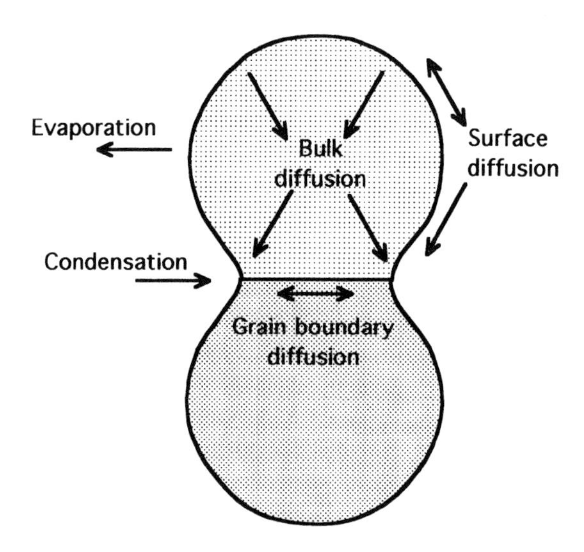


**Figure 5.6.** Transport paths of matter during sintering of two grains.

Mass transport (Figure 5.6) usually takes place over the surface of the grain, and occasionally through the gas phase by evaporation and condensation. There is little shrinkage and distances between the grain centers remain the same. There may also be some bulk diffusion. If a porous product for ceramic filters or membranes is wanted the sintering process is stopped after this stage.

The relative neck size is defined by $x/r$, where $x$ is the diameter of the neck and $r$ the diameter of the particles. The kinetics of neck growth are modeled by the expression

$$\left(\frac{x}{r}\right)^n = \frac{B}{r^m}\, t$$

Expressions for $n$, $m$, and $B$ have been set up for different sinter mechanisms.

The linear shrinkage $\Delta L/L$ at this stage, which covers the first 3% of the total densification, follows the kinetic law

$$\left(\frac{\Delta L}{L_0}\right)^{n/2} = \frac{B}{2^n r^m}\, t$$

in which $n$ has a value between 2.5 and 3. In this equation $B$ depends exponentially on the temperature: $B = B_0 \exp[-Q/kT]$.

2. Second stage: Densification and pore shrinkage. If grain boundaries are formed after the first stage, these are a new source of atoms for filling up the concave areas, which diminishes the outer surface area of the powder compact. Now there is a porous agglomerate consisting of particles sintered together with grain boundaries between them and a surface that is the boundary with the gas phase in the pores. This surface area can decrease by the diffusion of atoms from the grain boundaries to the pore surface. The driving force for this process is less than in the case of surface

diffusion because the atoms in the grain boundary are more firmly bound than on a flat or convex surface. In this stage the total surface area (and concomitantly the porosity) decreases; the distances between the grain centers decrease and shrinkage occurs because of bulk diffusion. Atoms now diffuse from the grain boundaries to markedly concave sites on the surface as long as such sites are still present. The driving force is still the difference in atomic bonding energy between the grain boundary and the pore surface, which is the only mechanism that contributes to densification as soon as the pores are closed.

The densification at this stage follows a logarithmic time dependence. Let $\rho(t)$ be the sintered fractional density at time $t > t_i$, where $t_i$ is the time of onset of the second stage, and $\rho_i$ is the green density at this threshold. Then,

$$\rho(t) = \rho_i + B \ln(t/t_i)$$

where $B$ is a temperature-dependent parameter as in the first sinter stage. The mean grain size $r$ increases with time according to

$$r^3(t) = r_0^3 + Kt$$

where $K$ is again a thermally activated parameter. In this stage the pore structure remains interconnected. The exposed surface area is now less than 50% of the initial surface area of the ceramic powder.

3. THIRD AND FINAL STAGE: Grain growth. At 8% porosity the pores break up and become closed spherical bubbles. Grain surfaces have a specific surface energy, which varies with the face orientation. The crystal surface comprising the boundary that is the most stable will accept atoms from its less-stable opposite partner. The orientation of the crystal planes in the grain boundary may (depending on the temperature) determine the direction of its movement during sintering. Another driving force for grain growth is the size of the grains. The large ones tend to grow, while the smaller grains shrink and disappear. Thus grain boundary curvature is important, although the popular explanation (Figure 5.7) based on equilibrium thermodynamics (curvature and surface-tension-dependent chemical potential) is incomplete. Kinetic effects based on surface roughness are more consistent with

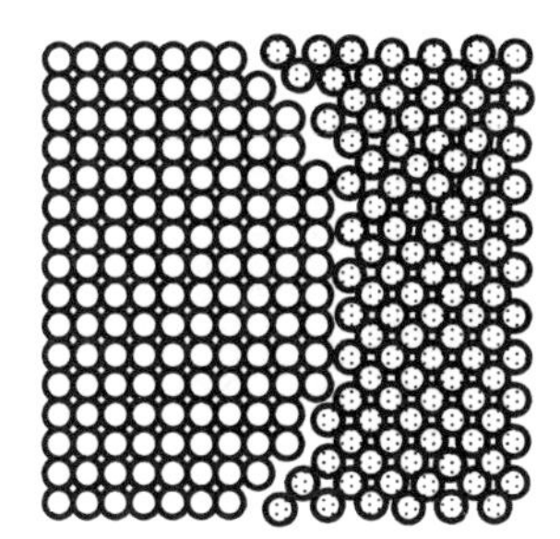

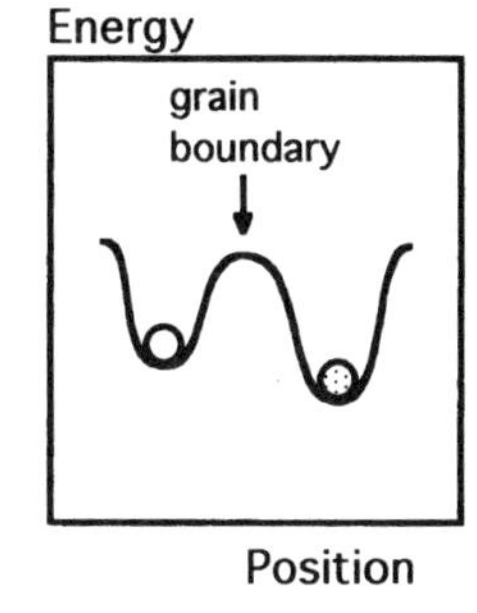


**Figure 5.7.** Convex grain boundary and energy of the atoms on either side. The slight difference in average coordination number means a different energy, which drives the boundary during grain growth.

sintering data and with recently established crystal growth kinetics. If the grain growth rate is determined exclusively from the grain size $r$ as a driving force, the size increases in this third stage as

$$r(t) - r_0 = (2k)^{1/2} t^m$$

in which $m = \frac{1}{2}$. Usually $0.1 < m < 0.5$ is observed.

Pore bubbles appear to behave like knots tying grain boundaries together; they can be swept along with the moving boundaries during grain growth. As long as the pores are on the grain boundaries, they can shrink by diffusion of atoms or vacancies through the attached grain boundaries. This diffusion (by a vacancy mechanism) would be considerably more difficult through the bulk of the grains. The grain boundaries also serve as vacancy sinks. Curved grain boundaries move through the material during this third phase in the direction of the center of curvature: small grains disappear and neighbouring large grains become larger. Fast grain growth is deleterious for densification since fast-moving grain boundaries leave the pores behind in the interior of the grains. In this position they cannot fill up by emitting vacancies and the resulting rest porosity remains too high (Figure 5.8). As long as the pores are able to move with the grain boundaries densification can continue (Figure 5.9). Sintering conditions are adapted to the grain size of the powder in order to enhance densification (Figure 5.10). During sintering grain growth should be prevented as long as there are large pores. When these have become small enough some grain growth can be tolerated as long as the small pores remain attached to the moving grain boundaries.

Extreme grain growth from fine powders is a phenomenon that, although carefully avoided in sintering, is exploited in growing single crystals by solid-phase epitaxy. Large and small crystals have different rates of growth. Within a limited temperature window large crystals are able to grow while Ostwald ripening (growth of larger particles at the expense of small ones) is still too slow to compete with the growth. The size of the powder particles strongly affects sinter rates, as was shown in Figure 5.10. The rest porosity may affect the properties of the sintered product according to the functions listed in Table 5.3. A percolating open porosity may remain, with a permeability for gases that may be intentional or unacceptable. The ratio of open to closed porosity is given in Figure 5.11 for a representative case.

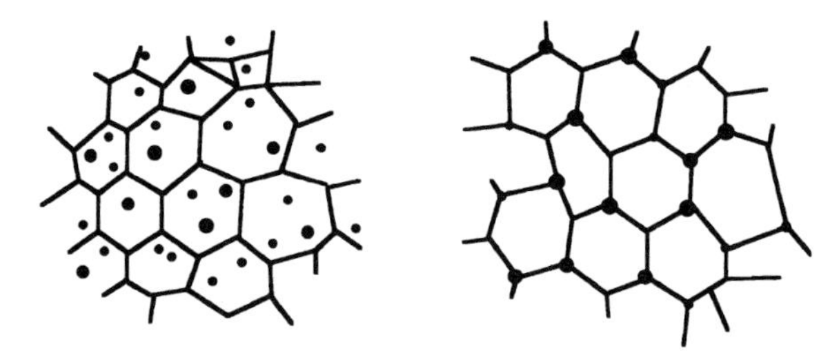

**Figure 5.8.** Grain boundary configurations after sintering.[6] At left fast grain growth has left rest pores in the interior of the grains; at right grain growth is slow enough to keep the pores on the boundaries and densification can continue. With permission from *Engineered Materials Handbook, Vol. 4, Ceramics and Glasses* (1991). ASM International, Materials Park, OH.

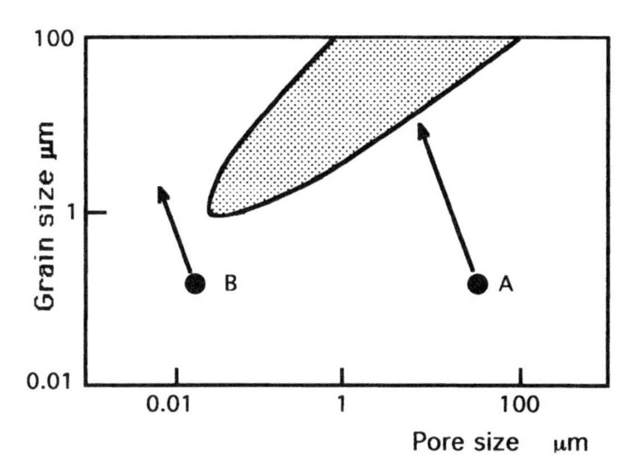


**Figure 5.9.** Approximate plot of grain size vs. pore size in a sintering ceramic.[6] Arrows indicate sinter paths, the shaded area showing conditions for breakaway of pores from moving grain boundaries. The sinter path of ceramic A crosses the breakaway area and has some rest porosity after sintering. Ceramic B with the same grain size but finer initial pores in the green compact attains theoretical density. With permission from *Engineered Materials Handbook, Vol. 4, Ceramics and Glasses* (1991). ASM International, Materials Park, OH.

**Table 5.3. Some Changes of Properties Owing to Porosity**[a]

| Property | Equation |
|---|---|
| Electrical resistivity ($\Omega$) | $\Omega = \Omega_0 P f^2$ |
| Magnetic saturation ($B$) | $B = B_0(P - Qf)$ |
| Elastic modulus ($E$) | $E = E_0 f^{3/4}$ |

[a]Bulk values have subscript zero, $P$ and $Q$ are adjustable parameters, and $f$ is fractional density.

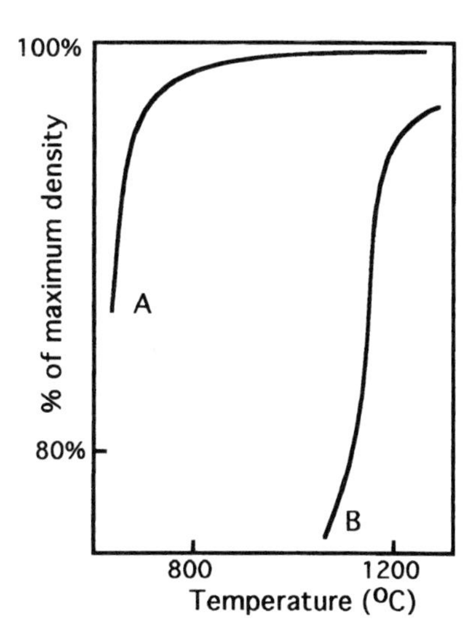


**Figure 5.10.** Densification during sintering of two titania compacts.[6] Ceramic A is sol-gel made and ceramic B is made of commercially available milled powder. The former sinters at lower temperature and to a higher final density. With permission from *Engineered Materials Handbook, Vol. 4, Ceramics and Glasses* (1991). ASM International, Materials Park, OH.

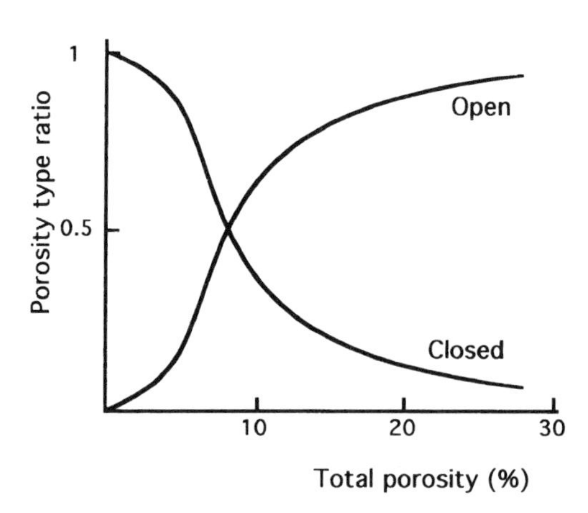


**Figure 5.11.** Variation in the type of porosity with total porosity in a fictive ceramic sintered from particles made by milling. More pores are open in highly porous materials.

Conventional sintering models found in the older literature are based on equilibrium considerations. According to this approach grain boundaries behave like soap bubbles. A number of results of model calculations have been reported to be consistent with measured data. Observations suggest that the microstructure of a ceramic material is kinetically determined in certain cases. Some grain boundaries, for instance, are seen to roughen during sintering instead of becoming smoother as the soap bubble model would have it (Figure 5.12). Therefore, as anywhere else, the thermodynamic potentials are by themselves of limited use in sintering models.

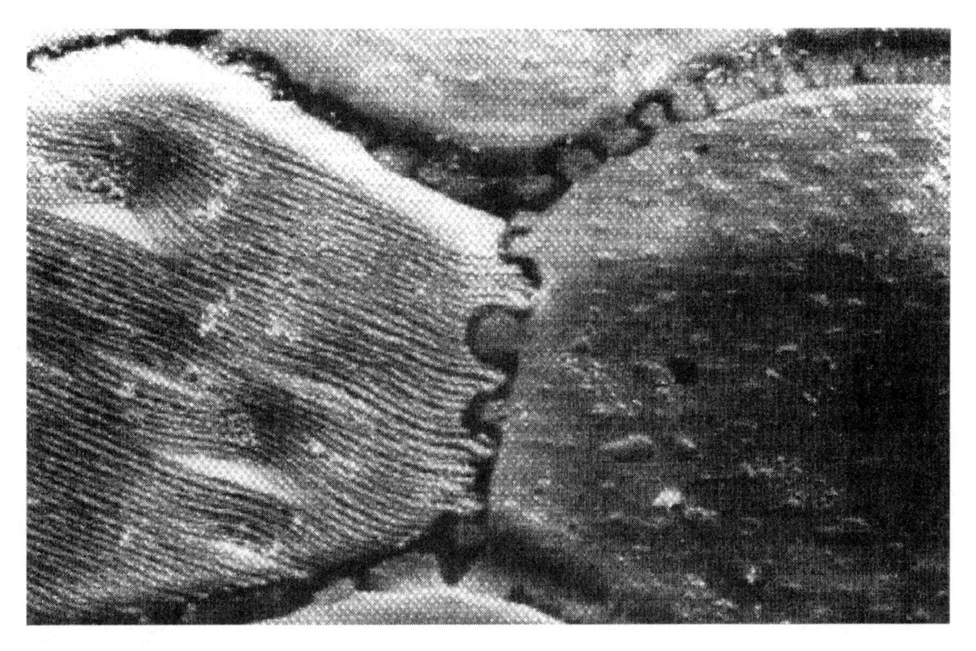

**Figure 5.12.** Grain boundaries of a sintered zirconia alloy (from S. Meriani. Thirteenth International Conference on Science of Ceramics, p. 485. Orleans (1985). With permission from *les éditions de physique.*

### Sinter Aids

Inorganic minority additives are often used to improve the densification of the powder compact during sintering. Five main categories of sinter aids are distinguished at present:

1. Those that temporarily form small amounts of a liquid phase at the sintering temperature by alloying or reacting with the majority phase. Phase diagrams can be a source of inspiration while searching for alternative sinter aids of this type. This form of sintering is liquid-phase sintering, described below.
2. Those that inhibit grain boundary motion by second-phase formation at grain boundaries; second-phase inclusions (pores are a special case) may retard the moving grain boundaries if they remain attached to them, as discussed above.
3. Those that affect diffusivity at grain boundaries by a suitable modification of boundary point defect concentrations.
4. Those that inhibit crystallite growth kinetically by blocking growth sites and by doing so poison the crystallite facets.
5. Those that inhibit grain growth thermodynamically by lowering the driving force. In certain cases surface enrichment in very small particles lowers the surface energy even to negative values. This removes the driving force for physical sintering.

Table 5.4 lists examples of dopants for a few ceramic materials. Often two or more dopants are combined that mutually reinforce the desired effect such as solubility or ambivalent diffusivity.

### Role of Ambient Atmosphere

The gas atmosphere during the sintering process can be modified to improve the sinter rates in two ways:

1. Mass can be transported in the initial phases of sintering through the vapor phase, which is faster than through the solid. Thus if the metal ion diffusion

**Table 5.4. Additions to Several Ceramics to Improve Densification during Sintering**

| Ceramics | Crystallite growth inhibitors | Other densification aids | Modifiers, enhancers |
|---|---|---|---|
| BeO | — | $Li_2O$ | — |
| MgO | $Cr_2O_3$, $Fe_2O_3$ | LiF, NaF | MnO, $B_2O_3$ |
| $Al_2O_3$ | MgO, ZnO, $ZrB_2$ | LiF | MnO, $TiO_2$, $H_2$ |
| $ZrO_2$ | $TiO_2$, $Cr_2O_3$, $H_2$ | — | — |
| $BaTiO_3$ | $TiO_2$, $Al_2O_3$, $SiO_2$ | — | — |
| PZT | $Al_2O_3$, $La_2O_3$ | — | — |
| SiC | — | B, Al, $Al_2O_3$ | — |
| $Si_3N_4$ | — | MgO, $Y_2O_3$ | — |
| TiC, TaC, WC | — | Mn, Fe, Ni | — |
| $TiB_2$, $ZrB_2$ | — | Cr, Ni | — |

is fast compared to the oxygen ion diffusion (which is then rate-limiting), any additional oxygen transport through the gas phase would increase the reaction rate. Sometimes the metal moiety can also be made volatile: in reactive sintering involving chromite products gaseous oxygen may form small amounts of the volatile $CrO_3$ intermediate. Diffusion of this molecule through the gas phase also increases the chromite formation rate.

2. The oxygen partial pressure affects the defect concentrations in an oxidic solid and therefore also the diffusivity in nonstoichiometric oxide ceramic compounds. The oxygen gas pressure then affects the apparent reaction rate through the solid state diffusion rates even when there is no mass transport through the gas phase during sintering.

#### *Differential Sintering*

Actual random packing of the powder particles in the green form (which concept differs from that of the atomic random close packing used in models for amorphous metals) inevitably implies density fluctuations. On sintering the powder compact shrinks but not at the same rate everywhere. Particles at sites of high packing density tend to draw together while those at sites of low packing densities are pulled away by their neighbors at denser sites. Thus locally low-packing-density sites develop into flaws in the second sintering stage, which weaken the product. Applying high pressures, either isostatic (HIP) or unilateral (HUP), during sintering will oppose this tendency and improve the strength of the product.

Locally dense parts in the green form initially have a higher sinter rate than the less dense parts, where the less dense stacking renders the material more porous. During sintering the powder particles in a packing density gradient (i.e., the average coordination number on the dense side is higher than on the less dense side) tend to move to the dense side. Thus the more porous areas bordering on the locally denser areas are under tension while the denser parts are under compression. These stresses, which result from local differences in sinter rates, might be partially relaxed by viscous creep.

Differential sintering is a case of positive feedback: initial fluctuations in the packing density (e.g., owing to agglomeration) affect the local densification rate, which causes differential sinter stresses. The initial fluctuations are amplified by the differential sinter stresses to such an extent that they may lead to the formation of sinter flaws if viscous creep does not relax the resulting strains that develop during sintering. Thus because particle packing fluctuations are unstable under sintering nonagglomerated powder particles which pack well are of vital importance for ceramics.

### *5.5.2. Liquid-Phase Physical Sintering*

Sintering in the presence of small amounts of a liquid phase comprises several processes simultaneously or sequentially. There is viscous creep, the movement of the liquid and the particles under capillary forces, which induces shrinkage and pore-filling from the beginning. The grains may disintegrate owing to attack by the liquid phase (liquid salts are good solvents and therefore usually quite corrosive). Finally the grains may grow by ripening, which consumes the smallest particles.

**Table 5.5. Examples of Application of Liquid-Phase Sintering**

| Material | Application |
|---|---|
| Co/WC; TiC/Mo/Ni | Cutting tools |
| $BaTiO_3$/LiF/MgO | Capacitors |
| $SrTiO_3/SiO_2$ | Capacitors |
| $Fe/Al_2O_3/C$ | Friction materials |
| $Al_2O_3$/glass | Grinding material |
| $Si_3N_4$/MgO; SiC/B | Refractories |
| $Al_2O_3/MgO/SiO_2$ | Refractories |
| $Si_3N_4/Y_2O_3$ | Metal working tools |
| Clay/feldspar/flint | Porcelain |
| Fe/P, Fe/Si | Soft magnets |
| $Fe/Cr_3C_2$ | Cermets |

Again in the thermodynamic model the particle size is assumed to determine the solubility in the liquid phase and therefore the sinter rate. The rates of liquid-phase sintering are usually much more rapid than those of solid state sintering because diffusion through a liquid is faster than in the solid state. Table 5.5 lists several composites that are densified by liquid-phase sintering. Cases are known where the liquid phase solidifies as an intergranular glass on cooling. In other cases, the liquid or glass-phase crystallizes or is dissolved in the solid phases as an alloy. The best conditions for liquid-phase sintering can be read from the phase diagrams (Figure 5.13). Densifying aids promote liquid-phase formation by reaction with one or several of the solid phases.

During vitrification there is a partial melting of the ceramic powder mass by mutual solution. On cooling after sintering, a composite results consisting of a glass matrix in which the remaining solid particles are embedded. This resembles liquid-phase sintering but the function of the liquid phase differs in these cases.

## 5.5.3. Reactive Sintering

In the case of reactive sintering chemical potentials are the driving forces for the sintering processes.[24] The ceramic compound is actually formed during sintering by

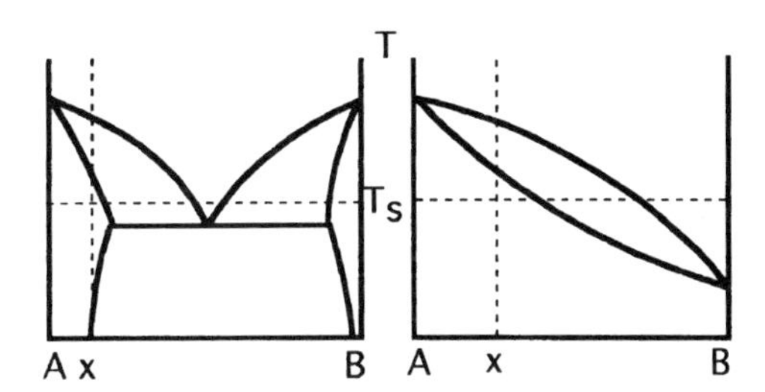


**Figure 5.13.** Schematic phase diagrams of two mixtures of ceramic components indicating the conditions for liquid-phase sintering that ends with a solid of composition $x$ at the sinter temperature $T$. Initial compositions $x$ of A and B may develop liquid phases at sinter temperatures $T_s$ even in cases where the components or the product are not liquid at that temperature.

the reaction between the solids or between the solid and liquid or gaseous reactants. Examples of reactive sintering are the formation of ternary oxides (such as spinels, perovskites, or garnets) from two simple solid oxides. Again phase diagrams are useful for choosing the conditions for liquid-phase sintering. Sometimes reaction heat evolved during reactive sintering could melt the green form if the temperature is not controlled. Figure 5.14 shows two reaction paths for the formation of $MoSi_2$, one adiabatic and the other nonisothermally cooled. Reactive formation of $MoSi_2$ from its elements is a case in which the mixture should be cooled once the reaction starts.

If the reaction enthalpy during reactive sintering is high enough, once the reaction starts it may self-propagate through the green powder mass. These are SHS processes (self-sustaining high-temperature synthesis). It is usually difficult to make dense ceramics with SHS as the reactions are, as a rule, too violent. The final density may be increased by adding inert phases (fibers or a certain amount of the end product itself) or phases that form liquids during the reaction. Table 5.6 lists a few ceramic compounds made by SHS.

Silicon oxynitride ($Si_2ON_2$) is a ceramic formed from a reaction between silicon and nitrogen in a liquid glass of CaO or MgO with $SiO_2$. This is an example of reactive sintering. In the final ceramic the oxynitride ($Si_2ON_2$) forms a composite with a glassy phase of calcium silicate. A density of 90% of the theoretical density can be obtained. Silicon oxynitride is refractory and chemically inert even in strongly oxidative melts of salts. For example, at 700°C the material is resistant to corrosive attack by a cryolite melt containing chlorine.

RBSC (reaction-bonded silicon carbide) is another example of a material formed by a chemical reaction between a solid and a liquid. In this case liquid silicon and solid carbon are the reactants. The sintering temperature (1410°C) is slightly above the melting point of silicon and because the reaction with carbon is strongly exothermic, the rates are high and thermal stresses may develop. If carefully controlled, however, the product has an acceptable density and high strength.

In the Lanxide process for the synthesis of fiber-reinforced ceramic composites a liquid alloy (Al based), impregnated in a whisker felt or fiber preform, is gradually oxidized in air to form an oxide matrix.

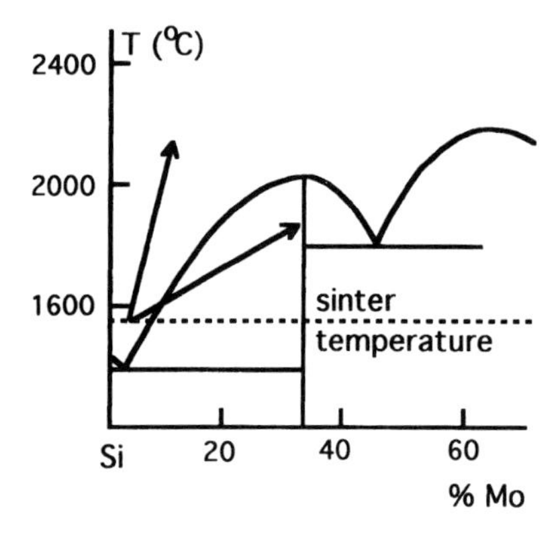


**Figure 5.14.** Part of the phase diagram of Si/Mo indicating possible reaction paths (arrows) of reactive liquid phase sintering of molybdenum disilicide. If the reaction heat increases the temperature the solid does not form during sintering and the whole mixture melts.

**Table 5.6. Materials Prepared by Self-Sustaining High-Temperature Synthesis**

| Reactants | Product | $T_{adiab}$(K) |
|---|---|---|
| $Ti + 2B$ | $TiB_2$ | 3190 |
| $Zr + 2B$ | $ZrB_2$ | 3310 |
| $Ti + C$ | $TiC$ | 3200 |
| $Al + \frac{1}{2}N_2(g)$ | $AlN$ | 2900 |
| $SiO_2 + 2Mg + C$ | $SiC + 2MgO$ | 2570 |
| $3TiO_2 + 4Al + 3C$ | $3TiC + 2Al_2O_3$ | 2320 |
| $3TiO_2 + 4Al + 6B$ | $3TiB_2 + 2Al_2O_3$ | 2900 |

## Exercises

1. When solid CoO and $TiO_2$ react to form the two possible mixed oxides, the mechanism of a reaction in an oxygen ambient differs from that in an inert gas ($N_2$) or in vacuum. Why? Derive the position of the marker threads in the reaction zone of a diffusion couple of these oxides.
2. Solid state reactions involving chromia and sintering of $Cr_2O_3$ are helped by an oxygen atmosphere. Why? Give the reaction mechanisms.
3. When an excess of boron is reacted with a transition metal M the solid state reaction rate is rather high up to the compound $MB_2$. Continued boration rates are low and the resulting material becomes amorphous. Explain.
4. What is the advantage of deriving the activation energy of a reaction by means of the method described in Section 5.4 compared to a least-squares curve fitting?
5. Why are solubilities better and existence regions in phase diagrams wider at higher temperatures?
6. How does the reaction mechanism affect the Avrami coefficient?
7. Arrange the following three oxides in a series of increasing oxide ion conductivity: $V_2O_5$, $Nb_2O_5$, $Ta_2O_5$.
8. How would one distinguish experimentally grain boundary diffusion from bulk diffusion in an Arrhenius plot (see Figure 10.15)?
9. Why is the iron ion diffusivity of FeO so high?
10. Why can glassy cation conductors only be used at low temperatures?
11. In transition metal carbides, carbon diffuses faster than the metal atoms. Why?
12. When two solid reactants can form several compounds, some of them are not found in a diffusion-couple experiment. Give three possible reasons for their absence and suggest experiments to determine the correct one.
13. What is the use of the phase rule in describing solid state reactions in diffusion couples?

## References


1. F. P. Glasser and P. E. Potter. *High Temperature Chemistry of Inorganic and Ceramic Materials.* The Chemical Society, London (1977).
2. Y. Chen, W. D. Kingery, and R. J. Stokes. *Defect Properties and Processing of High-Technology Nonmetallic Materials.* MRS, Pittsburgh (1986).
3. L. G. Harrison. The theory of solid state kinetics. In: C. H. Bamford, and C. F. H. Tipper (eds). *Comprehensive Chemical Kinetics. Vol 2.* Elsevier, Amsterdam (1969), Ch. 5.

4. T. Kudo and K. Fueki. *Solid State Ionics.* VCH, Weinheim (1990).
5. A. L. Laskar, J. I. Bocket, G. Brebec, and C. Monty (eds). *Diffusion in Materials.* NATO ASI E 179, Kluwer, Dordrecht (1990).
6. S. J. Schneider (ed). *Engineered Materials Handbook, Vol. 4: Ceramic and Glasses.* ASM International, Materials Park (1991).
7. S. Engels. *Anorganische Festkörperreaktionen.* Akademie Verlag, Berlin (1981).
8. H. Schmalzried. *Chemical Kinetics of Solids.* VCH, Weinheim (1995).
9. A. M. Alper (ed). *Phase Diagrams in Advanced Ceramics.* Academic, San Diego (1995).
10. E. S. Machlin. *An Introduction to Aspects of Thermodynamics and Kinetics Relevant to Materials Science.* Giro, Croton-on-Hudson (1991).
11. J. H. Westbrook and R. L. Fleischer (eds). *Intermetallic Compounds: Principles and Practice. Vol. 2.* Wiley, Chichester (1995).
12. E. J. Mittemeijer, Liu Cheng, P. J. van der Schaaf, C. M. Brakman, and B. M. Korevaar. Analysis of non-isothermal transformation kinetics; tempering of iron–carbon and iron–nitrogen martensites. *Metall. Trans.* **19A**, 925 (1988).
13. I. J. McColm. Special ceramics for modern applications: which? why? how? In: R. A. Terpstra, P. P. A. C. Pex, and A. H. de Vries. *Ceramic Processing.* Chapman and Hall, London (1995).
14. D. W. Richersen. *Modern Ceramic Engineering: Properties, Processing, and Use in Design.* Marcel Dekker, New York (1982).
15. R. J. Brook (ed). *Processing of Ceramics: Materials Science and Technology, Vol. 17*, VCH, Weinheim (1996).
16. H. V. Swain (ed). *Structure and Properties of Ceramics: Materials Science and Technology, Vol. 11.* VCH, Weinheim (1994).
17. A. G. Evans. Considerations of inhomogeneity: Effects in sintering. *J. Am. Ceram. Soc.* **65**, 497 (1982).
18. A. G. Evans. Inhomogeneous sintering: Stresses, distortion, and damage. In: Y. Chen, W. D. Kingery, and R. J. Stokes (eds.). *Defect Properties and Processing of High-Technology Nonmetallic Materials.* MRS, Pittsburgh (1986), p. 63.
19. Y. M. Chiang, J. S. Haggerty, R. P. Messner, C. Demetry. Reaction-based processing methods for ceramic-matrix composites. *Cer. Bull.* **68**, 420 (1089).
20. R. J. Brook. Controlled grain growth. In: F. F. Y. Wang (ed). *Treatise on Materials Science and Technology, Vol. 9. Ceramic Fabrication Processes.* Academic, New York (1976), p. 331.
21. C. H. Hsueh, A. G. Evans, R. M. Cannon, and R. J. Brook. Viscoelastic stresses and sintering damage in heterogeneous powder compacts. *Acta Metall.* **34**, 927 (1986).
22. A. A. Chernov. *Modern Crystallography III: Crystal Growth.* Springer, Berlin (1984).
23. B. C. Kellett and F. F. Lange. Thermodynamics of densification: 1. Sintering of simple particle arrays, equilibrium configurations, pore stability, and shrinkage. *J. Am. Ceram. Soc.* **72**, 725 (1989).
24. M. E. Washburn and W. S. Coblenz. Reaction-formed ceramics. *Ceram. Bull.* **67**, 356 (1988).

*Chapter 6*

# THE CHEMISTRY OF INORGANIC SURFACES

*But perhaps more important than the physics behind these processes is the fact that they have been observed.*

*SYLVIA CEYER (1990)*

## 6.1. Surface Chemistry

The materials chemist has to deal with three different types of chemistry: solid state (in homogeneous solids), molecular (in homogeneous liquids or gases), and surface. Of these the last is the most important for materials chemistry because[1]:

1. Solids react with fluids on their surface; catalysis and corrosion are surface reactions of solids.
2. When solid products are made, they grow by surface reactions. Syntheses of crystalline and amorphous solids and of nanostructured materials are surface processes.
3. Many material properties are surface properties, which are easier to tune by modifying the surface than by changing the solid.

Solid, molecular, and surface chemistry have much in common: they share most types of bonding and the driving forces are the thermodynamic potentials. Heterogeneous reactions differ from homogeneous processes in kinetics, in the morphology-dependence of the reaction rates, and in the possibility of electrical control.

For there to be chemistry there must be movement of the reacting species. Some surface reactions occur between adsorbed species that diffuse over the surface. The number of degrees of freedom for moving reacting species is lower on surfaces than in three-dimensional space (in the fluid or the interior of a bulk solid). Transport is anisotropic and the diffusion rate on the surface is between that in the solid and in fluids. The dimension of the reaction space is a very significant factor in the chemistry and for diffusion rates in that space.

Another difference between homogeneous and heterogeneous process rates concerns the role of morphology. Because of *reconstruction*, which is a rearrangement of surface atoms, the surface structure is not as the interior of the crystal lattice would suggest. This reconstruction affects the chemical properties of the materials. The smaller the grain size, the larger the ratio between the number of atoms on the surface and in the interior bulk. This ratio is sometimes called the dispersion. The dispersion affects many properties and a material that is finely divided differs from the same material that is coarsely crystalline. The atoms on the outer surface or interface also have different neighbors than those in the bulk. Surface atoms are bound on one side to atoms of the solid and on the other side to atoms in the liquid or gas phase or to species adsorbed on the surface. This bonding strongly affects their reactivity. Surfaces can have a varying roughness and that also has consequences for the rates of reactions on the surface.

Solids have two types of properties: bulk and surface. The bulk properties, which are connected with the interior of the solid, are, e.g., compressibility, strength and toughness up to a point, refraction index, color, melting point, and bulk conductivity and magnetism. Some properties change with the morphology of the material, e.g., with the grain size, such as the conductivity for electrons and ions, the heat conductivity, and the toughness. The second and largest of the two groups of properties are the surface properties, which include corrosion resistance, catalytic activity, reflectivity, work function, adhesion, and hardness. These surface properties can be easily changed with a surface treatment that leaves the bulk properties virtually unchanged.

The mechanical properties of solids that are usually related to the interior of the solid also depend partly on crystallite surfaces or grain boundaries. For example, tough material has a microstructure that has a high fracture surface area while the interatomic bonding forces that keep the surfaces together need not necessarily be very strong, as can be seen in the case of tenacious glue, which is held together only by dispersion forces. Paradoxically, high toughness implies comparatively weak interphase and strong intraphase interactions, so that any crack starting in a material deviates from a straight line and a lot of energy is needed for it to continue. All inorganic materials are already strong by themselves but they are also usually brittle. The art of toughening brittle materials is in creating weak interactions in the right places, which is done in hybrid composites, such as by polymers in cerpols (ceramic–polymer composites) and by metals in cermets (ceramic–metal composites).

The three types of strong bonds between a solid and adsorbed molecules, radicals, or atoms are all relevant for surface chemistry. The two types of weak bonds between a solid surface and adsorbed molecules can also be observed. A reaction between a surface and a molecule that involves strong bonds is called chemisorption. Reactions under low driving forces are called physisorption. Chemisorption refers to strong bonding as between atoms and physisorption to weak bonding as between molecules.

After being physisorbed molecules may break open and their fragments can chemisorb. Chemisorption means strong reaction with the surface; e.g., hydrogen or nitrogen gas with metallic surfaces (Figure 6.1). The atoms that were covalently bound to each other in molecules break off and form separate metallic bonds with the atoms of a metallic substrate surface such as platinum, nickel, or iron. There also can be a covalent bond between a surface atom and a nonmetal atom of an adsorbate or a $\sigma$-bound organic moiety. Ionic bonds

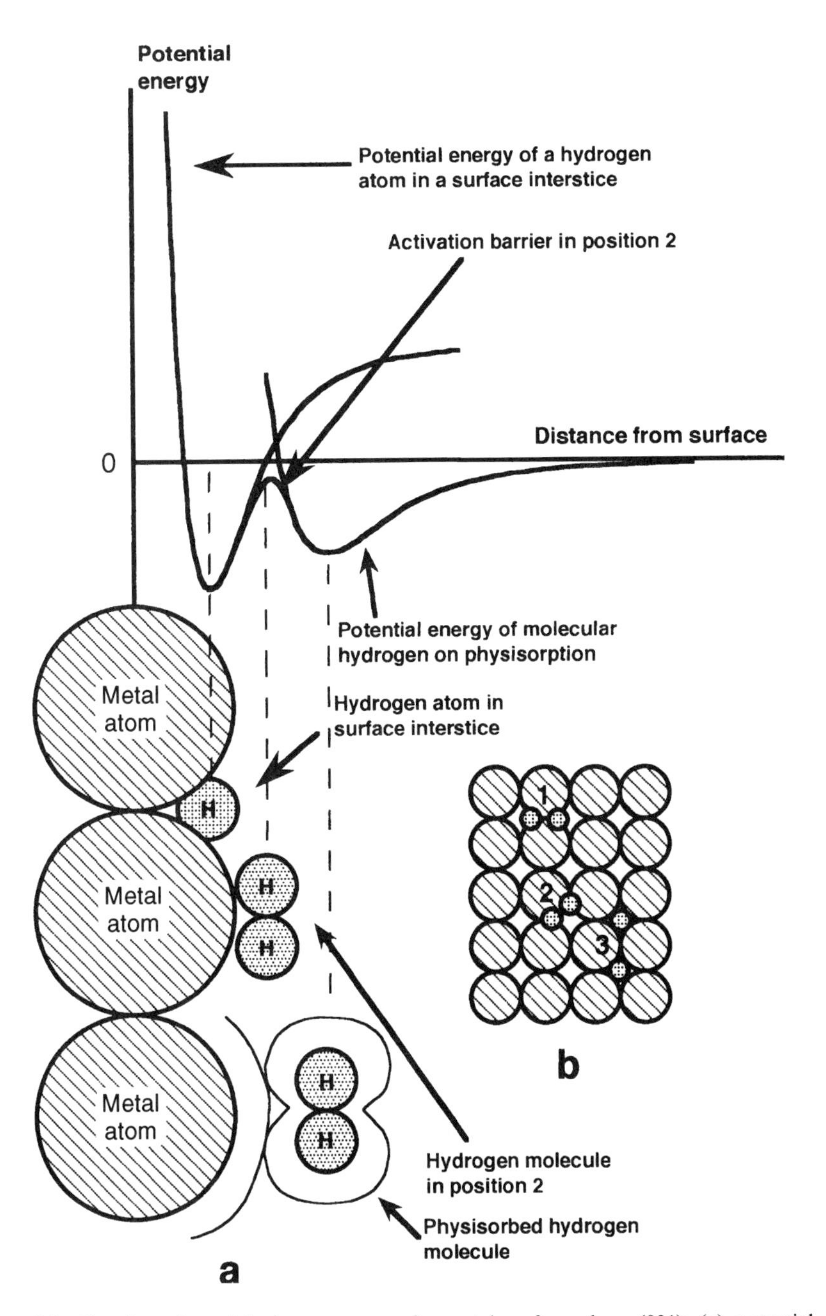


**Figure 6.1.** Chemisorption of hydrogen on an fcc metal surface plane (001): (a) potential energy representation of adsorption of hydrogen on a metal surface; (b) front view of hydrogen adsorption on a nickel crystal face. The large spheres represent the metal atoms, the small circles the hydrogen atoms. In the first step of the sorption process the hydrogen molecule might be expected to straddle two neighboring metal atoms as indicated in the top row of the front view (b). The phlogiston model predicts an asymmetrical position for the hydrogen molecule as shown in the center because bonding is more favorable. This configuration corresponds with the activation state between physisorption and chemisorption. When hydrogen is dissociated the atoms are bonded to nickel and they occupy interstitials at the surface.

between substrate surface ions and adsorbates (cations or ligands) can be formed in solution. The sol-gel process in ceramic synthesis is a well-known example of a process that is controlled by using ionic surface bonding.

The presence of bonds of a certain type determines the structure of the substrate. The structure of the diamond lattice is "stabilized" at the surface by chemisorbed hydrogen atoms, while the diamond surface without chemisorbed hydrogen relaxes to conjugated double, graphite-like bonds. If spontaneous formation of this graphite-like surface is not sufficiently countered with atomic hydrogen, the substrate growth will change to graphite instead of diamond when carbon is deposited.

The bond between covalent lattices such as silicon and diamond and chemisorbed gases such as hydrogen, nitrogen, and carbon monoxide is covalent. Species that are bound to the surface with covalent bonds generally have low mobilities over the surface. If they have a metallic bond or an ionic or coordination bond with surface atoms they tend to have higher mobilities. The mobility of surface species can be important for the reaction. High surface mobilities result in growth of coarsely crystalline solids, and the reaction rates have conventional kinetics.

For this chapter on surface chemistry a few examples have been selected to illustrate the type of surface chemistry with which a materials chemist should be familiar. First in Section 6.2 surface bonding and the effect of surface charge on the properties of inorganic materials is described. The surface charge of colloidal particles is the subject of Section 6.3. Heterogeneous catalysis is sketched in Section 6.4, not because a materials chemist will ever be asked to develop a novel catalyst (specialists who do nothing else will take care of that) but because there is useful information there for materials synthesis and for changing the chemical properties of solids. Section 6.5 summarizes the growth of solids from solutions or melts. Making and shaping solids simultaneously in the same process from the gas phase is increasingly done in manufacturing. The characteristic gas-phase techniques are the subject of Sections 6.6 and 6.7. The reverse of growth, corrosion, is described in Section 6.8. Finally Section 6.9 provides an example of how to use molecules to modify existing surfaces to change their chemical properties.

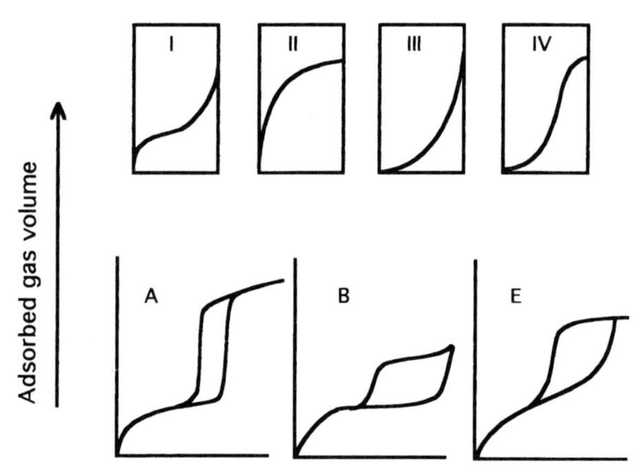


**Figure 6.2.** Gas adsorption isotherms of amount adsorbed vs. the vapor pressure. Several types are observed. Certain pores show hysteresis (A, B, E) in the isotherms, depending on the form and dispersion of the pores.

Surface physics is an old and well-established science but when ultrahigh vacuum equipment became available it was apparent that the previous results were inconsistent with the new data and everything had to be measured over again. The analytical techniques developed in surface physics were a great help for surface chemistry as the surface reactions in catalysis were analyzed with them.[2,3] One of those techniques is the BET (Brunauer–Emmett–Teller) technique to characterize surfaces: Figure 6.2 shows the adsorption isotherms that are used to measure specific surface areas of heterogeneous catalysts and characterize their pores.

## 6.2. Physical Properties of Inorganic Solid Surfaces

Many of the surface properties of a solid depend on the electrical and thermodynamic potentials that develop at its surface, and control of those potentials is basic for the design of functional materials. The consequences of these potentials are quite different for an ionic solid, a covalent semiconductor, or a metal.

Particles of ionic insulators suspended in solution can have charged surfaces because of an excess there of ions of one type as a result, e.g., of adsorption from the solution. The surface charge depends on the concentration in the solution of ions that can adsorb. The sol-gel technique is discussed in the next section as an example of a synthesis of nonmetallic inorganic polycrystalline materials that makes use of surface charge control of colloidal particles.

Covalent or ionic semiconductors generally have either *n*-type of *p*-type conductivity, depending on the sign of the majority charge carrier. Semiconductors have a depletion layer at their surfaces, which means that the majority charge carrier has a concentration at the surface that is lower than in the interior of the solid. The sign of the majority charge carrier is determined by the type of dopant or by the type of nonstoichiometry. The surface charges are crucial for the functioning of solid state diodes and electron transfer (redox) reactions of a solid.

Redox behavior of surfaces is always discussed in terms of the band model. The edges of the valence and the conduction bands are curved just below the outer surface. A crystallite lattice ends at the surface and the bonds of the surface atoms "dangle" in space. The (initially unpaired) electrons in those dangling bonds can do several things. They can be taken up by the lattice, making the surface positive with respect to the interior. Or they can pair with neighbors and form special bonds between the surface atoms; in other words there is surface reconstruction. The dangling bonds can also attract other electrons from the interior and form a surface charge that is negative with respect to the interior. If the excess charge is positive with respect to the interior the band edges curve downward; if the surface charge is negative, the top of the valence band and the bottom of the conduction band curve upward at the surface. The level of the band edges at the surface does not depend only on the nature of the compound but varies somewhat with the particular crystal plane of that surface. The position of the energies of acceptor or donor levels of molecular adsorbates with respect to the levels of those band edges at the surface determines the possibility of charge transfer between surface and molecule. Photoelectrochemical solar cells that convert the energy of light into chemical energy as for fuel (such cells operate without the Carnot losses of thermal machines) are semiconductors with the proper band levels and band bending at the surface.

Another electrical characteristic of semiconducting solids is the Fermi level. This level, which describes the thermodynamic potential of the valence electrons, is central to any discussion of potentials of electron transfer. The work function is for solids what the electronegativity is for molecules. Potentials in metals are schematized in Figure 6.3. The work function (energy to get a valence electron out of the solid) of two different facets is $\phi_1$ and $\phi_2$, the inner potential $\Phi_{\text{inner}}$ is the result of net charge on the metal lattice, $\mu$ is the chemical potential of the electrons, $E_F$ is the Fermi level, $\chi_1$ and $\chi_2$ are the surface potentials of the two facets, and $\Phi_{\text{outer}}$ is the potential difference of an electron between a position just outside of the solid and infinity, where the potential is $E_{\text{vac}}$. There is a contact potential between two different planes, which is equal to the difference between the work functions of those planes.[4]

The energy of valence electrons in a metal and on the surface depends on nonlocal factors such as the existence of certain facets elsewhere on the crystal. The bonding energy of valence electrons of adsorbed molecules also depends on the potential at the surface where the molecules are adsorbed and that potential depends on planes elsewhere on the crystal. The absolute bonding energy of electrons of adsorbed molecules can only be measured with respect to the Fermi level of the solid.

The large effect of bulk doping and Fermi level on surface reactivity is seen when *n*-type and *p*-type silicon single crystals are etched. A high Fermi level in the silicon (upward curling band edges) etches with formation of clean crystal planes, while the surface of anodically etched *p*-type silicon is ramified and highly porous.

The chemical activity of a metal is strongly influenced by the potentials at its surface. This has also been observed in heterogeneous catalysts that consist of metal particles on an ion-conducting ceramic substrate. Such a substrate makes it possible to change the potential of the metal particles. The reactions studied were the

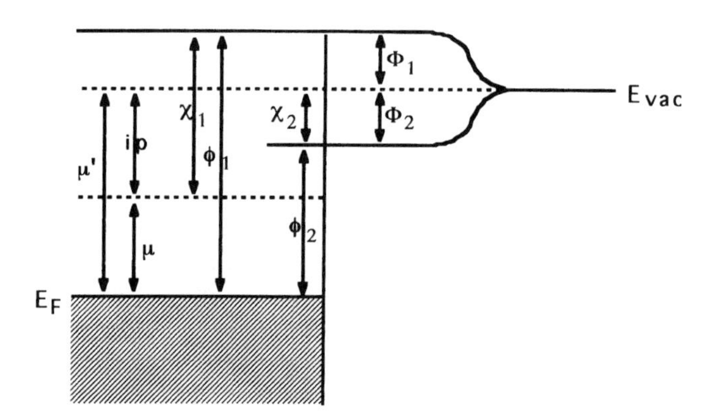


**Figure 6.3.** Potentials in a metal with two crystallite faces, 1 and 2. The work functions are $\phi$, $\mu$ is the chemical potential of the electrons, $\mu'$ the electrochemical potential, $\chi_i$ the surface potentials of the two faces, ip the inner potential owing to charge on the metal, and $\Phi$ the potential difference between an electron at the surface and far removed (at the vacuum potential).[4]

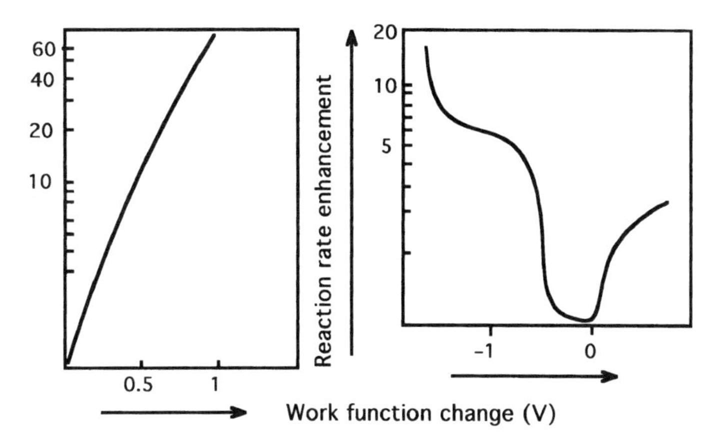


**Figure 6.4.** The rate enhancement of catalytic reactions with catalyst potential. At left the oxidation of methane on platinum to carbon dioxide; at right the oxidation of carbon monoxide on silver particles. Reprinted from C. G. Vayenas, S. Bebelis, I. V. Yentekakis, and H. G. Lintz. *Catalysis Today* **11**, 303. Copyright (1992). With permission from Elsevier Science.

oxidation of CO on silver and the oxidation of $CH_4$ on platinum (Figure 6.4). The effect of the potentials called NEMCA (non-Faradaic electrochemical modification of catalytic activity) was also observed for many other oxidation reactions on metals.[5] The NEMCA effect must be attributed to the presence of potentials because the oxidation activity is much greater than can be accounted for by transfer of oxide ions from the substrate. It is an activation of the surface by the electric effect that a few oxide ions (leaked from the substrate to the metal) have on the metallic surface. This electric effect requires a polarized interface between the metal and the ceramic support.

## 6.3. Inorganic Colloids

Colloids are suspensions of solid particles that are so small that a large proportion of the atoms in the particles are part of the surface layer and not of the interior. Ultrafine particles can be said to have more surface than bulk atoms. Colloid chemistry is surface chemistry.[6,7]

Inorganic colloids are made by growing solid particles from a solution of molecules and stopping the process when the particles are big enough. One method that allows extensive control of the microstructure of the solid deposit is the sol-gel technique, which is based on colloidal particles of oxides in a solution (sol) made by nucleation and growth from dissolved molecular reactants. After formation the particles are agglomerated to form a gel, a semisolid composite of a network of interlinked particles in a solvent matrix. Colloidal methods such as the sol-gel variants permit synthesis of sinter-active small particles of mixed oxides that have the right composition for subsequent sintering to the wanted mixed compound, alloy, or composite. Colloid shaping processes have been in use for a long time in ceramics.

Small particles in the gas phase or in solution easily agglomerate on collision because of the van der Waals–London or dispersion interaction, which is very effective in small particles (this interaction is attractive and has no activation barrier for aggregation). When a precipitate is formed in solution the particles grow gradually under surface reaction from adsorbed molecular precursors. In order to control the size of the particles it is necessary to prevent them from aggregating and keep them separate in solution during growth. The dispersive attractive forces that are always present can be compensated by giving the particles a surface charge that creates a barrier preventing collision and agglomeration when two particles meet. A sufficient surface charge causes the particles to repel one another and allows them to stay in suspension. The colloid is stable and it does not agglomerate and precipitate even at high particle concentration.

The Coulomb part of the interaction between two particles having the same surface charge is screened by an exponential factor as the result of the surrounding

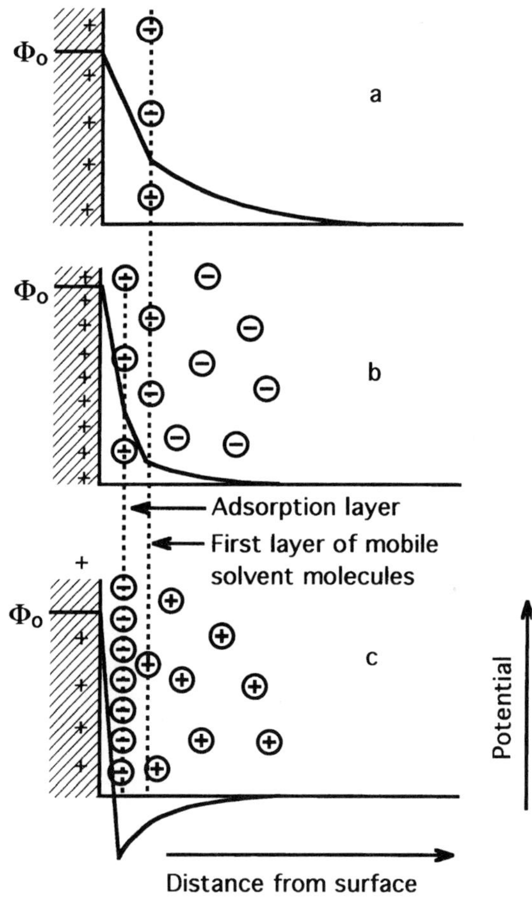


**Figure 6.5.** The potential near the surface of a positively charged colloidal particle: (a) without adsorption; (b) with some adsorption of negative ions; (c) more adsorption of countercharge than initial surface charge reverses the potential.

cloud of counter ions in the solutions around the charged particle:

$$E_{\text{rep}} = \zeta^2 e^{-\lambda(x-d)}$$

in which $\zeta$ is the potential on the surface of the particle, $\lambda^{-1}$ is the so-called Debye length (size of the cloud of counter ions around the particle), $x$ is the distance, and $d$ is the shortest possible distance between the surfaces of the colloid particles. This expression for the repulsion gives an exponential screening factor that is the result of the presence of the counter ions. When the distance between the particles is greater than the Debye length the particles do not repel each other as the surface potential is entirely screened by the ion cloud in the solution.

The surface charge on particles in stable colloids attracts counter ions in the surrounding solution and the particle has a double layer around it (Figure 6.5). This layer (lyosphere) moves with the colloidal particle as it diffuses through the solution. The electrical potential at the liquid side of the double layer, which is called the $\zeta$-potential, is strongly affected by adsorption of counter ions.

The shortest distance between the two particles can be larger than the sum of their radii. If the particle surface is covered with polymers there is steric hindrance, which prevents surface contact, so polymers are sometimes used as antiflocculation agents. The sum of the attraction and repulsion potentials is given in Figure 6.6 as a function of the interparticle distance. Once the particles are agglomerated they can only be dispersed again if the attraction between them can be lessened, e.g., with counter ions in the solution. This redispersion is called deflocculation or sometimes repeptization.

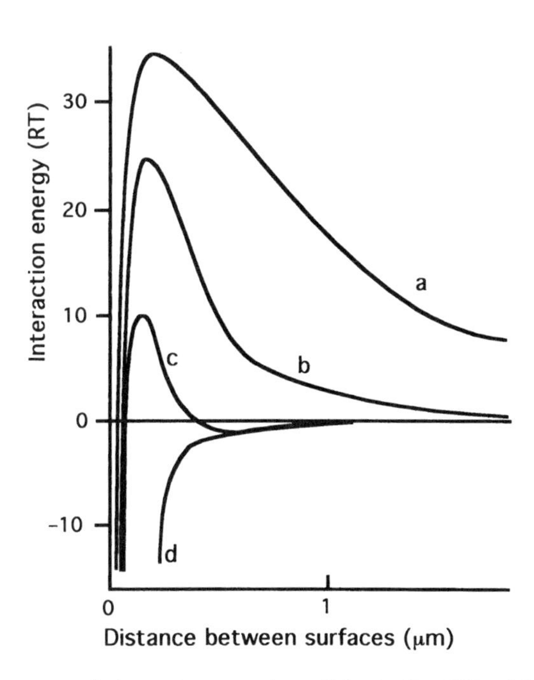


**Figure 6.6.** Interaction energy between two ceramic particles (radius 100 nm) in different monovalent electrolyte concentrations. Double layer thicknesses: (a) 10 μm; (b) 3 μm; (c) 1 μm; (d) 0.1 μm. The electrolyte concentration increases from a to d.

**Table 6.1. Isoelectric Points of Aqueous Colloids of Oxides**

| Compound | pH | Compound | pH |
|---|---|---|---|
| $SiO_2$ | 2 | $Cr_2O_3$ | 6–7 |
| Soda lime glass | 2–3 | $Fe_3O_4$ | 7 |
| Cordierite | 2–3 | $\beta$-$MnO_2$ | 7.3 |
| Quartz | 3.7 | Mullite | 7.5 |
| K-feldspar | 3–5 | AlOOH | 8–9 |
| Apatite | 4–6 | $\alpha$-$Fe_2O_3$ | 8–9 |
| $SnO_2$ | 5.6 | $Al_2O_3$ ($\alpha, \gamma$) | 8.5–9 |
| Kaolin (edge) | 5–7 | ZnO | 9 |
| $TiO_2$ (rutile) | 5–7 | $\alpha$-FeOOH | 9 |
| $TiO_2$ (anatase) | 6.2 | $CaCO_3$ | 9.5 |
| $ZrO_2$ | 6.5 | MgO | 12 |

The electrical charge of the particles is the result of both intrinsic surface charge and that of adsorbed ions (metals or ligands) or ionization of adsorbed water. The surface charge originally present can increase or decrease by adsorption of counter ions (e.g., protons). Thus oxide particles have a negative surface charge at high pH, a positive charge at low pH, and no charge at the point of zero charge: the acidity (pH) at which the surface charge becomes zero is called the point of zero charge (PZC). Table 6.1 shows the values of the PZC of some oxides used in ceramics. Near these pH values the colloids precipitate because there is not enough electrostatic repulsion between the particles to keep them apart. Instead of the point of zero charge the term isoelectric point is sometimes used (in colloid physics it has a slightly different meaning).

Figure 6.7 illustrates the way in which the pH of the solution changes colloid surface charges and electrophoretic mobility of an oxide colloid. The surface charge of colloidal particles is seen to change by adsorption of ions from the solution. Divalent or trivalent cations are effective in reducing negative surface charges and may act as flocculants.

The interaction between particles in solution can be controlled with the surface charge, the cloud of counter ions, and through steric hindrance by adsorbed polymers. Flocculants and deflocculants are additives that are used to control the stability of a colloidal suspension. Flocculants increase agglomeration and deflocculants decrease it. The $\zeta$-potential determines the thickness of the diffuse double layer and the strength of the repulsion between the particles. The plasticity of clay, e.g., is the result of the form and size of the particles, the amount of water bonded, and the $\zeta$-potential. The surface charge is always negative on the long flat sides of clay platelets but may be different on the short sides. At a high pH the side faces are also negatively charged and the clay suspension is stable. At sufficiently low pH the side faces are positive (the flats remain negative) and the particles may agglomerate and stack like a card house.

Flocculated powders have ramified clusters, which do not stack well in the green form which is then not dense enough. A sintered alumina ceramic will not become fully dense after sintering if it starts out from a green form made near the isoelectric pH.

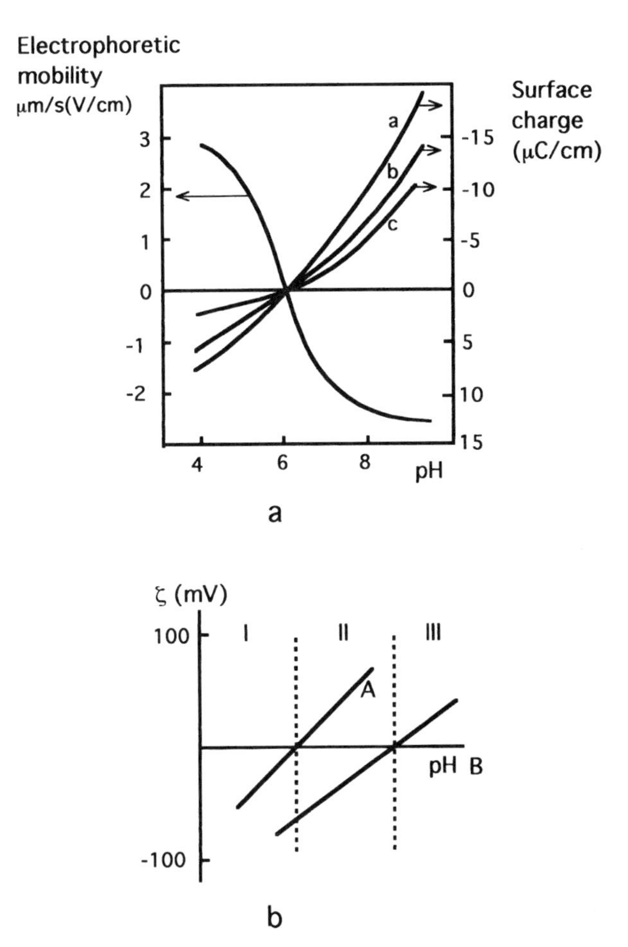


**Figure 6.7.** Some basic features of colloid chemistry for ceramic oxides: (a) relation between surface charge density and electrophoretic mobility of a titania colloid in an aqueous potassium nitrate solution. From a to c the concentration of potassium nitrate increases; (b) the dependence of the ζ-potential on the pH of two different colloids, A and B. At a pH in area I the particles in both colloids have a positive surface charge. In area III the surface charge is negative and at intermediate pH values (in area II) the colloids have surface charges of different signs.

The significance of coordination chemistry for colloid chemistry is obvious. Surface-adsorbed or lattice ligands or metal ions can react or exchange cations, ligands, or electrons with species in solution. The ligand exchange reactions at the surface of the particles form complexes that determine the surface potentials, the colloid stability, and the optical and electrical properties of solids. The stability constants and kinetic parameters of many mixed-ligand complexes in solution have been compiled by coordination chemists and these data can be used to control the nature of the surface complexes on colloidal particles with the right ions or ligands in the solution.

The colloidal particles in inorganic sols are usually oxides, sulfides, and stable metallic particles such as gold. Metallic clusters can be made in small amounts by reducing dissolved salts or by condensing metal vapors in an inert gas. Although this is an expensive process, very fine oxide and nonoxide particles are made on an industrial scale by gas-phase nucleation, e.g., silica, titania, and metal nitrides. Examples of all these techniques are described in Chapter 8.

## 6.4. Heterogeneous Catalysis

The type of bond between the surface of the solid and adsorbate molecules determines the kind of surface processes that can take place: crystal growth, growth inhibition, nucleation, corrosion, catalytic activity, and chemical passivation. Sometimes there are two types of surfaces involved in the reaction: metallic and ionic (many heterogeneous catalysts consist of very small metal particles on oxidic carriers).

Surface reactions are used in chemical technology to convert molecules by heterogeneous catalysis.[2,3,8] Heterogeneous catalysts are solids that have a surface that participates in reactions of molecular precursors to molecular products. After the reaction the surface of the catalyst is returned to its initial state and can react again. The solid catalyst is a reactant that is not consumed in a catalytic reaction. It is sometimes said that catalysts accelerate reactions by lowering the activation barriers. It must be realized that a reaction with a catalyst is altogether another reaction than one without a catalyst, with different reaction coordinates, although reactants and end products may be the same. The intermediate reaction steps are completely different and the two reaction paths cannot be strictly compared in a single graph since the two processes have different $x$-axes.

The reactions with heterogeneous catalysts can be categorized according to the nature of the surface bonds.[9] Thus there are reactions on metal surfaces, reactions on ionic surfaces, and reactions on covalent solids. The observed kinetics are specific[10] and proposed reaction mechanisms vary widely. "Reaction mechanisms come and go, and their ephemeral existence is often disconcerting. By contrast, the results of good chemical kinetics remain unchanged, whatever may be the future revisions of their underlying mechanism."[11]

The morphological state of a surface is usually decisive for its reactivity as can be seen in the following examples:

OXIDATION OF CARBON MONOXIDE ON PLATINUM: The adsorption reaction of oxygen on platinum is

$$O_2(g) + 2e' \rightarrow 2{*}O^-$$

and the subsequent surface oxidation of gaseous carbon monoxide,

$${*}O^- + CO(g) \rightarrow CO_2(g) + 2e'$$

sometimes has an oscillating reaction rate while at the same time the platinum has an oscillating surface potential. However, if the oxygen pressure over the catalyst is greater than the equilibrium decomposition pressure of $PtO_2$, there is no oscillation.[5] At high oxygen partial pressures the surface is not metallic, as it would be at low oxygen pressures, but rather is an oxide that does not sustain oscillations.

OXIDATION OF HYDROGEN BY OXYGEN ON PLATINUM BETWEEN 0 AND 100°C: A gaseous mixture of hydrogen and oxygen is catalytically oxidized at low temperatures. If there is an excess of oxygen in the mixture the reaction is rather morphology-independent and the reaction rate is first-order in hydrogen and does not depend on the oxygen pressure. In an excess of hydrogen the reaction is first-order in oxygen and zeroth order in hydrogen. The combustion rate then increases with increasing dispersion (the ratio of the number of surface atoms to bulk atoms in the catalytic crystallites). The observed kinetics indicate that as in the case of the oxidation of carbon monoxide, the reaction at high oxygen partial pressures occurs on a platinum oxide surface while the metallic active sites cease to have any effect on the rate.

SYNTHESIS OF $NH_3$ ON IRON: The reaction $N_2(g) + 3H_2(g) \rightarrow 2NH_3(g)$ is endothermic, and low temperatures, high pressures, and catalysts have turned out to be beneficial for the synthesis of $NH_3$ from nitrogen and hydrogen. The catalyst is the surface of iron particles spiked with a number of oxidic impurities. Not all iron is equally suitable. The catalytically active form of iron is made by reduction of magnetite ($Fe_3O_4$) in order to get the right particle morphology and dispersion. The reduced catalyst contains some potassium oxide and aluminum oxide. Alumina prevents the iron particles from sintering, keeping them small and chemically active. Potassium oxide is an electronic promotor that affects the potential of the iron surface. The rate-determining step in the complex set of intermediate reactions is assumed to be the surface reaction between adsorbed nitrogen atoms and adsorbed hydrogen.

The models of these surface reactions (more details are given in Section 6.7) imply a number of assumptions: (1) there is equilibrium in adsorption and desorption; (2) there is only one rate-determining intermediate reaction between adsorbed species; (3) the species on the surface are well mixed; and (4) there is a thermal probability of a transition of the physisorbed to the chemisorbed state before subsequent reaction and diffusion. These assumptions are not independently proved but are justified by the degree of success of the models in predicting the kinetics.

There is a large difference between the reaction rates observed for the same system in university laboratories and in factories: the rates measured in ultrahigh vacuum equipment in the laboratories were much lower than those of the same reactions at high pressure in the industrial setting. This phenomenon is known as the pressure gap[12] and is an example of the mutual incompatibilities that keep popping up between science and technology. University researchers use low pressures because it simplifies the system and enables analysis with *in situ* techniques. Chemical engineers use the high pressures that are necessary to make something fast. Experiments with molecular beams have now explained the mysterious pressure gap. The catalytic process is not merely the reaction between adsorbed species. There appears to be also a large contribution to the reaction by the impact of molecules from the gas phase, even of an inert gas, when they hit adsorbates. Collision-induced reaction at high pressures is the source of the pressure gap.

Catalytic reactivities of oxides have been associated with surface defects such as steps and kinks on crystallite faces and, more importantly, with surface vacancies.[13] Surface vacancies are basic or acidic, depending on which ion is missing. Cation vacancies can be highly basic; anion vacancies are acidic. *F*- and *V*-centers are examples of vacancies that have become paramagnetic by having trapped or lost a single electron. Electrons can be trapped in very acidic vacancies and form an $F$-center or they can break loose from a highly basic cation vacancy. Such defects can reduce or oxidize adsorbates.

Defects in solid oxides are generally made by doping the solid with a compound in which the metal ion has a valence other than the metal in the host oxide (Chapter 10). Lithium oxide increases the oxygen vacancy concentration in magnesium oxide when it is used as a dopant. The surface of doped MgO catalyzes the oxidation of methane. A steam treatment at high temperature regenerates catalytic action in poisoned catalysts by removing adsorbates that cover the active sites. The methods of preparation of the solid strongly affect the number of such active sites at the surface. MgO can be made by burning Mg metal or by decomposing $Mg(NO_3)_2$. Both methods result in faceted crystallites of noncatalytic MgO of low specific surface area. If $MgCO_3$ or colloidal $Mg(OH)_2$ is calcined a highly reactive and catalytically active porous magnesium oxide is formed with a high specific surface area.

## 6.5. Growth of Crystalline Solids from Liquids

Solid compounds can be made by reacting mixtures of precursor powders at high temperatures (reactive sintering), a method often chosen for large-scale processes because of its simplicity (Section 5.5). The preparation of solids is easier to control if the solids are made from molecules or atoms in solution, or from melts. When solids are made from a liquid the formation reaction occurs at the surface of the solid. Crystal growth and etching thus have much in common with catalysis. The practical aspects of making solid compounds from liquid phases are discussed in Chapter 8. Here some generalities are given on the formation of solids from melts or liquid solution.

Solidification of molten compounds can be considered a physical process and is merely a transfer of molecules or ions from a melt into a lattice while the bonds between the ions in the solid lattice remain approximately the same as those in the melt. This also occurs in a ternary oxide such as $BaTiO_3$ or $NiCr_2O_4$ made from a melt of the two component oxides. Solidification has consequences for the microstructure: the crystallites may be faceted. Facets are flat surfaces of a single crystal that have angles between them with values that are characteristic for the lattice structure of the crystal. The surfaces of single crystals are the facets having low Miller indexes because atoms in such surfaces are strongly bound to their nearest neighbors in the plane of the facet. The atoms in facets with a high Miller index are less strongly bound. Which particular facets form the outer surface of the crystal depends on the conditions during formation (e.g., the presence of impurities or the nature of the solvent) but also on the crystal structure itself. Perfectly stacked atomic planes do not grow or break up easily during dissolution. Such facets can only grow in modest supersaturation if there are enough surface defects such as screw-dislocations (Figure 6.8).

Planar crystal surfaces that survive during crystal growth are the slowest growing facets; the fastest growing ones disappear entirely in the evolving crystal

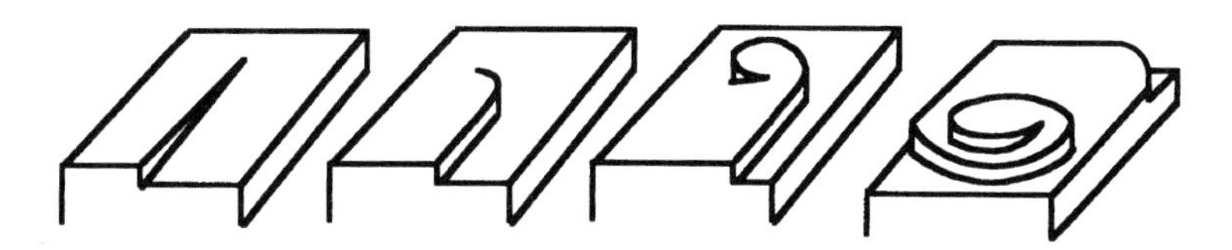

**Figure 6.8.** Growth of a crystal face in a spiral around the epicenter of a line dislocation.

(Figure 6.9). Impurities in the solvent affect the growth rate of different facets and these rates determine the habit or outer shape of the crystal. There have been some efforts to develop an equilibrium model for predicting observed crystal habits from the crystal structure alone. According to that model every crystal face has a roughening temperature that is a function of the strengh and anisotropy of the bonds between the lattice atoms. When the crystal is grown below this critical temperature, the facet exists and is smooth and flat.

When it is formed above the roughening temperature of all facets of the crystal no facet is found in the shape of the crystal. Single crystals such as silicon that have been grown by slow solidification of their melts are not faceted at all because the melting point of the solid at which the crystal is made is above the roughening temperature of all facets. Amorphous solids have no facets because they have no crystal structure and no regular anisotropy in their bonding. The grains in sintered ceramics do not have facets either for the same reason that crystals grown from the melt are unfaceted. The flat grain boundaries that are so often seen in a sintered ceramic are therefore rarely related to crystal planes.

Dendrites form if solidification is limited by diffusion. The dissipation of latent heat can be the rate-determining factor. As heat conduction is better in the liquid than in the solid phase, there is spatial anisotropy in the solidification rate and the solid assumes a ramified shape. Latent heat is more easily disposed of at the end of the dendrites than in depressions on the growing solid surface, and the tops of dendrites grow faster than the depressions in the surface. This positive feedback yields the same type of morphology as diffusion-limited aggregation, which is

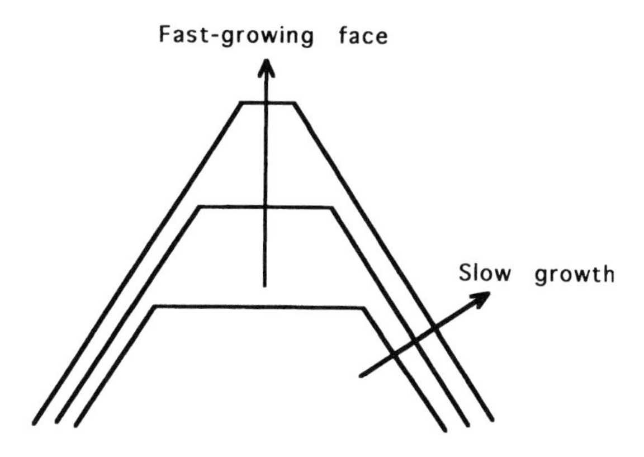


**Figure 6.9.** Growth stages of a crystallite. Slowly growing faces survive, while those that grow rapidly disappear.

discussed in Chapter 7. Eutectic or directed solidification also makes use of heat diffusion limitation to make a composite of anisotropic stacked layers from an eutectic melt, as described in Chapter 8.

## 6.6. Converting Solids by Reaction with a Gaseous Reactant

Conversion coatings are surfaces formed by chemical reaction between a reactive gas and the substrate material as a reactant. In this process not all reactants are gaseous; one of the precursors is the original solid substrate itself. Examples of conversion coatings are nitrided or carbided steel and anodized aluminum; the metal substrate is converted into a metal nitride, carbide, or oxide. *Pack cementizing* is an old form of chemical vapor deposition, developed for chromatizing. Steel objects to be hardened are embedded in an inert powder mixed with some solid reactants (Cr, Si, $NH_4I$) and heated for a while in a closed molybdenum box. The volatile chromium or silicon iodides that are formed at high temperature diffuse through the powder and react at the metal surfaces to form metal chromides or silicides. Another example of conversion is nitrided steel: a steel surface is hardened with iron nitride by reacting steel with nitrogen radicals from a plasma (a glow discharge at low gas pressure).

Other industrial examples of gas–solid reactions are reaction-bonded ceramics. Certain ceramics, such as silicon nitride and titanium nitride, have attractive properties but are difficult to sinter because they have very high melting points. These ceramics can be made by first shaping the solid precursor into a porous green form and then reacting the green form with a gas to convert the original solid into a ceramic. To make a compact of reaction-bonded silicon nitride (RBSN) the ceramic green form is made from silicon powder that is lightly sintered at 1200°C to make a porous green form, which can easily be machined to the right shape.[14] The green form of porous silicon is then converted to silicon nitride in nitrogen at 1250 to 1450°C for several days. As the reaction $3Si + 2N_2(g) \rightarrow Si_3N_4$ is exothermic it is carried out slowly in order to avoid melting and coarsening the silicon. If the silicon particles are too big they are passivated by a surface layer of silicon nitride and stop reacting. As silicon powder has a 3-nm-thick native oxide layer on the surface that has to be removed, the reaction needs an iron or iron oxide catalyst and some hydrogen to proceed. The end product is porous (porosity 15%), the grain size is 1 to 20 $\mu$m, and the reaction is not entirely complete; it contains some unreacted silicon that remains inaccessible for nitrogen. The ceramic is a mixture of crystallites of $\alpha$-$Si_3N_4$ (made by a vapor-phase mechanism involving gaseous Si and SiO) and $\beta$-$Si_3N_4$ (which grows from partially molten silicon with dissolved nitrogen). The reaction temperature determines the relative amounts of the two forms in the end product. Reaction bonding is a "near net shape" process, which means that there is no external size change of the product during the reaction although the reaction from silicon to silicon nitride means a molar volume increase of 22%. The pore volume of the green form decreases during nitridation. The shape of the ceramic is limited to a thickness of at most 1 cm because the accessibility of the interior of the product for nitrogen drops too much with thicker objects.

RBSN is not fully dense and often the product is given an additional HIP or HUP treatment after reaction. The morphology and modification of the product depends again on the processing conditions. Higher sinter temperatures (above 1410°C) result in coarser microstructures of predominantly the $\beta$-modification while a low (less than 1350°C) sinter temperature makes for a high content of fine-grained $\alpha$-$Si_3N_4$. Rates of heating also affect the $\alpha/\beta$ ratio in the ceramic. High nitrogen pressures give $\beta$-RBSN with a small grain size but a relatively high hydrogen partial pressure increases the $\alpha$ content.

Reaction-bonded titanium nitride (RBTN) ceramics are like RBSN made from a porous green shape of titanium powder that is reacted with nitrogen to titanium nitride (TiN) at temperatures up to 1000°C.[15] Here the titanium hardly increases in molar volume when nitrided and the initial porosity remains the same but the gas permeability of a pressed titanium tablet is increased after it has been converted to titanium nitride. If the titanium powder particles are too large, the reaction stops after passivation of the metal surfaces: the TiN formed at the surface is a diffusion barrier that stops the reaction. A fractal powder morphology of the starting metal (such as can be obtained from gas-phase preparation) is a very suitable reactant for complete reaction at modest temperatures.

## 6.7. Chemical Vapor Deposition

Chemical vapor deposition (CVD) is chemical synthesis of solids from gaseous precursors or reactants. The solid product is obtained in the form of a coating or a powder. It is used for synthesis of coatings that modify electrical, optical, and chemical properties of materials or improve their surface properties, such as hardness, wear, and corrosion resistance.[16–21] The combination of metals with ceramic CVD coatings is a particularly successful one. CVD is the technique used for improving the durability of tools and machine parts. It will be discussed here in more detail than other methods of synthesis because it is representative.

Table 6.2 lists properties modified by coatings and from it one can get an idea of the changes in properties that can be achieved with coatings. Thin layers suffice for vastly improved performance and if scarce resources are required for modifying or improving surface properties, they have the biggest effect and function most efficiently when applied as coatings. Applications for CVD coatings are mainly found in electronics, energy technology, mechanical engineering, optics, and telecommunications.

Apart from its use for coatings or films on tools or on silicon wafers for integrated circuit manufacture, CVD is also used in the synthesis of high-quality powders. Large-scale industrial synthesis of oxide powders such as $SiO_2$ or pigments such as $TiO_2$ are best made from gaseous precursors.

Composite powders, i.e., powder particles that are built up of alternating layers of different materials or particles with surfaces of a material other than the one of the interior are occasionally necessary as components in composites. Powder particles with modified surfaces can easily be made by means of CVD in a fluidized bed, which is of considerable interest because the nature of the interface between component phases in composites determines their performance. Whiskers and fibers

Table 6.2. Properties of Materials Affected by Coatings

| | | | |
|---|---|---|---|
| Electrical | Resistivity | Mechanical | Wear |
| | Superconductivity | | Friction |
| | Dielectric constant | | Hardness |
| | Magnetism | | Adhesion |
| | Work function | | Toughness |
| | | | Ductility |
| Optical | Refraction | | Strength |
| | Emissivity | | |
| | Reflectivity | Chemical | Diffusion |
| | Photoconductivity | | Corrosion |
| | Selective absorption | | Oxidation |
| | | | Catalysis |
| Morphology | Grain size | | Electrochemistry |
| | Porosity | | Wetting |
| | Surface area | | |

having the proper surface composition for their use in composites are also made by means of CVD.

CVD is one of the most universal techniques for solid state synthesis: virtually any material (even the most refractory and inert ones) can be synthesized with CVD under unusually mild process conditions.[22] Important solids made by CVD include the elements Si, C, B, W, Al, other refractory metals, and the transition metals. Binary compounds deposited by CVD are:

- The fluorides of Li, Na, and Ca
- Refractory carbides of silicon, boron, and group 4–6 elements
- Borides of group 4–6 transition elements and rare earths
- Silicides of group 4–6 metals, in particular Ti, Mo, W
- Simple oxides of Al, Si, Sn, and group 4–6 metals
- Mixed oxides such as perovskites
- Other chalcogenides such as the 12–16 semiconductors, $TiS_2$, $WS_2$
- Nitrides of B, Al, Si, and the group 4–6 metals
- Other pnictides such as 13–15 semiconductors, $B_xAs_y$, $Zn_2P_3$.

We will see below that a CVD coating deposited under the right conditions has a good step coverage, i.e., the layer thickness is the same everywhere on a rough or profiled surface.

A vacuum technique closely related to CVD is PVD (physical vapor deposition), in which a solid (metal or compound) is evaporated in a vacuum by heating or by a plasma (called sputtering) and condensed on a substrate to form a coating. Often there is no chemical reaction during deposition—hence the name. PVD is a line-of-sight process, and unlike CVD it suffers from shadowing on profiled surfaces if the substrates are stationary with respect to the source.

CVD is comparatively expensive because it is run batchwise and the size of the objects in the reactor is restricted to less than 1 m. There are some exceptions:

considerably longer tubes can be coated internally and large glass panes are covered on-line by CVD with optical coatings.



### 6.7.1. Operational Aspects and Equipment

The schematic of the gas line (in this case for deposition of $TiN/TiB_2$ composite coatings) is shown in more detail in Figure 6.10. Hydrogen is purified by removing traces of oxygen and water with a suitable copper or palladium catalyst and a zeolite, respectively. The purified gas is bubbled through liquid precursors (in this case $TiCl_4$ and $BCl_3$) or passes through a packed bed with a solid but volatile reactant. In the reactor, which may be of the hot-wall or cold-wall type, the gas mixture reacts at high temperatures to form the wanted product (in this case TiN or $TiB_2$ or both) on the substrate and waste gas (HCl), which is removed in the gas stream and neutralized in a scrubber. If necessary a vacuum pump will lower the deposition pressure.

Deposition parameters determine the nature and morphology of the product. By changing gas concentrations, pressure, or temperature during deposition, depth-

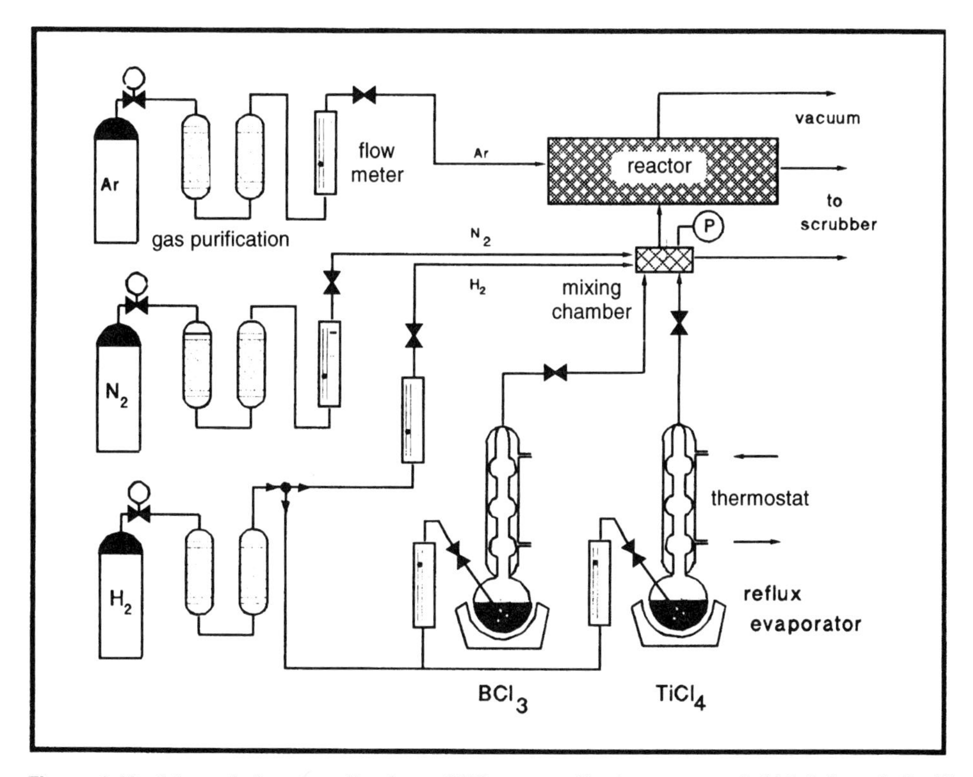


**Figure 6.10.** Schematic for a gas line for a CVD reactor. Carrier gases are bubbled through liquid precursors and the concentration of the saturated vapors reduced by a reflux condenser kept at the temperature of the wanted vapor pressure. Laminar composites are made by changing the reactant gas mixture during the deposition.

graded coatings or multilayer composites are formed in a single process. Thus prior to deposition of a wear-resistant coating of titanium boride on steel, an interface coating of TiN is first deposited on the bare steel, which prevents reaction of the substrate with the aggressive boriding ambient during boride deposition. TiN coatings are excellent diffusion barriers and are used as such on tools as well as in semiconductor technology to prevent diffusion of aluminum from the metallized conducting parts into the silicon part of integrated circuits.

Another possibility is integrated device formation in a single process: alternating electrode–electrolyte layers can be deposited by switching to the proper gas concentration during deposition. CVD processes are being developed for the manufacture of entire thin-film fuel cells or sensors in one process. In graded or layered products, expansion coefficients must match for there to be adhesion as the use temperatures are lower than those during CVD. The development of thermal stresses and the concomitant adhesion problems can be prevented by paying due attention to the development of proper interface layers between the different components during deposition.

### 6.7.2. Physical Chemistry of Chemical Vapor Deposition

The growth rate and morphology of a deposit are determined by the gas flow rate, the diffusion rate in the gas phase, the reaction rate at the surface of the substrate, and thus by the process parameters pressure, temperature, and gas-phase composition. This dependence is shown below for CVD but similar behavior is also observed in deposition techniques where solids are made from precursors in a fluid phase.

The oldest description of CVD uses the boundary-layer model, which assumes that between the bulk gas phase (uniform in composition) and the substrate there is a stagnant boundary layer in which gradients develop in temperature (cold wall reactors) and in the partial pressures of the reactants and the gaseous reaction products. The boundary-layer model presents some difficulties. Stagnant gas layers have a variable thickness in the reactor as long as unmodified bulk gas concentrations exist, if they exist at all. Moreover the flow is always laminar. Conceptually, however, the model has its advantages.The boundary-layer thickness is an effective parameter by which to characterize the deposition regime. This model will be used here for a simple overview of reaction conditions.

Figure 6.11 shows the eight groups of elementary steps in the CVD process. (1) and (8) bulk reactant gas flow into the reactor and flow of excess reactants and gaseous reaction product out of the reactor; (2) reactions between precursors in the gas phase to form either other reactive gases or solid products (homogeneous nucleation); (3) and (7) diffusion of reactants and gaseous waste products through the boundary layer (Stefan flow); (4) and (6) adsorption of reactants on the surface and desorption of product gas molecules; and (5) diffusion and chemical reaction of surface species to form the product, nucleation and crystallite growth, and solid state reactions in the layer and between layer and original substrate.

CVD comprises many steps, both reactive and diffusive. Rates are predominantly determined by the slowest reaction step in a series of consecutive reactions or by the fastest step of a set of parallel reactions.[23,24] Process conditions can

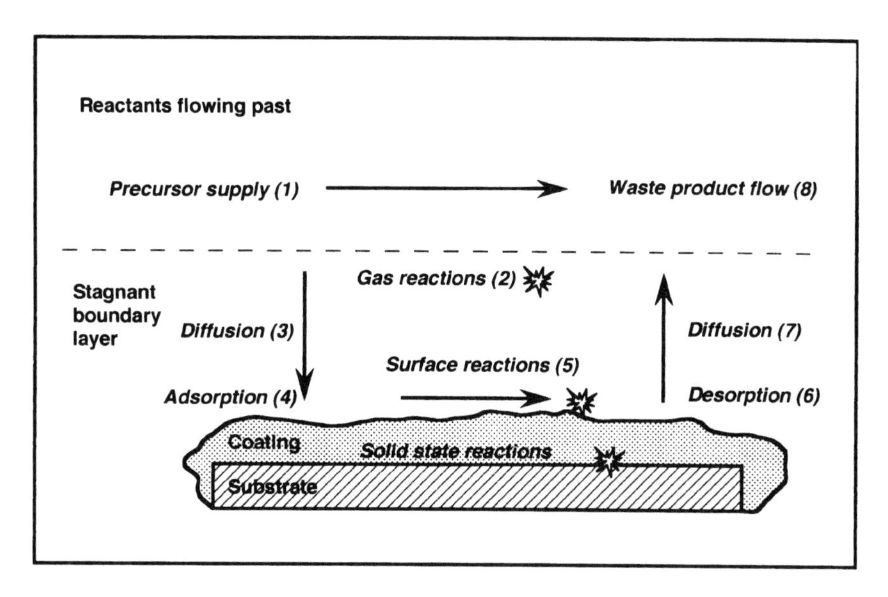


**Figure 6.11.** The processes that occur during CVD, gas supply flow, diffusion, and reaction of precursors. Chemical reactions are indicated with an asterisk.

be grouped in three regimes, the gas-flow-limited, the diffusion-limited, and the reaction-limited regimes, depending on which process determines the overall deposition rate. High deposition rates and fast reactions occur in the two transport-limited regimes and low reaction rates prevail in the reaction-limited regime. The consequences of operating in these regimes are discussed below.

Transport limitation is usually supposed to mean that the diffusion coefficient [in step (3) or (7)] is low compared to the slowest surface reaction step; deposition is then said to be in the diffusion-limited regime because the diffusion determines the growth rate. If reactant gas supply [(1) or (8)] is the slowest step (feed-rate limitation) and surface reaction rates and diffusion are much faster, deposition is sometimes said to be in the thermodynamic regime because mobility is high and equilibrium supposed to be more likely than in the other regimes. Since reactant gas flow is another form of transport some authors regard this regime (here called the pyrolytic regime) as being also transport-limited. The two transport-limited conditions (flow rate and diffusion limitation) have similar effects on the Arrhenius plot (log growth rate vs. $1/T$): they flatten the curve.

Reaction limitation obtains if the rate coefficients of the surface reactions [(4), (5), (6)] are very low compared to the transport steps. Chemical reactions are thermally activated and rates of simple reactions follow Arrhenius behavior. Thus a plot of the logarithm of the deposition rate vs. $1/T$ is a straight line in the temperature range corresponding to the reaction-limited regime. The slope of this line is determined by the adsorption heats of the reactants and the activation energy of the rate-determining surface reaction step.

How the regimes are identified in an Arrhenius plot is indicated in Figure 6.12 for deposition of silicon using various precursors having different reactivities (but similar slopes in the reaction-controlled regime). The reaction-limited (or catalytic)

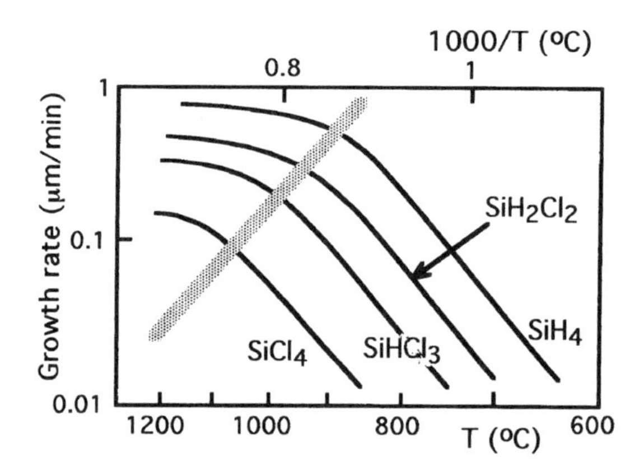


**Figure 6.12.** Arrhenius plot for silicon deposition from several precursors. The shaded area indicates a change from the catalytic to the dialytic regime on increasing temperature.

regime is recognized by a straight line in the Arrhenius plot and the onset of a diffusion-limited (or dialytic) regime is indicated by a flattening-out of the curves in this plot. The boundary temperature between the regimes depends not only on the nature of the precursors but also, and strongly so, on the pressure that affects mean free paths of molecules in the gas phase and hence the diffusion rates.

Deposition rates, deposit morphology, and the possibilities for simultaneous codeposition of several phases in composite coatings vary considerably with the process parameters and therefore with the deposition regime. However, codeposition of different phases from gases remains difficult. Usually codeposition is possible only under brute force conditions, where reaction rates are high and do not depend on the nature of the substrate.

Physical modeling of CVD processes means solving the Navier–Stokes equations, partial differential equations for mass and heat transport in fluids given the constant boundary conditions of the reactor. These processes affect the uniformity of the deposit in all parts of the reactor. The proper name for CVD reactor physics is chemical vapor technology (CVT) and it is a subject of some significance for industrial reactor design. (This branch of continuum physics is outside the scope of this book, which is concerned with materials rather than machinery.)

Chemical modeling of gas-phase and surface reactions, nucleation, and growth is based on concepts developed for surface reactions on heterogeneous catalysts. In order to understand the reaction rates and the microstructure of the products observed during CVD, the driving forces for the reaction, the kinetics in the pyrolytic regime, powder formation at high supersaturation, and the uses of thermodynamic calculations have to be discussed briefly.

### *6.7.3. The Pyrolytic Regime*

If diffusion and chemical reaction rate coefficients are high with respect to the flow of the bulk reactant gas supply (similar to the situation of a well-stirred reactor in which the contents are in thermodynamic equilibrium), so-called CVD diagrams or stability diagrams can be used to predict the process conditions necessary for desired reaction products. These diagrams, which are produced from the published thermodynamic data of the reactants and reaction products, indicate the conditions under which certain simple solid products are in equilibrium with the gas phase given an initial input reactant gas mixture. These process conditions represent the pyrolytic regime.

The pyrolytic regime is distinguished from the (diffusion-limited) dialytic regime by the effect of the input gas flow rate on the deposition rate. Both transport-controlled regimes occur at high temperatures, where the slopes of the Arrhenius plot decrease (Figure 6.12). In the case of small equilibrium constants, the slope of the Arrhenius plot in the pyrolytic regime equals the reaction heat, as will be shown below.

Thermodynamic calculations allow estimates of the species in the gas and solid phases that are in equilibrium at the conditions specified. This is useful for determining the choice of process conditions necessary for ensuring a positive driving force for the reaction.

For the calculation of thermodynamic equilibria the Gibbs energy of the system is minimized at given temperature and pressure as a function of the composition of all the reactants and reaction intermediates. Boundary conditions are the amounts of the reactants that enter the reactor. Also the stoichiometric coefficients of the reactions are constraints in the minimization. Commercial programs for routine thermodynamic calculations are SOLGASMIX, CHEMSAGE, or HSC.

Figure 6.13 shows the results of an equilibrium calculation for a combined deposition of SiC and $HfB_2$. It prescribes the partial pressures and temperatures necessary for having the indicated solids in equilibrium with the (reacted) gas mixture of initial composition indicated in the graph. If thermodynamic equilibria were the main factor determining deposition, CVD diagrams would give the necessary deposition parameters. However, SiC is a semiconductor and $HfB_2$ is a metallic conductor and they have different surface chemistries and surface reaction rates for growth and nucleation. Although these thermodynamic calculations were published many years ago, the results have not yet been experimentally shown to be usefully related to what actually occurs in the reactor under given conditions.

One case in which equilibrium calculations have been compared to experimental deposition observations is SiC/TiC codeposition. Figure 6.14 shows a calculated CVD diagram for deposition of mixed silicon and titanium carbides by reduction of a gas mixture of silicon and titanium tetrachloride precursors with hydrogen ($CCl_4$ was the carbon source). The solid phases that are actually found experimentally under these conditions are given in the second CVD diagram in Figure 6.14. Codeposition of SiC and TiC is predicted to be thermodynamically allowed only at temperatures below 1400°C, but it is actually found experimentally only above 1530°C. This example also illustrates the point that even at high reaction temperatures equilibrium estimates are of doubtful reliability as a predictive tool in CVD. Although knowledge of driving forces might seem very useful at first sight, chemical resistances and kinetics dominate CVD processes far more than driving forces, even if the system is likely to be near equilibrium.

The deposition rate in the pyrolytic regime is proportional to the equilibrium constant $K$ for small values of $K$; for large $K$ the reaction rate does not depend on $K$, but only on the mass flow rate. This will be shown for the case of a simple fictitious pyrolysis equilibrium: $AB(g) \rightleftharpoons A(s) + B(g)$. The equilibrium constant is

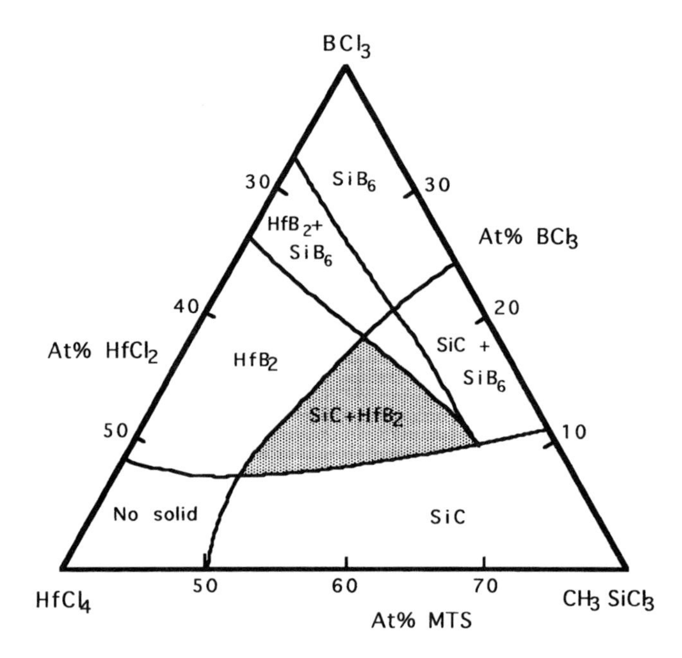


**Figure 6.13.** The calculated stability diagram at 1125°C of a mixture of hafnium chloride, boron chloride, and methyl trichlorosilane. The solids expected to be in equilibrium with a gas of the stated composition are noted in the different concentration ranges. The composition of the gas mixture at which both silicon carbide and hafnium boride are stable is shaded. From W. J. Lackey, A. W. Smith, D. M. Dillard and D. J. Twait. *Proc. 10th Intern. Conf. CVD*, Honolulu (1987), p. 1008. Reproduced by permission of The Electrochemical Society, Inc.

(considering [A] = [B]):

$$K = \frac{[B]}{[AB]} = \frac{[A]}{[AB]_0 - [A]}$$

Hence,

$$[A] = [AB]_0 \frac{K}{1 + K}$$

Here the concentrations are those at equilibrium and $[AB]_0$ is the initial concentration of the precursor AB introduced into the reactor.

If the time derivative is taken, the expression for the deposition rate in this regime becomes

$$\frac{d[A]}{dt} = \frac{d[AB]_0}{dt}\left\{\frac{K}{1 + K}\right\}$$

Figure 6.15 shows the plot of the reaction rate for this case and for the bimolecular case of two reactants. The reaction rate is proportional to the gas input flow and to the equilibrium constant $K$ if $K \ll 1$. In CVD equilibrium constants are usually much larger than that and therefore rates do not depend on $K$. Although shown here for the case of CVD the lack of dependence of the rates on driving forces (if these are large) holds quite generally in chemistry.[25]

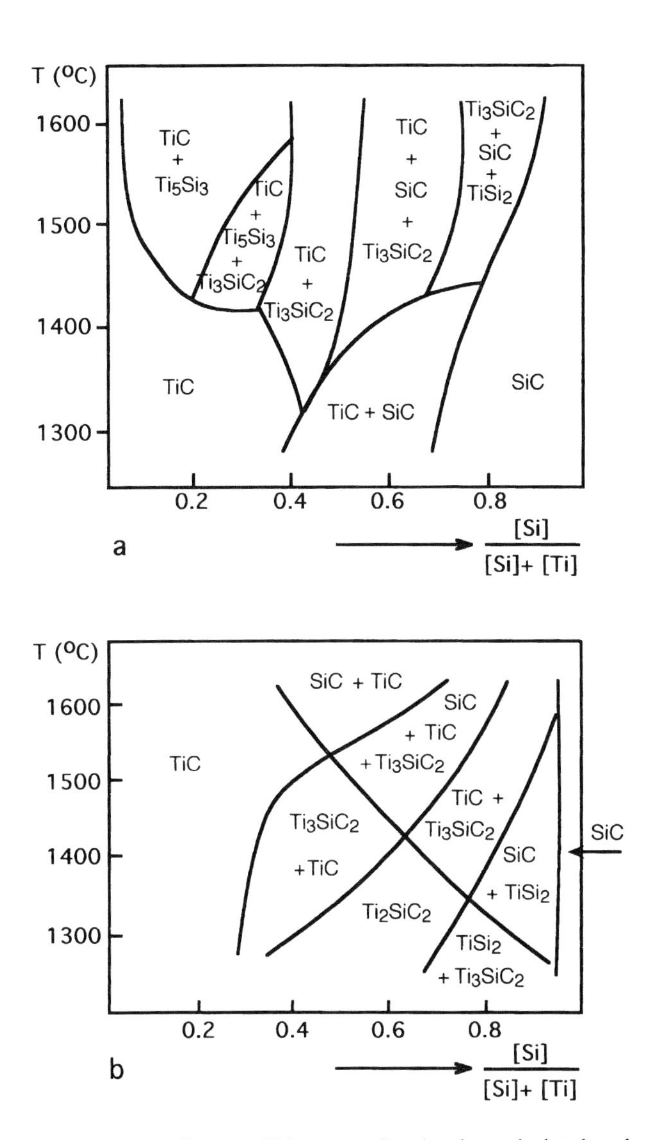


**Figure 6.14.** Stability diagrams for a carbide composite showing calculated and actually observed depositions at the same conditions. (a) Calculated stability diagram. The indicated compounds are expected to be in equilibrium with a gas mixture of the tetrachlorides of silicon and titanium (with hydrogen and methane) at different concentration ratios. A composite of SiC and TiC is predicted to be stable only below 1440°C. (b) The solids deposited under different reaction conditions (the experimentally observed stability diagram). A composite of SiC and TiC is formed only at temperatures over 1550°C. From T. Goto and T. Hirai. *Proc. of the 10th International Conference CVD* (1987), p. 1070. Reproduced by permission of the Electrochemical Society, Inc.

Codeposition in CVD means deposition of several solid phases simultaneously from the same precursor gas mixture. Making composite coatings by codeposition requires relatively high temperatures, i.e., working in the pyrolytic regime. The reason is that the deposition rates of the phases must be high and approximately comparable and must not depend on the substrate too much. This last condition is

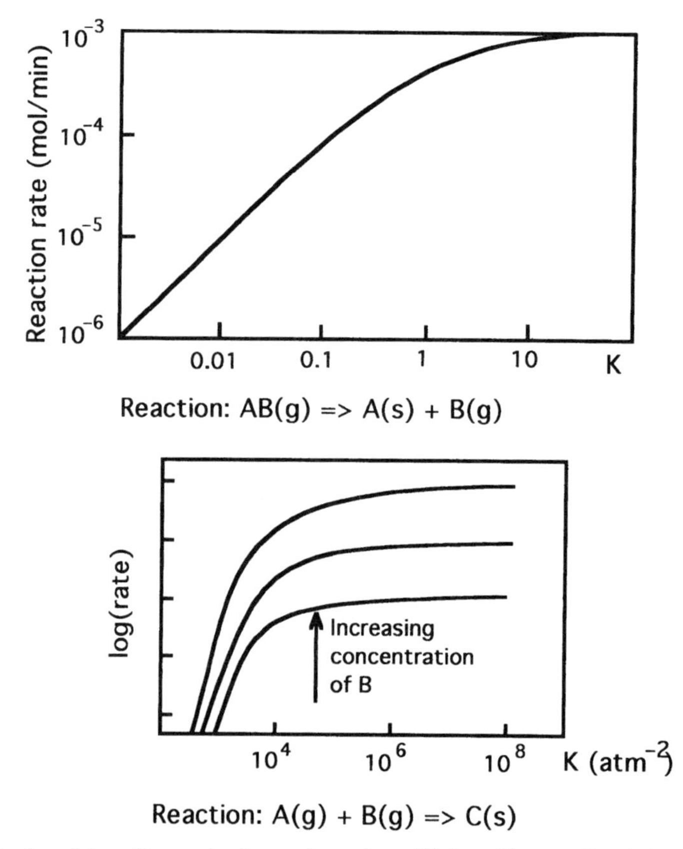


**Figure 6.15.** Kinetics of deposition under thermodynamic equilibrium. The growth rate is proportional to the equilibrium constant $K$ when $K$ is small but does not depend on $K$ when $K$ is large.

most likely to occur in the pyrolytic regime. When very reactive precursors are used, e.g., most metal–organic reactants, this regime will start at lower temperatures than with more stable precursors.

Thermodynamic calculations that predict negative $\Delta G$ values (decrease of Gibbs free energy) for the reaction do not guarantee that the reaction is possible at the conditions considered. Nor do positive $\Delta G$ values imply that CVD is impossible. Plasmas may act catalytically and may also shift equilibria. The thermodynamic driving forces for deposition of TiC and TiN are shown in Figure 6.16. Threshold temperatures can be lowered by as much as several hundreds of degrees centigrade by a weak glow discharge. The plasma shifts the equilibria in this case. An even more dramatic case of plasmas countering thermodynamic restrictions is the deposition of TiN and AlN using only nitrogen, which is thermodynamically impossible at all accessible temperatures, as is also shown in Figure 6.16. However, a glow discharge permits deposition above temperatures similar to those in the presence of hydrogen. The catalytic action of plasmas is suggested by the data presented in Table 6.3, which lists a few reactions that are thermodynamically "allowed" (have a negative $\Delta G$) but which will not occur without plasmas.

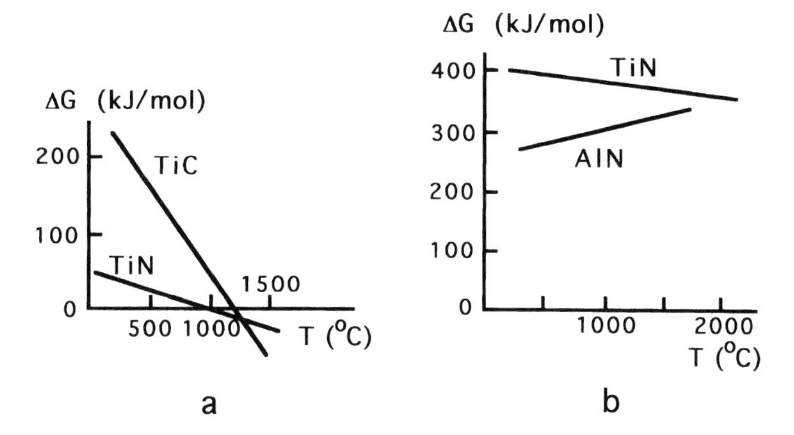


**Figure 6.16.** Gibbs energies of a few CVD reactions. Deposition occurs if $\Delta G$ is negative: (a) Nitrides are made from titanium chloride, and methane or nitrogen, with hydrogen at high temperatures as indicated. In a glow discharge, TiC and TiN form in reactions with hydrogen already present at 430 and 250°C, respectively. (b) Nitrides made from aluminum and titanium chlorides and nitrogen without hydrogen. Although $\Delta G$ is very positive AlN and TiN can be deposited from this gas mixture in a glow discharge at substrate temperatures of 530 and 850°C, respectively.

**Table 6.3. CVD Reactions that Despite Sufficient Driving Force Do Not Occur Thermally at 230°C[a]**

| Reaction | $\Delta G$ (kJ/mole metal) |
|---|---|
| $SiCl_4 + O_2 \rightarrow SiO_2(s) + Cl_2$ | −77 |
| $TiCl_4 + O_2 \rightarrow SiO_2(s) + Cl_2$ | −150 |
| $AlCl_3 + \frac{3}{4}O_2 \rightarrow \frac{1}{2}Al_2O_3(s) + \frac{3}{2}Cl_2$ | −200 |
| $\frac{1}{2}B_2H_6 \rightarrow B(s) + \frac{3}{2}H_2$ | −64 |
| $\frac{1}{2}B_2H_6 + NH_3 \rightarrow BN(s) + 3H_2$ | −294 |
| $SiH_4 + \frac{4}{3}NH_3 \rightarrow \frac{1}{3}Si_3N_4(s) + 3H_2$ | −272 |

[a] In a weak glow discharge the formation rate is up to 0.5 μm/h at this temperature.

## 6.7.4. Powder Synthesis

Very high supersaturations (defined below) lead to homogeneous nucleation, i.e., solid powder formation in the gas phase. Usually CVD is applied to make coatings by heterogeneous nucleations and growth on surfaces. If the supersaturation is sufficiently high, the deposition reaction takes place in the gas phase (homogeneous nucleation) and an aerosol or powder forms instead of a coating on a substrate. This phenomenon is made use of in the synthesis of ultrafine, monodisperse ceramic powders, and for ultrarapid deposition of thick films and monoliths from the gas phase.

The chemical supersaturation $S$ of a reaction: $aA(g) + bB(g) \rightleftharpoons cC(s) + dD(g)$ could be defined as:

$$S = \frac{([A]^a[B]^b/[D]^d)_{reactor}}{([A]^a[B]^b/[D]^d)_{equil.}}$$

where $S$ is the ratio of two reaction quotients. The numerator is the reaction quotient determined by the gas concentrations actually present in the reactor. The denominator is the reaction quotient expected for thermodynamic equilibrium and equals the reciprocal of the equilibrium constant $K$ for the reaction. The supersaturation expressed in the actual partial pressures of the reactants and the equilibrium constant is:

$$S = K\frac{[A]^a[B]^b}{[D]^d}$$

In this equation the concentrations are not the equilibrium values but the actual values in the reactor. This definition corresponds to that of the supersaturation used with physical condensation. Physical supersaturation is a special case of chemical supersaturation as defined above: if the condensation "reaction" is $A(g) \rightarrow C(s)$, the definition given above reduces to: $S = [A]_{reactor}/[A]_{equil.}$, which defines physical supersaturation. There is saturation when $S = 1$, and $p(A)$ can be used instead of [A].

Homogeneous nucleation is used in powder synthesis of ultrafine particles, in particular for covalent ceramics ($Si_3N_4$), in which the particles, owing to their size, have a high sinter activity. A disadvantage of the technique is particle agglomeration, which precludes achieving sufficiently high packing densities in the green form. To prevent such agglomeration the particles can be deposited as soon as they are formed. This is done in PP-CVD (particle precipitation CVD), which combines powder synthesis with ceramic-forming in one process. This eliminates the necessity to isolate the powder and make a green form of it, as in conventional ceramics processing.

Homogeneous chemical nucleation is similar both to stepwise polymerization reactions in fluid media and to spinodal decomposition in mixtures. The characteristic differences with a binodal-type nucleation reaction are a strong sensitivity to the presence of impurities, the absence of an activation barrier and a concentration threshold, and no incubation time.

Physical condensation models for molecular liquids based on the concepts of vapor pressure and surface tension have in the past been adapted to chemical nucleation but were not very successful in predicting nucleation rates from compound properties. The critical nucleus for chemical reactions is too small and so this is not a useful concept. Moreover surface energies are not defined for atoms, nor are surface areas defined for solids, and equilibrium thermodynamics only applies (if at all) to systems having a sufficiently large number of units, not on the scale at which nucleation reactions occur. However, homogeneous nucleation rates are observed to correlate roughly with chemical supersaturation. Table 6.4 lists CVD reactions with their equilibrium constants leading to powder deposition as well as examples of reactions in which powders will not form but only coatings.

**Table 6.4. Equilibrium Constants and the Possibilities of Synthesizing Powders of Nitrides and Carbides from Gaseous Reactants**[a]

| Reactants | Product | log $K$ (1500°C) | Powders |
|---|---|---|---|
| $SiCl_4$, $N_2$, $H_2$ | $Si_3N_4$ | 1.4 | − |
| $SiCl_4$, $NH_3$ | $Si_3N_4$ | 7.5 | + |
| $SiH_4$, $NH_3$ | $Si_3N_4$ | 13.5 | + |
| $SiCl_4$, $CH_4$ | SiC | 4.7 | −(pl+) |
| $CH_3SiCl_3$ | SiC | 6.3 | −(pl+) |
| $(CH_3)_4Si$ | SiC | 10.8 | + |
| $TiCl_4$, $H_2$, $N_2$ | TiN | 1.2 | − |
| $TiCl_4$, $H_2$, $NH_3$ | TiN | 5.8 | + |
| $TiCl_4$, $CH_4$ | TiC | 4.1 | −(pl+) |
| $TiI_4$, $CH_4$ | TiC | 4.2 | + |
| $TiI_4$, $C_2H_4$, $H_2$ | TiC | 3.8 | + |

[a](+) observed powder formation, (−) no powders formed, (pl+) powder formation in plasma.[26]

a

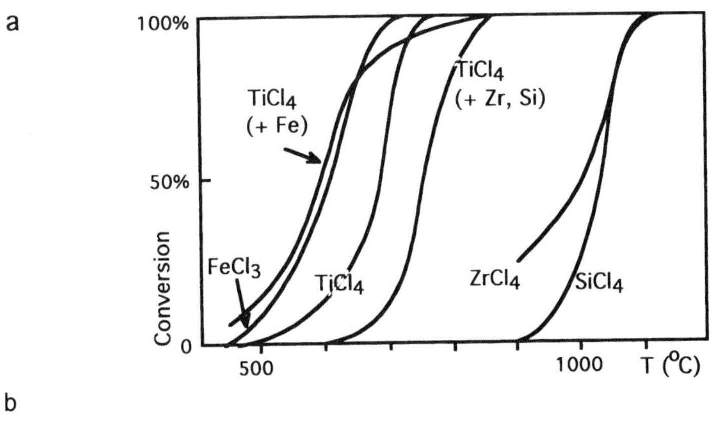


b

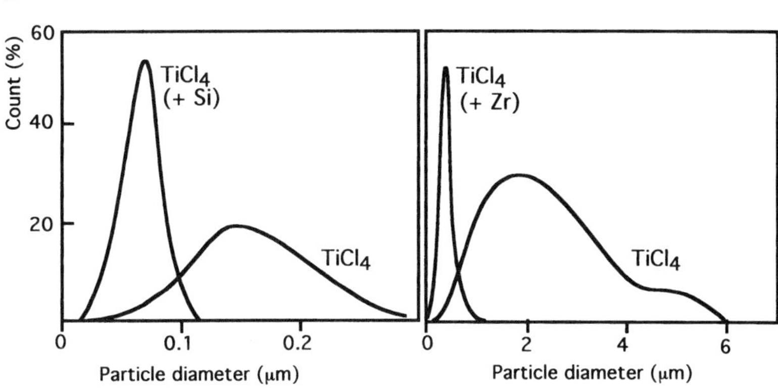


**Figure 6.17.** Powder formation in the gas phase by oxidation of gaseous metal halides at high temperature and the effect of impurities on the reaction: (a) Reactions of chlorides with oxygen to powders of the oxides. Impurities (in parentheses as chlorides) may accelerate or inhibit titania formation. (b) Particles size distributions of the titania products of the oxidation reactions. Impurities also affect the size distributions.[26]

Very small nuclei that might form in the gas phase can deposit on the substrate and contribute to heterogeneous deposition and accelerate growth. This accounts for blanket deposition (irrespective of substrate) at high supersaturation and also explains the morphological features of high-temperature deposition discussed below.

Homogeneous nucleation rates can be controlled by gaseous additives. Particle morphologies, particle size, and size distributions as well as temperature thresholds for nucleation may vary considerably with small amounts of nucleation inhibitors or nucleation boosters as is shown in Figure 6.17 for the case of $TiO_2$ formation by combustion of $TiCl_4$.[26]

### 6.7.5. The Catalytic Regime

If transport rates (bulk gas flow and diffusion coefficients) are high and the surface reaction rate coefficients comparatively low, the chemical reaction on the surface determines the overall kinetics of the process. The temperature dependence of a single reaction step is of the Arrhenius type. Increasing the temperature will increase the surface reaction rate exponentially whereas the diffusion rate increases according to a (low) power law with temperature. At sufficiently high temperatures the surface reaction rate is so high that diffusion takes over as the rate-limiting step and the process changes regime to diffusion or flow limitation. The temperature at which this occurs depends on the pressure. Lower pressures improve diffusion and lower temperatures decrease surface reaction. The reaction-limited regime is found at low temperatures and low pressures. A dimensionless number characterizing regimes will be given below.

Owing to the fact that under the conditions of the catalytic regime diffusion is easy compared to the surface reaction, reactant molecules can penetrate far into pores and are able to react there. The result is that the coating on a profiled substrate is uniform and the deposit has an excellent step coverage. Complicated forms and tubes can be covered with a uniform film in this regime even internally in the pores.

Chemical vapor infiltration (CVI) is a CVD variant capable of internally coating porous objects, e.g., an object made out of carbon fibers, with a ceramic material. Silicon carbide (SiC) or boron carbide ($B_4C$) are examples of ceramic matrix materials that are used in combination with carbon fibers. Strong, light, durable, wear-resistant, and biocompatible joint prostheses made of ceramic–ceramic composites are manufactured by means of CVI. Figure 6.18 shows how the degree of penetration is affected by temperature and pressure. Clearly, to get deposit deep in the interior of the porous object low temperatures are necessary for reaction limitation and low pressures for helping the diffusion. Under these conditions growth rates are low.

If information on the surface reaction is to be collected, for instance to study the reaction mechanism, the process parameters must be chosen in such a way that the deposition is in the catalytic regime. In this regime only the observed growth rate is determined by the surface reaction. In the other two regimes the growth rates do not supply information on chemical reactions since rates are determined by mass transport.

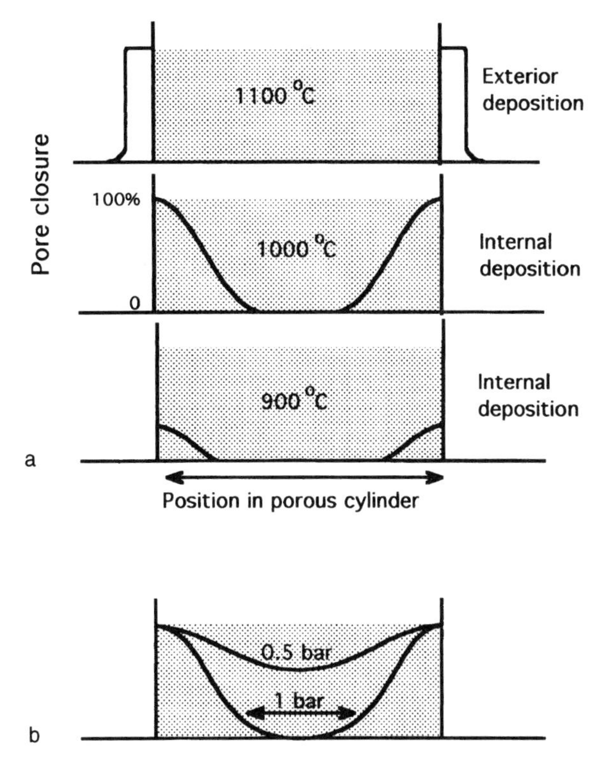


**Figure 6.18.** CVI of silicon carbide from methyl trichlorosilane and hydrogen in a porous substrate. The curves indicate the degree of pore filling after a deposition period (the same period in all cases): (a) Effect of temperature on the amount deposited by CVI in a porous cylinder. (b) Effect of pressure.

### *Determination of Reaction Kinetics with a Thermobalance*

Growth rates are determined by *in situ* measuring the weight increase of the substrate during deposition by means of a thermobalance from which a substrate is suspended in a vertical tubular reactor. Temperatures and all reactant partial pressures are varied in as wide a range as the catalytic regime allows during growth rate measurements. The results are presented: (a) in Arrhenius graphs: log growth rate vs. $1/T$; and (b) in reactant order graphs: log growth rate vs. log reactant partial pressure. The slopes in these graphs contain activation energies (combined with adsorption heats) and reaction orders, respectively.

Determination of reaction kinetics alone is not enough to get a reasonably complete picture of surface processes during deposition. Often the gaseous reactants and products are identified by mass spectroscopy and time-dependent concentrations are measured. Adsorbed species are identified by *in situ* spectroscopy (infrared, Raman, or laser-induced fluorescence). The morphology of the product as studied by electron microscopy also contributes to an understanding of CVD reaction mechanisms.

Models for surface reactions during CVD are based on those used in heterogeneous catalysis and are usually related to the Langmuir–Hinshelwood model,[11] which is illustrated by the following simplified fictitious case of a bimolecular reaction based on competitive monolayer occupation of surface sites by two adsorbed reactants. Adsorption and desorption steps are usually assumed to be in equilibrium, and there is assumed to be one surface reaction step which is very slow (hence irreversible), which determines the reaction rate.

Let A($g$) and B($g$) be reactant gas molecules that adsorb (and desorb) rapidly on the surface (surface species are labeled *A and *B). They react with free surface sites*:

$$A(g) + * \underset{k_d}{\overset{k_a}{\rightleftharpoons}} *A$$

in which $k_a$ and $k_d$ are rate coefficients for adsorption and desorption and $K_A = k_a/k_d$ is the adsorption equilibrium constant for A($g$). On the surface, these adsorbed species diffuse freely and on meeting each other react slowly (*A + *B → X) to form the substrate material X. If $\theta_A$ and $\theta_B$ represent the fraction of surface sites occupied by *A and *B, respectively (with a total of $N$ sites), the reaction rate $r$ is proportional to the concentrations of the reactive surface species:

$$r = k\theta_A\theta_B$$

This equation implies that the surface is assumed to be a well-stirred reactor: the diffusion rate is much higher than the reaction rate. The occupied site concentrations $\theta$ can be expressed in the partial reactant gas pressures and adsorption equilibrium constants.

In the stationary state the adsorption rate equals the desorption rate, or

$$k_a(1 - \theta_A - \theta_B)Np(A) = \theta_A N k_d$$

Hence $\theta_A = K_A p(A)\theta_0$, where $\theta_0 = 1 - \theta_A - \theta_B$ and $p(A)$ is the partial pressure of A($g$). Similarly $\theta_B = K_B p(B)\theta_0$. From these equations it follows that

$$\vartheta_A = \frac{K_A p(A)}{1 + K_A p(A) + K_B p(B)}$$

and a similar equation for B. These surface concentrations for $\theta_A$ and $\theta_B$ lead to the reaction rate

$$r = k\frac{K_A p(A) K_B p(B)}{\{1 + K_A p(A) + K_B p(B)\}^2}$$

Figure 6.19 shows the partial-pressure-dependent reaction rates for a unimolecular reaction (to be derived as an exercise), for bimolecular reaction (derivation given above), and for a Langmuir–Rideal mechanism (explained below).

There are two limiting cases. All three reactions described above are first-order in the reactant A at low reactant partial pressures. If one of the reactants is in excess, the order becomes zero except in the bimolecular case. The bimolecular process has a negative order in the excess reactant, which is interpreted as being due to surface "poisoning" by that reactant.

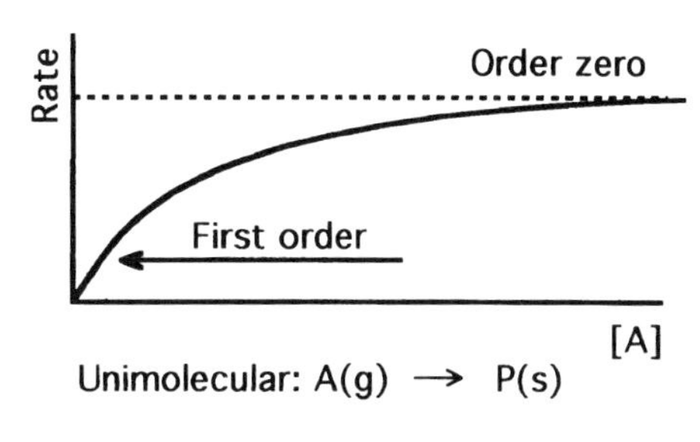


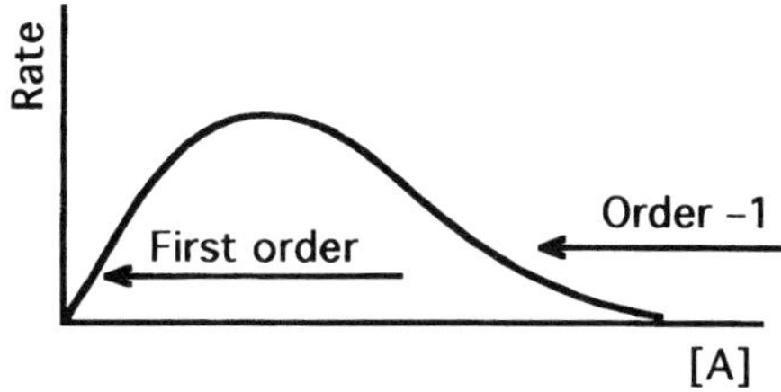


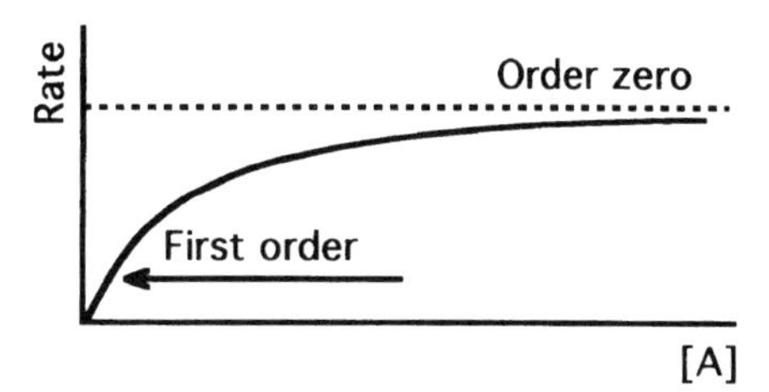


**Figure 6.19.** Reaction kinetics according to Langmuir-type models.

In the Eley–Rideal mechanism the surface reaction is between an adsorbed species (*A) and a gas-phase molecule B($g$): *A + B($g$) → $X$. The reaction rate is $r = kp(\mathrm{B})\theta_{\mathrm{A}}$. Substituting the value for $\theta_{\mathrm{A}}$ yields the rate

$$r = k\frac{K_{\mathrm{A}}p(\mathrm{A})p(\mathrm{B})}{1 + K_{\mathrm{A}}p(\mathrm{A})}$$

which turns out to be rather similar to the unimolecular growth rate.

In the Langmuir–Rideal mechanism (also shown in Figure 6.19), a molecule from the gas phase reacts with an adsorbed species although both reactant molecules are adsorbed and competitively occupy surface sites. In this model there is no surface reaction between adsorbed species.

The validity of the basic assumptions of the Langmuir-type models (which have been developed for heterogeneous catalysis) for CVD has been questioned. The Langmuir model is too restrictive and often cannot cover the whole parameter range in one mechanism. In a number of CVD mechanisms more than one irreversible step or gas-phase reaction had to be invoked in order to account satisfactorily for the observed kinetics in the entire parameter range studied.

CVD differs from heterogeneous catalysis (organic reactions on inorganic surfaces) in at least one important aspect: in CVD the inorganic surface is not only the catalyst (or reactant) for the reaction but also the reaction product. This feature leads to such phenomena as autocatalysis,[27] oscillations,[28] nonlinearities, and selective deposition, and these aspects are not implicit in the simple treatment given above. Another difference between CVD and catalysis is the strength of surface bond and the mobility of the surface species. For these reasons the Langmuir–Hinstelwood model is often combined with gas phase reactions in CVD.

Reports in the literature on experimentally determined activation energies (i.e., Arrhenius slopes) vary wildly from authors to author, even if the energies are measured under similar conditions; e.g., for the formation of TiN, activation energies have been reported that range from 25 to 309 kJ/mole. In this case the reaction order changes from positive to negative with increasing concentration of one reactant, similar to the bimolecular case described above. Such behavior combined with a certain unnoticed reactant depletion can easily yield slopes that are either too high or too low.[29] At low concentrations (positive order) depletion leads to a lower apparent slope, whereas at high concentrations (negative order) depletion increases the measured slope.

### *Kinetics of Titanium Nitride Deposition*

Chemical modeling procedures can be illustrated by the following analysis of TiN formation.[29] Deposition of TiN from $TiCl_4$, $N_2$, and $H_2$ at 800–1000°C shows Arrhenius behavior. The three reactants have different orders:

- The order in nitrogen is $+\frac{1}{2}$ under all conditions studied.
- The order in hydrogen varies between $+\frac{1}{2}$ to $+1\frac{1}{2}$.
- The order in $TiCl_4$ changes with increasing concentration from positive to negative values.

A mechanism consistent with these observations comprises the following steps:

- Reduction of $TiCl_4$ to $TiCl_3$ in the gas phase as well as on the surface (the latter is dissociative chemisorption of $TiCl_4$).
- Dissociative chemisorption of $N_2$ and $H_2$.
- Adsorption of $TiCl_4$ (dissociative) and $TiCl_3$.
- Formation of HCl: $*H + *Cl \rightarrow HCl(g)$.
- Intermediate surface reaction: $*N + *H \rightarrow *NH$.
- Rate-determining surface reaction: $*NH + *TiCl_3 \rightarrow HCl(g) + 2*Cl + TiN$.

The rate expression derived from this model has equilibrium and rate constants determined by being fitted to measured kinetic data, and the model describes the experiments well in the entire parameter range studied.

*Surface Chemistry of Nucleation and Growth*

The three types of solids, metals, covalent semiconductors or insulators, and ionic compounds (including oxides) have characteristic surface reactions. In organic catalysis only metals and ionics are considered (Table 6.5), while in CVD all three types of solid surfaces are of interest. Metals absorb hydrogen and nitrogen dissociatively while ionic substrates have redox reactions or acid/base reactions with molecules. Oxidation of gases is often catalyzed by the surface of metal oxides. So is deposition of oxides by oxidation and hydrolysis of metal-containing precursors. When mixed oxides (e.g., perovskites) are deposited care must be taken to ensure a sufficient availability of the separate components.

While metallic solids are deposited by reactions that involve metallic intermediates and ionic solids result from ionic reactions, the solids with covalent bonds grow by means of radical surface reactions. Examples of such materials are diamond, amorphous diamond-like carbon, silicon, and silicon carbide. Diamond and diamond-like carbon can be deposited if hydrocarbon and hydrogen radicals are available at the growing surface. Silicon carbide and boron nitride growth has also been modeled in terms of radical reactions at the surface.

Isonucleation (or homeonucleation) is the name given to heterogeneous nucleation on a surface of the same compound while allonucleation means nucleation of a solid on the surface of another material. The rates of both may vary considerably if the substrates have different chemistries. An incubation time (time delay before the onset of layer growth) indicates slow allonucleation and faster isonucleation or growth. An example of this effect is deposition of titanium nitride on alumina. Conversely, faster allonucleation than growth means a fast initial reaction rate followed by slow deposition (Figure 6.20), e.g., in tungsten deposition on silicon, where in the initial allonucleation silicon is the reducing agent instead of hydrogen.

**Table 6.5. Types of Reactions on the Surface of Heterogeneous Inorganic Catalysts**

| Types of surface | Examples of catalysts | Surface reactions |
|---|---|---|
| Metals | Fe, Ni, Ag, Pd, Pt | Hydrogenation<br>Dehydrogenation<br>Hydrogenolysis<br>Oxidation |
| Semiconductors | NiO, ZnO, $MnO_2$<br>$BiO_3$–$MoO_3$, $WS_2$<br>$Cr_2O_3$ | Dehydrogenation<br>Oxidation<br>Desulfuration<br>Hydrogenation |
| Insulators | MgO, $SiO_2$, $Al_2O_3$ | Dehydration |
| Acids | $SiO_2$–$Al_2O_3$<br>Zeolites | Polymerization<br>Cracking<br>Alkylation |

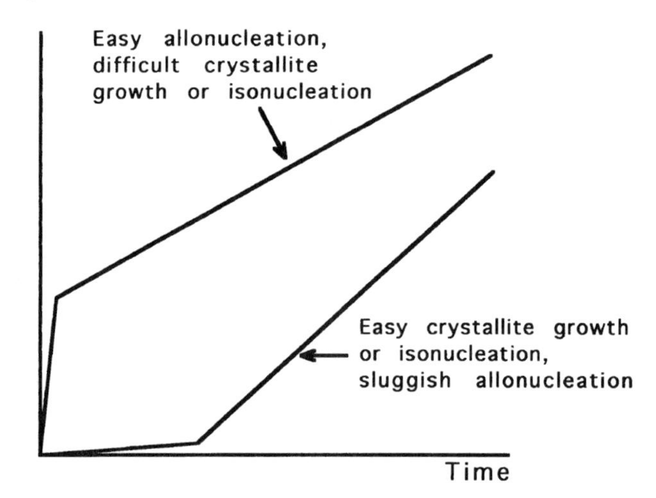


**Figure 6.20.** Effect of initial transient behavior on deposition.

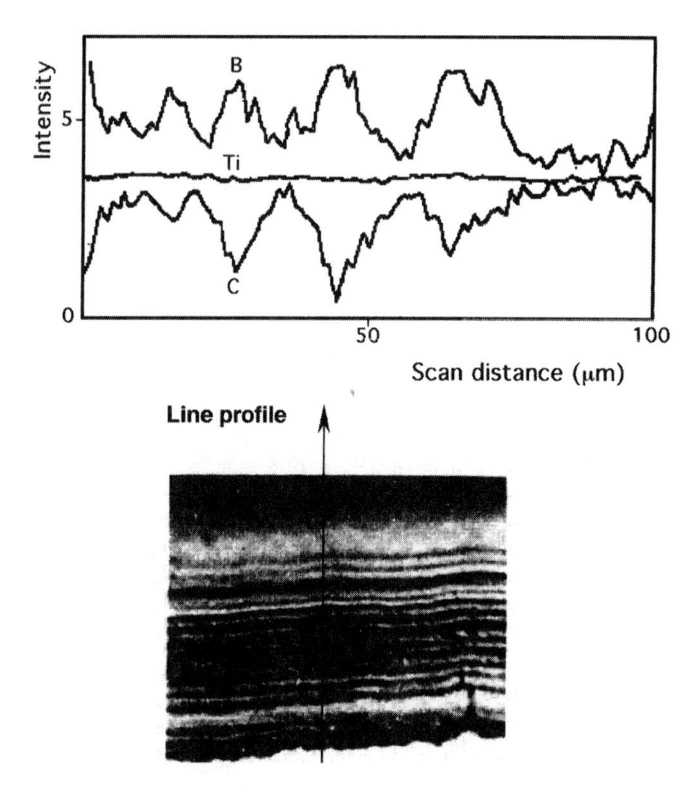


**Figure 6.21.** Oscillating deposition of TiC and $TiB_2$ from one gas mixture of constant composition.[28] A line profile of a cross section made with X-ray microprobe analysis shows that each compound preferably deposits on the other. With permission of *les éditions de physique*.

If no hydrogen is present in the gas phase the deposition is self-limiting, and the growth stops at a certain thickness of the deposited tungsten layer for lack of reducing agent (which was the initial substrate itself).

If surfaces are covered with a relatively nonreactive native oxide surface, etching or reduction by the ambient may be necessary before a proper deposition reaction can take over and form a layer on the bared surface. This may also be a cause for an initially slow deposition rate.

Another example of surface-dependent (selective) deposition is amorphous boron carbide on titanium and molybdenum surfaces. Nucleation on the titanium surface is much faster than on molybdenum and so is isonucleation or growth, with the result that boron carbide under certain reaction conditions will grow on titanium but not on molybdenum if they are both present in the reactor.

An example of fast allonucleation and slow growth is codeposition of titanium carbide and titanium diboride that shows oscillating behavior (Figure 6.21). The sensitivity of the surface reactions to the nature of the substrate in the reaction-limited regime is the reason for the difficulties with codeposition of several phases or alloy formation from a single gas mixture in this regime. Reaction rates may differ enough on the different component surfaces to cause preferential deposition of one component instead of the mixture or alloy intended, independent of the size of the driving forces involved.

### 6.7.6. Diffusion Limitation in the Dialytic Regime

The dialytic regime is characterized by high surface reaction rate coefficients and by rate-limiting diffusion. The Sherwood number (Sh) characterizes the regimes. Sh is defined as the ratio of the driving force for diffusion in the boundary layer to the driving force for surface reaction; alternatively, it is the ratio of the resistivity for diffusion to the resistivity for chemical reaction (reciprocal reaction rate coefficient). Diffusion limitation is the regime at $\mathrm{Sh} \ll 1$ and reaction limitation means $\mathrm{Sh} \gg 1$. The Sherwood number is closely related to the Biot, Nusselt, and Damköhler II numbers and the Thiele modulus. Some call it the CVD number. In the boundary-layer model it is a simple function of the thickness of the boundary layer, the diffusion coefficient, and the reaction rate coefficient. For simplicity a first-order reaction will be considered in the derivation below.

The diffusion current $J_d$ of a reactant from the bulk gas flow through the boundary layer to the surface is (according to an Ohm-like law):

$$J_d = \frac{D(c_b - c_s)}{d}$$

in which $D$ is the diffusion coefficient of the reactant in the gas phase, $d$ is the effective boundary layer thickness, $c_b$ the reactant concentration in the bulk gas, and $c_s$ the reactant concentration in the gas phase at the surface.

The reaction mass current $J_r$ at the surface from precursor to product is (first-order reaction):

$$J_r = k(c_s - c_e)$$

in which $k$ is the mass conversion constant [m/s] or the reaction rate constant, and $c_e$ the reactant surface concentration at thermodynamic equilibrium.

In the stationary state $J_d = J_r = J$, and eliminating $c_s$ yields

$$J = \frac{c_b - c_s}{(d/D) + (1/k)}$$

Again, this is an Ohm-like law: the reactant conversion mass current $J$ equals the driving force $(c_b - c_s)$ divided by the sum of two resistors in series. The resistance for diffusion is $d/D$ and the resistance for chemical reaction is $1/k$.

The Sherwood number equals the ratio of the two driving forces or the two resistances:

$$\text{Sh} = \frac{c_b - c_s}{c_s - c_e} = \frac{kd}{D}$$

The boundary-layer thickness $d$ is difficult to determine in a laminar flow. In practice the effective value of $d$ often has the same order of magnitude as the reactor diameter.

When there are high concentration gradients in the boundary layer the deposition may become nonuniform. On a strongly profiled substrate the solid will be preferably deposited at sites where the gradients are highest, i.e., more at the tops of protuberances and less at the bottoms of surface depressions. These tops will grow faster than the deep-lying parts, which are screened from the bulk reactant supply. Thus positive feedback amplifies fluctuations. In certain cases, however, solid state diffusion in the deposited layer rather than gas-phase diffusion in the boundary layer is rate-controlling. Solid state diffusion is a thermally activated process and reaction rates show Arrhenius behavior (exponential dependence on reciprocal temperature). One example is TiC deposition on carbon steels, where the carbon source is the substrate and titanium is deposited from the gas phase. If there is no carbon source in the gas phase, the titanium carbide (TiC) layer on the surface can only grow if carbon diffuses from the carbon steel substrate, through the titanium carbide coating already present, to the outer surface to react there with $TiCl_4$ and hydrogen to form TiC. Growth is faster in places where layer is thin than where it is thick. Diffusion limitation now stabilizes the surface growth, thickness fluctuations are equalized by negative feedback, and a smooth uniform coating obtains.

Another example of growth limited by solid state diffusion is electrochemical vapor deposition (EVD). EVD is a method for making very thin, yet gastight ceramic oxide coatings on a porous substrate. Such coatings are used in solid oxide fuel cells. The oxide is made from two precursors one on each side of the porous substrate, the appropriate metal chloride and hydrogen on one side and an oxygen/water mixture on the other. A hydrolysis reaction clogs the substrate pores with oxide. As soon as the pores are closed, the reaction can only proceed if one of the reactants diffuses through the solid coating to the reactant at the other side of the coating. The reaction is limited by this solid state diffusion and that means that the layer will have a uniform thickness (Figure 6.22): parts of the layer that happen to be thinner grow faster than thicker areas. In this case the oxide ions diffuse from the oxygen side of the coating to the metal chloride side. Layer growth is proportional to the square root of the time if diffusion in the solid is rate-controlling.

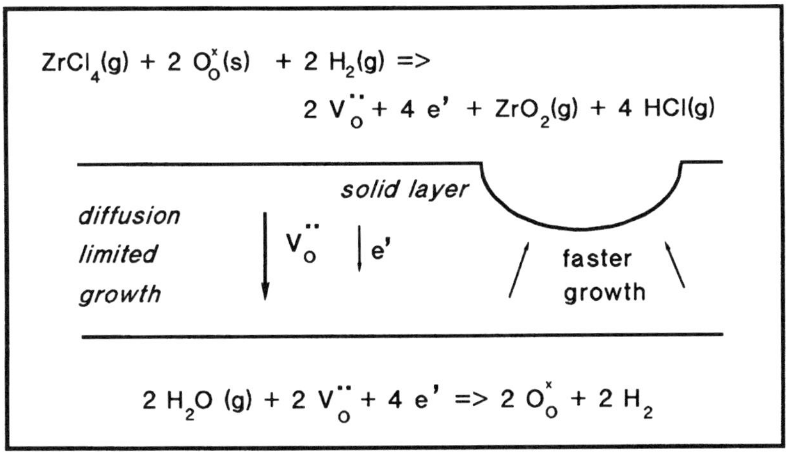

$$ZrCl_4(g) + 2\,O_O^x(s) + 2\,H_2(g) \Rightarrow 2\,V_O^{\bullet\bullet} + 4\,e' + ZrO_2(g) + 4\,HCl(g)$$

solid layer

diffusion limited growth

$V_O^{\bullet\bullet}$ $e'$

faster growth

$$2\,H_2O\,(g) + 2\,V_O^{\bullet\bullet} + 4\,e' \Rightarrow 2\,O_O^x + 2\,H_2$$

**Figure 6.22.** The principle of EVD: on both sides of a porous support the two different precursors are introduced. Initially they react to form zirconia in the pores of the substrate. When the pores are clogged and then closed the two reactants can react with each other only via oxide ions through the deposit. The growth reaction in this second phase becomes limited by diffusion of oxide ions in the solid deposit and the layer grows fastest where the deposit is thinnest.

The Sherwood number can be determined experimentally by measuring the thickness of the deposited layer in a deep standard pore (Figure 6.23). If $R$ is the radius, and $G(z/R)$ the layer thickness at depth $z$, then

$$G(z/R) = G(0)\exp[-(z/R)\sqrt{2\mathrm{Sh}}]$$

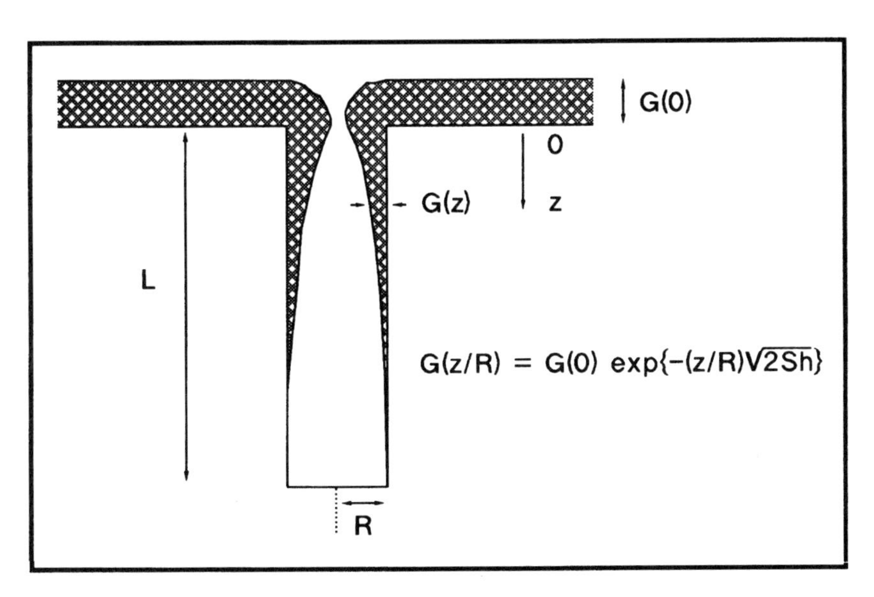


**Figure 6.23.** Deposition in pore. The Sherwood number affects the penetration depth.

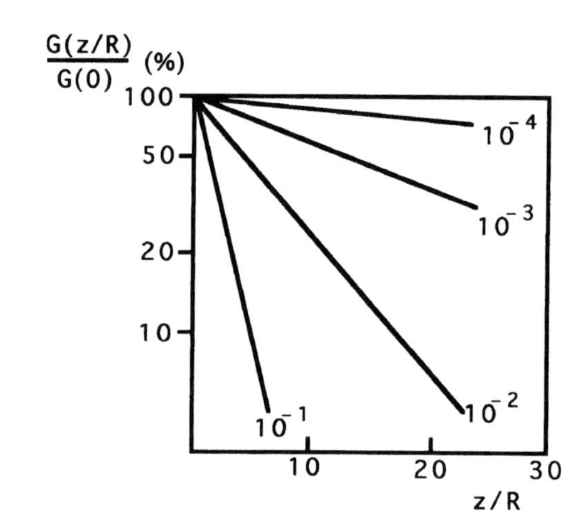


**Figure 6.24.** The calculated growth rate in a pore at different Sherwood numbers.

This relation allows experimental determination of the Sherwood number by measuring the deposited layer thickness in a deep standard pore. The layer thickness decreases exponentially with depth. A high Sherwood number implies low penetration, as shown in Figure 6.24.

Surface roughness can be the result of random spatial fluctuations in reaction rate. The characteristic length (or the surface roughness) for which $Sh = 1$ is amplified in the deposition process and the size of this characteristic roughness depends on the ratio of the time for diffusion over that characteristic length and the characteristic time for reaction. Hence the characteristic length of the roughness depends on the diffusion coefficient and the reaction rate. Thus the surface roughness may contain independent information on the rate of the surface reaction.

The extent to which gas-phase diffusion can be prevented from controlling the deposition rate is of considerable importance for chemical vapor infiltration (CVI). Low pressures and low temperatures (conditions in the catalytic regime) favor penetration. Both factors slow the deposition rate, however, and very long reaction times would be necessary for this way of doing CVI. Consequently, thermal gradients and forced reactant gas flows are sometimes applied to increase deposition rates.

### *6.7.7. Morphology Control*

The most important aspect of materials design is how to make a material with the right morphology (structure on the micrometer and nanometer scales). This section includes descriptions of a few general procedures for morphology control that are afforded by gas-phase techniques. As noted above, diffusion/reaction rate ratios affect solid growth and deposit morphologies have diagnostic value that is helpful for deriving reaction mechanisms.

By means of CVD products can be made with different morphologies by changing process conditions during the reaction. The grain size, size distribution, porosity, anisotropies, texture, roughness, and stoichiometry can be graded over the depth of the deposited layer by changing the gas concentrations, the pressure, the temperature, or the activation power levels during deposition. It is possible to make composites of nonequilibrium phases because CVD is a nonequilibrium process. Several morphologies can be formed in vapor-grown composites, such as spherical fine fibers or flakes in a matrix of another material. However, obtaining a flakelike composite with the plane of the platelets parallel to the surface of the substrate requires more care than the other morphologies.

In PP-CVD homogeneous nucleation (aerosol formation in the gas phase) is combined with thermophoretic particle precipitation in a temperature gradient and conventional CVD or reaction on the surface of the particles. Sometimes called thermophoretic CVD, and originally intended as a procedure to increase deposition rates two orders of magnitude, it has turned out to be a convenient way to control the morphology of a composite product. Whether a dense or a porous product is the result depends entirely on the imposed reaction conditions, such as the temperature gradient at the surface, the particle size, the pressure, the concentration of the reactants, and the rates of the different reactions. Figure 6.25 presents a schematic of one of the simpler reactors for the process. Two precursors are mixed in the hot zone of the reactor and react to form ultrafine, small, monodisperse, and very sinter-active particles by homogeneous nucleation. A mold in the reactor is kept at a temperature slightly below the gas temperature. The aerosol particles diffuse in the thermal gradient to the surface of the mold, where they precipitate and are sintered into the deposit by additional conventional heterogeneous CVD. At a high Knudsen number (Kn $\gg$ 1) the thermophoretic deposition velocity $v_{\text{th}}$ is proportional to the thermal gradient and the kinematic viscosity $v$ of the gas:

$$v_{\text{th}} = 0.55(v/T)\frac{dT}{dx}$$

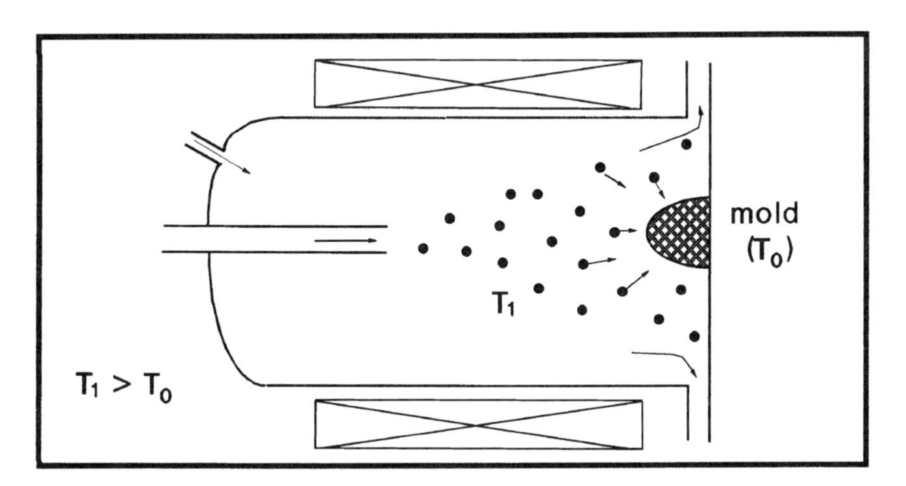


**Figure 6.25.** Schematic of a reactor for particle precipitation aided CVD: homogeneous nucleation, thermodiffusion, and precipitation, followed by thermophoretic CVD.

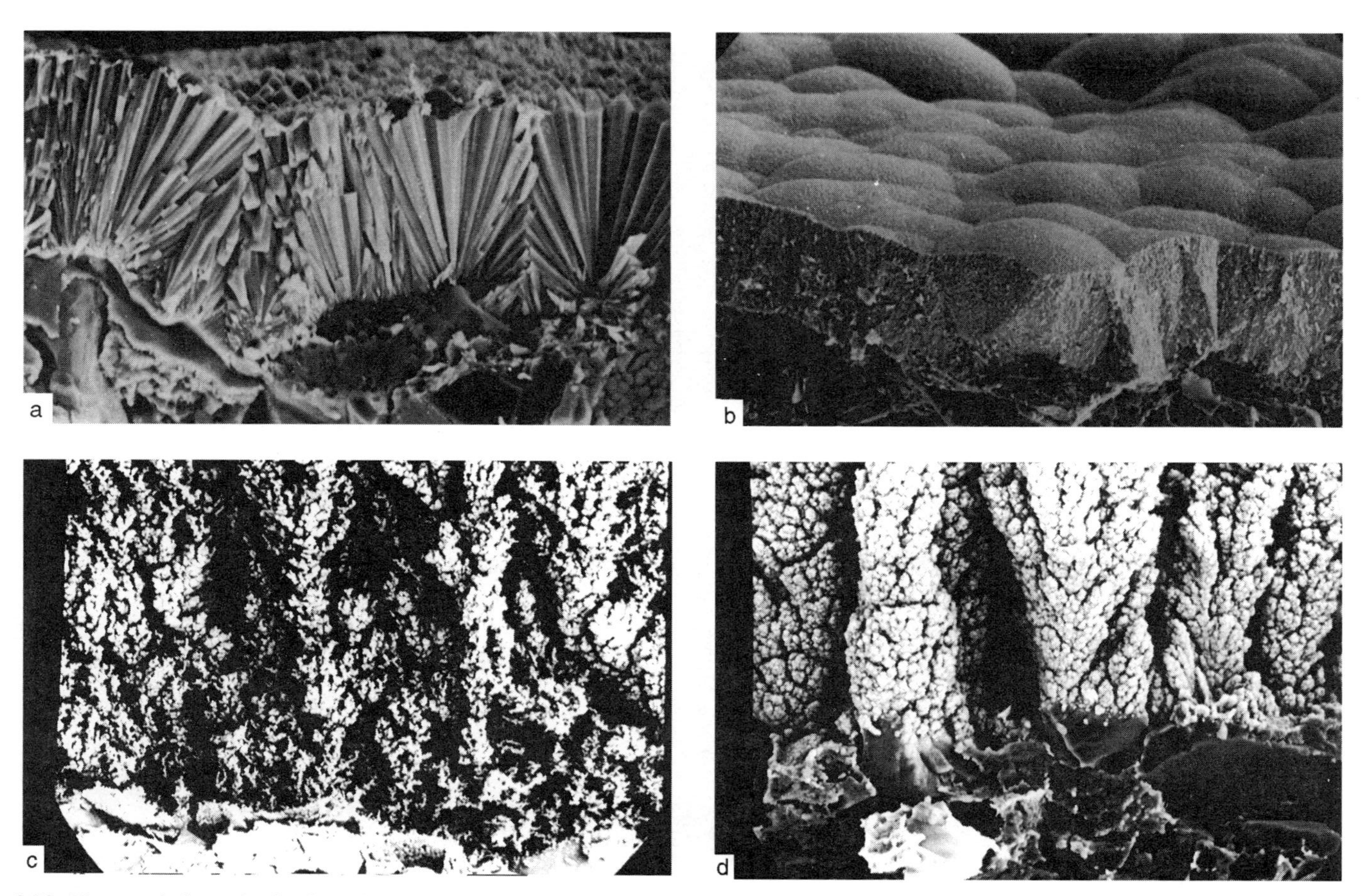


**Figure 6.26.** The morphology of a titanium nitride deposit made by thermophoretic CVD under different conditions: (a) a dense polycrystalline layer; (b) a fine-grained dense layer; (c) a porous deposit of fairly large particles sintered together; (d) a porous sintered deposit concentrated on isolated particles of the porous substrate.[29]

Figure 6.26 shows four very different layer morphologies obtained with PP-CVD of TiN from $TiCl_4$ by varying temperatures, concentrations, and thermal gradients.[29] The substrate is a porous alumina ceramic. The aerosol formation was forced through the use of ammonia as a nitrogen precursor, while the heterogeneous CVD reaction used $N_2$. In this case the reaction for homogeneous particle precipitation differs from the simultaneous heterogeneous CVD reaction although the same product (TiN) is formed. In a similar way composites of different phases have been made using PP-CVD by two simultaneous deposition reactions that form different solids.

*Faceted crystallite growth* in CVD requires a high surface mobility but growth temperatures lower than the corresponding roughening temperature; the crystallite habit of the solid depends strongly on the impurities present, which affect growth rates of the different facets (low-indexed lattice planes) to a different extent. Screw dislocations are necessary for efficient facet growth.

*Grain refinement* is realized in a deposit by having a high nucleation rate or by suppressing crystal growth or both. Several tricks are possible:

1. High temperatures increase heterogeneous nucleation rates. Apparently isonucleation is faster at high temperatures than regular crystallite growth

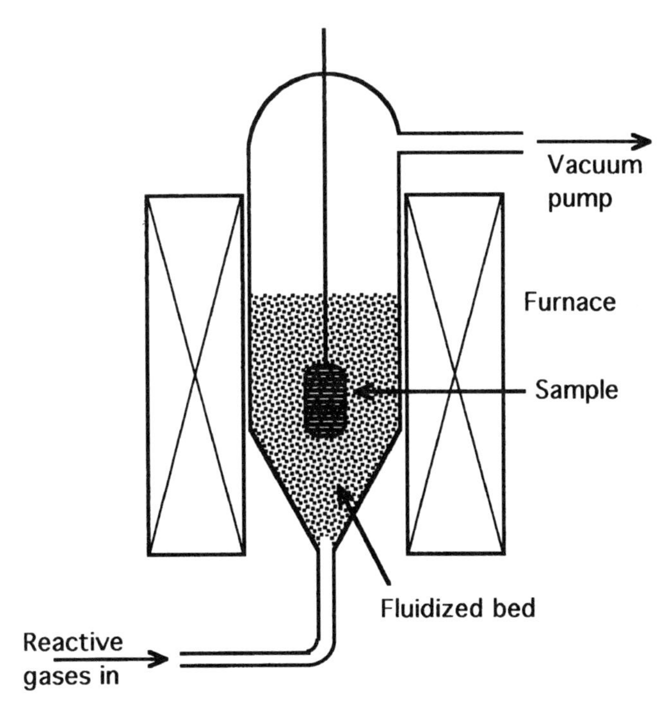


**Figure 6.27.** Schematic setup of a fluidized-bed reactor used either to cover a suspended sample with a fine-grained coating or to coat the fluidized particles to change their surface characteristics. Powder is fluidized in a precursor gas mixture at the reaction temperature. In addition to being the substrate for deposition the powder sometimes functions as a slowly evaporating precursor for deposition on the suspended substrate.

(which is favored by slow growth processes) and texture development is suppressed.
2. Fluidized-bed CVD is a version in which the substrate is suspended in a fluidized bed and yields fine-grained deposits because particle impact on the surface continuously creates new heterogeneous nucleation sites. Figure 6.27 shows a reactor schematic.
3. Crystallite face growth can be inhibited by poisoning the crystallite surface with an additional gas, which leaves heterogeneous nucleation as the main deposition mechanism. Addition of methane to the $H_2/CO_2$ mixture used for hydrolyzing $AlCl_3$ reduces the average grain size of the $Al_2O_3$ deposit. The proposed mechanism for grain refinement of TiC by the addition of small amounts of an aluminum-containing precursor to the reactant gas mixture is formation of a surface cover of very thin aluminum carbide that prevents growth but not nucleation.

*Texture* develops if the crystallite growth rate is anisotropic (if it differs for different crystal lattice directions), in which case a preferential crystallite orientation develops during growth with the fast growth direction perpendicular to the surface. This is a case of what might be called Darwin kinetics: some sites or directions grow at the cost of other sites or directions. In texture crystallite orientations with the fast axis perpendicular to the substrate surface survive, while those with their fast-growing axis oriented more parallel to the substrate surface cannot. As indicated above, texture can be suppressed if necessary by increasing heterogeneous (iso)nucleation rates by choosing high temperatures or by deposition in a fluidized bed.

## 6.8. High-Temperature Corrosion

At high temperatures materials start to react at their surfaces: they burn or react with their surrounding atmosphere, decompose, dissolve, are etched, melt, or evaporate; in short they corrode.[30] Corrosion is a surface process and corrosion of materials can be retarded by passivating their surfaces by means of a suitable treatment, e.g., by converting the surface or by applying a coating. Ceramics are very corrosion-resistant and so are used to passivate inexpensive metals such as iron. Abundant and cheap materials can then be used, for instance, in building chemical reactors to replace scarce and costly materials such as gold or tantalum and yet have the same or better durability in use.

There are several types of corrosion reactions and the specific type dictates the kind of protection needed against it:

1. Corrosion in oxidative atmospheres
2. Corrosion in reductive atmospheres
3. Corrosion in carburizing atmospheres
4. Corrosion by metal melts
5. Hot corrosion by salts
6. Electrochemical hot corrosion in fuel cells

**Table 6.6. Corrosion of Ceramics in Combustion Reactors under Cyclic Exposure to $H_2SO_4$ and HCl Vapors and $NO_x$ at Temperatures up to 300°C for 900 h**

| Material | Corrosion depth (μm) | Material | Corrosion depth (μm) |
|---|---|---|---|
| Cordierite | 9 | Mullite | 0.8 |
| Glass enamel | 1.4 | α-SiC (sintered) | 0.4 |
| RBSN ($Si_3N_4$) | 0.9 | RBSC (SiC) | 0.2 |

(Low-temperature aqueous corrosion is not covered here since it is a standard part of undergraduate physical chemistry courses for materials scientists and chemists.)

Inhibitors against all these forms of corrosion are basically coatings of materials that function as diffusion barriers and are simultaneously inert. Sometimes materials are selected for corrosion protection that are stable in the ambients under operating conditions but these do not always function as might be expected from the stability data. If solids are to be protected against corrosion, the corrosion reactions should be slowed down with diffusion barriers or inert ceramics rather than by a high stability. Stability is no guarantee of the absence of reaction as was discussed above for noble metals as catalysts and in Chapter 3 for the case of complex molecules that can be very stable yet can have fast ligand exchange (reaction).

A few high-temperature corrosive properties of ceramics are mentioned here as examples. The book by Samsonov and Vinitskii has many details on high-temperature reactions with various ambients.[31] Table 6.6 compares the corrosion resistance of several refractories in combustion gases.

Most metal oxides are stable in oxygen at high temperatures. Group 6 oxides, however, are much less stable as solids because their higher oxides are volatile. In reducing ambients nonoxide ceramics (borides, carbides, and nitrides) are preferred; $SiO_2$ is attacked in sufficiently reducing gases by reduction to the volatile SiO. However, many nonoxide ceramics are also resistant to oxidation in air at high temperatures.

Molybdenum silicide ($MoSi_2$) holds a record among nonoxides. It suffers a weight increase of only 0.3 mg/cm$^2$ after 4 h at a temperature of 1200°C in air and of 2.1 mg/cm$^2$ after 100 h. It is an electronic conductor that is used as a heating element in electric furnaces working up to 1700°C in air and as an electrode material in high-temperature soldering. The reason for its corrosion resistance is superficial oxidation and subsequent evaporation of volatile molybdenum oxides, which leaves a protective thin $SiO_2$ film on the surface. This film acts as a diffusion barrier for oxygen and protects the substrate from further oxidation. $MoSi_2$ has a large work function and catalytic (dehydration) properties. It also resists attack by a NaCl/$BaCl_2$ melt at 1100°C.

Silicon carbide (SiC) oxidizes slowly in air at a temperature of 1400°C. After 200 h the weight increase is 9.2 mg/cm$^2$. It is also highly resistant to metal melts and vapors, acids, and alkalis. It is often used in pump parts, heat exchangers, and abrasives, and is a useful high-temperature semiconductor.

Boron carbide ($B_4C$) is also surprisingly oxidation-resistant. Its weight decrease after 100 h in air at 1200°C is 11 mg/cm$^2$. It is an abrasive used in sandblasting nozzles, chemical vessels, ignitrons (semiconductor sparkers), thermoelectric energy converters, and nonlinear electronics. It is considered one of the most suitable

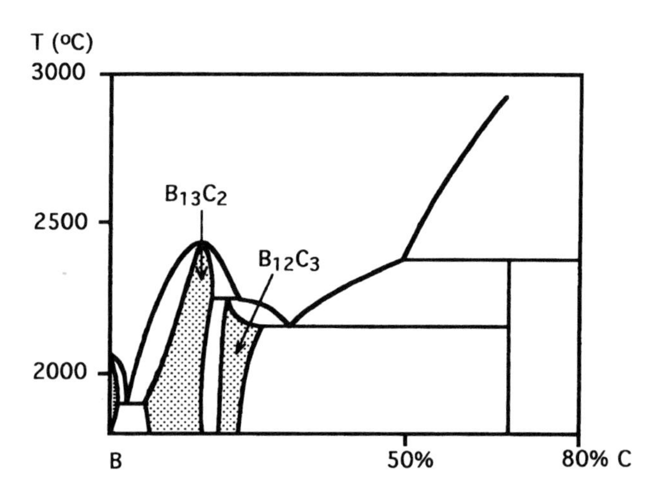


**Figure 6.28.** The phase diagram of the B/C system.

first-wall materials in fusion reactors. The boron–carbon phase diagram in Figure 6.28 shows the wide range of carbon solubility in the $B_4C$ phase, which implies a large structural disorder lowering the conductivity.

Zirconium diboride ($ZrB_2$), an electronic conductor (Figure 6.29 shows the Zr–B phase diagram) has a weight increase of 4 mg/cm$^2$ after 200 h in air at 1150°C. Besides being oxidation-resistant, it is not attacked by very corrosive metal melts such as aluminum at 1000°C, a property it shares with $TiB_2$ and $HfB_2$; the latter starts reactions with oxygen at 600°C. However, a composite of $HfB_2$ with SiC (Figure 6.13) is extremely oxidation-resistant at high temperatures. This example of synergy has no relation whatsoever to calculated thermodynamic stability diagrams.

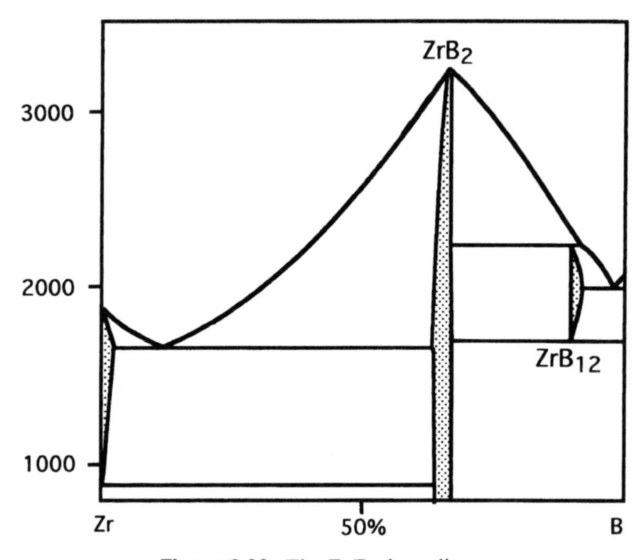


**Figure 6.29.** The Zr/B phase diagram.

Chlorine is very corrosive to virtually all metals at high temperatures but not to the binary ceramic compounds tungsten carbide, hexagonal boron nitride, and silicon nitride. WC shows no reaction up to 700°C. BN suffers a weight decrease of 0.55 mg/$cm^2$ after 40 h at 700°C. $Si_3N_4$ remains inert even after 500 h of exposure to chlorine at a temperature of up to 900°C.

Molten salts are excellent solvents but are for that reason difficult to handle as is apparent from the corrosion problems with the molten carbonate fuel cell. However, a few ceramics do not suffer attack by molten salts. ZrN is inert to cryolite melts at 1050°C; one of the chromium carbides ($Cr_3C_2$) does not react at all with a $BaCl_2$/NaCl melt at 1100°C; and $VSi_2$, $NbSi_2$, and $TaSi_2$ are electronically conducting ceramic compounds that are reported to resist an NaOH/$Na_2CO_3$ melt at 700°C. $Si_3N_4$ does not react with aluminum melts at 1000°C, acids, alkalis, salts, or slags. It is a good electrical insulator.

The perovskite lanthanum chromite ($LaCrO_3$) is one of the exceptional ceramic compounds that are chemically very resistant to both oxidizing and reducing ambients. Moreover, being an electronic conductor, it is eminently suitable as a bipolar connector in solid oxide fuel cells. It is evident that the thermal expansion coefficients of the different components in a fuel cell (electrolyte, electrodes, bipolar connector) must be closely matched. Doping the chromite with strontium or magnesium ions is necessary to increase its electronic conductivity as well as its sinter activity.

## 6.9. Surface Modification by Immobilization of Molecules

Surfaces of solid particles can be made to react in much the same way as certain molecules would by immobilizing those molecules on the solid surfaces. These surface molecules are attached with covalent bonds to the bulk solid and any reaction between the atomic bulk of the solid and molecules in the ambient gas or liquid can take place through the bound surface molecules instead of through the surface atoms of the bare solid. Immobilization can make a world of difference in redox behavior of solids or their adsorption of metal ions. Materials for chemical sensors can be activated this way and surfaces can be rendered catalytically active. Molecules that are biologically active need no longer be used in a homogeneous solution but can be made to act heterogeneously when they are bound to a solid substrate. Surfaces can be made antimicrobial and impossible for micro-organisms to attach to by immobilizing quaternary alkyl ammonium salts on them. If the alkyl groups are large, the surface layer is hydrophobic as well and difficult to remove.

Immobilization of functional organic groups is essential for adhesion between a reinforcing mineral phase and a plastic matrix in structural composites such as cerpols. In such composites the adhesion between the components is determined by the interface and strongly affects the performance for the same reason as it does in cermets. If there is no adhesion the components do not reinforce each other and the material fails. If there is too much adhesion the phases do not interact as intended and the composite becomes as brittle as the matrix would be without the reinforcement. Controlling the interface is the key to improving any property of a composite that depends on the interaction between the components.

Inorganic parts of cerpols are usually oxides such as $SiO_2$, $SnO_2$, $TiO_2$, or glass. Almost all metal surfaces are oxidized at the surface and behave somewhat like bulk oxides. The compounds that are used to attach molecules and organic plastics to inorganic surfaces are called coupling agents,[32] and very small amounts have large beneficial effects on mechanical properties. The most convenient coupling agents are substituted silanes such as $(H_3CO)_3SiZ$, where Z represents the functional organic group to be immobilized. The alkoxy group is hydrolized either by adsorbed water or by surface M–OH groups and the silylating agents form more or less covalent M–O–Si–Z bridges at the M surface. However, the interface is usually thicker than a monolayer. The coupling agents hydrolyze and form condensates in the same way and according to the same mechanisms as sol-gel processes do in making ceramics. When glass fibers are treated with silylating agents they form adherent polysiloxanes at the glass surface while the organic groups in the siloxanes link to the organic matrix with covalent bonds. The polysiloxane interphase is the reason for the necessarily nonrigid interaction between the matrix and reinforcement phases.

The quality of the bonding that is obtained with silylating agents between plastic and inorganic oxide varies with the substrate:

1. Bonding with silica, alumina, glass, SiC, and aluminum needles is very good.
2. With talcum, magnesium silicate, iron powder, clay, and hydrated alumina, bonding is acceptable.
3. Little bonding occurs with ZnO, $TiO_2$, and hydroxyapatite.
4. There is no bonding with $CaCO_3$, graphite, and boron using silylating agents.

Carbon electrodes have been grafted with complexes such as tetraaminophenylporphyrine and substituted cobaltocinium. The carbon surface is first treated with $SOCl_2$, after which the carbochloride groups at the surface are reacted with an amine-substituted complex.

It is common practice to modify organic polymers with carboxyl groups in order to improve adhesion to metals. However, the carboxyl link, being ionic, is not very resistant to hydrolysis and a much less ionic siloxane bridge is more inert toward water. Very high heat stability is obtained by using mixtures of silicon and titanium orthoalkoxides as coupling agents for silicone rubbers. Such mixtures form mixed Si–O–Ti–O chains. Adhesion with polyimides and silicones remains intact up to temperatures at which the plastic decomposes. In addition to the titanate esters, chromium carboxylates are nonsilylating coupling agents that are occasionally used for joining plastics with inorganic fillers but much less often than the silylating compounds. These chromium salts also partially hydrolyze and the colloidal agglomerate particle bonds to the surface M–OH groups as the silanes do. The organic acid groups remain attached to the chromium hydroxide and form the link to the organic phase.

Silylating coupling agents are used to attach inorganic complexes to $SnO_2$ and $TiO_2$ electrode surfaces in order to facilitate electron transfer and lower overvoltages. Carboxyl- or sulfonyl-substituted metal phthalocyanines are immobilized with aminosilyl agents, which form amide bonds. Such links have a much higher resistance against photolytic aging than physisorbed dyes. Cobalt phthalocyanine fixed on $SnO_2$ in this manner showed no activity loss after cycling 1000 times.

Chelating organic ligands can also be immobilized on $SnO_2$ surfaces with coupling agents $(CH_3O)_3Si{-}R$ or $Cl_3Si{-}R$ with R ethylene diamine or bipyridyl.[33] Cations from solution can be efficiently trapped on such surfaces for purification or for making heterogeneous catalysts from homogeneous ones. Proton conducting polymers such as nafion and electron-conducting polymers such as polypyrrole have also been immobilized with silylating coupling agents.

## Exercises

1. Give some examples of how surface properties dominate the material properties and suggest ways to modify them.
2. In what way can reaction mechanisms observed in catalytic processes be of practical use for materials synthesis?
3. The corrosion resistance of CVD layers and solids that are made out of (nominally) the same products fabricated with standard ceramic techniques is quite different. Glasses also are often more resistant against chemical attack than the polycrystalline modification of the same compound. Why?
4. Discuss the influence of gas pressure, temperature, and concentrations of reactants in solid state reactions. In what way do the kinetics of reactions involving solids differ from the rates of homogeneous reactions between molecules?
5. Unlike the usual types of fuel cell, kinetic fuel cells do not need two gas departments hermetically sealed from each other. Cells of the kinetic type develop power from surface reaction in a mixture of fuel and air (supplied simultaneously to both electrodes, which are made of different materials). Suggest a few possible explanations for the effect.
6. The reactivity of a heterogeneous catalyst depends on the work function of the solid. Design a catalyst surface that has an activity that can be electrically controlled.
7. List a few applications of oscillations in reactions. How can oscillations be suppressed?
8. Discuss the sequential reactions or processes in thermally activated (conventional) CVD of a ceramic compound. Which reaction conditions are preferred for determination of kinetics intended to study surface reaction kinetics? Which process conditions are required for the production of fine powders? Which conditions are necessary for good throwing power (uniform step coverage)?
9. What is a CVD diagram and what sort of information in it is useful for choosing process conditions? What are the limitations of CVD diagrams?
10. Three process conditions can be distinguished in CVD, but not in PVD. To which one of the three CVD regimes does the PVD regime correspond most closely?
11. Draw an Arrhenius plot of the temperature dependence of the deposition rate of a CVD reaction (at a sufficiently large temperature interval to cover both regimes) for an equilibrium constant small with respect to 1, and also for an equilibrium constant that is much larger than 1.
12. During the production of fiber-reinforced composites with CVI, the interior is poorly filled with the matrix material. How should the deposition conditions be changed to improve the density of the composite?
13. Let a two-phase composite coating be formed by CVD in the reaction-limited regime and assume that two different reaction steps having different activation energies are running in parallel at the surface: (a) how can the composition of the product be changed without changing the gas-phase composition? and (b) sketch the Arrhenius plot; how are the two activation energies determined separately?
14. Define chemical supersaturation. What is its temperature dependence? Explain the function of impurities for $TiO_2$-formation shown in Figure 6.17.

15. Fine uniform particles of SiC are made by CVD. In what sense must the reaction conditions be changed in order to make smaller particles? And larger powder particles? Explain.
16. During a certain deposition process for a HfN coating a higher process pressure was observed to lead to homogeneous nucleation and undesired powder production. How can this tendency be compensated for by changing other process parameters?
17. Rank homogeneous nucleation, heterogeneous allonucleation, and heterogeneous iso-nucleation is order of increasing supersaturation required. Explain.
18. Which morphological features identify the VLS process? What is the preferred growth direction in VLS deposition?
19. What morphological features do products that have been made by EVD (e.g., $ZrO_2$) have in common with conversion coatings (e.g., TiC on carbon-containing steel)?
20. What assumption is implicit in the derivation of the Sherwood number that is not valid at large driving force?

## References


1. P. Grange and B. Delmon. *Interfaces in New Materials.* Elsevier, London (1991).
2. G. A. Somorjai. *Introduction to Surface Chemistry and Catalysis.* Wiley, New York (1994).
3. G. Ertl. Self-organization in reactions at surfaces. *Surf. Sci.* **287/288**, 1 (1993).
4. J. Q. Broughton and D. L. Perry. Electron binding energies in the study of adsorption by photoelectron spectroscopy: the reference level problem. *Surf. Sci.* **74**, 307 (1978).
5. C. G. Vayenas, S. Bebelis, I. V. Yentekakis, and H. G. Lintz. Non-Faradaic electrochemical modification of catalytic activity: a Status Report. *Catalysis Today* **11**, 303 (1992).
6. J. Lyklema. Interfacial electrochemistry of disperse systems. *J. Mater. Educ.* **7**, 203 (1985).
7. J. T. G. Overbeek. How colloid stability affects the behaviour of suspensions. *J. Mater. Educ.* **7**, 393 (1985).
8. G. C. Bond. *Heterogeneous Catalysis: Principles and Applications.* Clarendon, Oxford (1987).
9. E. Shustorovich. Chemisorption phenomena: analytical modeling based on perturbation theory and bond-order conservation. *Surf. Sci. Rep.* **6**, 1 (1986).
10. R. Mezaki and H. Inoue. *Rate Equations of Solid-Catalyzed Reactions.* University of Tokyo Press, Tokyo (1991).
11. M. Boudart and G. Djega-Mariassou. *Kinetics of Heterogeneous Catalytic Reactions.* Princeton University Press, Princeton (1981).
12. S. T. Ceyer. New mechanisms for chemistry at surfaces. *Science* **249**, 133 (1990).
13. A. J. Bard. *Integrated Chemical Systems: A Chemical Approach to Nanotechnology.* Wiley, New York (1994).
14. W. E. Lee and W. M. Rainforth. *Ceramic Microstructures: Property Control by Processing.* Chapman and Hall, London (1994).
15. A. Pivkina, P. J. van der Put, Yu. Frolov, and J. Schoonman. Reaction-bonded titanium nitride ceramics. *J. Eur. Ceram. Soc.* **16**, 35 (1996).
16. M. L. Hitchman and K. F. Jensen. *Chemical Vapor Deposition: Principles and Applications.* Academic, London (1993).
17. H. O. Pierson. *Handbook of Chemical Vapor Deposition (CVD): Principles, Technology and Applications.* Noyes, Park Ridge, N.J. (1992).
18. C. E. Morosanu. *Thin Films by CVD. Elsevier, Amsterdam* (1990).
19. *W. S. Rees (ed). Chemical Vapor Deposition of Nonmetals.* VCH, Weinheim (1996).
20. T. Kodas and M. Hampden-Smith (eds.). *The Chemistry of Metal CVD.* VCH, Weinheim (1994).
21. F. S. Galasso. *Chemical Vapor Deposited Materials.* CRC, Boca Raton (1991).
22. J. T. Spencer. CVD of metal-containing thin-film materials from organo-metallic compounds. *Progr. Inorg. Chem.* **41**, 145 (1994).
23. M. E. Jones and D. W. Shaw. Growth from the vapour. In: N. B. Hannay (ed.). *Treatise on Solid State Chemistry: Vol. 5. Changes of State* (1975), p. 283. Plenum, New York.

24. K. E. Spear. High-temperature reactivity. In: N. B. Hannay (ed.). *Treatise on Solid State Chemistry, Vol. 4. Reactivity of Solids.* Plenum, New York (1975), p. 115.
25. P. Glansdorff and I. Prigogine. *Thermodynamic Theory of Structure, Stability and Fluctuations.* Wiley Interscience, London (1977).
26. A. Kato. Some common aspects of the formation of nonoxide powders by the vapor reaction method. In: *Materials Science Research, Vol. 17: Emergent Process Methods for High-Technology Ceramics.* R. F. Davis, H. Palmour III, and R. L. Porter (eds.). Plenum, New York (1984), p. 123.
27. P. Gray and S. K. Scott. *Chemical Oscillations and Instabilities: Non-linear Chemical Kinetics.* Oxford University Press, Oxford (1990).
28. K. Bartsch, A. Leonhardt, E. Wolf. Composition oscillation in hard material layers deposited from the vapour phase. *J. de Physique Coll. C2 Suppl.* II, Vol. 1. P. 563 Proc. 8th Eur. Conf. CVD, Glasgow 1991.
29. J. P. Dekker, P J. van der Put, H. J. Veringa, and J. Schoonman Particle-precipitation-aided CVD of titanium nitride. *J. Am. Ceram. Soc.* **80**, 629 (1997).
30. R. J. Fordham (ed.). *High-Temperature Corrosion of Technical Ceramics.* Elsevier, London (1990).
31. G. V. Samsonov and I. M. Vinitskii. *Handbook of Refractory Compounds.* IFI/Plenum, New York (1980).
32. E. P. Plueddemann. *Silane Coupling Agents.* Plenum, New York (1991).
33. P. Laszlo (ed.). *Preparative Chemistry Using Supported Reagents.* Academic, San Diego (1987).

*Chapter 7*

# INORGANIC MORPHOGENESIS

*Causality is an attempt to mesmerize the world into some sort of significance.*

LAWRENCE DURRELL, TUNC

## 7.1. Introduction to the Chemistry of Microstructure and Nanostructure

Traditionally and conventionally chemistry deals with the elementary composition and structure of molecules and crystal lattices. Books on chemistry sometimes also discuss how to run processes to make those structures, but the importance of secondary and higher structures for properties of solids and polymers is not often acknowledged. The information that impurities, synthesis conditions, grain size and form, and the dimension of surfaces in atomic compounds strongly influence the properties and the chemistry and how that influence can be put to use can only be found in articles and books on materials,[1,2] virtually never in chemical descriptions. Materials science deals with complex matter and its function and that is outside the scope of molecular and structural chemistry. Yet as polymer chemists, colloid scientists, and catalyst chemists know, morphology is a branch of chemistry. "In the area of solid-catalyzed reactions it has been well accepted that differences in the fine structure of solid catalysts yield totally different reaction mechanisms and rate expressions even in cases where the chemical compositions of the catalysts are exactly identical."[3] This is a statement from a well-known compilation of catalyzed reactions, and this chapter is an attempt to correct this neglect in textbooks, at least for the case of inorganics, by discussing forms (morphology or microstructure) and how synthesis creates them.

The word microstructure indicates form and spatial distribution on a scale having characteristic size $\lambda$, which is larger than molecular sizes ($\lambda > 0.5\,\mathrm{nm}$) but smaller than $1\,\mu\mathrm{m}$. Its realm is between those of crystal structures and microcrystalline forms. The word microstructure has gradually undergone a change in meaning with the increasing use of novel words such as nanostructure (with characteristic length in the nanometer scale) and mesostructure (between nanostructure and microstructure in the micrometer range). "Morphology" covers both microstructure and nanostructure but is sometimes used as a synonym for microstructure.

Morphogenesis is the chemical formation of morphology or formation of microstructure or nanostructure. The macroscopic shaping of materials to make a product belongs to materials technology but not to materials chemistry. Morphogenesis is a function of several driving forces. It may be the result of enthalpy or entropy differences (thermodynamic driving forces) or of diffusion because of anisotropic concentration differences. From looking at the forms it appears that in a reaction that creates a solid, the diffusion and reaction steps are sequential and that usually one step (the one having the highest resistance) determines the kinetics of the process and the form or morphology of the solid product. This means that two types of morphology can be distinguished: reaction-limited and diffusion-limited structures. In reactions between solids, diffusion is intimately linked to reaction and this chapter will focus on diffusion-limited morphogenesis (Sections 7.2 to 7.6), beginning with an overview and a brief discussion of reaction-limited morphogenesis.

Morphology indicates the distribution of matter in grains, grain size, size distribution, and anisotropy and the connection among them. The form of the grains in polycrystalline solids can be anything from spheres and needles to plates and polyhedra. These grains are assembled into compact materials or in ramified clusters called fractals when they have a symmetry of scale and a noninteger dimension. Fractal distributions are ubiquitous in materials[4] and will be discussed below.

In reality, the forms and boundaries of phases are not as ideal as our mathematically convenient models would have it. Particles are not spherical, surfaces are not flat, and pores are not cylindrical, although that is invariably assumed in models without taking account of the role of the form on the calculated kinetics. There is another problem in trying to calculate solid state reaction rates by solving partial differential equations that include mass transport numerically. The boundary conditions change during a reaction in a way that depends on the solutions of the equations, which means that heterogeneous processes are difficult to model with classical continuum mathematics. The results of calculated estimates based on idealized boundary conditions are of limited use for designing a synthesis for a solid state compound.

Morphogenesis consists of four processes:

1. Potentials or gradients in concentration or other driving forces determine the direction of the reactions. If the potentials are low and the mobility of the atoms is high enough, potentials affect rates. These effects have been well studied and dominate the formation of solids if it takes place under conditions not far removed from thermodynamic equilibrium, e.g., the growth of single crystals from a melt or from solution.
2. In many if not most formation processes diffusion is rate- and shape-determining, in which case the reaction path of the process rather than the driving forces determines the result. Diffusion of reactants may introduce fractality and fractal processes on fractal surfaces (see below). If diffusion is rate-limiting when solids are formed the temporal fluctuations are frozen in spatial structures. The direction of transport of the atoms or energy during the formation of the solid introduces anisotropy in the solid products.

3. The reaction resistances or their reciprocals, the chemical admittances, or reaction rate coefficients for phases that grow simultaneously on different facets of one solid phase determine which reactions occur and to what extent, and which do not. Rates are key factors for morphogenesis.
4. Time modulation of the process conditions during synthesis (or adding impurities) is used to control the microstructure and the composition of phases in the material.

Examples of morphology control during synthesis can be found throughout this book.

With the exception of single crystals, the materials that are synthesized with the methods of inorganic chemistry are random materials, i.e., glasses, composites, or polycrystalline compounds that have random elements such as defect sites or bond angles, distribution of different phases, or arrangement of grain boundaries. Materials are not made with tight external controls in the way electronic chips are made with planar technology or with the atomic force microscope (AFM) for moving atoms around on surfaces. Such manufacturing techniques are for building devices, not for making materials. When materials are synthesized, powders are mixed and sintered, or particles nucleate, grow, and deposit from fluid phases. The difference between a material and a device becomes clear if photographic paper is compared with a display screen. In a display each pixel is precisely localized, carefully composed, and separately addressed while photographic paper has a random distribution of pixels. The big difference in the fabrication processes reflects the basic differences in the two concepts.

Generally, neither the form nor the size of crystallites is the result of equilibrium during formation. The existence of grain boundaries already implies an absence of equilibrium because they represent extra energy. The grain size and form depend on the history of the solid and on the rates at which the different faces grow.

Texture or preferential growth in the crystallographic directions perpendicular to the surface of the substrate occurs whenever growth is faster than nucleation and there is a large anisotropy in crystal growth rates as discussed in chapter 6. A fractal dimension of the surface is adjusted with the ratio of the diffusion rate to the reaction rate (see below). Anisotropic grains or whiskers, even of cubic compounds, are easily made with the VLS (vapor-liquid-solid) technique or by eutectic solidification (Chapter 8). The average grain size can be controlled if the temperature for maximum nucleation rates differs significantly from the temperature for maximum crystal growth rate. The size distribution can be adjusted with the choice of those temperatures during synthesis. The morphology is controlled in particle-precipitation CVD with the temperature gradient and in the sol-gel method with the pH of the solution.

Surfaces can be reactants and, of course, their chemical activity depends on their atomic (crystal) structure. This is no surprise: the reactivity of molecules in chemical reactions also depends on their molecular structure. The fact that the chemical reactivity of surfaces depends on the microstructure of the solids is evident in catalysis and in growth reactions, as was discussed in Chapter 6. The art of making good catalysts as well as the art of creating the right microstructures (morphogenesis) consist of choosing the right preparation processes for the surface.

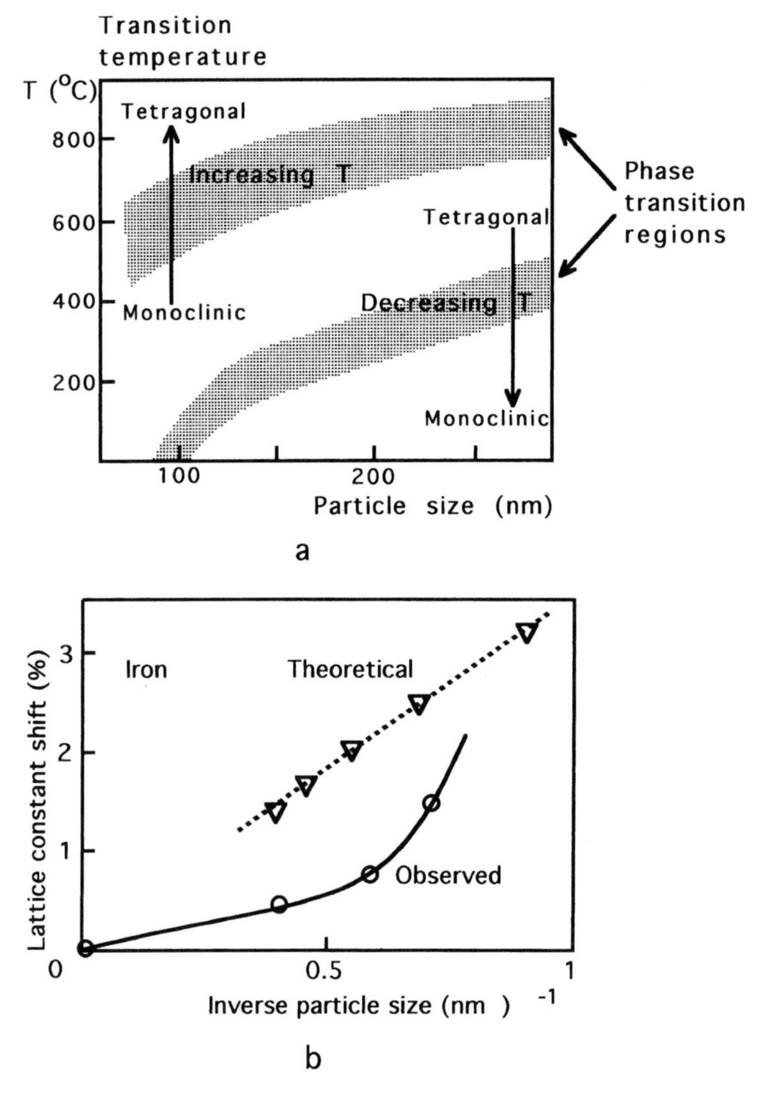


**Figure 7.1.** The influence of particle size on property: (a) The transition temperature of zirconia particles between the monoclinic and the tetragonal modification. There is a considerable hysteresis and the transition temperature drops with particle size. (b) The size dependence of the lattice constant of iron particles. (c) The particle size dependence of the melting point of a few metals. Their bulk melting points are given at the right-hand side of the graph.

Surface atoms are bound to their neighbors is a way that differs from atoms deeper in the solid, which has consequences for the mobility of atoms at the surface. The atoms at the surface are more mobile than defects in the lattice and the surface appears molten far below the melting temperature $T_m$ of the solid (at what is called the Tammann temperature $T \approx 2/3\,T_m$). Particles become slippery as a result of this enhanced mobility, which is why they sinter far below their melting point.

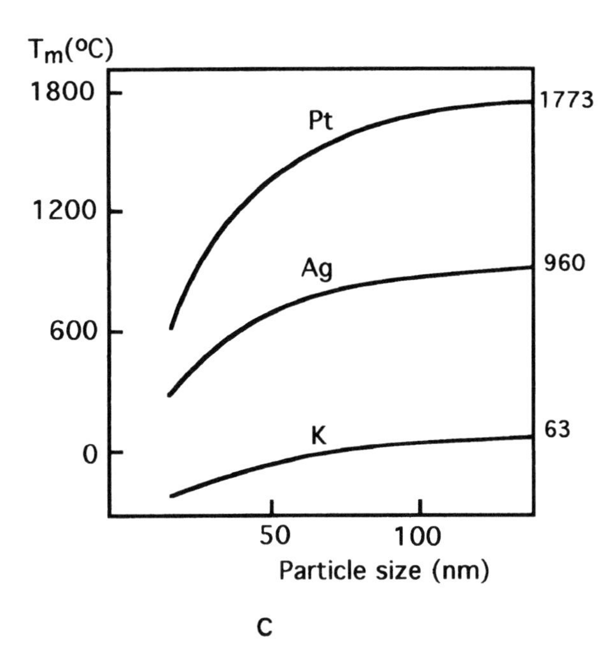


**Figure 7.1.** *(Continued)*.

Lead, e.g., is rather inert and is used on roofs because it does not corrode. However, finely divided lead is pyrophoric. There are other examples of the influence of particle size on properties. One is the behavior of electrons in very small particles, which cannot be easily described with delocalized orbitals because the electrons have discrete energy levels and do not show bandlike behavior. Very small particles can be considered to be large molecules that have novel shape-sensitive electric, magnetic, optical, and chemical properties that makes them interesting for use in new microelectronic devices and materials. Hence matter properties are related to the surface dimension, and this number (defined below) has replaced the concept of specific surface area. The dimension of the solid surface is generated during synthesis.

The particles of tetragonal zirconia ($t$-$ZrO_2$) are stable at room temperature if they are small enough and can be processed in toughened ceramics. Small particles of cobalt-doped $\gamma$-$Fe_2O_3$ are used in audio tapes if they are needle-shaped. A few other examples of how properties change with particle size are illustrated in Figure 7.1.

Polycrystalline solids develop a microstructure during their synthesis that can be very variable. Zinc deposited from an aqueous zinc sulfate solution by electrolysis at low voltages has an open fractal morphology. At high voltages the deposit is more dense and the growth front is uniform (it has a "random dense branching" morphology). An unexpected change in microstructure can be the result of a slight change in formation conditions (Figure 7.2). On the other hand, the same morphologies can arise as a result of very different processes, such as aggregation of particles, liquid infiltration, dielectric breakdown, and galvanic deposition, all of which result in the same ramified fractal form, as described below.

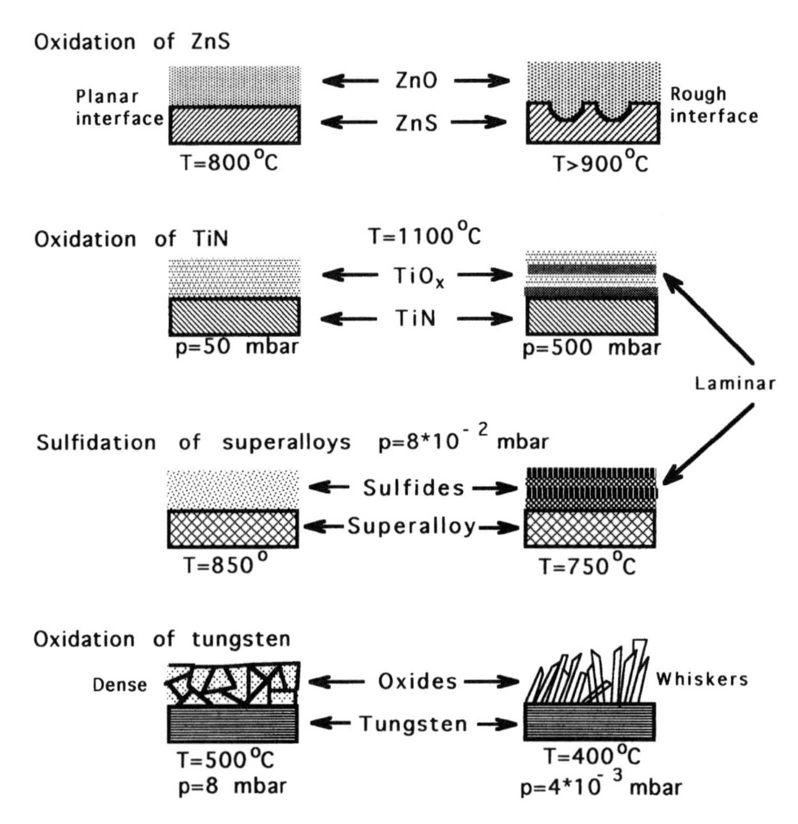


**Figure 7.2.** Some morphologies of product layers on the oxidized surfaces of materials under different conditions.

One example of morphology development during galvanic deposition is the overvoltage dependence of the microstructure (Figure 7.3). However, different observers see different things when they look at microstructures: Figure 7.3 illustrates how two authors describe the same structure differently. There appears to be a certain lack of reproducibility in reactions involving solids.

The properties of random solids made of several components have been a subject of intense interest for a long time because improved properties and entirely novel ones can be realized in composite materials. A survey has been done of how many physical properties of composites relate to properties of their components and their connection.[5,6]

In a solid state reaction at low temperatures a solid reactant may be entirely converted to a solid product that keeps the original structure of the reactant. In such a thermal process new nonequilibrium solids can be made that cannot form when the system would be close to equilibrium. The temperature cannot be too low so that the reaction remains possible (sufficient mobility), and it should not be so high as to prevent crystallization to equilibrium structures. This procedure is called "chimie douce" or soft chemistry.[7] An example of chimie douce from industrial chemistry is the oxidation of magnetite (a mixed valence form of iron oxide, $Fe_3O_4$) to the ferromagnetic $\gamma$-$Fe_2O_3$ (maghemite) in a microstructure that renders it suitable for

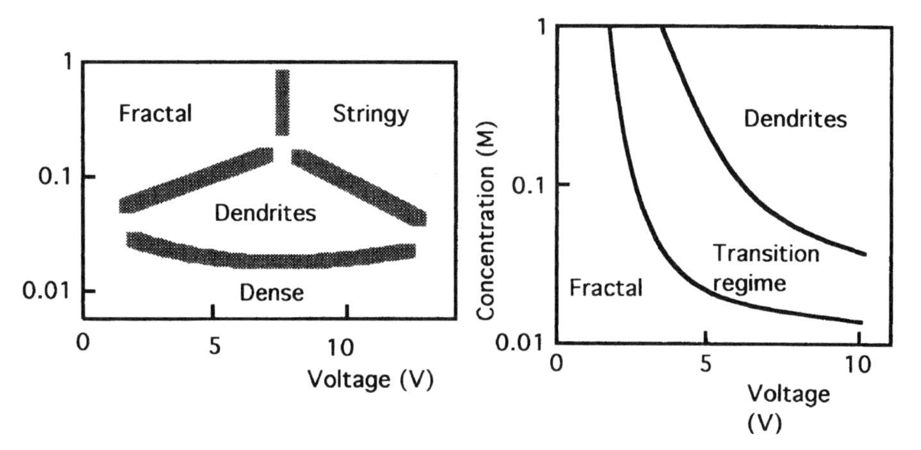


**Figure 7.3.** The morphology of electrodeposited metal and its voltage dependence as given by different observers. From J. Kertesz. Morphological transitions in pattern growth phenomena. In: *Random Fluctuations and Pattern Growth*, p. 42. With kind permission from Kluwer Academic Publishers.

use in magnetic tapes. Another example is the soft reduction of metal oxides with substituted boranes to colloidal metal particles. Intercalation recipes are also forms of chimie douce.

Recently a novel type of symmetry in random matter known as symmetry of scale was discovered. A material with this symmetry when viewed at small scales is seen to have the same microstructure (is not distinguisable) as when viewed at larger scales. For example, a photograph of the north wall of the Mont Blanc shows the same surface features as a grain from that wall in a scanning electron microscope; this system shows symmetry of scale. Figure 7.4 shows a colloidal agglomerate that has symmetry of scale (in a statistical sense, it is not an exact one). Use of symmetry is in general illuminating for discussions of properties and saves a considerable amount of work. Atomic and molecular spectra would still be a mystery without group theory, which is generally used in chemistry to exploit symmetry to the utmost.

Scale invariance is a symmetry that has provided a new view on random matter and suggests ways to improve processes, structures, and properties of solids. The concept of this symmetry has made it possible to understand and control reaction kinetics. The equations that are generally used for reaction rates were developed for three-dimensional idealized systems. With the help of the new symmetry they can be generalized and made applicable to random matter. Of course we must remember that application of this symmetry presupposes a great deal and models based on it are just models.

The discovery of a fractal symmetry in random matter was great news for the materials scientists. Until a decade or two ago a systematic study of the influence of synthesis conditions on morphology and therefore on the properties of random matter was difficult and therefore rarely done because a quantitative measure of random matter did not exist. The symmetry in random spatial distributions was recognized after the introduction of the concept of fractality by Mandelbrot. A fractal is a set that has a dimension that usually is not an integer. It is not 0 (a point), 1 (a line), 2 (a plane), or 3 (a volume) but a fractional number somewhere between

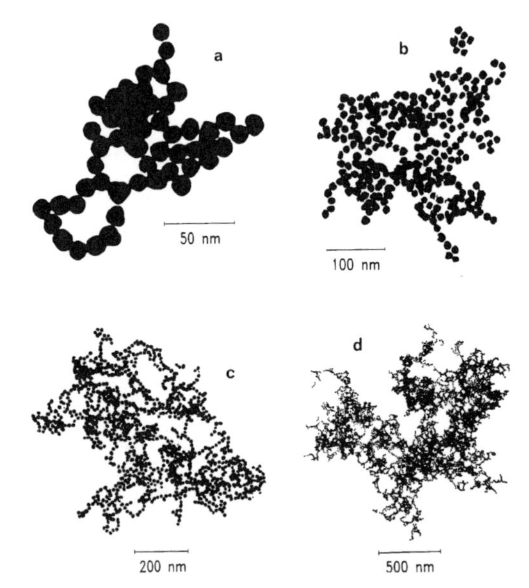


**Figure 7.4.** A cluster of particles in a gold colloid. Sizes are in nm. The agglomerate shows the same structure at different degrees of enlargement and differs only because the cluster size is evident at low magnifications and the primary dense and smooth particles become visible at high magnifications. From D. A. Weitz and J. S. Huang. In F. Family and D. P. Landau (eds). *Kinetics of Aggregation and Gelation.* North-Holland, Amsterdam (1984).

those values. The most remarkable feature of fractals is not that they have a fractional dimension, but that they have the property of scale invariance or self-similarity. This means that the form of the whole can be found in the parts or alternatively that the observed form does not depend on the scale or the enlargement of the image of the set. Mathematically designed geometric fractals such as the Koch curve or Menger sponge are regular and self-similar in an infinite range. In nature different types of fractals, random fractals, are found and not the simple geometric morphologies. These natural fractals are self-similar in a statistical sense and the range of self-similarity is limited to values between a lower and a high cutoff. Figure 7.4 shows both cutoffs: the size of the primary particles and the size of one cluster. Fractal forms can be found in biology, fracture mechanics, catalysis, corrosion, particle technology, amorphous solids, colloids, interface phenomena, chemical vapor deposition, and ceramics.

The morphology of random inorganic solids is often the direct consequence of local fluctuations in the formation process, e.g.,

1. Solidification of melts causes distributions that are used in certain solids. These distributions are fractals.
2. Spinodal decomposition of liquids or solids gives a distribution of the phases that has scale symmetry and these composites are fractals within a range of sizes.
3. Diffusion-limited aggregation of particles also yields fractal agglomerates that have scale symmetry within wide cutoff limits.

Synthesis involves a complex set of processes and complex systems often show chaos, in particular in autocatalysis. Chaotic processes diverge: two systems that follow the same process but have an unnoticed difference in start concentrations follow completely different paths. Chaos is not recognized if the system is locked into what is called a "stable attractor" or a "limit cycle." Then it seems to be simply causal and is usually easy to model. When the system is in a "strange attractor" it seems to run amok.[8] Chaos in synthesis can be recognized by several indicators:

- Nonlinearity, sudden irreversible changes, hysteresis, discrete chemistry.
- The breaking of imposed symmetry in a spatial or chemical sense; the occurrence of nonreproducibilities in synthetic reactions.
- Oscillations in reaction rates at constant average reaction conditions.

Fractals are solidified chaos. Usually chaotic behavior during synthesis is unwanted. Living nature however uses fractal structures that are very robust and comparatively insensitive to damage. The matter designer can make use of the typical properties of fractal structures. Heterogeneous catalysts also have a fractal surface but until recently very clean facets of idealized single crystals have been more popular in basic catalyst research.

The chaotic formation processes that have fractal morphologies are difficult to simulate with continuum mathematics, e.g., by solving partial differential equations that describe the diffusion and reaction steps. The boundary conditions of those differential equations change during the process and depend upon the solutions of those equations. Moreover considering nature as continuous (as Ostwald did) is an approximation that does not work at the nanolevel where statistical fluctuations have effects that cannot be described well by simulations with partial differential equations. There is, however, a more appropriate simulation technique that does incorporate fluctuations from the outset and has become possible with computers: the method of cellular automata, developed by Ulam and von Neumann. Its chemical use is described in Section 7.4. Cellular automata are pixel games played on the computer screen. The rules of the game simulate reactions and diffusion of the particles. Formally it has been proved that each dynamical problem can be simulated with a cellular automaton with the proper rules. Some of those simulations allow us to understand how diffusion, fluctuations, or subsequent or parallel chemical reaction steps affect chemically produced microstructures (chemomorphs). Cellular automata are suitable models to describe a novel type of chemistry that could be called digital chemistry. This is a discrete form of chemistry that produces sharp transitions in space or time. Examples are the chemical clock, Winfree spirals, nucleation with incubation time,[9] and spinodal decomposition.

The differential equation of Laplace ($\nabla^2 c = 0$) describes the formation of fractal distribution of solid matter that results from very different processes (Figure 7.5). Diffusion-limited aggregation ($c$ is the concentration), electrogalvanic deposition ($c$ is the electric potential), and viscous invasion ($c$ is the local pressure) are three Laplacian processes that produce similar fractal distributions. They all imply a strong positive feedback and have the same mathematics. The first two are of significance for materials synthesis.

In Section 7.3 the definition of fractal dimension is given and the experimental methods used to determine the dimension are described. Section 7.4 summarizes

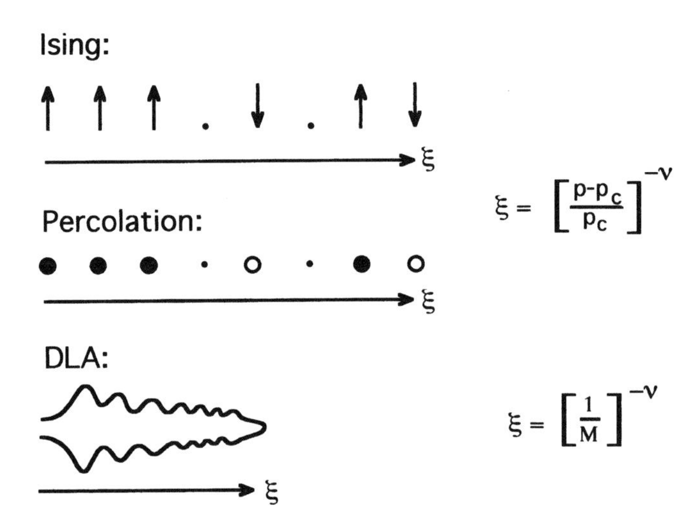


**Figure 7.5.** The analogy among three different phenomena: ferromagnetic ordering of lattice spins in clusters of size up to $\xi$ as calculated according to the Ising model; percolation clusters of size smaller than $\xi$; and diffusion-limited growth with size fluctuations up to $\xi$. All have the same power law behavior; $\nu$ is the reciprocal fractal dimension, $p$ is the occupation fraction, and $M$ the mass. From H. E. Stanley. In: *Random Fluctuations and Pattern Growth*, p. 1. With kind permission from Kluwer Academic Publishers.

recent experiments with cellular automata that have been used to model agglomeration of particles and clusters. Finally in Section 7.5 examples will be given of morphologies that have been made under transport limitation.

## 7.2. Extrinsic Properties of Materials

The extrinsic properties of materials are particularly interesting for materials technologists because these can be adjusted with the preparation. Examples of structural properties that depend on the morphology are toughness, yield strength, and wear resistance. The most important extrinsic properties by far of functional materials are mechanical, electrical, optical, and chemical. There are also many combination properties in composites such as thermoelectrical, thermochromic, electromechanical, electrochemical, optoelectric, and optochemical, and these are described in Chapter 9.

### Mechanical

Strength is a function of the microstructure and not of the crystal structure: the presence of macrodefects and their size and spread in a material are a result of processing during synthesis. Toughness in a material is somewhat related to the crystal structure (the possibility of dislocation transport) but is mainly a function of the microstructure: a very high or low crystallite aspect ratio and a high dimension of the fracture surface are signs of high toughness as shown by imitation shell (glued mineral platelets) or jade, a tough stone.

Hardness is a complicated property involving plasticity and compressibility, the latter being correlated with the density of bond energy and intrinsic. Adhesion is related to strength but does not depend on the strength of atomic bonding in the interface but entirely on the nature and size of the defects in the interface.

The mechanical functions are traditionally supposed to be the province of mechanical engineering but some mechanical properties (such as hardness and friction behavior) are related to chemistry. The high hardness and favorable tribological behavior of many ceramic compounds are used in hardcoatings on tools and machine parts to give them a longer lifetime, i.e., to make them mechanically and chemically inert (this is done chemically). They do not corrode and do not wear out easily. Examples of coating materials are diamond and diamond-like carbon, boron nitride, silicon carbide, alumina, borides, carbides, and nitrides of the early transition metals. Moving parts of precision instruments are protected against wear with a hardcoating. The friction coefficient of a material is not an intrinsic property either but is characteristic for its surface morphology.

Other illustrations of extrinsic mechanical properties are:

- An armor plate for individual protection consists of a plate of 6 mm of boron carbide, which is light and very hard, buffered by 6 mm of fiber glass. This combination will stop a 0.3-caliber antiarmor bullet of tungsten carbide.
- A nozzle opening of a sandblaster may suffer severe wear during operation. This can be countered by using hot-pressed boron carbide ($B_4C$), which is expensive but is very hard and has a long lifetime in this application. Sintered alumina or hardmetal (a cermet of tungsten carbide sintered with some cobalt or nickel metal) show more wear but are suitable as well and cheaper to make.
- A grinding stone is made up of hard sharp grains that should not be too tough because the grains should not become rounded during use. Small parts of the particles should break off leaving the remaining grain sharp. Porous silicon carbide is a suitable material for grinding wheels.
- Rubberlike ceramics have an electrically adjustable elasticity based on stress-induced electric effects in composites of mixed perovskites. This is an example of an active material, which is discussed below.

### *Electrical*

The electrical conductivity of semiconductors depends on their microstructure for several reasons. The grain boundaries can be conduction paths or barriers for electron transport either because the interphase material has a lower or a higher resistance than the bulk material or because there are grain boundary charges that generate depletion layers in the crystallites. This also holds for ion conduction: ionic conductors have a similar conductivity behavior.[10] The activation energy for ionic conduction in the intergrain phase usually differs from that in the grain itself, and therefore the temperature dependence of the ion conductivity in polycrystalline materials depends on the grain size. In the Arrhenius plots of two samples of the same material with different grain sizes the slopes change (if noticeable) at different temperatures.

### *Optical*

Light is scattered at grain boundaries if there is a change in refraction index and the degree of scattering depends strongly on the microstructure. Glasses are single-phased and there is only Rayleigh-type scattering by density fluctuations. If the wavelength of the light is chosen well, the degree of scattering is very low and fiber optics are used for information transfer by light at near-infrared wavelengths.

A remarkable optical effect of the microstructure is quantum confinement in very small (nanosized) particles. This size effect strongly modifies the optical behavior of semiconductors and metals. It is linked to the removal of the translation symmetry in crystalline solids.

### *Chemical*

Ceramic materials can accelerate chemical reactions in the chemical industry and in car exhausts to convert CO, $NO_x$, and left-over hydrocarbons. As was shown in Chapter 6, the microstructure is determinative for catalytic activity. So it is for the inverse of catalysis: slowing reaction rates in corrosion inhibition. Chemical passivity is strongly morphology-dependent: corrosive attack begins at the grain boundaries. Chemical passivity is as difficult to discuss in a systematic manner as catalysis. Both chemistries are descriptive, empirical, and not derivable from formal theory. Chemical properties of solids are mainly surface properties and therefore extrinsic. Ionic transport through solids depends on the morphology as well.[10] High-temperature corrosion of inorganic materials is described in Section 6.7.

## *7.3. Fractal Dimensions*

The dimension is one of the numbers that characterize the form of a point set or mass distribution.[11–14] There are mass distributions that do not have integer dimensions (0, 1, 2, or 3) but a noninteger one. Examples can be found in galaxies, coastlines, mountain surfaces, rivers, powders, colloidal agglomerates, polymers, heterogeneous catalysts, blood vessels, lungs, corals, cauliflowers, other plants, and many marine organisms.[15–17] Englands's coast is not a line that has a dimension 1 because the length of the circumference increases with the length of the measure used to determine it. With a measuring stick of decreasing length more capes and bays are included and the total measured length increases. The length observed with a measuring stick that approaches atomic dimensions becomes very large while a proper one-dimensional line has a finite limit for its length at decreasing size of measure. The dimension is the slope in a doubly logarithmic graph of measured length vs. size of measuring stick. The west coast of England has a dimension of 1.25 and that of Norway is 1.52: the circumference is not expressed in meters but in $m^{1.25}$ or $m^{1.52}$.

An example of a regular mathematical fractal (Figure 7.6) is the well-known Koch curve, which has a dimension 1.26 or the related Cesaro curve, whose dimension approaches 2 as the base of the pyramid gets smaller and the curve fills the plane. The fractal dimension of these curves is calculated with one of the definitions: $D = \log M/\log r$, where $M$ is the increase in the measure or weight of an $r$ times larger part of the set. For a Koch curve, $D = \log 4/\log 3 = 1.26$ and for the

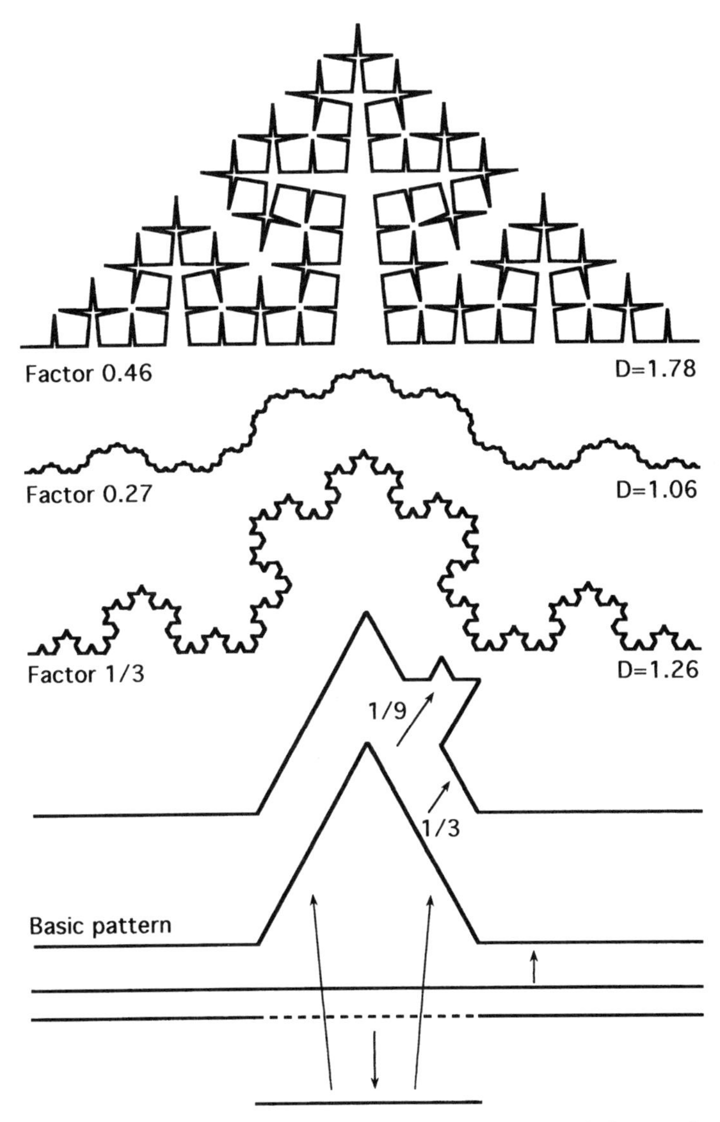


**Figure 7.6.** Fractal curves of infinite length, constructed by leaving out a middle part of a line segment and replacing it by two segments of the remaining length (the factor of the original line) as shown. This process is iterated on each new line segment and the line turns into the Koch curve. The size of the part replaced determines the dimension of the final curve or its space-filling capacity.

infinitely sharp Cesaro curve, $D = \log 4/\log 2 = 2$. These regular geometric curves are clearly self-similar.

Processes in nature often result in fractal-like forms that differ from the mathematical fractals such as the Koch curve in two ways: (a) the self-similarity is not exact but is a congruence in a statistical sense; and (b) the number of repeated splittings is finite and random "fractals" have an upper and a lower cutoff length. A spatial example of a random fractal is the colloidal gold particle agglomerate shown earlier in Figure 7.4.

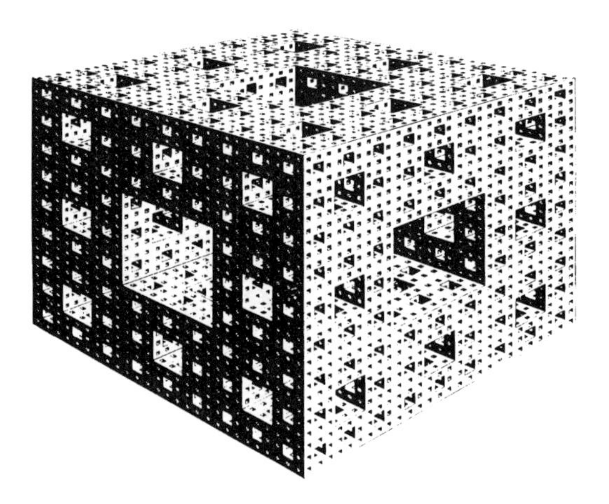

**Figure 7.7.** The Menger sponge, a geometric fractal made of a cube from which three central parts (cube stacks) have been removed. This leaves 20 of the original 27 subcubes in a cubic arrangement. All the remaining subcubes get the same treatment. If this is iterated an infinite number of times the Menger sponge of no weight and infinite surface area is formed. Its dimension is 2.73. From B. B. Mandelbrot. *The Fractal Geometry of Nature*. W. H. Freeman & Co., New York (1983). With kind permission from B. B. Mandelbrot.

Two definitions of random fractals can be applied in chemistry:

1. The dimension $D$ of a set indicates how a measure $M(r)$ of the set scales with the linear size $r$. If this size increases by a factor $a$, the measure changes according to $M(ar) = a^D M(r)$. Figure 7.7 shows another geometric example of a fractal, the Menger sponge, which, according to this definition, has a dimension $D = \log 20/\log 3 = 2.73$. Its inner surface is infinite and the volume and the weight zero. The same definition was used above in calculating the dimension of the Koch curve.
2. The minimum number of spheres of radius $r$ that can enclose the set is $N(r)$. If $r$ decreases with a factor $a$, then $N(r)$ increases with a value determined by the dimension $D$: $N(ar) \propto a^{-D} N(r)$. This definition is used for determining the surface dimension of solids using gas adsorption (Figure 7.8).

There are two different types of fractals in solid state chemistry: (a) mass fractals, sets of solid particles that form aggregates and have as measure their mass that scales as $L^D$ with $0 < D \leqslant 3$; and (b) surface fractals that consist of interfaces between solids and the vacuum and that have as measure the surface, which also scales as $L^D$ with $0 < D \leqslant 3$. The fractal dimension of an object, composite, or aggregate affects the values of the heat capacity, heat conductivity, electric conductivity, mechanical resistance against deformation, specific mass, and light scattering.

There are four methods of measuring the dimension $D$ of fractal solids:

1. Electrical conductivity at different frequencies, or impedance spectroscopy. This electrochemical method is used to measure the dimension of macroscopic surfaces of conducting solids. The surface is used as an electrode in a liquid electrolyte of a cell in which the impedance is determined at different

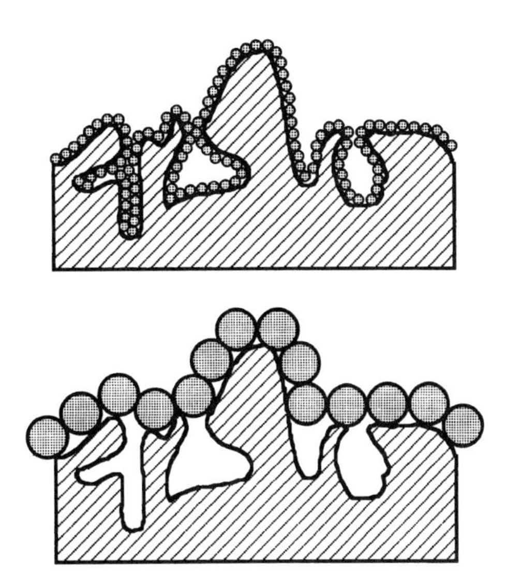

**Figure 7.8.** Schematic of gas adsorption on a rough surface by gases of different molecular size. The surface area measured with large molecules is smaller than when smaller molecules are used.

frequencies. The dimension of the surface can be derived from the impedance spectrum.

2. Scattering of light, X-rays, or neutrons at small angles. The intensity of the scattered light $I(q)$ is a function of the scattering vector $q$ [with $q=4\pi \sin((\vartheta/2)/\lambda]$, where $\lambda$ is the wavelength and $\vartheta$ the angle of the scattered light. $I(q)$ is also determined by the spatial distribution of the scattering centers, the particles in the agglomerate. Under suitably chosen conditions the scattering intensity $I(q)$ is proportional to $q^{-D}$ for mass fractals or particle aggregates and to $q^{D-6}$ for surface fractals. With this method one measures the dimension of agglomerates. If light or X-rays are used the electrons in the atoms scatter the light; if neutrons are used the cross section of the nuclei causes the scattering. If the three techniques are combined, they can cover a range of sizes from 0.1 to 1000 nm (Figure 7.9).
3. Adsorption of gases that have molecules of varying sizes (Figure 7.8). Monolayer gas adsorption can be used to measure the dimension of the outer surface layer of particles or aggregates. According to the second definition of the dimension of a surface, the specific surface area that is determined by the BET method (gas adsorption) will depend on the size of the gas molecules $\sigma$. If $n$ is the number of adsorbed molecules in a monolayer of surface, the surface dimension of the solid surface is proportional to the logarithms of number and size: $D \propto \log n/\log \sigma$. The slope in the graph of $\log n$ vs. $\log \sigma$ is half the surface dimension $D$. Table 7.1 gives a collection of surface dimensions of different powders determined by adsorption of different gases and measuring the change of adsorbed amount.
4. Sometimes digitized TEM (transmission electron microscopy) micrographs of an object are analyzed by computer to determine its dimension. In a TEM picture the intensity of a site in the image can be made proportional to the quantity of matter that the electron beam has traversed at that spot. The double-logarithmic graph of the size of an aggregate against its total

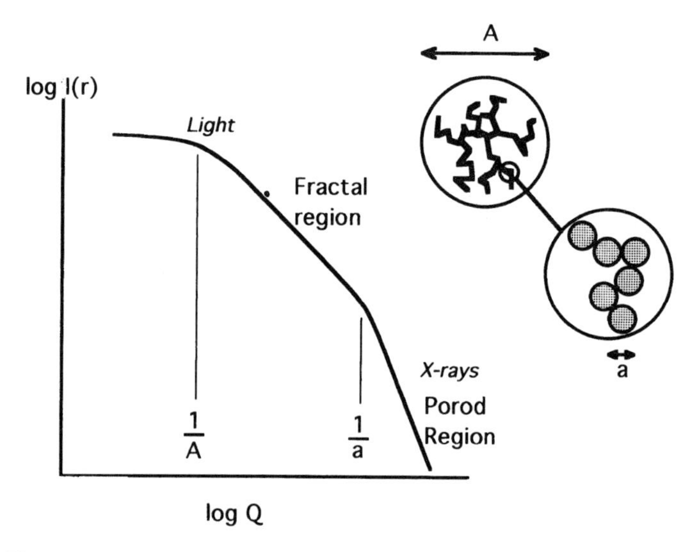


**Figure 7.9.** Scattering of electromagnetic radiation by clusters of particles.

integrated mass as determined in the micrograph for different aggregate sizes gives the dimension as the slope.

5. The fractal dimension of composed powder particles can be determined by measuring both the volume and the mass of the agglomerates of different sizes.[18]

The fractal dimension of an agglomerate is not the only number that characterizes the morphology. There are other numbers, related to the fractal dimension, that characterize properties such as diffusion coefficients and mechanical strength. In the case of nonfractal objects or sets, these numbers are equal to the dimension $d$ of the topological space ($d = 3$ mostly).

The tortuosity $\tau$ of a fractal is a scaling factor for the distance between two points in the fractal set. If $L$ is the shortest distance in a set between two points that have a topological separation distance $a$ (as the crow flies) then $L(ka) = k^{\tau}L(a)$, with

**Table 7.1. Surface Dimension of Powder Particles Determined with Gas Adsorption**

| Solid | Dimension | Solid | Dimension |
|---|---|---|---|
| A graphite | 2.07 | Magnetic $\alpha$-FeOOH pigment | 2.57 |
| Fused and milled periclase | 1.95 | A soil | 2.29 |
| An active charcoal | 2.04 | Dolomitic rock | 2.91 |
| Zeolite | 2.02 | Porous silicic acid | 2.94 |
| Aerosil | 2.02 | A granulated activated carbon | 2.71 |
| A carbon black | 2.25 | An alumina catalyst carrier | 2.79 |
| Porous charcoal | 2.54 | A soil | 2.92 |
| Crushed lead glass | 2.35 | | |

$\tau$ being also a measure of the elastic deformability of a set and numerically equal to its so-called backbone dimension $D_T$. The backbone dimension belongs to a set that obtains from the original set if the dead ends in the set are removed.

The chemical dimension $D_c$, also called the spreading dimension, is a measure of the ease by which a chemical reaction spreads on a fractal: $D_c = D/\tau$.

The spectral dimension $S$ is a function of the branching rate of the set and indicates how the whole aggregate can vibrate. The vibration density is proportional to $\omega^{S-1}$ at low frequencies $\omega$. The specific heat of the fractal aggregate is then proportional to $T^S$ (in which $T$ is the absolute temperature). The heat conductivity is a function of the distance $a$ and is proportional to $a^{-2(-1+D/S)}$. A typical value for $S$ is 4/3 in simple percolation clusters.

## 7.4. Simulations of Reaction–Diffusion Processes Using Cellular Automata

A cellular automaton is an algorithm that permits digital simulation of growth processes including those of the sort that cannot be described with partial differential equations.[19] Cellular automata are better adapted to the computer than continuum mathematics, the latter being a 17th century invention, more appropriate for life without computers. A cellular automaton is simple: space is covered by discrete cells that have a digital value that may also be a vector. The value or vector content of the cells evolves in time. Time in a cellular automaton is also discrete and is a series of instants. At each next instant the content of each cell is updated according to local rules that involve only its present content and the present content of its nearest neighbors. The rules of the automaton are chosen so that they give results that are similar to existing processes.

Three groups of experiments in cellular automata have been useful for understanding the formation of microstructures: (1) agglomeration of particles to form clusters and agglomeration of clusters; (2) percolation; and (3) reaction–diffusion processes in or on solids.

### Agglomeration

The dimensions of aggregates depend upon the path of the particle or cluster before they are incorporated in the aggregate.[20] There are several types of agglomerates (Figure 7.10), for example:

DIFFUSION-LIMITED AGGREGATION (DLA) OR THE WITTEN–SANDER MODEL. A pixel is fixed in the middle of the screen and a mobile pixel, a random walker (RW) is added on a circle around the fixed pixel. When the moving pixel hits the fixed pixel or other immobilized pixels on the screen it is stopped, locally fixed, and a new RW is activated on the circle. The fixed initial point in the middle grows in size with captured RWs but not as a dense blot. Local fluctuations are enhanced and the aggregate of pixels becomes strongly ramified because the outer branches of the growing dendrites prevent the incoming pixels from filling the indentations by catching them before they can reach the shielded interior parts of the aggregate. It is a fractal cluster that has a dimension of 1.71 on a screen with $d = 2$. In a

**Figure 7.10.** Simulation in two-dimensional space of three types of monomer–cluster agglomerations: (a) diffusion-limited aggregation (DLA), where clusters are ramified and have pores of all sizes between cutoff values; (b) reaction-limited aggregation with dense clusters; (c) ballistic aggregation when the paths of the particles are straight and the sticking factor is 1. From P. Meakin. Models for colloidal aggregation. *Ann. Rev. Phys. Chem.* **39** (1988) 237. Published with permission from Annual Reviews, Inc.

three-dimensional space a fractal agglomerate of pixels or particles that results from DLA would have a dimension of 2.52. One modification of the rules would be lowering the sticking probability on collision: there is a chance that the pixel that hits an immobilized pixel is not immediately fixed (sticking factor $= 1$) but continues its random walk until the next collision (sticking factor $< 1$). Such a rule would give thicker dendrites. Figure 7.10 a shows the pure DLA-type cluster that is formed.

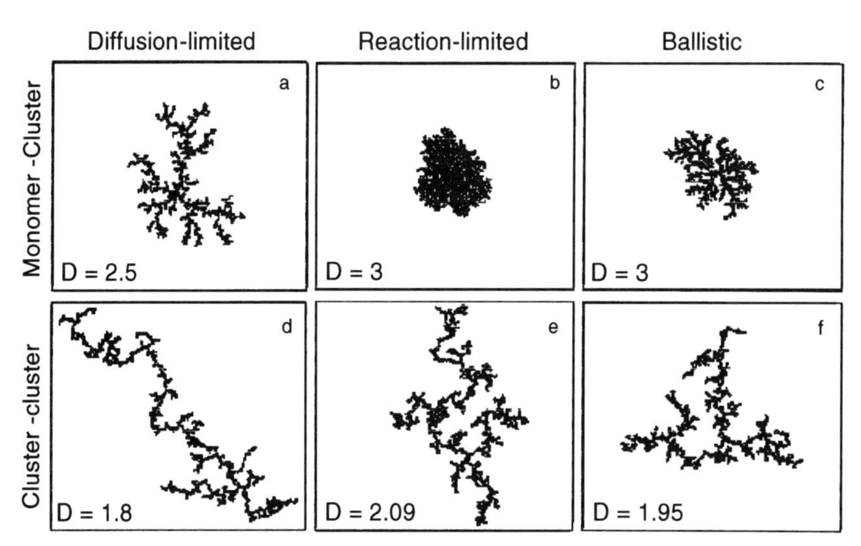


**Figure 7.11.** Simulated particle clusters and their dimensions. Morphologies of clusters grown by adding single particles and the dimensions ($D$) of the clusters are given in a, b, and c. Images are in two-dimensional space, and their dimensions $D$ are given for three-dimensional clusters. Case b is sometimes called Eden growth. When clusters themselves also agglomerate the dimensions of the aggregates are much lower, as shown in d, e, and f. Images are made in two-dimensional space, and dimensions are given for clusters that would form in a three-dimensional reaction space. Reprinted with permission from D. W. Schaefer. Polymers, fractals, and ceramic materials. *Science* **243**, 1023 (1989). Copyright (1989) American Association for the Advancement of Science.

REACTION-LIMITED AGGREGATION (RLA) OR THE EDEN MODEL (Figure 7.10b). An Eden cluster grows by a surface reaction (like cell colonies), not by collecting RWs from the outside with sticking factor 1, but rather by occupying at random an empty place next to an occupied one in the cluster at each instant. The resulting cluster is not fractal but dense, and has a dimension $d$, which equals that of the topological space of the reaction. Some variants that have site-dependent growth rates have been tried. If the sticking factor of a DLA process is made very low, it becomes an RLA process and an Eden cluster obtains.

BALLISTIC AGGREGATION (Figure 7.10c). This is similar to DLA but the incoming monomers that are trapped by the cluster do not follow a Brownian path but straight lines. The agglomerate that results is a bit more porous than in the case of RLA but it is not fractal; the dimension of the cluster is $d$ (3 in space and 2 on a flat screen).

CLUSTER–CLUSTER AGGLOMERATION. What has been described above for aggregation of moving primary particles and fixed clusters can be done with a dilute mixture of many pixels that move simultaneously and can connect to others to form moving clusters that also stick when they hit each other to form larger moving clusters. The gas of single pixels evolves to one supercluster in the end. Such aggregations can be made to occur ballistically, by a reaction-limited cluster aggregation process (RLCA) or a diffusion-limited growth (DLCA). Such aggregates have low dimensions. Sometimes variable sticking factors are used. The aggregation processes and the shapes of the resulting aggregates are summarized in Figure 7.11.

### *Percolation*

Percolation literally means flows oozing through a porous solid. Figure 7.12 illustrates the concept for materials. A vessel filled with marbles of which an increasing fraction $p$ is replaced by steel balls becomes electrically conducting if $p$ exceeds a critical fraction $p_c$ (by a percolation path of conduction). The percolation threshold is 0.3 for a primitive cubic three-dimensional lattice and 0.539 for a quadratic plane lattice. There is also a percolation threshold for the other component. Percolation in a composite of two types of particles can also be simulated with a cellular automaton. Here the cellular automaton is not a process in time but a microstructure that is the result of an occupation chance $p$ of each lattice site or particle position. The percolating cluster, the connected aggregate of one component at the percolation threshold, is a fractal. If some interaction between the particles is introduced in this automaton, other morphologies result.

Percolation is used to describe properties of composites. As percolating clusters are fractals some scale invariance holds for them. The particular composition of the different phases in the solid determines the path for transport of atoms, electrons, or phonons (heat) through the solid, which can be nonlinear and "anomalous." For example, the magnetic properties of composites that have one ferromagnetic phase are seen to be remarkably dependent on percolation (Figure 7.13). Section 9.3 shows the particle size dependence of the percolation threshold.

Another example of the use of percolation in inorganic chemistry is the SHS reaction as described in Chapter 6. The reaction between two solid powders (such as the reaction between titanium and boron particles) can propagate after local ignition only if the heat of reaction is high enough and loss of heat low enough.

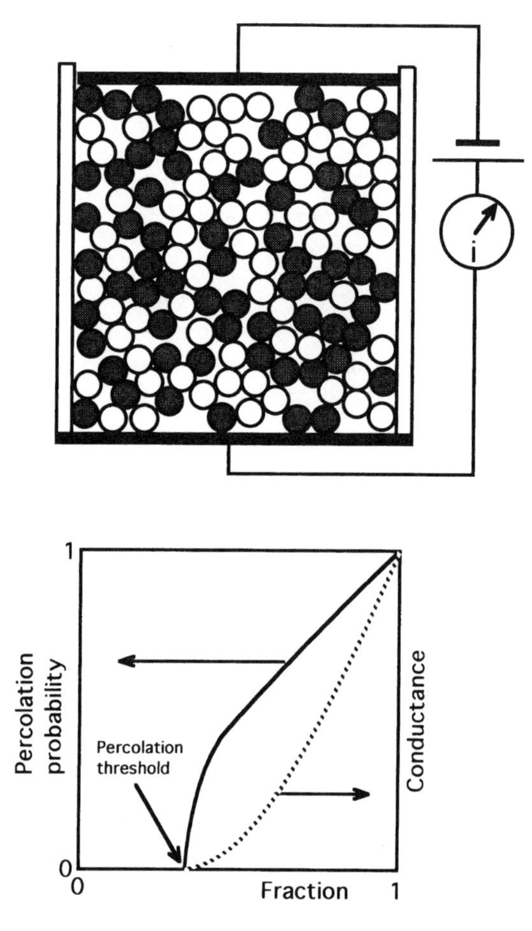


**Figure 7.12.** Percolation in a random mixture of equisized conducting and insulating particles. At a threshold concentration of conducting particles the mixture starts to conduct current. One of the coherent clusters of those particles then spans the vessel. The fraction of particles that are part of those conducting clusters is the percolation probability.

Moreover the reactant powders have to be mixed at a concentration between the two percolation limits. Mixing in ratios outside the percolation range means that there are only isolated small clusters that have a noncoherent boundary between the two types of reacting particles and thus the SHS reaction cannot propagate through the whole powder mass.

### *Reaction–Diffusion Processes*

A classical example of reaction-diffusion processes is the Belousov–Zhabotinsky reaction that forms moving Winfree spirals in a two-dimensional reactor. In this reaction the many sequential reaction steps comprise a very complex system, yet the spirals can be simulated with a very simple three-state cellular automaton. This model accounts for the morphology given certain initial conditions but it does not

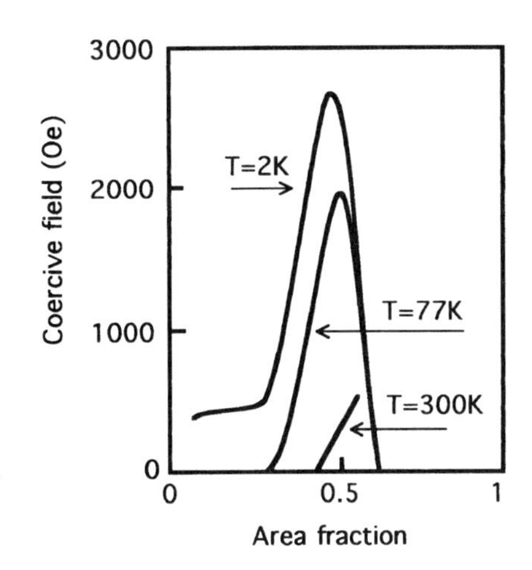


**Figure 7.13.** The coercive field of a granular film of iron silicate on a flat surface. It depends strongly on two-dimensional percolation of the particles.

provide a reductionistic explanation in terms of reaction steps as is usually attempted in chemical kinetics.

## 7.5. The Chemistry of Fractal Distributions

### 7.5.1. Processes that Generate Fractal Distributions

Two types of chemical behavior of fractal surfaces must be considered: how the fractal surfaces are formed by reaction with fluids (morphogenesis) and how the fractal surfaces of solids react with fluid reactants. The first will be addressed here, the second in the following section.

Surface reactions, agglomeration, and spinodal decomposition are known to yield fractal mass distributions but other processes, such as fracture, milling or etching, also form scale-invariant surfaces.[21–24]

As the name implies, fracture is a source of fractal surfaces. Fracture surfaces of broken steel have a surface dimension of 2.28, which has been measured by covering the rough fracture surface of the steel with a nickel coating and partially removing it by grinding and polishing and then analyzing the nickel–iron pattern that results. The surface and perimeter of the steel domains in the nickel matrix are measured and from these data the dimension of the steel fracture surface is derived. The dimension of the fracture surface depends on the toughness of the steel, which can be influenced with a heat treatment. A tough material has a fracture surface that has a high dimension. The relation between fracture toughness and dimension $D$ is $K_{IC} \propto \sqrt{D}$. In principle the toughness of a material can be read from the surface dimension of a milled powder of that material.

Another example of a fractal surface is anodically etched $p$-type silicon. It has a fractal porosity and shows photoluminescence and even electroluminescence, which would be impossible in compact bulk silicon. Electrons behave differently in fractally distributed solids than in three-dimensional extended lattices: they are supposed to be more localized. The density of states $\rho$ in a three-dimensional box is

$\rho(E) \propto \sqrt{E}$ (with $E$ the orbital energy), while in a fractal distribution, e.g., in a percolation cluster with a dimension four-thirds, the density of states is $\rho(E) \propto E^{-1/3}$. There are other differences between three-dimensional bulk solids and fractal solids. The bandgap or what passes for it in fractal solids is larger and the wavenumber $k$ is no longer a good quantum number because there is no translation symmetry. In short, there cannot be bands in fractal solids as we know them in infinite lattices, and this has consequences for the interpretation of optical and electrical properties. It is not surprising that fractally distributed semiconductors and metals are electrically very different from three-dimensional large blocks that have well-behaved two-dimensional surfaces.

There are two fundamentally different ways that a solid can precipitate from a homogeneous fluid (gas or liquid) or in a homogeneous bulk solid: by spinodal decomposition or as a nucleation and growth process. The morphologies of the products of these two processes are very different, as described in Section 4.2.

Spinodal decomposition occurs when the free energy curve in the graph vs. the composition $x$ is convex or, in other words, if the second derivative is negative:

$$\frac{\partial^2 G}{\partial x^2} < 0$$

The mixture is then unstable against any small fluctuation in composition and demixing occurs gradually, provided there is enough mobility. The atoms diffuse uphill, against the concentration gradient. Initially concentration differences are not sharply bounded, but the nucleation/growth process does have sharply bounded phases. Borosilicate glass demixing as described in Section 4.2 is an example of spinodal decomposition, while the formation of glass ceramics is a case of nucleation and growth. Vycor glass (and Raney nickel) has a morphology resulting from spinodal reaction. It has spatial density oscillations with a distribution of frequencies. In the short interval from 1 to 10 nm the surface of the $SiO_2$ phase has a dimension of 2.4. Above 10 nm the morphology becomes more involved, and the dimension is rather low.

When phases precipitate or separate in a solid, the particles do not agglomerate because they do not diffuse, but they grow until they touch each other. Solid particles that are formed in a fluid phase (colloidal or aerosol processes) move around and aggregate when they collide and sinter together.[24] The agglomerates may have a fractal morphology in some cases, depending on the reaction conditions. The primary particles have a very narrow size distribution but the clusters are fractal and often have a dimension of approximately 2.5, which points to diffusion-limited clustering and fast reaction between monomer and cluster as was discussed in Section 7.4.

Powder is preferably made from the gas phase in those cases where size, size distribution, shape, composition, and structure of particles should be within narrow limits, e.g., pigments. Nonoxides particularly are made more easily with gas-phase reactions than from liquids. Silicon nitride powders for engineering ceramics need to be fine and monodisperse. For a high final density after sintering, the primary particles must be dense; for a high sinter activity they must be in the nanometer range while they must be weakly agglomerated for handling. How this can be achieved can be derived in principle from computer simulations: ballistic agglomeration or Eden-type growth at high temperatures is preferred followed by low-temperature DLCA, which is needed in order to harvest the agglomerates. Care has to be taken to avoid hard agglomerates, which are ramified and cannot be well densified with colloidal techniques because they do not stack well and cannot be broken up (deflocculated).

Silica particles that have been made with the sol-gel method have a primary particle size and a cluster dimension that depends on the chosen conditions in the precipitation process. The measured dimensions of the polymeric silica particles are around 2, close to the computer-simulated value of 2.1 of the DLCA clusters. In polymerization of $Si(OH)_4$, the dimension of the product depends on the catalyst used. When a base is used as the catalyst to form a silica colloid the particles are compact because bases catalyze Eden-type growth of silica. These rather dense particles do not easily agglomerate since they have a negative surface charge at these high pH values. A green form of silica made from basic colloidal solutions is porous because it consists of stacked dense particles and has a considerable interparticle space. Acids catalyze the formation of less branched and more linear polymers that readily entwine even at the beginning of the polymerization reaction. From acid solutions of the precursors rather dense silica solids can be formed after removal of the solvent because the polymers stack with less interstitial space than the particles formed during base catalysis. On drying acid-catalyzed gels shrink more than base-catalyzed gels.

An aerogel also shows a fractal distribution in a certain size range, as can be seen from the small-angle light scattering behavior illustrated in Figure 7.14. This figure also shows the scattering plot of Vycor glass. Etching or attacking the surface by partial dissolution can generate a fractal surface. Figure 7.15 shows, e.g., the scattering behavior of a powder of colloidal dense silica particles that have a surface dimension that increases in time by etching in a basic solution.

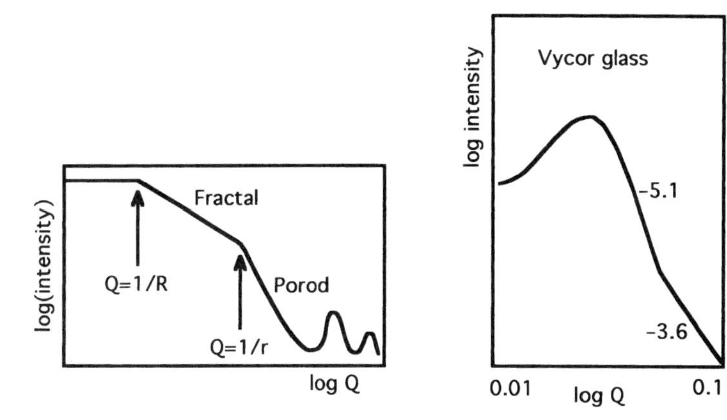


**Figure 7.14.** The small-angle diffraction pattern of a silica aerogel (left) that consists of clusters of size $R$ built up of dense smooth particles of size $r$. Within a size range bounded by $r$ and $R$ the clusters are fractal. At right the scattering pattern of Vycor glass is shown. From D. W. Schaefer, A. J. Hurd, and A. M. Glines. In: *Random Fluctuations and Pattern Growth: Experiments and Models*, p. 62. With kind permission from Kluwer Academic Publishers.

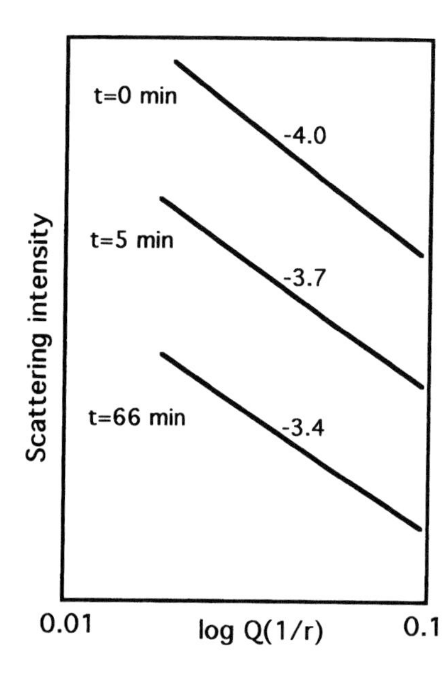


**Figure 7.15.** The gradual increase of the surface dimension of aerosil on etching. The initial Porod slope of −4 decreases in time. From D. W. Schaefer.[16] From D. W. Schaefer, A. J. Hurd, and A. M. Glines. In: *Random Fluctuations and Pattern Growth: Experiments and Models*, p. 62. With kind permission from Kluwer Academic Publishers.

Features of particle flocculation and gelation in colloids of considerable significance for ceramics are: (1) The kinetics of agglomeration. How fast do particles in suspension agglomerate? (2) The structure of agglomerates. How many particle aggregates of a given size are formed?

The study of the rates of aggregation of colloidal particles owing to Brownian motion has a long history and was initiated by von Smoluchowski in 1916. Several solutions to the Smoluchowski equation have been attempted since then and two will be sketched here. The problem is to establish the time evolution of the number and size of the agglomerates in terms of the number of monomer particles. The expressions are typical for simplified models as is usual in physical chemistry.

Starting with a number $N$ of monomers that form oligomers $M$ on reaction, which in their turn agglomerate to form higher polymers, the reaction equation is $M_i + M_j \rightarrow M_k$, with reaction rate constant $\kappa_{ij}$. The indexes $i$, $j$, and $k$ stand for $i$-, $j$-, and $k$-fold oligomers and $k = i + j$. According to classical kinetics (which is a mean field theory and ignores fluctuations), the reaction rate depends only on the number of reactant molecules (or concentration):

$$\frac{dn_k}{dt} = \frac{1}{2} \sum_i \sum_j \kappa_{ij} n_i n_j - n_k \sum_i \kappa_{ik} n_i$$

The first (double) summation is over $i$ and $j$ with $i + j = k$. The number of $i$-mer clusters is $n_i$. The use of numbers instead of concentrations implies a constant volume. The reaction rate constants $\kappa_{ij}$, which depend on the size $i$ and $j$ of the clusters, contain the chemistry of the agglomeration problem.

The first solution supposes size-independent reaction rates. Von Smoluchowski gave the solution for constant rates ($\kappa_{ij} = \kappa$ for all $i$ and $j$):

$$n_k(t) = \frac{N[(\kappa N/2)t]^{k-1}}{[1 + (\kappa N/2)t]^{k+1}}$$

For a given size $k$, $n_k$ goes through a maximum and tends to zero as $t^{-2}$ for large values of $t$. The mean cluster size $\langle k \rangle$ increases with time ($\langle k \rangle \propto t$) while the total number of clusters decreases ($N_c \propto t^{-1}$). For $t \gg 2/\kappa N$ the solution reduces to

$$n_k = \frac{4}{k^2 \kappa^2 N} \left(\frac{k}{t}\right)^2 \exp\left[\frac{-2k}{\kappa N t}\right]$$

The second solution assumes ballistic agglomeration. Friedlander[25] estimated cluster growth by aggregation of monomer particles: (a) assuming that the mean free path of monomers and clusters is much larger than the cluster size; (b) assuming spherical $i$-fold clusters of radius $R_i$, which radii vary as $i^{1/3}$; and (c) taking as the reaction probability the product of cross section and relative speeds: $\kappa_{ij} = \sigma_{ij} v_{ij}$. It follows: $\sigma_{ij} = \pi(R_i + R_j)^2$; $v_{ij} = (v_i^2 + v_j^2)/2$ and $v_{ij} \propto kT/m_i \propto i^{-1/2}$. The ballistic assumption leads to an average cluster size $\langle k \rangle$ increasing as $t^{6/5}$, while for large values of $t$ the number of clusters of size $k$ decreases as

$$n(t \to \infty) \propto \frac{1}{k^2} \left(\frac{k}{\langle k \rangle}\right)^{-\frac{1}{2} - 5k/(6\langle k \rangle)}$$

If the reaction constants scale as $\kappa_{\lambda_i, \lambda_j} = \lambda^\omega \kappa_{i,j}$, then several general conclusions concerning the solutions have been derived: (1) If $\omega < 1$ (flocculation regime), the size distribution always decreases with $k$. (2) If $\omega > 1$ (gelation regime), an infinite cluster (the size of the reactor vessel) appears after a finite time $t_g$. At $t_g$ the change in behavior of the suspension implies a gel point.

### *7.5.2. Reactions on Fractal Surfaces*

Formation processes often determine the dimension of the surface of the solid but the dimension of a surface also affects its chemistry.[26,27] They are closely linked: the growth reaction at a surface affects the surface morphology, which in turn affects the reactivity. This reciprocal influence can make the overall reaction oscillate, which was shown for two cases in Chapter 6. These are examples of positive feedback in surface reactions.

Another chemical consequence of broken dimensions in heterogeneous catalysts is that the parameter "specific surface area," expressed in square meters per gram, is not defined if the gas that is used to determine that parameter is not mentioned. One should add to the surface area that is always mentioned with catalysts the dimension that characterizes its fractal surface. The dimension does not depend on the size of the molecules that measure the surface area in the BET method but the specific surface area does. Concepts such as porosity and permeability as well as concentration and density should be redefined because dimensions no longer need to be integer.

Chemical reactions on surfaces often have a noninteger and time-dependent reaction order[3,21,26-28] for a simple reason. The surfaces of solids are themselves fractal because they have been formed by fracture, by dissolution, or by deposition under diffusion-limited conditions. The reactive surface is therefore fractal but the distribution of reactive sites on the fractal surface may itself be fractal-like with a different dimension. During heterogeneous deposition of solids a certain distribution of the active growth sites on the growing surface imposes the morphology on the solid product. During the deposition the surface dimension and the reaction dimension $Q$ (see below) can change, which means changing growth rates, morphologies, surface dimensions, and reaction dimensions during the course of the reaction.

Diffusion on fractal surfaces is less easy than in topological spaces with an integer dimension. As in the diffusion of atoms, the conductance properties also depend on the dimension. Reactions are so frequently "anomalous" that many reports on rate measurements, interpreted with two-dimensional surfaces, conclude with the note that the observed data invoke new questions that will have to be studied in the sequel.

The reactions of deposition or crystal growth are surface reactions. The reactants are adsorbed, more or less mobile molecules, e.g., *A and *B in the fictitious reaction $^*A + {}^*B \rightarrow 0$. These adsorbates form the substrate surface and growth is the annihilation reaction between the adsorbed reactants. The reaction rate $r$ is expressed as usual (Chapter 6) in the reactant concentrations as $r = k[^*A][^*B]$. This can be done if the surface (the reaction space) can be considered to be a well-stirred reactor. In other words, the mobilities of *A and *B are high compared to the rate of the growth reaction. If that is no longer true and there is diffusion limitation the reaction can still be fitted to the above rate equation except that the reaction rate coefficient $k$ is replaced by $k_1 t^{-h}$ in which $h = 1 - \frac{1}{2}S$ (with $S$ being the spectral dimension). A characteristic value for $h$ is $\frac{1}{3}$ for the case of a reaction on a percolation cluster with a spectral dimension of $\frac{4}{3}$.

The reaction rate between adsorbed species in a diffusion-limited reaction decreases in time by demixing, which is the result of thermal fluctuations in the initial concentrations. There are regions with an excess of adsorbed A and regions with a slight excess of adsorbed B. After a while there are domains over the surface with only leftover adsorbed A and regions with only B. The reaction between the adsorbed reactants can then only take place on the boundaries between the different domains and the global rate depends on the overall length of those boundaries. This demixing also results in the growth of scale-invariant cauliflower morphologies, which is often observed in gas-phase deposition.

Apart from the fractal dimension $D$ of the surface, there is a so-called reaction dimension $Q$ (sometimes also called $D_r$), which characterizes the way in which the rate of the reaction $r$ on the surface scales with the size of the particles $L$. The power law is $r \propto L^Q$. The value of $Q$ is usually between 1 and 3 but extremes such as 0.2 and 5.8 have also been observed. The distribution of chemically active sites on the surface determines the way in which the reactivity scales with the size of the crystallite.

To get an impression of the meaning of $Q$ the chemical activity of a few simple forms can be considered. If the active sites of a crystallite are only its corner points,

Table 7.2. The Reaction Dimension of Heterogeneous Catalysts



| Catalyst | Dimension | Reaction |
|---|---|---|
| Pt on $TiO_2$ | 1.4 | Photocatalytic decomposition of alcohol |
| W on C | 2.21 | Electro-oxidation of hydrogen |
| Pt on C | 1.85 | Electroreduction of oxygen |
| Fe on MgO | 5.8 | Ammonia synthesis |
| Fe on MgO | 3.0 | Carbon monoxide methanation |
| Ru on $Al_2O_3$ | 4.0 | Carbon monoxide methanation |
| Pt on $SiO_2$ | 1.95 | Carbon monoxide oxidation |
| Pt on $SiO_2$ | 1.58 | Propene hydrogenation |
| Pt on WC | 5.24 | Benzene hydrogenation |
| Ag on $SiO_2$ | 0.71 | Oxidation of ethylene |
| NiMo on $Al_2O_3$ | 0.1 | Coal liquefaction |
| Pt unsupported | 0.1–0.4 | Electroreduction of oxygen |
| BiMo | 4.0 | Dehydrogenation of butadiene |

$Q = 0$, because the number of those vertices does not depend on the size, and on a growing crystallite the chemical activity remains constant. If only the ridges are chemically active, $Q = 1$, because the length increase of the edges is proportional to the linear size of the crystallite. Similarly, if the faces are homogeneously active, $Q = 2$ and the chemical activity is proportional to the surface area of the crystallite. If all sites in a particle are active, $Q = 3$. If $Q < 2$ the active sites at the surface may form a fractal subset on the two-dimensional surface of the particle. A value of $Q > 2$ may indicate a fractal surface that is entirely reactive. Table 7.2 gives the reaction dimension of several catalyst surfaces. The growth rate $r$ of particles expressed in mass increase per unit of weight scales with their size $L$ as $r \propto L^{Q-3}$. Hence the reaction dimension of the process for making solid particles affects the size distribution of the particles that are formed.[28]

The reaction dimension $Q$ may be the same as the fractal dimension $D$ of the reactive surface or it may be different. If $Q < D$ there may screening or poisoning of reactive sites. A value of $Q$ larger than $D$ usually means that the reaction occurs only in the micropores. The value of $Q$ is not a constant and may change during the course of the reaction.

All this is valid only in the catalytic or reaction-limited regime (see Chapter 6). There is another effect of the fractal dimension of the surface on the growth rate and on the morphology of the resulting product. This is the ease of transport of the reactants to the surface before they adsorb and react. In the dialytic regime, e.g., in the case of an Eley–Rideal mechanism with slow diffusion, the reaction occurs mainly on the top of a fractal surface and less on the less accessible parts. The Eley–Rideal expression of the rate is then raised to a power dependent on the dimension of the surface.[29] Higher surface dimensions mean relatively higher reaction rates because the reactants have better accessibility to the surface.

Most surfaces encountered in practice are fractal (always in a bounded range of sizes) and that means that the usual way of discussing surface phenomena again need to be reconsidered for these ranges. Surfaces of solids and grain boundaries are singularities and therefore thermodynamic anomalies. In practice their behavior deviates from the simple rules based on minimalization of what, for convenience, is called "surface energy." Surface energy presupposes an equilibrium and that requires sufficient mobility. Liquid drops have mobile atoms and can therefore develop well-behaved two-dimensional surfaces if the driving forces are favorable for them. However, atoms on solid surfaces are usually not mobile and what holds for liquid droplets does not apply to solid particles. Most solids have a surface dimension larger than two and the value of energy per square meter is undefined and unrealistically high if a measuring scale is chosen that is much smaller than one that is relevant for the function of the material.

## Exercises

1. List a number of extrinsic properties of inorganic materials and show how they could be changed with conditions during synthesis.
2. Explain what a fractal mass distribution is and how its dimension is measured.
3. Describe the processes that form fractal solids.
4. Give the definition of fractal dimension and describe four methods to measure it.
5. Give the consequences of the dimension of a mass distribution on some of its physical and chemical properties.
6. List the factors that determine morphology.
7. Indicate the role of the Laplace equation for morphogenesis.
8. Name some indicators for chaotic processes in synthesis.
9. Describe the rules for three cellular automata to simulate morphologies.
10. Under which process conditions are dense agglomerates of particles formed and which processes lead to fractal agglomerates? Agglomerates of particles can have various dimensions when they are fractal. Describe some actual aggregation processes and give the dimension of the products of those processes.
11. Describe the role of percolation in explaining corrosion of stainless steel and brass.[30] Use the localized electron model of Linnett to account for the effect of boron and arsenic on corrosion inhibition.
12. From a small-angle scattering graph of a sample of agglomerates derive the dimensions of the agglomerates and the surfaces of the powder particles.
13. Describe the effect of fracture, etching, and spinodal decomposition on the surface dimension.
14. Derive the time dependence of the reaction rate coefficient of a reaction on a fractal surface.
15. Indicate the sources of anomalous reaction and diffusion rates on fractal surfaces.

## References

1. A. Mocellin. An introduction to the morphological characterization of sintered structures. *J. Mater. Educ.* **4**, 211 (1982).
2. E. Hornbogen. A systematic description of microstructure. *J. Mater. Sci.* **21**, 3737 (1986).
3. R. Mezaki and H. Inoue. *Rate Equations of Solid-Catalyzed Reactions*. University of Tokyo Press, Tokyo (1991).
4. A. Harrison. *Fractals in Chemistry*. Oxford University Press, Oxford (1995).
5. Ce Wen-Nan. Physics of heterogeneous inorganic materials. *Prog. Mater. Sci.* **37**, 1 (1993).
6. D. S. McLachlan, M. Blaskiewicz, and R. E. Newnham. Electrical resistivity of solids. *J. Am. Ceram. Soc.* **73**, 2187 (1990).

7. J. Rouxel, M. Tournoux, and R. Brec. *Soft Chemistry Routes to New Materials*–Chimie Douce. Trans. Tech Aedermannsdorf (1994).
8. M. Schroeder. *Fractals, Chaos, Power Laws.* W. H. Freeman, New York (1991).
9. S. K. Scott. *Chemical Chaos.* International Series of Monographs on Chemistry, Vol. 24. Clarendon, Oxford (1993).
10. N. J. Dudney. Composite electrolytes. In: M. Z. A. Munshi (ed.) *Handbook of Solid State Batteries and Capacitors.* World Scientific (1995), p. 231.
11. D. Avnir. Fractal geometry—a new approach to heterogeneous catalysis. *Chem. Ind.* **16**, 912 (1991).
12. T. Vicsek, M. Shlesinger, M. Matsushita. *Fractals in Natural Sciences.* World Scientific, Singapore (1994).
13. F. Family, P. Meakin, B. Sapoval, and R. Wool (eds). Fractal aspects of materials. *Mater. Res. Soc. Symp. Proc.* **367** (1994). MRS, Pittsburgh.
14. D. W. Schaefer. Fractal models and the structure of materials. *Mater. Res. Soc. Bull.* p. 22 (Feb. 1988).
15. H. E. Stanley and N. Ostrovsky (eds.). *On Growth and Form: Fractal and Non-Fractal Patterns in Physics.* Martinus Nijhoff, Dordrecht (1986).
16. H. E. Stanley and N. Ostrovsky (eds.). *Random Fluctuations and Pattern Growth: Experiments and Models.* Nato ASI Series E, Vol 157. Kluwer, Dordrecht (1988).
17. H. E. Stanley and N. Ostrovsky (eds.). *Correlations and Connectivity: Geometric Aspects of Physics, Chemistry, and Biology.* Nato ASI Series E, Vol. 188. Kluwer, Dordrecht (1990).
18. P. J. van der Put, R. A. Bauer, A. van den Assem, F. E. Kruis, B. Scarlett, and J. Schoonman. Determination of particle aggregation in ultrafine silicon nitride powders. In: G. L. Messing, S. Hirano, and H. Hausner (eds.). *Ceramic Powder Science, Vol. III.* American Ceramic Society, Westerville (1990), p. 259.
19. Y. Liu, T. Baudin, and R. Penelle. Simulation of normal grain growth by cellular automata. *Scripta Materialia* **34**, 1679 (1996).
20. P. Meakin. Fractals and disorderly growth. *J. Mater. Educ.* **11**, 105 (1989).
21. D. Avnir. *The Fractal Approach to Heterogeneous Chemistry, Surfaces, Colloids, Polymers.* Wiley, Chichester (1989).
22. A. L. Barabasi and H. E. Stanley. *Fractal Concepts in Surface Growth.* Cambridge University Press (1995).
23. P. J. Reynolds (ed.). *On Clusters and Clustering: From Atoms to Fractals.* North-Holland, Amsterdam (1993).
24. J. Zarzycki. Fractal Properties of Gels. *J. Non-Cryst. Sol.* **95/96**, 173 (1987).
25. S. K. Friedlander. *Smoke, Dust, and Haze.* Wiley, New York (1977).
26. R. Kopelman. Fractal reaction kinetics. *Science* **241**, 1620 (1988).
27. W. G. Rothschild. Fractals in heterogeneous catalysis. *Catal. Rev. Sci. Eng.* **33**, 71 (1991).
28. A. Bunde and S. Havlin (eds.). *Fractals in Science.* Springer-Verlag, Berlin (1994).
29. A. Bunde and S. Havlin (eds.). *Fractals and Disordered Systems.* Springer-Verlag, Berlin (1991).
30. R. W. Cahn. Percolation frustrated. *Nature* **389**, 121 (1997).

*Chapter 8*

# SYNTHESIS OF INORGANIC MATERIALS

*The place was always stiff with researchers trying to get to the bottom of it all and taking a very long time about it.*

*DOUGLAS ADAMS, So Long and Thanks for All the Fish*

## 8.1. Introduction to Inorganic Synthesis

The synthesis of inorganic compounds for use in materials[1,2] is about making solids and this chapter on synthesis describes the preparation of (mainly) nonmolecular solid compounds in solid state reactions, in reactions from liquids, or reactions from the gas phase. The synthesis of inorganic molecules[3] is not much different than that of organic compounds. Some methods of making coordination compounds and organometallics were given in Chapter 3. Their synthesis is not described here because, for the materials technologist, inorganic molecules are used mainly as precursors and not as part of materials.

Chemical processes in synthesis are difficult to generalize; they are *specific*, depending very much on which elements are used.[4] Although compounds all have their characteristic preparation recipes, they can be categorized according to the morphology of the product that is formed: powders,[5–8] coatings,[9] single crystals,[10] polycrystalline compounds,[11] or glasses.[12] Three groups of synthetic methods are described in this chapter:

1. Reactions between solids, such as reactive sintering.
2. Reactions in liquids, such as melts, solutions, and suspensions.
3. Reactions from the gas phase, such as in PVD, CVD, and vapor transport.

Several preparative tricks have been developed to increase the rate of inorganic reactions. They are based on:

1. Shortening the diffusion distance between the reacting atoms in solids: The atoms are brought together within atomic distances in the precursor, which shortens the diffusion time in solid state reactions.
2. Increasing the mobility of atoms in solids by increasing the diffusion coefficient, e.g., by using a different interphase or by increasing the vacancy

concentration: A higher mobility means faster diffusion and therefore higher solid state reaction rates.

3. Increasing the mobility by making use of phase transitions in solids during synthesis: the Hedvall effect as described in Section 5.2.
4. Using liquids such as molten salts or water under hydrothermal conditions as the reaction medium: Liquid-phase reactive sintering is also an example of the use of a liquid phase to increase the reaction rate.
5. Using gaseous precursors or reaction intermediates for rapid mass transfer.

When reactions must be slowed down one reverses these tricks if applicable.

If the compound is made under conditions of high atomic mobilities, which usually means high temperatures, the product is near equilibrium and thermodynamic estimates can be used to formulate the recipe. Application of thermodynamics to solids is less simple than to molecules in liquids because solid lattices may have stresses that have chemical consequences: structures are not perfect and nonstoichiometry and surface defects may considerably affect the driving forces. Phase diagrams are also useful for synthesis under these conditions because they give the phases that are at their thermodynamically lowest state at certain concentrations and sufficiently high temperatures. If the mobilities are high enough and there are no other obstacles, such phases may actually occur.

Other diagrams that are of some use for synthesis under equilibrium conditions are Ellingham diagrams, which are the graphs of the Gibbs free energy $\Delta G_f(T)$ of binary compounds vs. the temperature. Figures 8.1, 8.2, and 8.3 give the Ellingham diagrams for metal oxides, nitrides, and sulfides as examples. It should be remem-

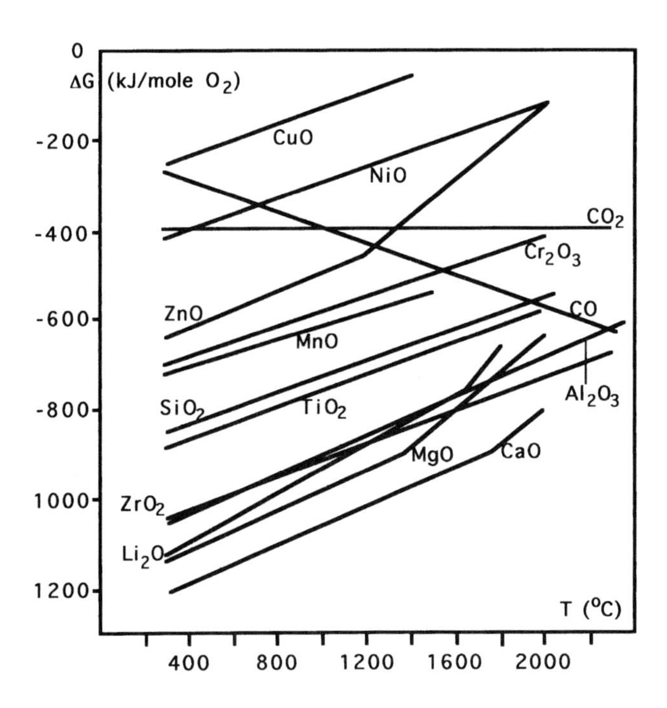


**Figure 8.1.** The Ellingham diagram for metal oxides.

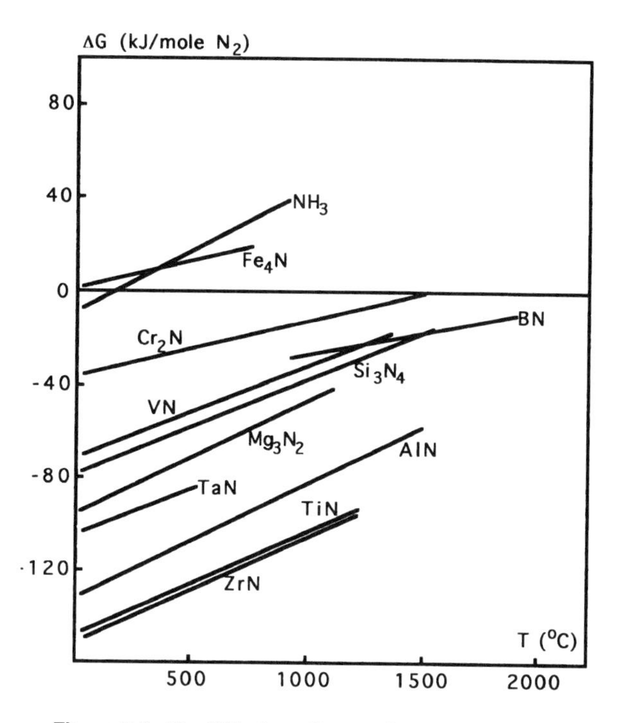


**Figure 8.2.** The Ellingham diagram for metal nitrides.

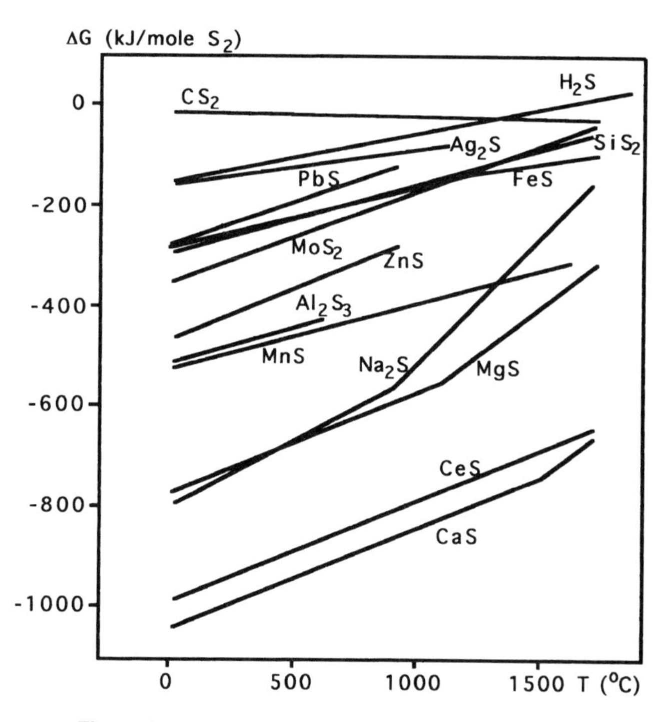


**Figure 8.3.** The Ellingham diagram for metal sulfides.

bered that these diagrams show equilibrium values. Only in exceptional cases is there a relationship between the activation energy and $\Delta G$ of a reaction.

Nonequilibrium structures are difficult to make in high-temperature processes when all the reactants are mobile. Compounds that are not in their lowest thermodynamic state are made under kinetic control.[13] Solvents often have to be used to obtain certain morphologies: micelles, liquid crystals, and adjusted acidity are used in recipes for nonequilibrium solids in solvents. High temperatures are used to increase rates, low temperatures inhibit mobility. In soft chemistry the reactivities and mobilities are selected by subtly adjusting the temperature and concentrations of reactants and catalysts in such a way that the wanted compound gets formed and reaction paths to other products are excluded kinetically. The alternatives to a nonselective "heat it and beat it" method involve other techniques for activation of the reactants: plasmas and light. Plasmas increase rates because they increase electron temperatures while the gas temperature remains low. Light generates active radicals or excited species. Often these nonthermal activation techniques are chosen if, for some reason, high temperatures must be avoided.

Sometimes in synthesis, special conditions are chosen to eliminate the effects of convection, which is the result of density differences in a liquid (owing to temperature or concentration differences) combined with gravity. Gels are solids and the liquid in them does no flow. Therefore the formation of crystals from solutions that are immobilized by gels is not affected by convection, and mass transport during the growth of solids is only by diffusion. Another way to prevent convection is to eliminate gravity. Reactions in fluids on board satellites in outer space are not affected by convection so it is thought that crystals grown under microgravity have fewer defects. A potentially more important application of microgravity in synthesis is the use of gaseous fluidized beds (suspensions of solid particles in a gas). Terrestrial fluidized beds are often used in chemical technology to let gases react with particles, usually heterogeneous catalysts. A gaseous fluidized bed consists of particles falling through an upward flow of gas. It can exist at high pressures (a few bars) and at low gas pressures (a few torrs). The mixture of particles and gas behaves as a fluid and can develop bubbles in a so-called emulsion phase. Thermal reactions can occur between the gas and the solid particles if the temperature is high enough, but a glow discharge (plasma), for lowering the process temperature, can be formed only in the bubbles, not in the emulsion phase. Plasmas do not penetrate the emulsion phase of terrestrial dense fluid beds because the particle density in that phase is too high and the mean free path of electrons in the gas is too low for a plasma. In the absence of gravity a suspension of particles in a gas is stable at any particle density and the particle concentration can be chosen so low that a plasma can be maintained. Microgravity therefore means more for synthesis than just suppression of convection: it also allows reactions in dusty plasmas. Increased gravity is used in the physical separation of nuclides in ultracentrifuges.

Physical phase transitions have much in common with solid state reactions. In both cases the bonding between the atoms changes and the heat of reaction is of the same order of magnitude. However, there are some clear differences. In a phase transition the atoms do not have to travel over large distances in the lattice to be brought to or removed from the reaction site. Mass transport limitation slows chemical reaction rates significantly and phase transitions are usually faster than solid state reactions. By mixing the reactant atoms prior to the reaction on an atomic scale chemical solid state reactions start to behave like phase transitions in solids.

Crystalline solids are usually made from liquids (melts or liquid solutions). The solid product may be a single crystal or a powder. Nucleation, or the formation of very small solid grains, is distinguished from crystal growth, in which the crystallites that are initially present increase in size by growth of outer planes or facets. As was described in Chapter 7, nucleation can be either homogeneous or heterogeneous. Homogeneous nucleation means formation of nuclei in a homogeneous phase (a liquid or gas), while heterogeneous nucleation occurs on surfaces.

Another concept in synthesis is epitaxy. Epitaxy is the continuation of the crystal orientation of the monocrystalline substrate in the deposited crystalline product, which may be the same compound as the substrate or a different solid that has the same crystal orientation as the monocrystalline solid. Epitaxial layers are essential for microlithography in the electronic industry: carefully formed epitaxial layers do not have localized electronic interface states, which are deleterious for the functioning of the device. The process conditions for epitaxy by molecular beam epitaxy (MBE) are very low process pressure, comparatively high temperatures, and a low growth rate. MBE is a form of CVD, which was described in Chapter 6. Liquid phase epitaxy (LPE) is a form of growth of single crystals from a melt.

There are three types of growth on surfaces of solids, depending on the morphology of the deposit (Figure 8.4):

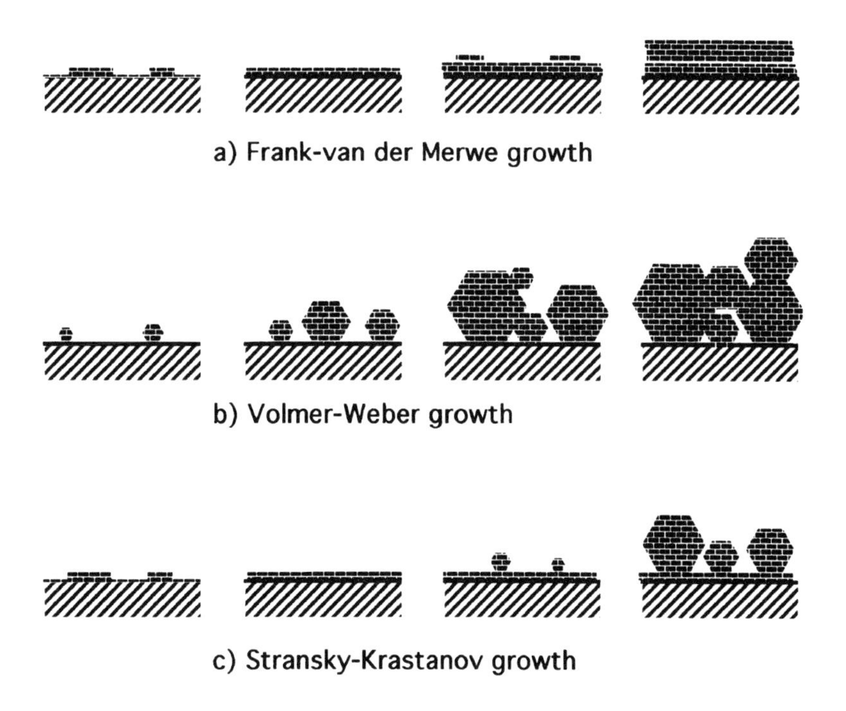


**Figure 8.4.** Three types of growth of solids, their kinetic origin, and morphology. (a) Frank–van der Merwe growth: easy allonucleation and easy homeonucleation, slow growth rate; (b) Volmer–Weber growth: low allonucleation and homeonucleation rate coefficients and relatively high growth rate; (c) Stransky–Krastanov growth: fast allonucleation rate and low homeonucleation rate coefficient, high growth.

1. *Frank–van der Merwe:* Polycrystalline solids grow in layers. There is fast allonucleation, fast homeonucleation, and slow crystallite growth. This combination means strong adhesion between the deposit and the substrate.
2. *Stransky–Krastanov:* There is fast allonucleation while crystallite growth is faster than homeonucleation. Adhesion is also strong owing to fast allonucleation although fast growth may introduce some macrodefects in the layer.
3. *Volmer–Weber:* Slow allonucleation and homeonucleation and fast crystallite growth. The adhesion between the deposit and the substrate is generally poor because this kind of growth is likely to occlude macrodefects and pores at the interface.

The three types of growth are summarized as follows

| Growth type | Allonucleation | Homeonucleation | Crystal growth |
|---|---|---|---|
| F–vdM | Fast | Fast | Slow |
| S–K | Fastest | Slow | Fast |
| V–W | Slow | Slow | Fast |

Good adhesion of a deposit to the underlying substrate usually means a deposition of nuclei that is faster than the growth of the nuclei themselves. It does not necessarily mean a high bonding enthalpy between the atoms of the substrate and the deposit. Good adhesion can occur only if deposition reactions for layers are also fast. Reaction limitation is favorable for adhesion because diffusion limitation (the alternate regime) may introduce defects.

The following example shows how the surface reaction rate is adjusted to improve the adhesion of a coating of a ceramic nitride to two different substrates:

1. Titanium nitride (TiN) adheres well to an alumina substrate only if the $Al_2O_3$ surface is partially nitrided in $NH_3$ before deposition of TiN. This surface has a high allonucleation rate for TiN.
2. A TiN deposit does not stick well to a nitrided steel surface, which has many different types of surface crystallites such as $Fe_3N$. Deposition of a very thin interlayer of titanium that is low in nitrogen concentration improves the allonucleation of TiN and hence its adherence to nitrided steel.

A high nucleation rate and low crystal growth rate imply a fine-grained product. The crystallite growth rate can be lowered with adsorbed foreign atoms or molecules that adsorb but do not dissolve in the solid. In the formation of titanium carbide from gaseous precursors, the crystallite growth can be prevented by adding a low concentration of aluminum-containing gases to the reactants. As noted in Section 6.7 aluminum carbide obstructs crystallite growth but not nucleation of TiC.

The growth of nuclei in powder formation often results in monodisperse crystallites or particles with a narrow size distribution. The LaMer model[14] explains this remarkable phenomenon as a decrease in the supersaturation after formation of an avalanche of nuclei from the solution or the gas phase. The concentration is then not high enough to continue nucleation but it can increase the size of the crystallites that are already present (Figure 8.5). The nuclei, which were formed at the same instant, all grow during the same time period; hence their equal size. This explanation is not correct in all known cases. Nucleation also appears to continue during growth of monodisperse particles. It has also been shown that the nuclei grow by aggregation and form composite particles, perhaps owing to oscillating surface charges during growth (oppositely charged primary particles attract each other). The agglomerated particle clusters are often monodisperse, which means that the smaller aggregates grow faster than the large clusters (why that is so remains unexplained). Aggregation of nuclei is not a generally valid mechanism for particle growth because certain particles (such as $SiO_2$ in colloidal solutions) are known to grow from monomers by polymerization. Monodisperse silica particles then have a higher growth rate for small particles than for large ones because small particles are rougher and have more reactive surfaces. The induction time, which is often observed in this case, is the result of the necessity to build up a sufficiently high concentration of hydrolized monomers in the solution.

One example of a carefully controlled synthesis of shaped powder particles is the preparation of magnetic pigments for recording information.[14,15] The microstructure of the particles is a determinant for the applicability and such properties as remanence, coercive force, anisotropy, orientability, and mechanical strength. Cobalt-doped $\gamma$-$Fe_2O_3$ (maghemite) is used in audio and video tapes but the lattice is cubic and the isotropy of $\gamma$-$Fe_2O_3$ precludes growing it from a solution as needles, the form needed for magnetic recording. It is made in a roundabout way by dehydration of Goethite (FeOOH) needles by heat, reducing with hydrogen to $Fe_3O_4$, and then oxidizing again to form $\gamma$-$Fe_2O_3$ particles that have the original Goethite particle shape. Sometimes the final powder particles, which consist of a

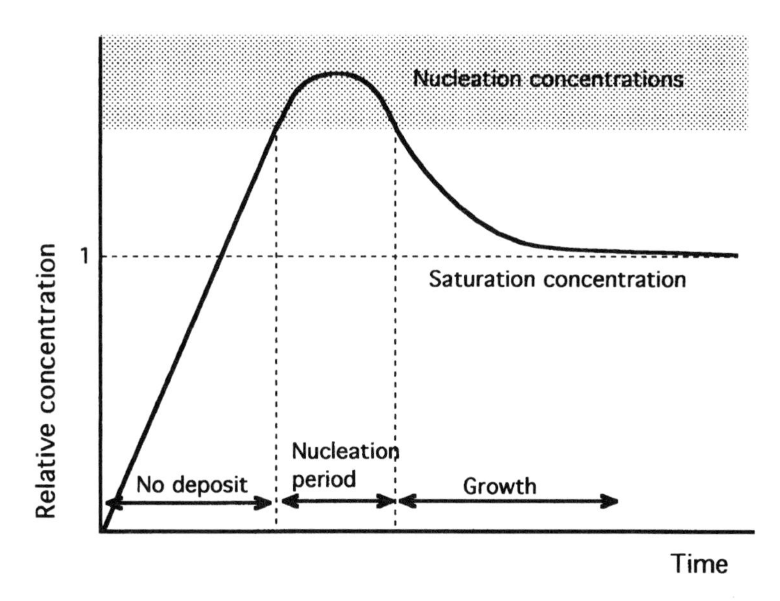


**Figure 8.5.** The development of supersaturation in a solution that nucleates and grows monosized particles according to the LaMer model. When the solution gets oversaturated there is a short period in which nucleation occurs. The burst of nucleation lowers the concentration to a value where no more nuclei form but all the existing ones start to grow simultaneously.

single magnetic domain (0.3 to 0.03 $\mu$m), are covered with a thin layer of colloidal $SiO_2$ or $CoO_{1-x}$ to increase the coercive force.

If the reduction of dehydrated Goethite is continued to metallic iron and silver or cobalt compounds are added to prevent sintering of the small particles, the magnetic particles have a doubled coercive force compared to cobalt-doped $\gamma$-$Fe_2O_3$. These metallic particles are very suitable for information recording but less corrosion-resistant than the oxide. Reduction of Goethite in the presence of ammonia yields useful $Fe_4N$ powders that have a high saturation magnetization. Chromium dioxide ($CrO_2$) pigments are made by oxidation of chromium (III) oxide $Cr_2O_3$ with an excess of $CrO_3$ under hydrothermal conditions. The reactant is toxic and expensive and for that reason $CrO_2$ has been virtually replaced by cobalt-doped $\gamma$-$Fe_2O_3$. Colloidal iron oxide particles have easily measured magnetic properties that vary strongly with their form and so such particles are used as tracers in environmental processes e.g., (labeled oil spills can be identified).

Inorganic solids are often made as powders. The average particle size can be varied over many orders of magnitude (Figure 8.6). Ultrafine nanosized particles are best made from the gas phase or if they are oxides, from colloids. The physical properties of the particles, such as density and magnetism, are used to separate the larger particles from the fine fractions. A hydrocyclone, shown in Figure 8.7, is a typical device used for this purpose.

This chapter on synthesis discusses solid state reactions (Section 8.2), reactions in liquids (Section 8.3), and gas reactions (Section 8.4), with and without the use of electric fields. Syntheses from liquids are: (a) melt processes (such as growth of single crystals or directed solidification); (b) processes from solution (flux or hydrothermal methods); and (c) colloidal methods. Synthesis from melts (Section 8.3) usually requires high temperatures but these methods are commercial and designed for products with a high added value, such as silicon for the semiconductor industry, laser hosts, or singly crystalline alumina crucibles. Table 8.1 summarizes growth methods for single crystals.

Molten salts (called fluxes) and also water at high pressures and temperatures are excellent solvents for many inorganic solid compounds. Inorganic synthesis from flux is discussed in Section 8.3.2, while water as a solvent for synthesis of oxides under hydrothermal conditions is the subject of Section 8.3.3. Colloidal methods are of increasing importance and the synthesis of mixed oxides by polymerization of hydrolized metal–organic molecules is the fourth subject in the group of synthesis from liquids.

Almost every conceivable solid compound can be made from the gas phase by chemical reactions between gases. The synthesis of single crystal by vapor transport is the subject of Section 8.4.1. The preparation of powders, layers, and fibers from gaseous precursors is dealt with in Section 8.4.2–8.4.4. The following four methods are described.

1. Vapor transport: solids are transported through the gas phase in a closed vessel using a reactive gas that is not consumed in the process. This method is used for crystal growth, reactive sintering, and synthesis of solids.
2. Physical vapor deposition (PVD), a vacuum method: Solids are sublimed or sputtered at low pressures using heat or a gas discharge. This technique is

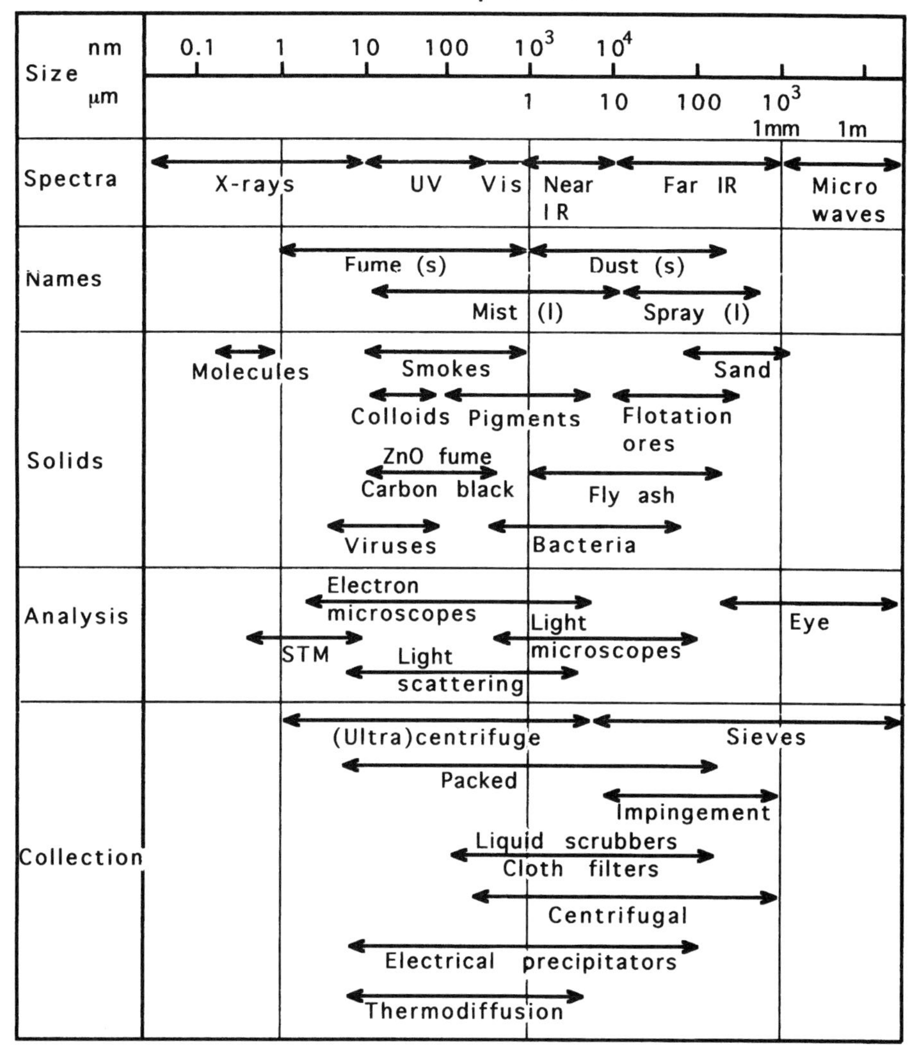


**Figure 8.6.** The size of particles in several dusts and aerosols and the way they are characterized and manipulated.

often used for low-temperature deposition of thin films or ultrafine powders. Fullerenes and their derivatives are made with PVD, not with molecular chemistry.

3. Chemical vapor deposition (CVD): Precursor gases react to form solids as a powder, a coating on a surface, or a bulk material. The method affords good control of the microstructure and virtually all inorganic solids that are useful as materials can be made by means of CVD.
4. Plasma-enhanced CVD (PE-CVD), a CVD technique that activates gases by

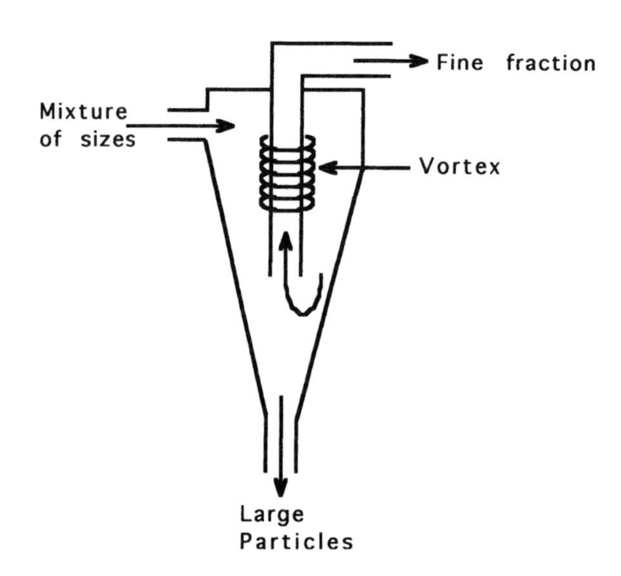


**Figure 8.7.** Schematic of a hydrocyclone for removing large particles from a fluid.

**Table 8.1. Methods for Growing Single Crystals**

| From a melt | From solution | From the gas phase |
|---|---|---|
| Bridgeman–Stockbarger | Evaporating solvent | Vapor transport |
| Czochralski | Cooling solution | Vapor-phase epitaxy |
| Kyropoulos | Liquid-phase epitaxy | |
| Zone-melting | | |
| Flame fusion | | |

means of a gas discharge (a plasma): This nonequilibrium method enables a kind of high-temperature chemistry in a gas that remains at low temperature. Plasma activation is used in synthesis of diamond and other nonequilibrium compounds at low temperatures and pressures.

Upscaling all the processes discussed here is a subject on its own. Details can be found in standard works on chemical technology.[16,17] As upscaling is in large part applied physics it has not been included here.

## 8.2. Solid State Reactions

When green forms of ceramics are sintered many solid state reactions occur.[11] In the end product different phases coexist and a sintered ceramic is in general a composite material.[18–21] The basics of the reactions that occur during sintering were discussed in Chapter 5. The recipes for solid state reactions are usually very simple: powders of solid reactants are mixed and heated or calcined. The reaction between two powder particles of the separate solid reactants means that the reactant atoms have to diffuse over large distances of the order of the particle size ($\mu$m) in order to react. The reaction between CoO and $Fe_2O_3$ powders at 1200°C to $CoFe_2O_4$ (cobalt ferrite) is sluggish and the reaction is incomplete (see Section 5.3). The reaction products must be milled, pressed into compacts, and recalcined many times before the conversion is complete. This is a simple preparation process as it does not involve difficult process control or expensive reactants. Owing to their simplicity, solid state reactions are very popular in industry.

Reactions involving gases are less slow. Decomposition reactions that produce gases are faster than purely solid state reactions because the evolving gas breaks up the precursor lattice. Use of reactants other than oxidic powders (e.g., carbonates) may also reduce the reaction time required for the formation of oxides.

As noted in the introduction to this chapter, the temperature needed for the solids to react fully can be lowered considerably if the diffusion distance for the reacting atoms is kept low by having them close to each other in the precursors. In the mixed salt (iron cobalt oxalate) that is the precursor for the corresponding spinel, the metal ions are already mixed on an atomic scale in the proper stoichiometric amounts and a complete decomposition of the oxalate (indicated with the abbreviation Ox) to the mixed oxide (the spinel) takes place at 700°C within 3 h:

$$2FeSO_4(aq) + CoSO_4(aq) + 3(NH_4)_2Ox(aq) \rightarrow CoFe_2Ox_3(s) + 3(NH_4)_2SO_4(aq)$$

$$CoFe_2Ox_3(s) \rightarrow CoFe_2O_4(s) + 4CO(g) + 2CO_2(g) \qquad \text{at } 700^\circ C$$

Many stoichiometric double salts (citrates, oxalates) are suitable as precursors for mixed oxides. If stoichiometries are needed in an oxide for which no double salts are available, coprecipitation of hydroxides can yield a precursor that has the metal ions atomically mixed in the proper amounts so that no excessive reaction times are necessary in the calcining step.

A method used in ceramics that is similar to coprecipitation is spray-calcining. A liquid solution (usually aqueous) that has the metal salts for the components in the right concentration is sprayed in hot air. The solvent evaporates from the droplets so fast that the dissolved solids are unable to form separate crystallites. A well-mixed coprecipitate forms that is a suitable precursor for a mild process to form a mixed oxide.

Nonoxides can be formed from polymeric precursors that are already mixed in the right proportion on an atomic scale. The atoms in silicon carbide are bonded with covalent bonds and are not mobile. Similar to most other carbides and nitrides SiC is difficult to sinter and so has slow solid state reactions even at high temperatures. A good precursor for solid state formation of SiC is the polymer polycarbosilane, which is made at 450°C from polysilanes formed from dimethyldichlorosilane and sodium:

$$Me_2SiCl_2 + Na \rightarrow NaCl + (Me_2Si)_n \text{ (polysilanes)}$$

The polycarbosilane in which the silicon and the carbon atoms are already bound together is pyrolytically decomposed to SiC after the precursor polymer has been shaped into a green form. Such a decomposition means evolution of hydrogen and excess hydrocarbons as gases.

## 8.3. Synthesis from Liquids

### 8.3.1. Preparation from Melts

Single crystals such as quartz and silicon are important functional solids. Figure 8.8 shows three methods for synthesizing single crystals from a melt[10] that yield cylindrical single crystals. Large crystals of semiconductors such as silicon, gallium arsenide, and indium phosfide are drawn from the melt using the Czochralski method. The method of Kyropoulos is also used for making large single salt crystals from their melts and the Verneuil technique is used to make single crystals of refractory oxides such as alumina.

Synthetic ruby ($2\frac{1}{2}\%Cr_2O_3$ in $\alpha$-$Al_2O_3$) for lasers is made industrially by the Verneuil method as are single crystals of corund itself, sapphire, rutile, spinels, and strontium titanate. In this method the oxide particles are dropped through a flame,

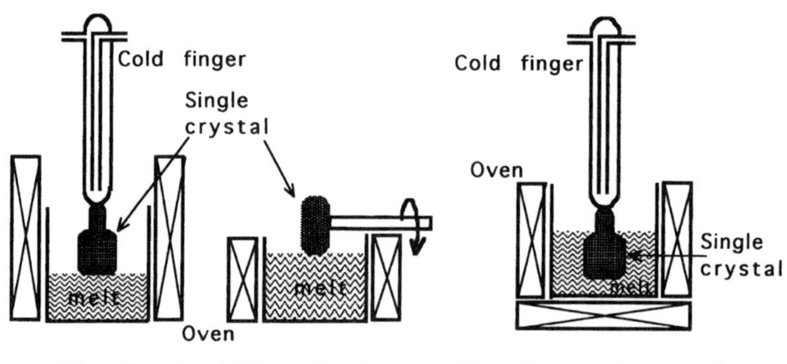


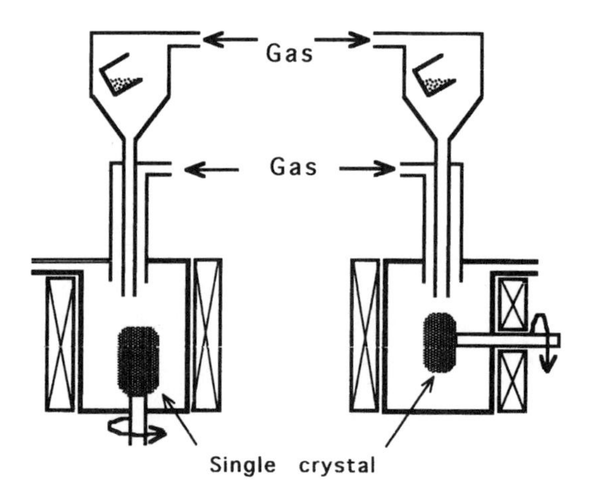


**Figure 8.8.** Techniques for growing single crystals.

**Table 8.2. Process Conditions for Growing Single Crystals**

| Single crystal | Method | Crucible | Atmosphere | Growth rate (mm/h) |
|---|---|---|---|---|
| Copper | Stockbarger | Graphite | Inert | 40 |
| Silicon | Czochralski | Graphite | Vacuum | 10 |
| $Al_2O_3$ | Verneuil | | | 13 |
| | Stockbarger | W, Mo | Inert | |
| | Edge-feed | W, Mo | Inert | |
| $MgAl_2O_4$ | Stockbarger | W, Mo | Inert | 4 |
| $Y_3Al_5O_{12}$ (YAG) | Stockbarger | W, Mo | Inert | 2 |
| $Gd_3Ga_5O_{12}$ | Czochralski | Ir | Air | 2 |
| $LiNbO_3$ | Czochralski | Pt | Air | 2 |

which melts them, onto the growing single crystal. During growth, single crystals can be shaped into crucibles for use in the manufacture of microelectronics. In the edge feed method, shaped objects of alumina that are made of one single crystal are drawn from the melt. In the Bridgeman–Stockbarger method, a tapered vessel containing the melt is slowly taken out of an oven that is kept at a temperature slightly above the melting point of the crystal. The crystal grows from the melt starting at the tapered end, where it first nucleates when the temperature is low enough there. Table 8.2 gives an idea of the process conditions for growing crystals from their melts and the crucible material that is used to hold the melts during growth.

Phase diagrams ($T$ vs. $x$) are used to control the solid product (the crystal) with the composition of the melt. If a binary melt is cooled the composition of the solid that may form can be derived from the phase diagram (Figure 8.9), the temperature,

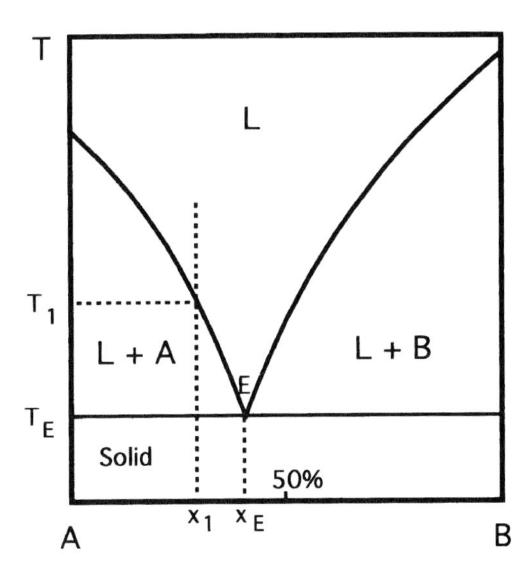


**Figure 8.9.** Phase diagram of a two-component system without mutual solubility or compound formation. A liquid mixture of concentration $x_1$ will deposit solid phase A when cooled to temperature $T_1$. A liquid mixture with the eutectic concentration $x_E$ can be directionally solidified to a laminar crystalline 2–2 composite of phases A and B if the eutectic temperature $T_E$ is not too low.

and the composition $x_1$ of the melt. From a melt that has a composition $x_1$, pure A will crystallize at a temperature $T_1$, and on cooling further the melt will get higher in B as more A is removed from the system as a solid (crystallized). Phase diagrams are also used when salts are the solvents. A melt having the eutectic composition $x_E$ at a temperature $T_E$ is in equilibrium with the solids A and B.

In a eutectic point, the morphology of the solid product depends on the direction of the temperature gradient when the melt is cooled to a temperature below the melting point of the solid. The two solid phases that crystallize during solidification of the eutectic mixture form alternating layers or columns with a direction perpendicular to the solidification front. If in the phase diagram the composition $x_E$ of the eutectic mixture lies close to the value of the composition of one of the two phases (say close to $x_A$ of phase A) then columns or fibers of the minority phase (B) are formed in a matrix of phase A. This composite is anisotropic and the direction of the parallel fibers is the same as that of the heat flow during solidification. If $x_E$ is not close to the composition of one of the two solid phases that crystallize on cooling but is roughly halfway between them, the two phases crystallize in the form of alternating parallel plates. The temperature gradient is parallel to the surface of the plates. The rate at which heat is removed during solidification also determines the morphology: the slower the cooling rate, the thicker the plates or columns. With this directed or so-called eutectic solidification it is possible to make composite materials that have anisotropic electrical and mechanical properties. Figure 8.10 shows the conditions for the formation of columns or plates and the characteristic distance between the phases in the solid composite. The value of $x_E$ in the mixture determines whether directional solidification will result in fibers in a matrix or alternating stacked layers.

### 8.3.2. Liquid Salts as Solvents

Molten salts are often used as solvents because they have a number of properties that make them very attractive for use in inorganic synthesis[22–24]:

1. Their dissolving capacity is large: they can easily dissolve large quantities of metals, nonmetals, covalent compounds, metal oxides, other salts, water and many gases.
2. The temperature range in which they can be used is wide: 100–1000°C.
3. The acidity and the oxidation/reduction properties of molten salts can be adapted to the required values with the composition of mixtures of molten salts.
4. The electrical and thermal conductivity of molten salts is very high.
5. The conversion efficiency in syntheses in molten salts is high.

Molten salts are not only useful as solvents in chemical synthesis, electrolysis, soldering, enameling, de-enameling, metal recycling and preparation, coal gasification, and desulfuration, but they are also reactants, catalysts, and ambients for heat storage and heat transfer, as well as electrolytes in fuel cells (molten carbonates). The solvent can participate in the reaction that is carried out in fluxes. $BaTiO_3$ is made in molten $TiO_2$ as a solvent and $Bi_8(AlCl_4)_2$ can only be made in a very acidic cryolite ($NaAlCl_4$).

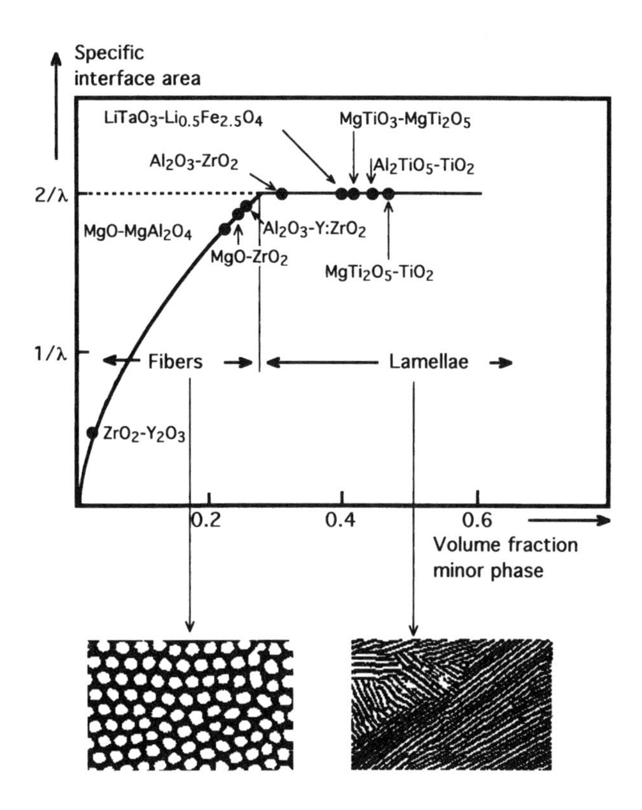


**Figure 8.10.** Morphologies in directed solidification. From V. S. Stubican, R. C. Bradt, F. L. Kennard, W. J. Minford, and C. C. Sorrel. Ceramic eutectic composites. In: R. E. Tressler, G. L. Messing, C. G. Pantano, and R. E. Newnham (eds.). *Tailoring Multiphase Composite Ceramics*. Plenum, New York (1984). Morphologies from D. Michel, Y. Rouaux, and M. Perez y Yorba. *J. Mater. Sci.* **15**, 61 (1960). With kind permission of Kluwer Academic Publishers.

Molten salts as solvents are categorized in two groups, covalent and ionic salts: (a) Covalent salts form molecules in their melts and suffer partial autoionization as water does. $HgBr_2$, e.g., is partially ionized in the melt:

$$2HgBr_2 \rightarrow HgBr^+ + HgBr_3^-$$

KBr is a base and $HgSO_4$ is an acid in molten $HgBr_2$ because they increase the anion or the cation concentration of the solution. These ions result from the autoionization reaction when the acid or base is dissolved in the molten salt. (b) Ionic salts such as NaCl are fully ionized in their melt. Often eutectic mixtures of salts are used to lower the melting point of the solvent. The eutectic point of a mixture of LiCl and KCl has a temperature of 450°C.

The acidity or the base character of the solvent is central to its use in synthesis. In mixtures of NaCl with $AlCl_3$ (or $ZnCl_2$) the acidity can be adjusted with the composition. Figure 8.11 shows the melting points and the chloride ion activity (read basicity) of mixtures of this system. An excess of $AlCl_3$ makes the solvent acid through

$$Al_2Cl_6 + Cl^- \rightarrow Al_2Cl_7^- \qquad \text{or} \qquad AlCl_3 + Cl^- \rightarrow AlCl_4^-$$

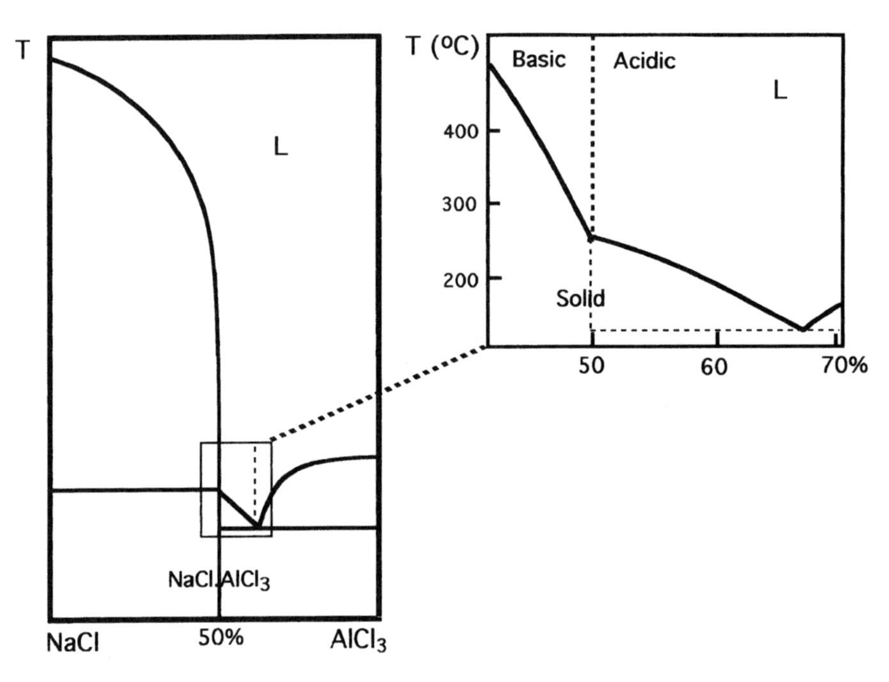


**Figure 8.11.** Phase diagram of the $AlCl_3/NaCl$ system. A melt of a mixture of the two components is a basic solvent if it has an excess of NaCl. The solvent becomes acid if there is more than 50% $AlC_3$ in the mixture.

The free chlorine ions that are present are the bases, but when the chlorine ions are bound in the aluminum complexes, they do not act as bases. The concentration $[Cl^-]$ determines the basicity of the salt mixture. Solids such as metals and metal oxides, which are difficult to dissolve in the usual solvents, may dissolve well in molten salts. Metals often dissolve well in their halides, e.g., Na in NaCl, Cd in $CdCl_2$, and rare earths in $LnCl_3$. A molten mixture of $Na_2CO_3$ and $B_2O_3$ is a good solvent for $\alpha$-alumina (ruby, sapphire). All metal oxides with the exception of $ThO_2$ dissolve in a melt of alkali hydroxides (a popular solvent in analytical chemistry used to analyze minerals) with some water and oxygen. For instance, $MnO_2$ dissolves as $K_2MnO_4$ in KOH.

Ionic melts are characterized by: (a) a comparatively high degree of order slightly above the melting point, and only a quarter of the full amount of melting entropy is accounted for at the melting point with three-quarters being contributed at temperatures above the melting point; (b) a lower average coordination number of cations by anions and vice versa in the melt than in a crystalline solid (Table 8.3); (c) a lower average distance between the ions in the melt than in the crystalline solids; (d) a density decrease on melting (in the case of halides even up to 25%), and a melt that is characterized by many vacancies, which is the reason for its dissolving power (Figure 8.12).

The eutectic of a NaOH/KOH mixture is at 175°C and this mixture is also a good solvent for synthesis of oxides. The acid/base equilibrium is

$$2OH^- \rightleftharpoons H_2O\,(\text{acid}) + O^{2-}\,(\text{base})$$

**Table 8.3. Some Parameters in Ionic Melts**[a]

| | Solid | | Melt | | |
|---|---|---|---|---|---|
| | CN | ID (nm) | CN | ID (nm) | AD (%) |
| LiCl | 6 | 0.266 | 3.7 | 0.246 | 7.1 |
| LiBr | 6 | 0.285 | 5.2 | 0.268 | 6.0 |
| LiI | 6 | 0.312 | 5.6 | 0.285 | 8.6 |
| NaI | 6 | 0.335 | 4.0 | 0.315 | 6.0 |
| KCl | 6 | 0.326 | 3.6 | 0.310 | 4.9 |
| CsCl | 6 | 0.357 | 4.6 | 0.353 | 1.0 |
| CsBr | 6 | 0.372 | 4.6 | 0.355 | 4.4 |
| CsI | 6 | 0.394 | 4.5 | 0.385 | 2.3 |

[a](CN = average coordination number; ID = average interatomic distance; AD = decrease of average interatomic distance of the compound on melting).

The strongest acid in this solvent is water, which dissolves almost all of the metal oxides:

$$MO(s) + H_2O \rightarrow M^{2+} + 2OH^-$$

If the solution is then neutralized, the oxides precipitate.

Doping of $BaBiO_3$ or $La_2CuO_4$ with Na or K to give them a *p*-type conductivity is difficult to do with solid state reactions but easy in a NaOH or KOH flux.

The following are examples of chemical synthesis in a flux:

1. Metal cluster compounds and complex ions that have a number of ligands that is lower than usual (e.g., $NiL_3$ instead of $NiL_6$) would disproportionate or coordinate with bases, and most solvents are Lewis bases. Such compounds that are susceptible to basic attack can be made in melts that are sufficiently acid. Thus, $Cd_2^{2+}$ compounds have been made in acid cryolite mixtures ($NaCl/AlCl_3$). In $BiCl_3$ the compounds $Bi_5(AlCl_4)_2$ and $Bi_8(AlCl_4)_2$ have been made from Bi metal.
2. The synthesis of hydrides. Silane ($SiH_4$) is made in a LiCl/KCl eutectic mixture:

   $$LiH + SiCl_4 \rightarrow SiH_4 + 4LiCl$$

   The LiCl that is formed in this reaction is electrolytically regenerated to LiH.
3. Halides are easily made from oxides in liquid salts. Boron trichloride is prepared in a $NaCl/CaCl_2/BaCl_2$ mixture:

   $$B_2O_3 + 3C + 3Cl_2 \rightarrow 2BCl_3 + 3CO$$

   At the process temperature all products are gases and easily removed from the melt.
4. Metal chlorides are methylated in melts:

   $$SiCl_4 + 12CH_3Cl + 8Al \rightarrow Si(CH_3)_4 + 8AlCl_3$$

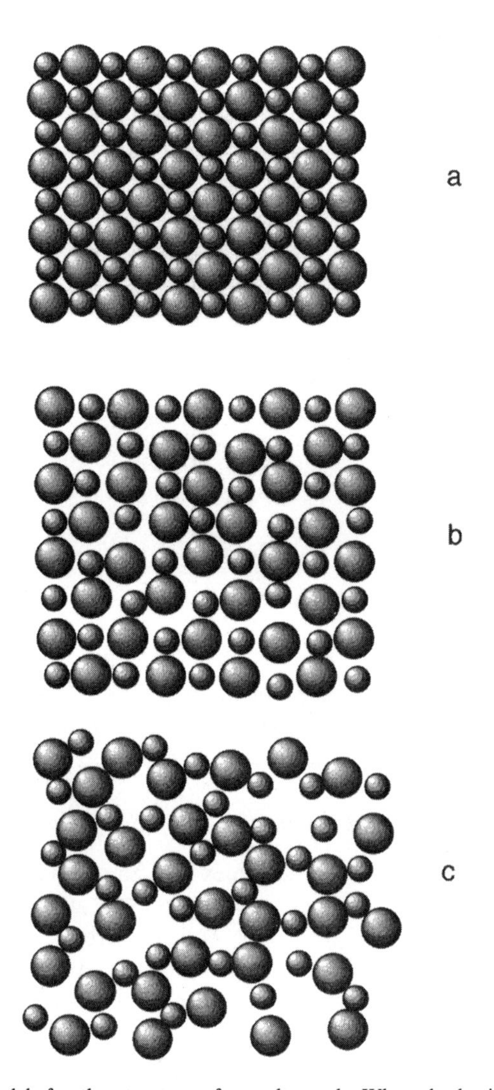


**Figure 8.12.** Two models for the structure of a molten salt. When the lattice (a) melts the ions form roughly equal cells according to schematic (b) as expected to minimize electrostatic energy. In model (c) the average interionic distance and the coordination number of the ions are lower than in (b). This model also has more and larger vacancies than the cellular model.

5. Ore processing, Volatile tantalum and niobium chlorides are formed in ores treated with molten chlorides:

$$Fe(TaO_3)_2 + 6C + 6\tfrac{1}{2}Cl_2 \rightarrow 2TaCl_5 + FeCl_3 + 6CO$$

6. Boron nitride is synthesized in a $NaOH/NaNH_2$ eutectic melt:

$$B_2O_3 + 3NaNH_2 \rightarrow 2BN + NH_3 + 3NaOH$$

7. The very hard and inert ceramic titanium carbide is made in a melt of alkaline halides:

$$CaC_2 + TiO_2 \rightarrow TiC + CaO + CO$$

**Table 8.4. Process Conditions for Growth of Single Crystals from Solutions**[a]

| Compound | Solvent | Conditions |
|---|---|---|
| $Y_3Fe_5O_{12}$ (YIG) | PbO | $T_0 = 1300°C$, CR = 3 deg/h |
| $Y_3Al_5O_{12}$ (YAG) | $PbO/PbF_2$ | $T_0 = 1500°C$, CR = 4 deg/h |
| $BaTiO_3$ | $TiO_2$ | $T_0 = 1200°C$, CR = 0.3 deg/h |
| $MgFe_2O_4$ | $PbP_2O_7$ | $T_0 = 1310°C$, CR = 4 deg/h |
| $YVO_4$ | $V_2O_5$ | $T_0 = 1200°C$, CR = 3 deg/h |
| $BeAl_2Si_6O_{16}$ (beryl) | $Li_2O/MoO_3$ | $T_0 = 975°C$, CR = 6 deg/h |
| $Al_2O_3$; $YCrO_3$; $TiO_2$ | $BiF_3/B_2O_3$; $PbF_2$ | $T = 1300°C$, evaporation |

[a] $T_0$ is the temperature at the beginning of the process; CR = cooling rate.

8. Single crystals of oxides are made from fluxes, as summarized in Table 8.4. Some typical process conditions are also mentioned.
9. Metals in groups 4–6 of the periodic table are galvanically deposited from FLINAK (a eutectic mixture composed of LiF, NaF, and KF with a $T_E$ of 450°C). With the exception of chromium, these metals cannot be precipitated at room temperature from either aqueous solutions or organic solvents.
10. Carbides such as $W_2C$ and $Mo_2C$ are also galvanically deposited from a FLINAK solution.
11. FLINAK is also an appropriate solvent to deposit $TaS_2$ and $TiS_2$ electrolytically from a solution of $K_2SiF_6$ and metal sulfides.
12. Zirconium diboride ($ZrB_2$), a very hard and inert ceramic, is galvanically deposited from a melt of $K_2ZrF_6$ and $KBF_4$.
13. Similarly, titanium diboride ($TiB_2$) is electrolytically precipitated from a borax melt with titanates and $TiO_2$ at 900°C.

### *8.3.3. Hydrothermal Processes*

Water is an excellent solvent for many ionic compounds (salts). At high pressures and at temperatures far above its boiling point at 1 bar pressure (hydrothermal conditions) it can also dissolve nonionic covalent compounds such as certain otherwise insoluble oxides.[25] Many oxidic gem stones have been formed hydrothermally in inorganic nature and it is not difficult to grow rather large single crystals of oxides in a hydrothermal way in autoclaves. Often a dissolving or mineralizing aid e.g. (NaOH) is added to the water. In particular the crystals that are made hydrothermally are those that cannot be made from the melt for some reason, e.g., because it is a nonequilibrium solid or it suffers phase transitions on cooling. Thus sphalerite (cubic ZnS) can be made hydrothermally at temperatures of 300–500°C but not from the melt because at the high fusion temperature of 1080°C the hexagonal Wurtzite structure is the stable modification. Single crystals of quartz are made hydrothermally in industry for the same reason: a low growth temperature is needed for a low-temperature modification. Not only single crystals but also powders are made hydrothermally on an industrial scale, e.g., zeolites, apatites, ceramic precursors such as $Al_2TiO_5$, mullite, stabilized zirconia, and magnetic pigments such as $\gamma$-$Fe_2O_3$ and $CrO_2$.

The process temperatures in hydrothermal synthesis (up to 400°C) are generally lower than the temperatures in flux methods. Sometimes the reaction temperature is even kept below room temperature, where the driving force for formation of low-entropy and low-symmetry compounds is comparatively high.

Another example of hydrothermal methods is the synthesis of aerogels. These very-low-density solids are formed by making a gel from a colloidal solution (sol) in an autoclave (filled to more than 32% with the sol) and then removing the solvent supercritically (Figure 8.13). The result is a material that has a very high porosity (>95%). An aerogel is a gel in which the solvent has been replaced by air while the colloidal particles have kept the same distribution and connectivity as they had in the liquid gel. The network of connected solid particles in the gel remains intact on removing the solvent. Although much is possible with synthesis in hydrothermal solvents, the most frequent technological use of hydrothermal techniques is in growing large single crystals for the electronics and telecommunication industries: silica, $\alpha$-$Al_2O_3$, $AlPO_4$, $KTiOPO_4$, ZnO, calcite, silicates, perovskites, garnets, bronzes, chalcogenides, and apatites. Large single crystals of quartz are made at a temperature of 300°C at a pressure of 70 bars in steel reactors of several cubic meters from an aqueous solution of NaOH or $Na_2CO_3$. Such single crystals are more suitable than most mineral crystals found in geological deposits because they have fewer defects. Zinc oxide, the 12–16 semiconductor that has the highest piezoelectric effect, is also preferably made hydrothermally. Very pure ZnS is made at a rate of 0.15 mm/day from a concentrated aqueous solution of KOH or $H_3PO_4$.

An autoclave for a hydrothermal reactor is a thick-walled steel reactor able to contain high pressures at elevated temperatures. Powders dissolve in the lower and hotter part of the reactor and crystallize in the form of single crystals in the upper part. Crystals can also be grown by gradually lowering the temperature of the entire reactor vessel; large PbO and $Na_2CoGeO_4$ crystals have been made this way.

Sodalite, $Na_8Al_6Si_6O_{24}(OH)_2 \cdot nH_2O$, is a piezoelectric low-density solid that can be easily made photochromic with suitable doping from solution. Sodalite crystals are grown hydrothermally from a concentrated NaOH solution at 100 bars and a temperature of 360–400°C. It has been shown that crystals can be grown epitaxially even if their facets have been covered with a gold coating.

Zeolites, metastable aluminosilicates, also called molecular sieves because of the molecular-sized pores these structures have, are made hydrothermally by building up the silica network around certain template molecules (amines) that are later removed from the product by oxidation. This is a kind of chimie douce. Similar structures of other oxides can be made hydrothermally. Molecular sieves of zinc silicon oxides (called VPI 70), molybdenum phophorus oxides, and double helixes of vanadium phosphorus oxides are made by reacting large organic molecules hydrothermally with the appropriate number of inorganic cations needed for charge compensation. The organic molecular template around which the lattice is built up is carefully burned away at the end. We will see in the following section that micelle structures in the solvent can also be used as an organic mold to make novel, open, catalytically active, inorganic structures that have a very high specific surface area.

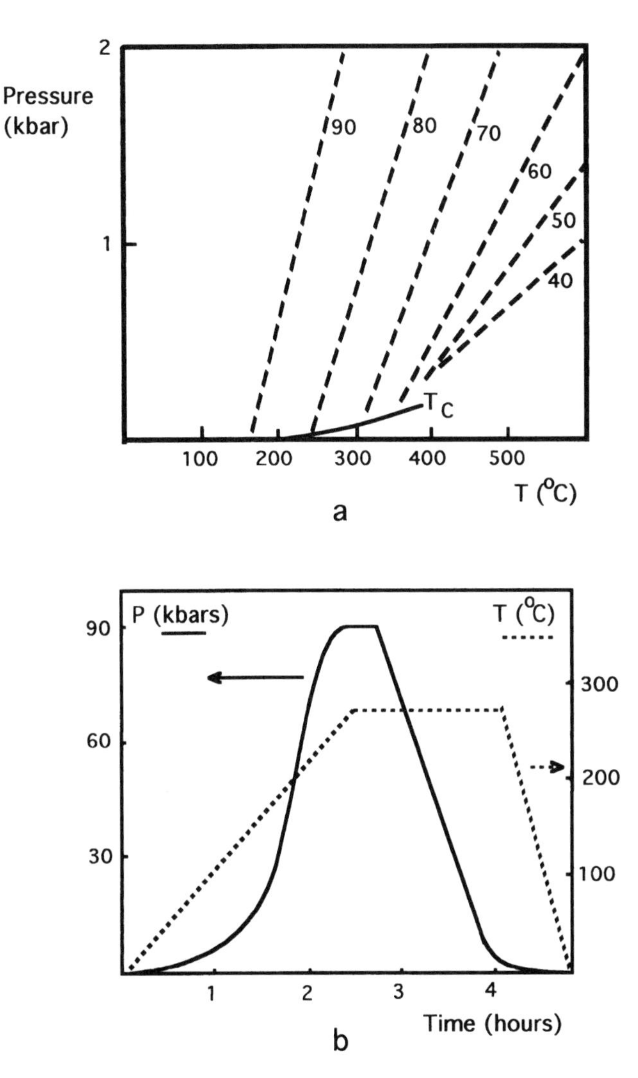


**Figure 8.13.** Pressure and temperature in an autoclave: (a) pressure–temperature relation for different fill ratios (in % of volume) of the autoclave in percent (drawn curve indicates the equilibrium relation of water); (b) process condition in a typical run during a hydrothermal process.

## *8.3.4. Sol-Gel Method*

The sol-gel method is very convenient for making oxides as the technique make it possible to form widely different morphologies and particle sizes with modest means.[26–30] The process conditions are temperature near room temperature, ambient pressure, and water or alcohol as a solvent. The method is as follows: Very small particles of a solid are precipitated as a colloidal sol from an aqueous or alcoholic solution of suitable molecules, after which the particles in the sol are agglomerated to a hydrogel or alcogel that can be shaped or otherwise processed; the gel can be conventionally dried to form a xerogel or the solvent can be supercritically evacuated to form an aerogel. Growing the primary sol particles and aggregating them to the gel is often controlled with the pH of the solution. Powders made with the sol-gel technique are precursors for making glasses or ceramics. Compounds made with this method include the oxides of silicon, germanium, aluminum, and the other metals. Mixed oxides and oxidic alloys can also be made with mixed precursor solutions precipitated with the sol-gel method.

Colloids or sols can be made: (a) by peptizing small particles in a solvent, which means that the suspended solid particles are given a surface charge that prevents them from coagulating and keeps them in a stable suspension, a sol; or (b) by growing small solid particles from a solution of molecules, e.g., hydroxides from hydrolized halides, or by polymerization from a solution of monomers.

The first method yields finally a green form with a wide distribution of pore sizes, rapid drying with a small drying shrinkage, and low cost but difficult doping possibilities. The second produces a green form with more uniform pore size, slow drying with large shrinkage and therefore increased possibility of cracking, and ease of doping. Sinter temperatures are lower than in the first method.

On gel formation the sol particles form a cohesive network by agglomeration that stretches over the entire volume of the reaction vessel. Precipitation or flocculation is a local form of particle aggregation in which the agglomerates do not interlink and extend over the entire solution volume.

The so-called alkoxide process is convenient because it is generally applicable to a large number of oxidic compounds and because the microstructure and properties of the product are in general easily changed with the process conditions.

In the alkoxide process, metal alkoxides are partially or completely hydrolized and the hydrolized monomers polymerized to form colloidal particles in a sol. Subsequently these particles can be made to aggregate in these sols to eventually form a gel. The hydrolysis and polymerization are catalyzed with acids or bases and the choice between them strongly affects the morphology of the product. Figure 8.14 shows this for the formation of silica gel. The two reactions, hydrolysis of the alkoxide (TEOS or tetra ethyl ortho silicate) and polymerization of the hydrolyzed monomers determine the particle size, form, and their connectivity, in short the morphology of the product. Base catalyzed polymerization yields highly branched polymers while acid catalysis yields linear polymers. The microstructure after sol and gel formation of silica under different conditions are shown schematically in Figure 8.15. High concentrations of the alkoxide precursor in the starting solution are favorable for cross-linking and therefore less shrinking on drying but the gels are then more porous.

The sol-gel method is used for making "nonexistent" (i.e. thermodynamically unstable) ternary oxides such as $CaO \cdot 4SiO_2$ and $CaO \cdot 8SiO_2$ after sintering at 800°C. These are compounds of a composition that cannot be made from the melt. $LiAl_5O_8$ is another example. Ultrafine sinter-active ceramic powder particles can be made with the sol-gel method into whiskers, fibers, spheres, coatings, or monolithic products. Aerogels can be prepared hydrothermally, while other materials that have been made with this technique are PLTZ (Lead lanthanum titanate zirconate, PSZ (partially stabilized zirconia), $\alpha$-alumina, and perovskites such as $KTaO_3$, $SrZrO_3$, and $LaCrO_3$.

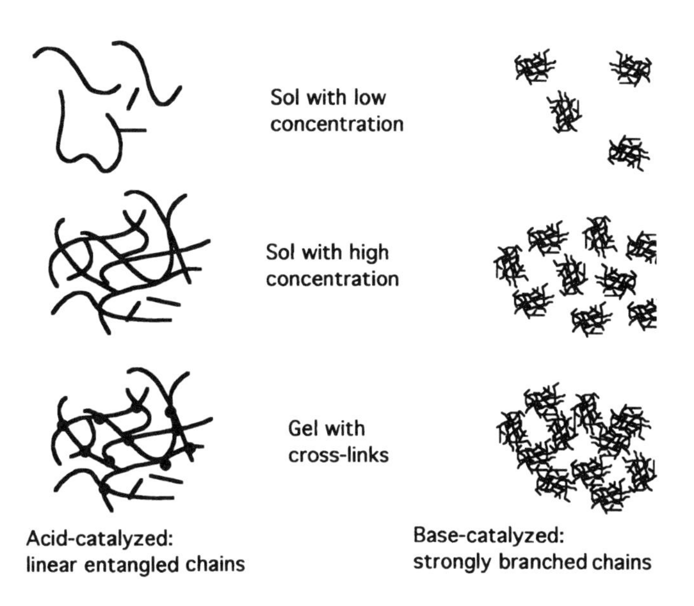


**Figure 8.14.** Growth of silica colloids and their morphology during hydrolysis of precursor molecules and condensation in a sol-gel process.

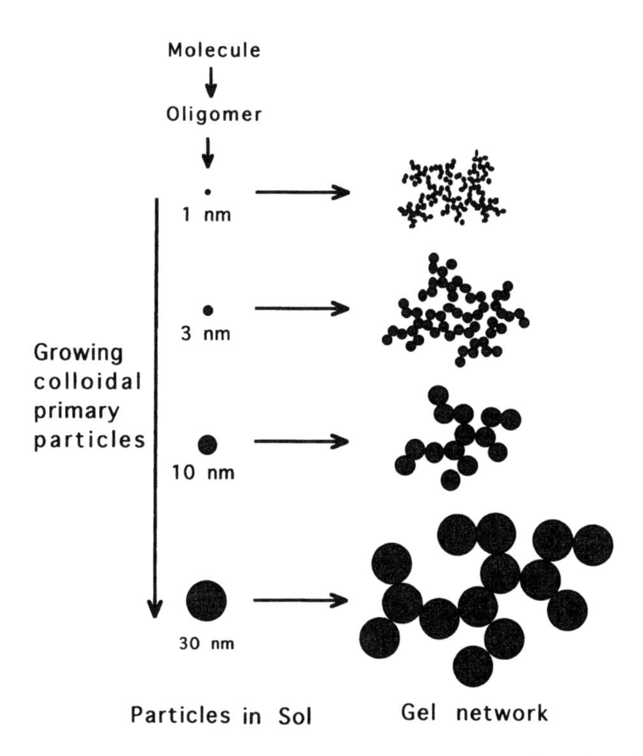


**Figure 8.15.** Schematic of particle growth and agglomeration in sols. Charged particles in a stabilized sol (far from the isoelectric point) grow as long as reactant molecules are present. With highly charged ions or a pH near the isoelectric point the particles are agglomerated. If the particle concentration is high enough a gel forms.

The sol-gel method is also used to make very fine spherical particles of oxides. By structuring the solvent with surface-active solutes, other forms can also be realized during condensation of the monomeric reactant molecules to form a solid particle. Figure 8.16 shows that normal or inverse micelles or liquid crystals (liquids having long-distance order) can be formed in such solutes. Micelles are small domains in a liquid that are bounded by a layer of surface-active molecules. In these domains the solid is condensed and the microstructure of the precipitated solid is affected by the micelle boundaries. Monodisperse colloidal metal particles (as model catalyst) have been made in solvents that have been structured with surfactants. In the concentration domains where liquid crystals obtain highly porous crystalline oxides can be condensed. After calcination such solids can attain specific surface areas up to 1000 $m^2/g$. Micro-organisms use structured solutions when they precipitate calcite, hematite and silica particles.

The sol-gel method is a convenient technique for making oxidic coatings. There are many applications for such coatings, including protective and optical coatings, ferroelectric films, electrically conducting layers, photocatalytic films, and catalysts. In the case of silica gels, the composition of the mixture of TEOS, water, and ethanol is chosen to optimize the final form, as is shown in Figure 8.17.

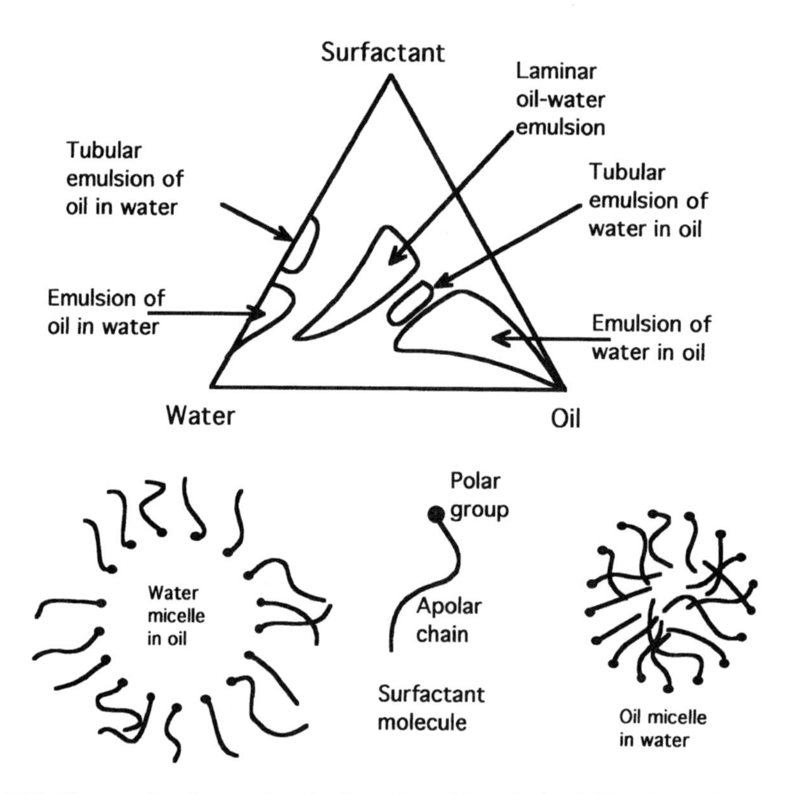


**Figure 8.16.** The use of surfactants in sol-gel reactions. Mutually insoluble solvents for reactants are structured by surfactant molecules which collect at the surface of solid or liquid phases in dispersions. The size of the solid particles grown from such emulsions is restricted by the structured emulsion. Dispersions are droplets, tubes, or platelets depending on the concentration as the triaxial phase diagram indicates.

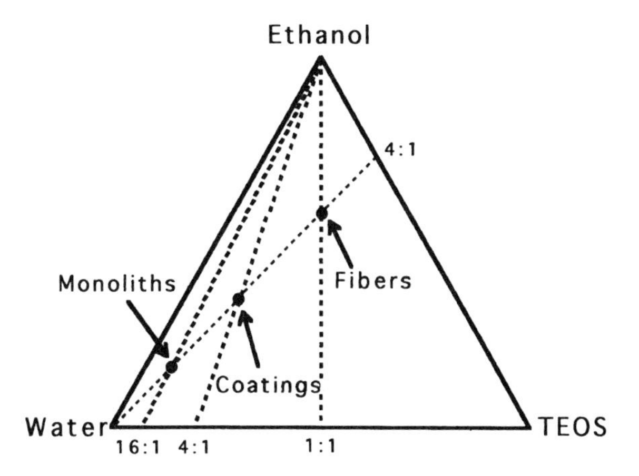


**Figure 8.17.** Concentrations of the three reactants used for making different silica products with the sol-gel method.

## 8.4. Gas Phase Techniques

### 8.4.1. Vapor Transport

Vapor transport is a technique in which solids react with gases and form volatile compounds that deposit the original solid reactant or a different solid elsewhere in the reactor at other temperatures. A temperature difference generates an equilibrium difference and polycrystalline solids can be said to dissolve and precipitate (recrystallize) in the reactive gas ambient. When the product deposits at the end of the reaction the original carrier gas is liberated and diffuses back to the solid reactant. Halogens, HCl, HF, and sometimes oxygen or hydrogen are gases used for transport. Figure 8.18 shows schematically a reactor used in vapor transport. The vapor transport method for making solid compounds is faster than solid state reactions because the mass transport (always necessary for reaction or recrystallization) is rate-limiting and it is faster through the gas phase than through solids. Table 8.5 lists some compounds that have been synthesized in this way. A few examples of the formation of volatile intermediates are:

$$Pt(s) + O_2(g) \rightarrow PtO_2(g)$$

$$Cr(s) + I_2(g) \rightarrow CrI_2(g)$$

$$WO_2(s) + I_2(g) \rightarrow WO_2I_2(g)$$

$$Cu_2O(s) + 2HCl(g) \rightarrow CuCl(g) + H_2O(g)$$

Single crystals of ternary sulfides such as $ZnAl_2S_4$ or selenides such as $Cu_3TaSe_4$ are made with iodine as a transport medium. Ternary oxides such as $ZnWO_3$ are made with chlorine. A versatile transport gas for the synthesis of single crystals is $GeI_4$.

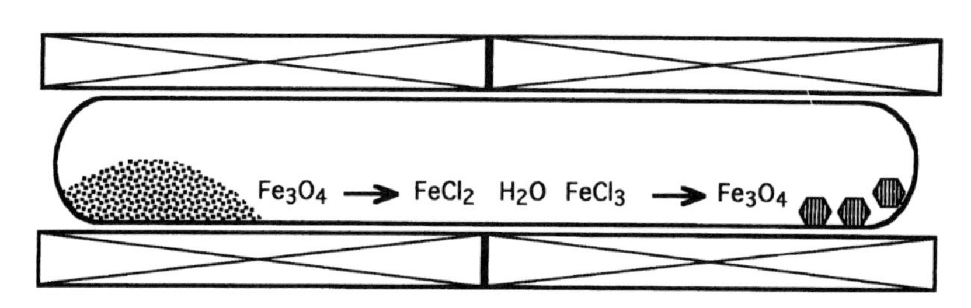


**Figure 8.18.** The vapor transport method for the growth of magnetite crystals. At a high temperature powder reacts with a reactive gas (HCl) at low pressure to volatile compounds. These decompose at a lower temperature in another part of the closed reactor. HCl is formed there and diffuses back to react again with the powder.

The volatile intermediate is used to increase the rate of reactive sintering in advanced ceramics. Examples of solid state reactions that are carried out with vapor transport are:

STANNATES: The reaction $2CaO + SnO_2 \rightarrow Ca_2SnO_4$ is accelerated with carbon monoxide as the transport gas because it reduces one of the solids to a volatile oxide that is easily transported:

$$SnO_2(s) + CO(g) \rightarrow SnO(g) + CO_2(g)$$

which reacts with CaO particles as

$$2CaO(s) + SnO(g) + CO_2(g) \rightarrow Ca_2SnO_4$$

CHROMITES: The reaction between NiO and $Cr_2O_3$ is faster with oxygen:

$$Cr_2O_3(s) + \tfrac{3}{2}O_2(g) \rightarrow 2CrO_3(g)$$

which forms the spinel with nickel oxide:

$$NiO(s) + 2CrO_3(g) \rightarrow NiCr_2O_4(s) + \tfrac{3}{2}O_2(g)$$

SILICIDES: Niobium does not react with $SiO_2$ at 1100°C but the reaction is easy if traces of reductive gases are present. Hydrogen "dissolves" silica and forms SiO gas:

$$SiO_2(s) + H_2(g) \rightarrow SiO(g) + H_2O(g)$$

**Table 8.5. Selected Crystals Grown with the Vapor Transport Method**

| Reactant | Product | Temperatures (°C) | Transport gas |
|---|---|---|---|
| $SiO_2$ | $SiO_2$ | 200–500 | HF |
| $Fe_3O_4$ | $Fe_3O_4$ | 1000–750 | HCl |
| $Cr_2O_3$ | $Cr_2O_3$ | 800/600 | $Cl_2 + O_2$ |
| $NbSe_2$ | $NbSe_2$ | 850/800 | $I_2$ |
| $Fe_2O_3 + MgO$ | $MgFe_2O_4$ | | HCl |
| $Nb_2O_3 + Nb$ | NbO | | $Cl_2$ |

and after that

$$8Nb(s) + 3SiO(g) \rightarrow Nb_5Si_3(s) + NbO(s)$$

Iodine is also helpful in forming niobium silicide from Nb and $SiO_2$:

$$Nb(s) + 2I_2(g) \rightarrow NbI_4(g)$$

and

$$11NbI_4(g) + 3SiO_2(s) \rightarrow Nb_5Si_3(s) + 22I_2(g) + 6NbO(s)$$

The halogen incandescent lamp is an example of vapor transport that increases the temperature and the lifetime of the filament. In a halogen-containing lamp bulb the evaporating tungsten does not precipitate on the interior of the bulb wall but reacts with the gas to volatile halides that pyrolize on the filament. Thus the evaporated tungsten does not obscure the transmission of light of the bulb walls but is returned to the filament (Figure 8.19). A high energy efficiency requires that the filament temperature be as high as possible; 6000°C would provide good efficiency and the right color. The highest known melting points (TaC, HfC) are around 4000°C but the high vapor pressure and temperature gradient limit the lifetime and applicability of most materials at temperatures far below the melting points. Evaporation is countered by returning the evaporated metal as a volatile halide to the filament by means of vapor transport. From thermodynamically estimated vapor pressures it can be seen that at high temperatures (on the surface of the filament) the tungsten halides decompose to the metal and halogen while at lower temperatures (at the bulb wall), the evaporated tungsten reacts to form volatile halides. There is vapor transport of evaporated tungsten back from the walls to the filament. Oxygen gas behaves as halogens with tungsten but the $W/O_2$ system is not used in incandescent lamps because $WO_2$ is a solid at the lower temperatures. Some oxygen, hydrogen, and carbon (as CO) are added to improve the equilibria and transport in the lamp.

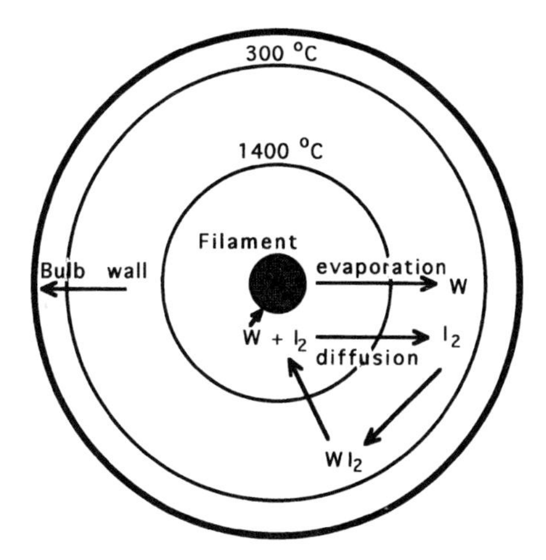


**Figure 8.19.** Principle of the halogen-filled incandescent lamp. This is a case of vapor transport. The tungsten filament can be kept at a very high temperature because evaporating tungsten is returned to the filament: at lower temperatures near the bulb wall the evaporated atoms do not deposit on the bulb wall but react to a volatile component that decomposes to tungsten on the filament.

The latest type of incandescent lamp is the cluster lamp. A tungsten bromide vapor in a bulb is excited by a plasma with external electrodes. Light is emitted by 2-nm tungsten particles that form in the center of a plasma, where they reach a temperature of 3300°C. At a lower temperature near the bulb wall the particles react with bromine and form the gas that decomposes again at the higher temperatures in the plasma.

### 8.4.2. Physical Vapor Deposition

Physical vapor deposition (PVD) is a technique for making thin films at low temperatures and is widely used in planar technology in electronics.[31] It consists of evaporating or sputtering a solid, such as a metal, an alloy, or a mixture of solids, in a vacuum and condensing the compound on the substrate to be covered. In certain variations the vapor is reacted with gases introduced in the vacuum. That variation is reactive evaporation or reactive sputtering. The product can be a polycrystalline deposit or a powder.

A solid is sublimed by heating it to a temperature at which it has a vapor pressure high enough for a sufficient mass transport rate. Figure 8.20 gives the vapor pressures of several metals that can be evaporated. To evaporate the solids heat can

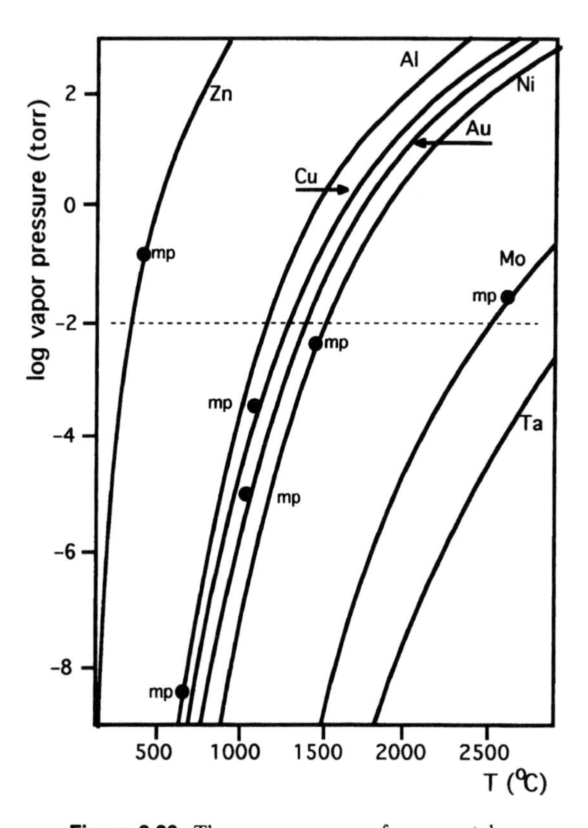


**Figure 8.20.** The vapor pressure of some metals.

be applied by resistive heating, by an electron beam impact (EB-PVD), or a laser beam (laser ablation) directed toward the target, which is the solid to be sublimed. The advantage of using a laser to heat the target is that evaporation is very fast and the evaporation rate is nonselective; it does not depend very much on the equilibrium vapor pressure of certain components of a mixture. Thus, with laser ablation mixtures of solids can be brought into the gas phase with roughly the same composition as that of the target.

For metals a sublimation temperature is chosen that corresponds to a vapor pressure of $10^{-2}$ Torr. This vapor pressure is not sufficient for all cases. Platinum and boron, e.g., have a vapor pressure of $10^{-2}$ Torr at 2100°C, yet platinum evaporates four times faster than boron at this temperature. The same difference in evaporation rate between metals and nonmetals is observed for osmium and carbon. Both have a vapor pressure of $10^{-2}$ Torr at 2650°C but their evaporation rates are as different as in the case of platinum and boron. Apparently the activation energy for evaporation of nonmetals is higher than for metals. Ease of evaporation or a high vapor pressure does not guarantee fast deposition rates even for metals. Although magnesium and zinc are volatile, they are difficult to deposit because they do not condense easily as their closed outer electron shell confers helium-like properties on their gaseous atoms.

Vapor phase epitaxy (VPE) is a PVD technique in which a compound is formed by shooting the evaporated atoms in an ultrahigh vacuum toward a substrate, where they react with each other. VPE is applied in the synthesis of superlattices. These are monocrystalline semiconductors that consist of alternating epitaxial layers several lattice cells thick of different semiconductors having slightly different bandgaps but almost equal lattice parameters. VPE has also been used in attempts to make $C_3N_4$, a theoretically predicted very hard compound that has a silicon nitride structure. An amorphous deposit that has approximately the right composition has been made with a VPE-like technique.

Reactive sputtering means hitting the solid (or target) with noble gas cations from a plasma followed by a reaction of target atoms as they deposit on the substrate. The target has a negative potential and is eroded by the impacting noble gas ions. The target atoms that are removed by the ion bombardment are deposited elsewhere in the reactor and react with added gaseous precursors to form compounds on all exposed surfaces, including the reactor wall. The vacuum reactors used in sputtering have electrode arrangements similar to those in plasma-CVD reactors discussed in Section 8.4.4. Although during reactive sputtering or reactive evaporation chemistry occurs these methods belong to PVD because the reactor is PVD equipment.

Not only polycrystalline or epitaxial layers of metals and many ceramics are made with PVD methods but also metal powders. Metal is evaporated (by resistive heating) in an inert gas at a pressure in which the mean free path length of the atoms is small with respect to the size of the reactor. The evaporated metal atoms combine on colliding with each other and form small particles that can be harvested on a cold finger by thermodiffusion.

The development of a novel technique for making known solid compounds often leads to the discovery of novel solids, e.g., carbon. Notwithstanding the enormous amount of information about molecular compounds containing carbon that has been collected by many generations of organic chemists, the molecular

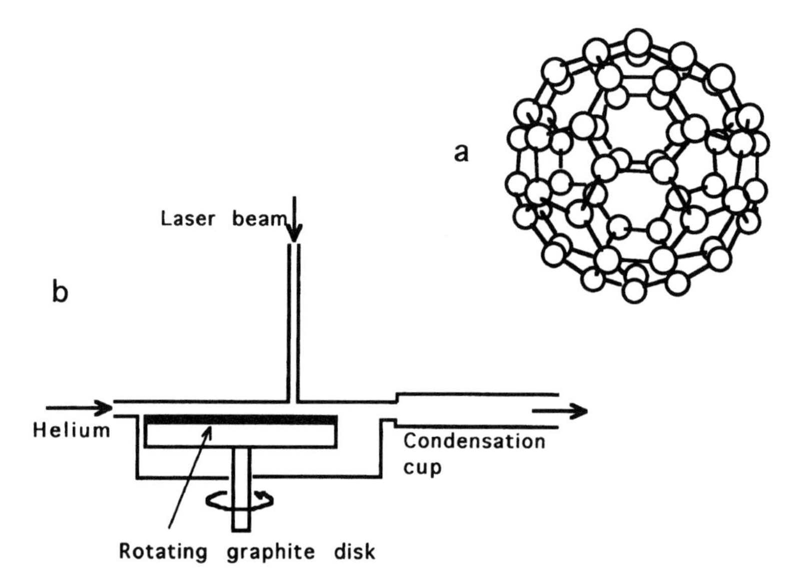


**Figure 8.21.** (a) The structure of one of the molecular forms of carbon, buckminsterfullerene, also called buckyballs. (b) Equipment for preparation of molecular carbon by laser ablation.

modification of carbon was discovered only recently as the result of the application of a new, nonmolecular method of synthesis to carbon.

Fullerenes or buckyballs are carbon molecules that were discovered in 1985. Now there are molecular modifications of carbon besides the usual diamond, graphite, and lonsdaleite. Their stoichiometries are $C_{60}$, $C_{70}$, $C_{76}$, $C_{84}$, and $C_{90}$ (Figure 8.21). They are not made by molecular synthesis as are most organic molecules but by means of PVD: carbon is evaporated by a laser or an arc discharge between carbon electrodes in a helium atmosphere ($T = 10{,}000°C$). From the condensation products the fullerenes are extracted with toluene, in which the buckyballs dissolve. $C_{60}$ is a semiconductor with a bandgap of 1.6 eV; it sublimes at a temperature of 350°C. If a layer of $C_{60}$ is exposed to a potassium vapor the metallic compounds $K_3C_{60}$ or, at a later stage, $K_6C_{60}$ are formed. If during the synthesis of fullerenes electropositive metals are present in the gas, the molecules can wrap around the metal atoms in a kind of template synthesis. In this manner, ($Li@C_{60}$), ($K@C_{60}$), ($La@C_{60}$), ($La@C_{82}$), ($Y_2@C_{82}$), and ($La_3@C_{100}$) have been synthesized (the symbol @ indicates enclosure of the metal ion or ions by the carbon cage). In spite of several efforts it has not been possible to make ($Fe@C_n$) or ($Cu@C_n$), probably because the transition metals are not electropositive enough compared to the carbon shell. Boron and nitrogen doping of fullerenes is possible. Figure 8.21 also shows a sketch of the equipment for making fullerenes by laser ablation.

### *8.4.3. Chemical Vapor Deposition*

Chemical vapor deposition (CVD) is a synthetic technique for making inorganic solids from the vapor phase by reacting gaseous precursors in a flow reactor,[32–35]

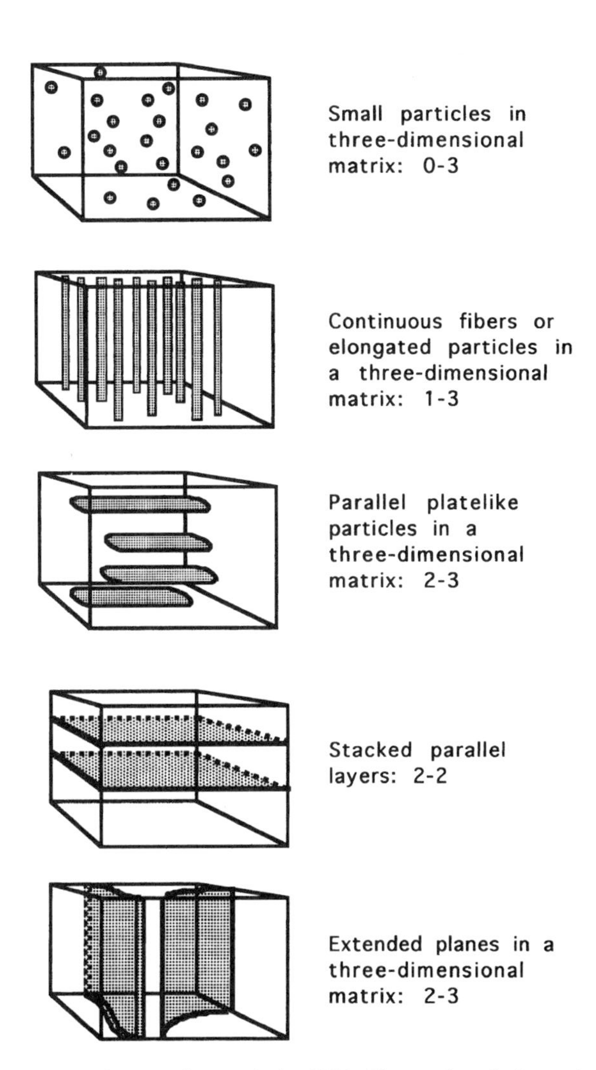


**Figure 8.22.** Morphologies of composites made by CVD. The numbers indicate the dimensions of the component phases. From T. Hirai and T. Goto. In: R. E. Tressler, G. L. Messing, C. G. Pantano, and R. E. Newnham (eds.). *Tailoring Multiphase and Composite Ceramics.* Plenum, New York (1984).

a method described in Section 6.7. The reactant vapors are usually excited by heat but in certain variants, light or electrical energy in plasmas are also used to activate the precursors. The morphology of the reaction product can be controlled by adjusting the nucleation and growth rates with the process conditions (Figures 8.22 and 8.23). Composites can be made in a single process step in some cases but not easily.

CVD is a form of vapor transport in an open reactor in which all the reactants are gases. Virtually all metals, ceramics, and some polymers have been deposited by means of CVD. The method is much used for making inorganic coatings in order to improve the performance of materials that are exposed to severe conditions. It has mainly been developed in the technology of electronic devices.

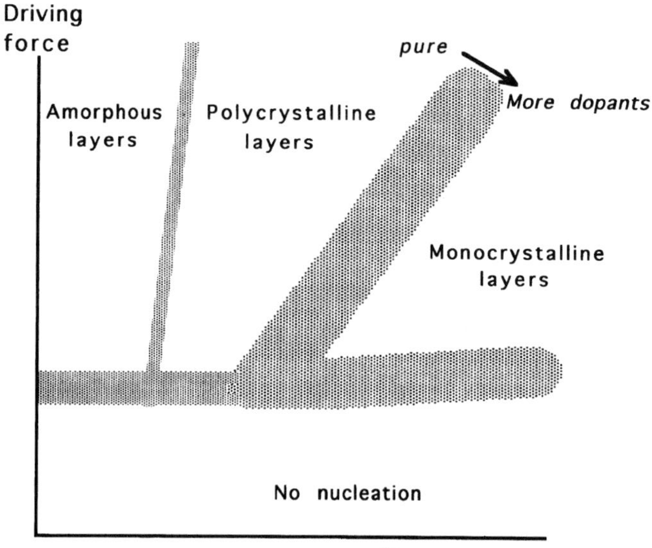


**Figure 8.23.** The effect of process conditions on the morphology of the deposit. Both the chemical driving force and the temperature affect the morphology as indicated in this qualitative scheme. The strength of the driving force is varied with the choice of reactants. For temperature read surface mobility. A minimum lowering of free energy is required for deposition, but its threshold value for the reaction depends on the impurities present. There are four domains. Amorphous growth needs low mobility and the growth of monocrystalline solids presupposes a high mobility of the surface atoms and a relatively low driving force for the reaction.

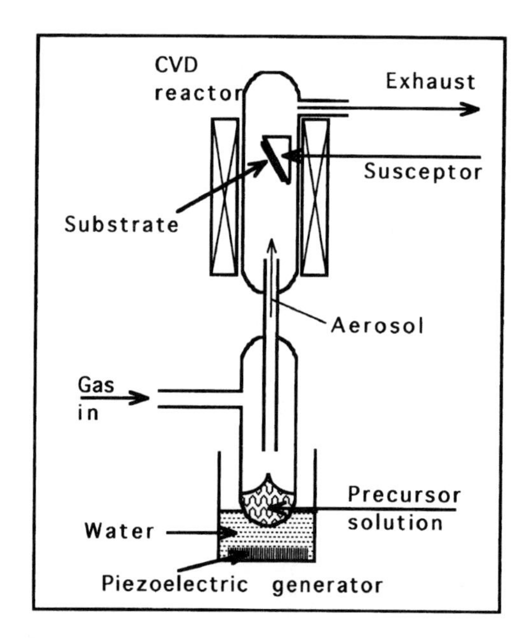


**Figure 8.24.** Equipment for producing a mist of a precursor solution in a carrier gas if the precursor itself is not volatile enough or if it decomposes on evaporation. From M. Langlet and C. Joubert. The pyrosol process or the pyrolysis of an ultrasonically generated aerosol. In: C. N. R. Rao (ed.). *Chemistry of Advanced Materials.* Blackwell, Oxford (1993), p. 55. With permission from Blackwell Science Ltd.

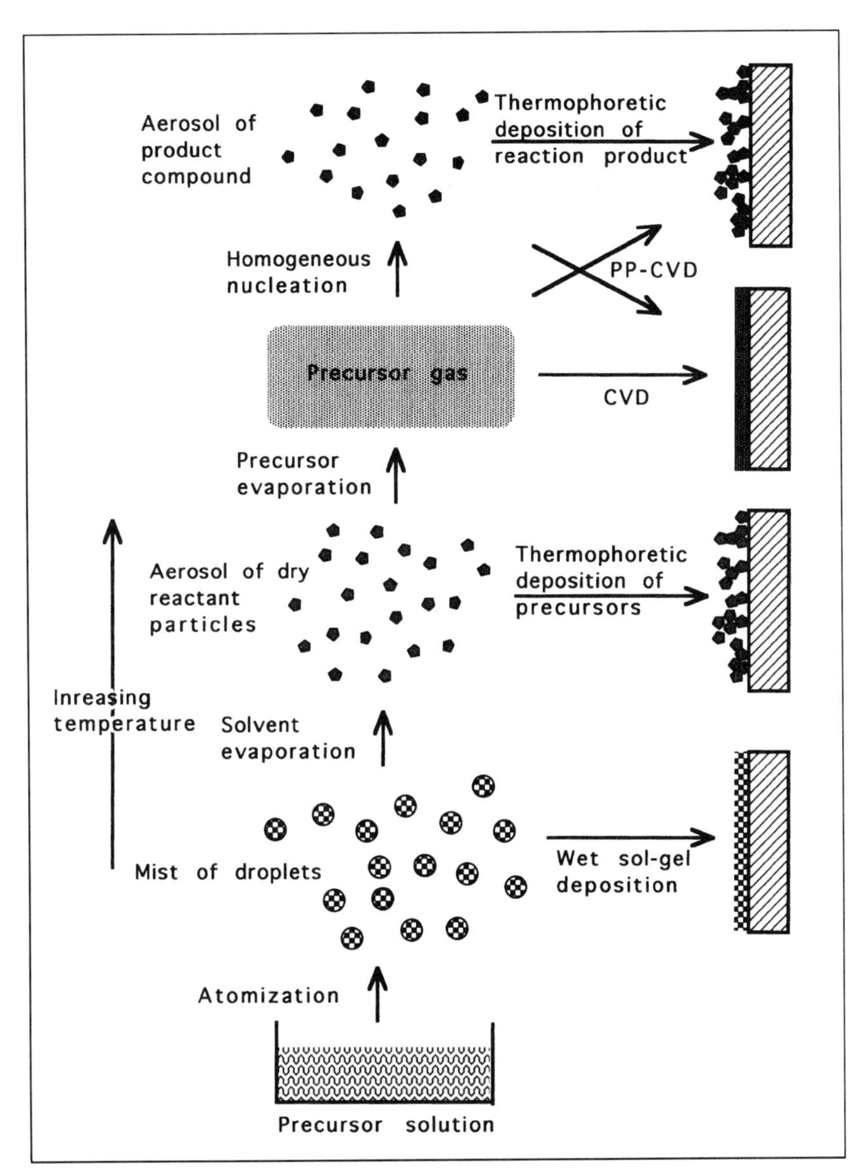


**Figure 8.25.** Application of nonvolatile precursors for synthesis of solids. Ultrasonic atomization of a solution leads to droplets from which the solvent evaporates in the reactor. At still higher temperatures the aerosol evaporates and the usual forms of CVD are possible.

The reactants in CVD are gases. Metal-containing precursors are liquids or solids that are sufficiently volatile. These precursors are evaporated, mixed with phases such as hydrogen, and fed into the reactor. Precursors should not decompose too early (on evaporation). The metal–organic compounds discussed in Chapter 3 are suitable precursors as are some metal halides and carbonyls. Solid reactants that are difficult to evaporate or are unstable but can be dissolved in a noninterfering solvent can be used with a special technique.[36] Figure 8.24 shows how to get a solid, nonvolatile precursor into the CVD reactor. An aerosol is made by dispersing a solution of the reactant in a carrier gas. Several types of deposit are possible as indicated in Figure 8.25. If the deposit is still wet, the solvent evaporates after deposition and the precursor that is left behind decomposes and reacts on the surface. If the solvent evaporates from the droplets before deposition, dry particles of the reactant are deposited and decompose and perhaps react with gases if they have not already done so as an aerosol just before deposition. The morphology of the product is most likely to be very different all these cases because the precursor distribution is not the same for wet and dry deposition. At high temperatures in the reactor the aerosol evaporates or decomposes or both and common CVD can occur.

The study of aerosol technology has so far been largely concentrated on its physical aspects.[37–39] The terminology in the aerosol world is physical: cluster formation by colliding particles that stick together after collision is "coagulation" and surface chemistry of particle growth is called condensation. Figure 8.26 shows schematically the various processes that contribute to the formation and growth of particles in the gas phase. In the customary physical discussions the chemical steps

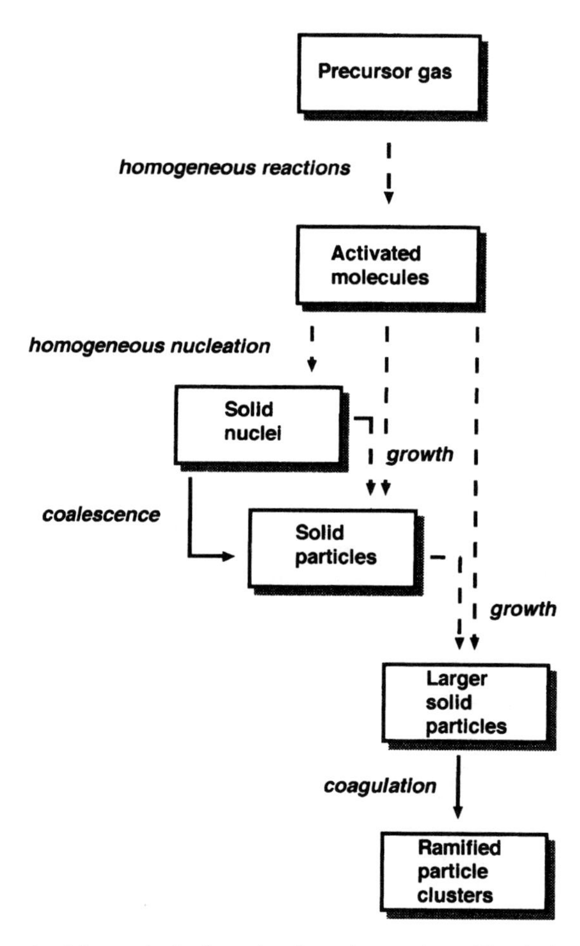


**Figure 8.26.** Schematic of the synthesis of powders from the gas phase. Chemical processes are indicated with broken arrows and physical processes with solid arrows.

are taken to be simplified first-order reactions to facilitate developing mathematical analytical expressions.

Whiskers (small powder particles having a high aspect ratio) are made with very small droplets of a metal or (eutectic) salt melt (e.g., a sulfide) that show some sort of catalysis. Two mechanisms of such vapor–liquid–solid (VLS) catalysis have been proposed, one for dense monocrystalline whiskers and the other for hollow polycrystalline carbon whiskers. Massive monocrystalline whiskers grow from a supersaturated droplet of a solution in which the concentration of the product is kept high by the reaction at the surface of the droplet. The product of the reaction is forcibly dissolved in the oversaturated droplet because the driving force for the surface reaction is high. The solute has to crystallize on the interface of the droplet on the whisker. If the droplet evaporates during growth it becomes smaller and the whisker tapers. If the surface tension depends on the concentration, the contact angle between the liquid droplet and the whisker may oscillate during growth and so may the whisker diameter. If the solid that forms on the surface of the droplet does not dissolve in the liquid substrate (carbon is an example) then hollow whiskers form. Whisker formation of silicon, carbon, silicon carbide, and silicon nitride (from gold or nickel droplets) have been used in industrial processes. Figure 8.27 shows how SiC whiskers that are used for toughening composite ceramics are made in principle.

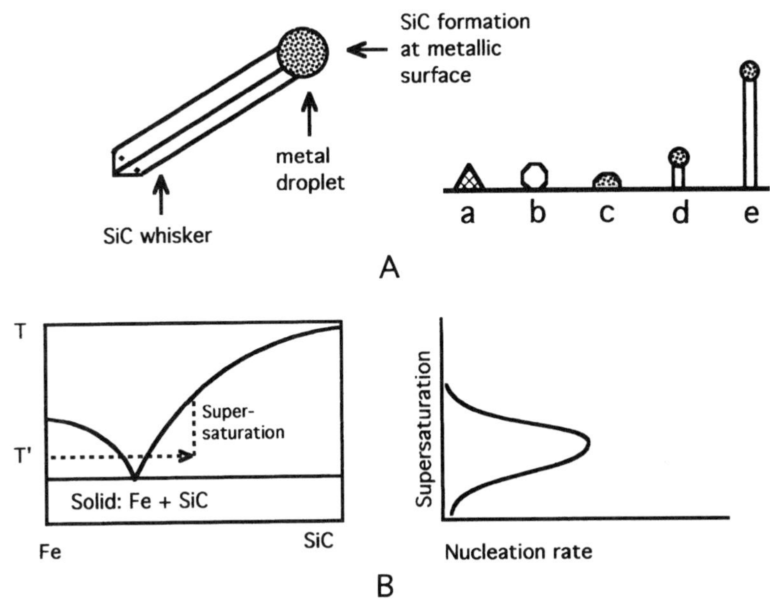


**Figure 8.27.** Growth of silicon carbide whiskers from gaseous precursors. (A) The VLS-mechanism for the growth of whiskers, which occurs when solid impurities (a) react to form metal particles, e.g., iron (b), which melt at the reaction temperatures (c). At the surface of these droplets gases react to form a product (e.g., SiC), which dissolves in the droplet. When the oversaturation in the metal melt is high enough the solute nucleates and a whisker grows out of the droplet (d, e). (B) The iron/SiC phase diagram. At the whisker growth temperature $T'$ the SiC concentration in the droplet follows the path indicated by the broken arrow. At a certain supersaturation the crystallite growth is rapid enough to keep up with the VLS production rate at the metallic surface of the droplet.

One of the advantages of CVD as a preparative technique is that materials that are very inert and have a high melting point, e.g., W, TaC, and $Si_3N_4$ can be deposited under comparatively mild conditions (300–1100°C). A "brute-force" method for synthesis of such materials (solid state reactions or melt growth) requires temperatures in the range of 1800–3500°C.

A few representative examples of CVD reactions are listed here.

1. ELEMENTS AND THEIR ALLOYS

(*a*) *Pyrolysis:*

$$ZrI_4(g) \rightarrow Zr(s) + 2I_2(g)$$

$$CrI_2(g) \rightarrow Cr(s) + I_2(g)$$

$$Ni(CO)_4(g) \rightarrow Ni(s) + 4CO(g)$$

$$SiH_4(g) \rightarrow Si(s) + 2H_2(g)$$

$$CH_4(g) \rightarrow C(s) + 2H_2(g)$$

$$AlR_3(g) \rightarrow Al(s) + nR'(g) \text{ (R is an organic group)}$$

$$(C_8H_{10})_2Cr(g) \rightarrow Cr(s) + nR'(g)$$

(*b*) *Reduction of halide precursors by hydrogen:*

$$2AlCl_3(g) + 3H_2(g) \rightarrow 2Al(s) + 6HCl(g)$$

$$SiCl_4(g) + 2H_2(g) \rightarrow Si(s) + 4HCl(g)$$

$$2TaCl_5(g) + 5H_2(g) \rightarrow 2Ta(s) + 10HCl(g)$$

$$x\,MoF_6(g) + (1 - x)WF_6(g) + 3H_2(g) \rightarrow Mo_xW_{1-x}(s) + 6HF(g)$$

(*c*) *Reduction by metals:*

$$TiCl_4(g) + 2Mg(s) \rightarrow Ti(s) + 2MgCl_2(g)$$

$$TiI_4(g) + 2Zn(s) \rightarrow Ti(s) + 2ZnCl_2(g)$$

2. SEMICONDUCTORS:

$$GaEt_3(g) + AsH_3(g) \rightarrow GaAs(s) + 3C_2H_6(g)$$

$$BBr_3(g) + PBr_3(g) + 3H_2(g) \rightarrow BP(s) + 6HBr(g)$$

3. OTHER CERAMICS

$$BCl_3(g) + NH_3(g) \rightarrow BN(s) + 3HCl(g)$$

$$4BCl_3(g) + CO(g) + 7H_2(g) \rightarrow B_4C(s) + H_2O(g) + 12HCl(g)$$

$$4BCl_3(g) + CH_4(g) + 4H_2(g) \rightarrow B_4C(s) + 12HCl(g)$$

$$4BCl_3(g) + CCl_4(g) + 8H_2(g) \rightarrow B_4C(s) + 16HCl(g)$$

$$3SiCl_4(g) + 4NH_3(g) \rightarrow Si_3N_4(s) + 12HCl(g)$$

$$SiCl_4(g) + CH_4(g) \rightarrow SiC(s) + 4HCl(s)$$

$$CH_3SiCl_3(g) \rightarrow SiC(s) + 3HCl(g)$$

$$Si(OEt)_4(g) + 2H_2O(g) \rightarrow SiO_2(s) + 4C_2H_5OH(g)$$

$$2AlCl_3(g) + 3CO_2(g) + 3H_2(g) \rightarrow Al_2O_3(s) + 3CO(g) + 6HCl(g)$$

$$Al_2Cl_6(g) + 3H_2O(g) \rightarrow Al_2O_3(s) + 6HCl(g)$$

$$AlCl_3(g) + NH_3(g) \rightarrow AlN(s) + 3HCl(g)$$

$$Al(CH_3)_3(g) + NH_3(g) \rightarrow AlN(s) + 3CH_4(g)$$

$$[(CH_3)_2AlNH_2]_3(g) \rightarrow AlN(s) + xR(g)$$

$$ZrCl_4(g) + 2CO_2(g) + 2H_2(g) \rightarrow ZrO_2(s) + 2CO(g) + 4HCl(g)$$

$$TiCl_4(g) + 2BCl_3(g) + 5H_2(g) \rightarrow TiB_2(s) + 10HCl(g) \text{ (also Zr, Hf, Ta)}$$

$$Zr(OPr)_4(g) + 2H_2O(g) \rightarrow ZrO_2(s) + 4C_3H_8OH(g)$$

$$2ZrCl_4(g) + N_2(g) + 4H_2(g) \rightarrow 2ZrN(s) + 8HCl(g) \text{ (also Ti and Hf)}$$

$$WCl_6(g) + CH_4(g) + H_2(g) \rightarrow W(s),\ W_2C(s),\ WC(s) + xHCl(g)$$

$$TaCl_5(g) + CH_4(g) + H_2(g) \rightarrow TaC(s) + 5HCl(g) \text{ (also Nb)}$$

$$2Mo(s) + CO(g) + H_2(g) \rightarrow Mo_2C(s) + H_2O(g) \text{ (conversion)}$$

The overall reaction equations of the type given above are mere mass balances that give the reactants and reaction products and are necessary for calculating the equilibrium constants or the driving forces for the processes. Such reaction equations do not contain information on kinetics or mechanisms. The precursors always react through a large and complex set of reaction steps involving many reaction intermediates that usually remain unidentified. Deposition rates and reaction orders are mainly determined by the slowest of a set of consecutive reaction steps.

To illustrate this point, we consider the reaction for tungsten deposition:

$$WF_6(g) + 3H_2(g) \rightarrow W(s) + 6HF(g)$$

The equilibrium constant for this reaction is

$$K = \frac{[HF]^6}{[WF_6][H_2]^3}$$

The order of the reaction in the reactants however has been shown experimentally to be zero in $p(WF_6)$ and one-half in hydrogen partial pressure.

### *Precursors and Methods for Their Evaporation*

Precursors molecules containing metal atoms are liquids or solids and, in a few rare cases, gases. Gaseous precursors if available are preferred because their manipulation in the gas line is easy. Volatile liquid precursors also are convenient but solids, particularly those having low vapor pressures, require special treatment to evaporate (Figure 8.24).

The main criteria for a suitable precursor are: (a) a sufficiently broad stability range between evaporation and decomposition temperatures; (b) a sufficient vapor pressure at those temperatures; (c) clean decomposition to the product wanted during reaction, i.e., leaving no contaminating species in the solid deposit. The choice of precursor is determined by many factors such as substrate, temperature permitted,

Table 8.6. Thermal Properties of Selected Halides

| Halide | Melting point (°C) | Boiling point (°C) | Halide | Melting point (°C) | Boiling point (°C) |
|---|---|---|---|---|---|
| $AlCl_3$ | 190 | 182.7 | $NbCl_5$ | 204.7 | 254 |
| $AlBr_3$ | 97.5 | 263 | $TaCl_5$ | 216 | 242 |
| $BCl_3$ | −107.3 | 12.5 | $CrCl_2$ | 824 | 1300 |
| $CCl_4$ | −23 | 76.8 | $MoF_6$ | 17.5 | 35 |
| $SiCl_4$ | −70 | 57.6 | $MoCl_6$ | 194 | 268 |
| $TiCl_4$ | −25 | 136 | $WF_6$ | 2.5 | 17.5 |
| $ZrCl_4$ | 437 | 331 | $WCl_6$ | 248 | 275.6 |
| $VCl_4$ | −28 | 148.5 | $ReF_6$ | 18.8 | 47.6 |

side reactions, codeposition, and resulting morphology. The most popular reactants are halides, carbonyls, carbonyl halides, hydrides, metal–organics (alkyls, arenes), alkanolates, substituted acetylactonates, and miscellaneous other molecules (amides, boranes).

1. HALIDES. A few relevant data are collected in Table 8.6. Their disadvantage for CVD is that they produce corrosive acid vapors.
2. CARBONYLS of certain refractory transition metals and of a few noble metals are suitable precursors. They easily decompose and usually are rather volatile. Examples are the mononuclear $Cr(CO)_6$, $Ni(CO)_4$, $Fe(CO)_5$, $Mo(CO)_6$, $W(CO)_6$, $Ru(CO)_5$, and $Os(CO)_5$, and the polynuclear compounds $Fe_2(CO)_9$, $Mn_2(CO)_{12}$, $Re_2(CO)_{10}$, $Ir_2(CO)_8$, $Ru_3(CO)_{12}$, $Os_3(CO)_{12}$, $Co_4(CO)_{12}$, $Rh_4(CO)_{12}$, and $Ir_4(CO)_{12}$. Several of them (Ni, Fe) are very poisonous and are best prepared *in situ*. The carbonyls react at comparatively low temperatures, usually in a hydrogen atmosphere.
3. METAL CARBONYL HALIDES are preferred for the noble metals because of their stability temperature range. Examples are $Ru(CO)_2I_2$, $Os(CO)_3Cl_2$, $RhCl_2$, $RhO(CO)_3$, $IrCl_2(CO)_2$, $Pt(CO)Cl_2$, and $Pt(CO)_3Cl_2$. Metal carbonyl nitrosyls are also useful: Cobalt is deposited from liquid $CoNO(CO)_3$.
4. HYDRIDES of the nonmetals are used extensively for deposition of semiconductors. These are arsine ($AsH_3$), diborane ($B_2H_6$), phosphine ($PH_3$), silane ($SiH_4$), germane ($GeH_4$), stibine ($SbH_3$), and hydrogen selenide ($SeH_2$). Hydrogen sulfide is used to deposit sulfides at low temperatures and use of $NH_3$ instead of nitrogen with hydrogen increases the driving force for deposition of metal nitrides. Several of the hydrides (B, P, As) are very toxic.
5. METAL ORGANIC COMPOUNDS are precursors for MO-CVD. The metal atom is bound to a carbon atom, either an alkyl group or an aromatic group such as cyclopentadienyl or aryl. Examples are $Al(CH_3)_3$, $As(CH_3)_3$, $In(CH_3)_3$, $Mg(C_2H_5)_2$, $Mg(C_5H_5)_2$ [or $Mgcp_2$], $Pb(CH_3)_4$, $Si(CH_3)_4$ [TMS], $Sn(CH_3)_4$, $Sncp_2$, and $Zn(C_2H_5)_2$. Most metal–organics are very reactive.
6. ALKOXIDES OR ALKANOLATES are useful for depositing oxides. Examples are $Si(OC_2H_5)_4$(TEOS), $Ti(OC_2H_5)_4$, and $Zr(OC_3H_7)_4$.
7. ACETYLACETONATES (acac) and derivatives are very versatile as precursors and many metals, even very electropositive ones such as the rare earths can be made volatile with (substituted) acetylacetones as ligands. Commercially available are

those of Ba, Ca, Ce, Cr, Co, Cu, In, Fe, Pb, Li, Mg, Mn, Ni, Pd, Pt, the rare earths, Rh, Sc, Sr, Ag, V, Y, Zn, and Zr. They are volatile solids. The vapor pressure can be increased by fluorine substitution of the alkyl hydrogen in the acac ligand.

Evaporators for precursors are bubblers equipped with a reflux condenser (Figure 6.10) for liquids or simple packed-bed evaporators for solids. If vapor pressures are too low the solid is evaporated in the CVD reactor, either by positioning the solid precursor upstream of the deposition site or by generating an aerosol of the precursor in the reactant gas. *In situ* synthesis of the precursors is also practiced, for instance, by reacting a metal with a halogen gas at high temperature in the CVD reactor itself shortly before deposition from the formed halide.

### *Reactor Types*

Reactors for conventional thermally activated CVD are of two types: cold-wall and hot-wall reactors, respectively internally and externally heated. The disadvantage of a hot-wall reactor is deposition on the wall and partial depletion of reactants leading to nonuniform coatings. A correct reactor geometry and gas inlet manifold can compensate for gas depletion in hot-wall reactors. There is no limit to the form of the objects to be coated, but sizes are restricted. In a cold-wall reactor the substrates to be coated are heated by a graphite susceptor that is inductively heated by an rf generator. Only the hot parts are coated and not the reactor walls, which remain relatively cold.

### *Other Methods of Activation*

Conventional CVD uses thermal activation, with the reaction taking place at elevated temperatures. The activation energy necessary for the reaction can also be supplied in the form of very hot electrons in a plasma (electrical gas discharge) or by absorption of infrared or ultraviolet electromagnetic radiation. Eximer lasers are used for photolysis of gas molecules into reactive fragments that are the activated precursors for the CVD reaction at low substrate temperatures.

High-power infrared carbon dioxide lasers selectively heat those molecules having a vibration absorption band at the laser frequency. The absorbing precursor molecule is heated and by collisions the vibrationally excited state is relaxed and heats the other (nonabsorbing) precursor molecules if present. A sensitizer (radiation absorbing molecule that does not itself react) is used to pass the laser energy on to the reactants if none of the precursors has a suitable absorption band at the laser frequency. Infrared lasers are also used to heat the substrate locally to a temperature high enough for common CVD.

Mercury vapor in the reactant gas mixture is sometimes used as a sensitizer for activation by light from a high-pressure mercury lamp.

### *Variants of CVD*

Depending on process conditions, we label the many possible CVD variants by prefixes:

- MO-CVD uses metal–organic precursors and has relatively low process temperatures.

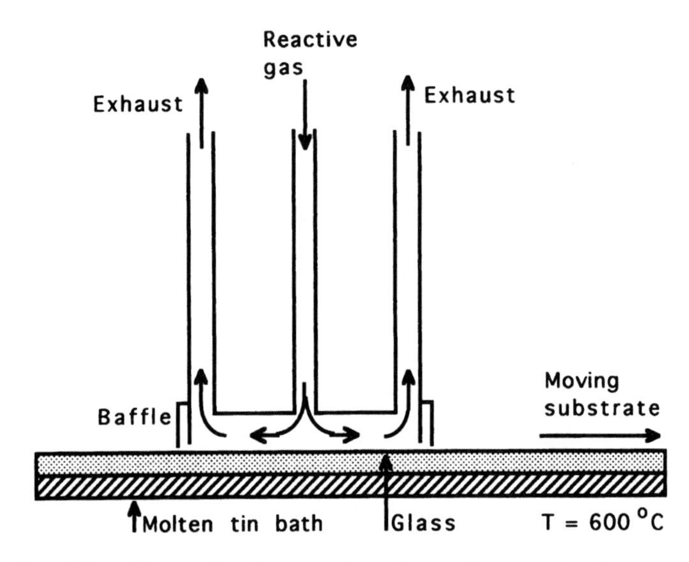


**Figure 8.28.** Deposition of inorganic solar control coatings on large panels of float glass by means of atmospheric pressure CVD.

- PE-CVD applies a plasma for activation and is also a low-temperature process.
- L-CVD uses laser activation.
- LP-CVD has low pressures in the reactor to improve mass transport by diffusion.
- AP-CVD occurs at atmospheric gas pressures, which is convenient and permits large substrate size; a reactor scheme is given in Figure 8.28.
- E-CVD or EVD (electrochemical CVD) relies for growth on ion transport in the deposited coating.
- FB-CVD (fluidized bed-CVD) is deposition on particles in a fluidized-bed reactor.
- CVI (chemical vapor infiltration) is CVD in the interior of a porous solid.

### *8.4.4. Plasma Synthesis*

Plasmas in inorganic synthesis are ionized gases. In inorganic nature they are found in the ionosphere and in the stars (Figure 8.29). In technology, plasma activation is used extensively for depositions at relatively low-temperatures.[40] At low pressures an electrical glow discharge by dc, rf, or microwave excitation allows high-temperature chemical processes to occur in the gas phase although the gas temperatures remain low. Low-pressure plasmas are used in sputtering, ion plating, and in plasma-enhanced CVD. High electric fields accelerate free electrons in the plasma, which on collision excite or ionize reactant molecules. A plasma contains radicals, ions, and free electrons that are chemically very active. The gases that are activated by a gas discharge can sometimes react to form useful solids even at low temperatures. This is called plasma-activated (PA-CVD or plasma-enhanced CVD (PE-CVD).

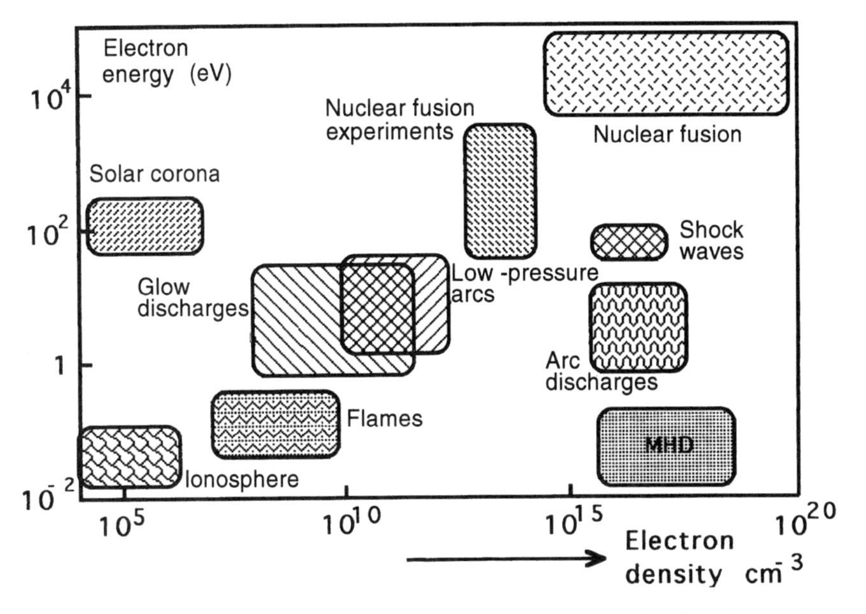


**Figure 8.29.** Types of plasma according to the density and kinetic energy of the free electrons in the gas.

Making solids with plasmas is done preferably at low pressures (0.1–10 mbar). At those pressures the plasma, an ionized gas, is not in thermodynamic equilibrium and the kinetic energy of the free electrons in the plasma in much higher than the translational energy of the molecules (Figure 8.30). Plasmas permit a kind of high-temperature chemistry in a low-temperature gas because the colliding high-energy electrons create more ions, reactive radicals, and excited states than would correspond with equilibrium at the molecular gas temperature. It has been shown in Section 6.7.3 that thermodynamic restrictions can be circumvented by using plasma

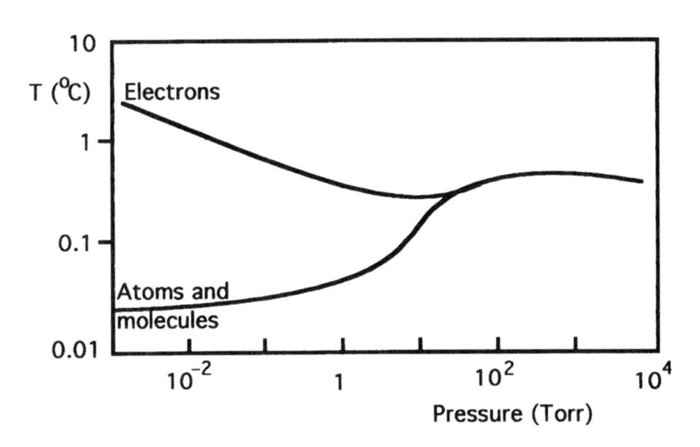


**Figure 8.30.** The temperatures in an air plasma. At low pressures there is no equipartition of the kinetic energy over the atoms and free electrons and they have different kinetic energies or different effective temperatures.

excitation in order to allow "impossible" reactions to occur at a reasonable rate. In plasma etching, the radicals in the gas react with the surface of the substrate in a gas discharge and remove atoms from the solid. Besides PA-CVD there is also a related technique known as surface conversion by a plasma of which one example is plasma nitridation of steel: a metallic surface reacts with nitrogen species from a plasma and is converted to a nitride.

The free electrons in the plasma are reactant or catalyst in chemical reactions. Plasmas shift equilibria or they help thermodynamically allowed reactions that have rates that are too low at the low gas temperature (Table 6.3). Table 8.7 lists selected compounds that have been made with PE-CVD at low temperatures with their precursors and the conditions for their synthesis. The amorphous products of PE-CVD are rarely stoichiometric and almost never pure.

Diamond is made by means of PE-CVD at low substrate temperatures (800°C) and low pressures (10 mbar) using hydrocarbons, hydrogen, and some oxygen in a microwave reactor. Other reactors use radicals made at atmospheric pressure with a heated tungsten or tantalum filament or an acetylene torch. The presence of hydrogen atoms and carbon radicals in the gas phase seems to be a necessary condition for deposition of diamond. Hydrogen radicals "stabilize" the diamond structure of carbon at the surface, which means that four-coordinated carbon (as in diamond) forms more readily than three-coordinated carbon (as in graphite). Hydrogen and carbon radicals react with dangling bonds at the surface, which are prevented from forming double graphitic links between surface carbon atoms. Figure 8.31 shows the schematics of several reactors that are used in PE-CVD.

In PE-CVD all reactants are gases but in plasma sprays a coating is made with a high-pressure plasma (at 1 bar the gas temperature is very high) in which powder particles are molten and sprayed on a cooled substrate. This is similar to the Verneuil method for growing single crystals. Plasma or flame spraying is a form of enameling with solids that have high melting points. It is a physical shaping process and there is no plasma chemistry involved; the atmospheric plasma is merely used in this case for making the temperatures very high.

**Table 8.7. Typical Reaction Conditions and Deposition Rates of Plasma CVD**

| Solid | T (°C) | Reactants | Rate (nm/s) |
|---|---|---|---|
| Silicon | 250–400 | $SiH_4$ | 0.1–1 |
| a-Carbon | 250 | Alkanes | 0.1–100 |
| Boron | 400 | $BCl_3/H_2$ | 0.1–1 |
| $SiO_2/GeO_2$ | 1000 | $SiCl_4/GeCl_4/O_2$ | 3,000,000 |
| $Al_2O_3$ | 250–500 | $AlCl_3/O_2$ | 0.1–1 |
| $TiO_2$ | 200–400 | $TiCl_4/O_2$ | 0.1 |
| AlN | 1000 | $Al_2Cl_6/N_2$ | 10 |
| TiN | 250–1000 | $TiCl_4/N_2/H_2$ | 0.1–50 |
| $Si_3N_4$ | 300–500 | $SiH_4/N_2$, $NH_3$ | 0.1–1 |
| SiC | 200–500 | $SiH_4$/alkanes | 0.1 |
| TiC | 400–600 | $TiCl_4$/alkanes/$H_2$ | 0.5–10 |
| $B_4C$ | 400 | $B_2H_6/CH_4$ | 0.1–1 |

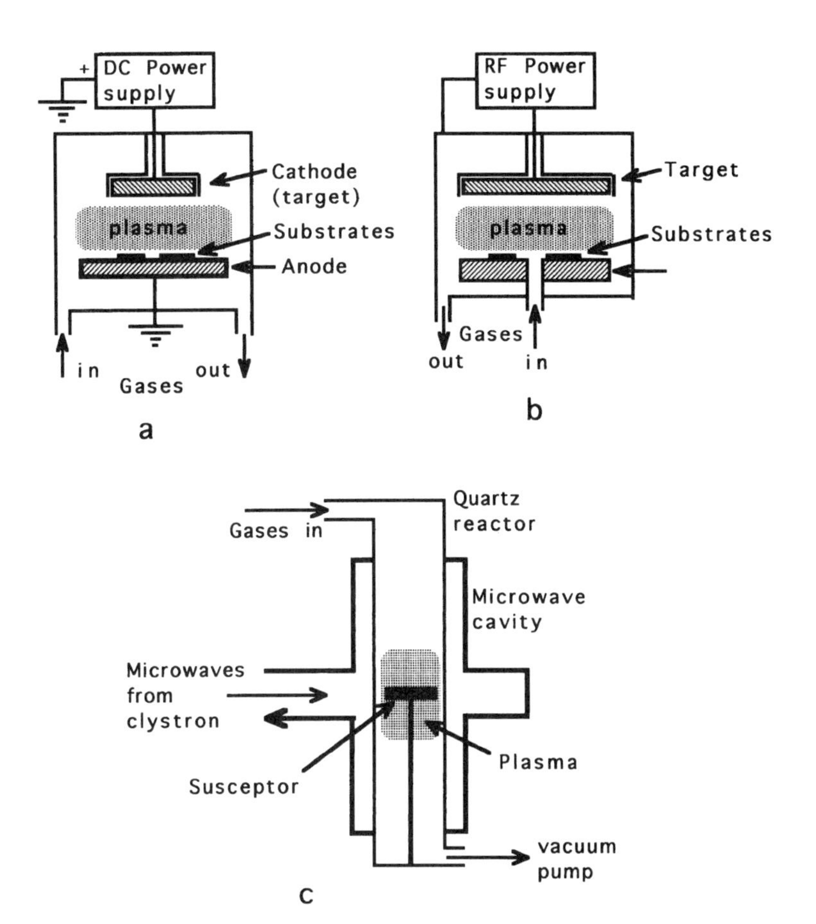


**Figure 8.31.** Three reactors for plasma-CVD: (a) dc-reactor; (b) rf-reactor; (c) microwave reactor.

## *Exercises*

1. Describe the usual methods for growing single crystals. Choose any solid compound from the text and devise the most suitable method for making a single crystal of it.
2. Explain how to shape single crystals during growth.
3. Name the three types of epitaxy (growth of solids). Why do they arise, what are their consequences for the product, and how does one control them?
4. Choose a metal and use the Ellingham diagrams to design a synthesis for a fractal powder of the metal from one of its compounds.
5. Give an example of controlling the particle shape with the synthesis conditions.
6. Describe a number of compounds that are preferably made using molten salt as the solvent. Give the conditions of the synthesis and its precursors.
7. Describe how one adjusts the acidity of certain salt mixtures and why that is done.
8. Explain why molten salts are good solvents.
9. Describe the hydrothermal procedure for synthesis of solids.
10. Give a few examples of the sol-gel method (and the conditions) for making mixed oxides.
11. Indicate the conditions that affect the morphology of oxides made with the sol-gel method.

12. Explain the use of structured solvents in inorganic synthesis.
13. Give a few examples of solids made with vapor transport and explain the mechanism.
14. Choose a combination of a metal and an oxide and give the process conditions for making a coating of a composite of them with PVD.
15. Describe several examples of making powders from gaseous precursors and discuss the function of supersaturation in homogeneous precipitation.
16. Explain the characteristics of the VLS mechanism for the growth of whiskers.
17. Design a recipe for making some given oxidic compound as a dense coating or a porous one having a specified porosity.
18. Indicate how precursors that have a low vapor pressure can nevertheless be employed in CVD.
19. What is the function of plasmas in gas-phase deposition? Explain their effectiveness.
20. Will the presence of oxygen in the gas phase increase or decrease the rate of the sintering reaction $ZnO + Cr_2O_3 \rightarrow ZnCrO_4$?
21. Powders that are made from the gas phase or with the sol-gel process are often monodisperse (all the primary particles have the same size). Why? Explain how the aggregation model for particle growth can suggest formation of monodisperse powders.
22. Arrange the following metals according to an increasing reaction rate with nitrogen to form the nitride: Al, B, Ti, Si. Do the same for an expected increasing heat formation at the nitridation reaction.
23. Can aluminum metal be used in principle to reduce the following oxides to their metals: MgO, $TiO_2$, $ZrO_2$, $SiO_2$, or ZnO?
24. CeS and MgS are very stable. Why can CeS be used as a refractory material for crucibles while MgS is unsuitable?
25. Is $CS_2$ a suitable reactant for the synthesis of metal sulfides from metals?
26. The Kroll process uses magnesium for the synthesis of metals from their chlorides. Which metals might be formed that way?
27. Arrange the three growth mechanisms, Frank van der Merwe, Stranski–Krastanov, and Volmer–Weber, in the order of increasing adhesion between substrate and coating. How would they be arranged in a series according to increasing texture?
28. Which process conditions help grain refinement? Which lead to texture?
29. Explain why immobilizing radioactive wastes in insoluble glasses and then burying them is an unsuitable method for disposal.
30. What happens at eutectic (or directed) solidification of the following mixtures: Sn/S, $Si_2O/SiO_2$, Borax/$SiO_2$, $ZrO_2$/MgO, $TiO_2/Al_2O_3$. What is the morphology of the product in each case, fibrous or lamellar?

## References


1. J. W. Evans and L. C. de Jonghe. *The Production of Inorganic Materials.* Macmillan, New York (1991).
2. C.N.R. Rao. *Chemical Approaches to the Synthesis of Inorganic Materials.* Wiley, New York (1994).
3. J. D. Woollins (ed). *Inorganic Experiments.* VCH, Weinheim (1994).
4. Schwertmann, U. and R. M. Cornell. *Iron Oxides in the Laboratory: Preparation and Characterization.* VCH, Weinheim (1991).
5. Y. Arai. *Chemistry of Powder Production.* Chapman and Hall, London (1996).
6. K. J. Klabunde. *Free Atoms, Clusters, and Nanoscale Particles.* Academic, San Diego (1994).
7. J. J. Moore and H. J. Feng. Combustion synthesis of advanced materials: Part I. Reaction parameters; Part II. Classification, applications, and modeling. *Progr. Mater. Sci.* **39**, 243, 275 (1995).
8. T. A. Ring. *Fundamentals of Ceramic Powder Processing and Synthesis.* Academic, San Diego (1996).
9. D. L. Smith. *Thin-Film Deposition: Principles and Practice.* McGraw-Hill, New York (1995).

10. A. A. Chernov and E. I. Givargizov (eds.). *Modern Crystallography, Vol. 3: Crystal Growth*. Springer, Berlin (1984).
11. W. E. Rhine, T. M. Shaw, R. J. Gottschall, and Y. Chen. Synthesis and processing of ceramics: scientific issues. Mat. Res. Soc. Symp. Proc. **249** (1992).
12. A. K. Varshneya. *Fundamentals of Inorganic Glasses*. Academic, Boston (1994).
13. A. Stein, S. W. Keller, and T. E. Mallouk. Turning down the heat: design and mechanism in solid state synthesis. *Science* **259**, 1558 (1993).
14. E. Matievic. Preparation and properties of uniform size colloids. *Chem. Mater.* **5**, 412 (1993).
15. M. Ozaki. Preparation and Properties of Well-Defined Magnetic Particles. *MRS Bull.* December 1989, p. 36.
16. J. R. Backhus, J. H. Harker, and R. K. Sinnet. *Coulson and Richardson's Chemical Engineering*, Vols. 1, 2, and 6. Pergamon, Oxford (1990).
17. W. L. McCabe, J. C. Smith, and P. Harriott. *Unit Operations of Chemical Engineering*. MacGraw-Hill, New York (1993).
18. D. Segal. *Chemical Synthesis of Advanced Ceramic Materials*. Cambridge University Press, Cambridge (1989).
19. S. C. Deevi (ed.). Manufacturing of covalent ceramics by exothermic reactions. *Mater. Res. Soc. Symp. Proc.* **327**, 171 (1994).
20. Y. M. Chiang, J. S. Haggerty, R. P. Messner, and C. Demetry. Reaction-based processing methods for ceramic-matrix composites. *Ceram. Bull.* **68**, 420 (1989).
21. H. Yanagida, K. Koumoto, M. Miyayama, and H. Yamada. *The Chemistry of Ceramics*. Wiley, Chichester (1996).
22. G. Mamantov and R. Marassi. *Molten Salt Chemistry: An Introduction and Selected Applications*. Reidel, Dordrecht (1987).
23. G. Mamantov and A. I. Popov (eds.). *Chemistry of Non-aqueous Solutions: Current Progress*. VCH, New York (1994).
24. T. A. O'Donnell. *Superacids and Acidic Melts as Inorganic Chemical Reaction Media*. VCH, New York (1993).
25. R. A. Laudise. Hydrothermal synthesis of crystals. *C&EN* **28**, 30 (1987).
26. C. J. Brinker and G. W. Scherer. *Sol-Gel Science: The Physics and Chemistry of Sol-Gel Processing*. Academic, Boston (1990).
27. O. A. Gzowski. Gels. *J. Mater. Educ.* **8** (5), 671 (1986).
28. L. L. Hench and J. K. West. *Chemical Processing of Advanced Materials*. Wiley, New York (1992).
29. R. J. Ayen and P. A. Iacobucci. Metal oxide aerogel preparation by supercritical extraction. *Rev. Chem. Engin.* **5**, 157 (1988).
30. L. C. Klein (ed). *Sol-Gel Technology of Thin Films, Fibers, Preforms, Electronics, and Specialty Shapes*. Noyes, Park Ridge, NJ (1988).
31. D. L. Smith. *Thin Film Deposition*. McGraw-Hill, New York (1995).
32. M. G. Hocking and V. Vasantarasree, and P. S. Skidky. *Metallic and Ceramic Coatings*. Longman, Harlow (1989), Ch. 4.
33. A. C. Jones and P. O'Brien. *CVD of Compound Semiconductors*. VCH, Weinheim (1997).
34. D. S. Rickerby and A. Matthews (eds.). *Advanced Surface Coatings: A Handbook of Surface Engineering*. Blackie, Glasgow (1991).
35. H. K. Pulker, E. Bergmann, and H. M. Gabriel. *Wear and Corrosion Resistant Coatings by CVD and PVD*. Expert Verlag, Ehningen (1989).
36. M. Langlet and J. C. Joubert. The pyrosol process or the pyrolysis of an ultrasonically generated aerosol. In C.N.R. Rao (ed.). *Chemistry of Advanced Materials*. Blackwell's, London (1992), p. 55.
37. S. K. Friedlander. *Smoke, Dust, and Haze: Fundamentals of Aerosol Behavior*. Wiley, New York (1977).
38. W. C. Hinds. *Aerosol Technology: Properties, Behavior, and Measurement of Airborne Particles*. Wiley Interscience, New York (1982).
39. T. Kodas (ed). Aerosols in materials processing. *J. Aerosol Science* **24**, 273 (1993).
40. S. Veprek. *Application of Low Pressure Plasmas in Materials Science—Especially CVD*. Current Topics in Materials Science, Vol. IV. E. Kaldis (ed.). North-Holland, Amsterdam (1980).

*Chapter 9*

# THE DESIGN OF INORGANIC MATERIALS

*It is useless to ask if design is an artistic activity or a scientific one; the etymology of the word technology or techne (art) provides a sufficient response.*

*A. A. MOLES, 1992*

## 9.1. Introduction to Materials Design

This chapter is in a sense the culmination of the line of this book, the goal of chemical technology of inorganic materials. The main professional activity of materials technologists is the designing of materials. They make and optimize materials for products with a wide range of structures and functions. The lack of suitable materials is always the main bottleneck in the introduction of new products or processes. Materials technology has a supporting or "enabling" role: all hardware is made of materials. Basically materials are what all our prostheses are made of, both inside the body (pacemakers, implants, artificial joints) and outside (cars, ships, computers, tools).[1]

The design of materials with specified properties is not an explicit and institutionalized discipline although many are in fact engaged in it. The books that discuss materials design[2,3] describe specific cases and do not provide a general treatment. Nobody can. Materials for different purposes do not have much in common; each material comes with its own processes, tricks, and possibilities. Some rules have been formulated for the materials design process itself and are offered here, but they do not provide a mechanism for making recipes for successful designs. Rather, creativity in working out each step is the basis for success.

The design of a material is only one part of the design of a product. The other parts are shaping, combining, marketing, dealing with environmental load, and maintenance. The design procedure for materials includes setting up a list of goals, listing available materials that come close to being a solution in view of one or more of the goals, devising composites and fabrication, and adjusting all this in view of other constraints. These generalities are discussed in Section 9.2.

Market pull means that there is not much chemistry in the first step (setting up the objectives) but the chemistry in later stages is strongly related to the original requirements and constraints set out in the first stage. When a new technology, e.g., in electronics, solar energy conversion, fuel cells, robot sensors, and exploring space, is conceived, generally a demand outside of chemistry sets the chemists to work to deliver the proper material for it. However, many cases are known where chemical findings created a demand. Ceramic superconductors and diamond coatings are instances of the original technological push coming from inorganic chemistry.

Designing means different things to different people because different parts of the overall process of product design are emphasized. Generally design means creating something, a chemical fabrication process, the form of a consumer product, a device or a novel material.[4,5] The term design for chemical engineers means working out a fabrication process and setting up a factory for making tons of a commodity at the lowest possible cost. Industrial designers use design as a word to mean creatively shaping and combining items to form a novel product that is the answer to a demand. They do not concern themselves much with making up new materials; rather, they choose the most suitable material from a list of those available. However, a totally new product needs a new material. It is unlikely that existing materials are the best possible way to meet the need because they have usually been optimized for another product with different requirements. Designing novel materials means creating new solids and new processes to make them and the design of a material cannot be separated from the end product or its intended function.

The fabrication processes are only part of the entire design framework but have a key role because they fix the properties. The intrinsic properties are determined by the chemical nature and the crystal structure (see Chapter 4) but the extrinsic properties are related to the morphology and the defects that arise during the synthesis (see Chapters 5 and 7). In the design of a new material both types of properties have to be considered and the nature of the material as well as its fabrication processes are adjusted to the requirements.

The logic of using composites in design is based on the idea that combining materials increases the range of their use, even for simple requirements.[6–8] The combination is more than the sum of its parts and if its components are chosen well, the material can be vastly more suitable than any individual component. Hybrid composites and active materials that will be discussed below illustrate this point.

New materials envisaged for new applications are rarely designed and made from first principles. The procedure in practice is different: existing materials are tried first; those that are provisionally chosen from a list of existing materials inevitably are not the best conceivable. If in spite of the less-than-suitable material that is chosen to make it, the product turns out to be successful, the material is gradually adapted to the requirements by small changes and improves. While materials continually need improvement, manufacturers are reluctant to replace established procedures and materials that work by better ones unless continuity of production is certain and increased sales are virtually guaranteed. An example is given in the next section. Therefore the materials engineer will be asked to improve the properties of an existing material without disrupting a fabrication process that is already in place, and this, of course limits the designer's freedom.

The trend in postindustrial technology is toward fewer material products and more action. The center of attention moves from making products to shifting symbols around. That trend toward dematerialization has consequences for the

## Biomimesis

Organisms are capable of breathtaking performances. Trees are constructions that can hardly be improved by civil engineers and sequoias have been pumping water up to heights higher than engineers were capable of until recently. The design of the circulatory system in organisms is gradually being appreciated by physical technologists. Vast amounts of information are routinely processed and transported by cells and compared to pattern recognition by those cells the performance of our computers is pathetic. Control techniques of very complex processes in biology are much more advanced than those we make for our factories or wars. Navigation and transport in fish and birds are legendary. Even refined electrical and chemical warfare is common in biology as is biotechnological exploitation of one organism by another. All these techniques are on the highest level conceivable, given the materials that are available to the organisms. The fascinating possibilities seen in biology raise the question of whether biology might be a source of inspiration for materials design. Trying to imitate biological behavior in materials (biomimesis) is a recent trend in technology[9] that employs the types of materials used by organisms, which are primarily organic with some inorganics but not the most durable ones.

Biomimesis, however, has turned out so far to be much less useful than expected in spite of the marvelous properties of biological tissues. Biological processes and the properties of biological materials are usually recognized only after similar materials or processes have been realized *in vitro*: plywood was recognized in the microstructure of abalone shells and sonar in bats. In inorganic technology processes and materials can be better and more efficient than in biology because the range of available materials is much wider.

Nature prefers fragile organics as basic building blocks because they are easier to overhaul and being alive means unceasing repair and replacement. Living is costly and imitation of organisms means loss of efficiency from the technology viewpoint. Conversion of solar energy to fuel is an example. Green plants have been specialists in this conversion for billions of years. Photosynthesis is carried out in chloroplasts, organic solar cells that are coupled to a complicated set of enzymes to make fuel, which is stored to be used to maintain life and reproduce. The highest efficiency shown for conversion from sunlight to fuel (glucose) is less than 3%. The biochemical conversion of glucose to fuel that the cell can use directly (ATP) means an additional energy loss of 60%, an amount comparable to that of an old-fashioned coal power plant. Thirty years after they were invented, run-of-the-mill silicon solar cells that are used for energy supply in ships, buoys, and houses far from utility nets had a conversion efficiency four times higher than the highest biological value. The biological way of photosynthesis is technologically unacceptable and there is little point in trying to imitate chloroplasts. Other examples of differences in efficiency are information transfer in nerves or glass fibers and transport by flight. Organisms have optimized their materials for other requirements and constraints than those in technology and that makes biomimesis in general of questionable significance for materials technology. Getting inspiration from biology is a different story.

design of materials. Dematerialization needs a solid material base. The material base of advanced technology requires maintenance, which is a heavy burden if the materials are not durable. The material side of an immaterial civilization must therefore be extremely reliable. The task of the designer is to ensure reliability in his

Table 9.1. Contrasting Features of Industrial and Postindustrial Materials Design

| Materials | | Processes | |
|---|---|---|---|
| Industrial | Postindustrial | Industrial | Postindustrial |
| Simple | Composite | Monocausal | Multicausal |
| Disposable | Durable | Deterministic | Chaotic |
| Passive | Active | Extremely controlled | Self-controlled |
| Strategic (vital) | Replacable | Wasteful | Environmentally clean |
| Standardized | Customized | | |

products by developing materials that are impervious to corrosion, wear, and aging. Some other consequences of recent trends on materials design are collected in Table 9.1.

In the search for novel materials it is sometimes pointed out that biology is far ahead of chemistry and that we chemists can find inspiration and exemplary materials in biology. An imitation of living tissues and processes is called biomimetics (see box). However, there are three essential differences between biomaterials and synthetic ones:

1. Organisms are restricted to making organic materials, which have a limited scope. Moreover organics are much less durable than the materials made with the wide range of elements and combinations that are available in inorganic technology.
2. Functioning organisms have to expend considerable effort to stay alive.
3. Organisms have to expend considerable effort to repair themselves continuously during their lifetime.
4. Any alleged teleology in organisms is aimed at genetic continuity while artificial materials are dedicated exclusively to a function.

Thus far because of these differences the performance of biomimetic materials when they are used out of context in technology has been disappointing. Biomaterials are not unique and most of them can easily be bettered by artificial ones. It is true, however, that organisms can do things such as organic synthesis of certain complex molecules much more cheaply than technology can, and single-celled organic chemists are competing increasingly with the human chemists. Biotechnology or the use of organisms for our purposes is a continuation of a long tradition in agriculture, which is basically applied organic technology; there have been hardly any inorganic developments in biotechnology.

## 9.2. Requirements and Constraints

The designers of a new material follow a checklist. First they look at why the new material is needed: what is its purpose, what functions are required, which properties must it have and not have, what fabrication methods are permitted, what resources are necessary and their cost, what is the availability of necessary elements, how much waste is produced in its fabrication and what is its durability. The search for a novel material narrows down all conceivable options in steps as all the constraints are considered successively. At every step the decision to reject some and keep other possibilities is determined by the requirements. Designers also have to deal with trade-offs: when a property cannot be improved or its cost lowered without concomitant undesirable changes in other features compromises are made. Materials are specific and it is difficult if not impossible to set up general rules to cover all activities in materials design. Each material has its own ideosyncrasies, advantages, functions, and recipes.

In design, there are steps that gradually narrow down the number of possible materials by imposing additional constraints. In each of them the success of the final design depends on originality, not on mechanically following a checklist, although that may help:

1. The first step is to survey the requirements. A list of purposes is created and constraints derived from product requirements and uses envisaged. Which features are wanted in the material and what is unwanted?
2. Then a list is made of all existing materials that have intrinsic properties that come close to the ones wanted. It should also include materials that only meet some of the requirements. The undesirable features of each possibility must be explicitly noted and perhaps also some seemingly irrelevant features that might come in useful later.
3. The third step is to make a list of the extrinsic modifications (morphology, purity) that would improve the performance of the materials that come out of step 2. Materials are combined in composites to form so-called sum and product properties that improve properties or eliminate unwanted characteristics. The result of this step is a list of potentially suitable materials and composites, including the compounds that make them up and their microstructures.
4. The next step is setting up fabrication techniques for the materials remaining in the list of possibilities. These production techniques further limit the choices arrived at in the previous stages. This fourth stage is the key part of materials design because the extrinsic properties of a material are fixed during its synthesis. Synthesis is inextricably bound up with properties. This production step further decreases the number of candidate compounds: some compounds might be suitable if they have the right properties but when they are to be combined in a composite they may be incompatible with one another during synthesis.
5. The final step is optimizing the production process of the material. Chemical engineers call this step design.

Of the five steps in the design procedure steps 1 to 3 are the easiest. For these steps the designer needs to know the properties and behavior of compounds and candidate materials. To design materials successfully, one must have a thorough familiarity with the many elements, compounds, and commodities as well as access to data reported in the literature. Chemical and ceramic abstracts help in building up the necessary body of knowledge. Although a knowledge of properties is necessary for parts 2 and 3 it is not sufficient for the design question. It is nice to know the attractive properties of materials that make them suitable for the objective but it is a different story to make those properties cheaply, consistently reproducibly, and reliably. The fourth step on fabrication is at least as important for the technologist or materials designer as the other steps: after all is said and done, making things is the basis of all technology and synthesis needs the attention it deserves.

One requirement that is always added (or should be added) to the functional constraints such as conductivity, color, elasticity, and toughness is chemical and mechanical passivity. The material must not corrode, sinter, wear, or react irreversibly during use. High durability is a nontrivial design constraint but it is not always mentioned explicitly.

The designer of materials needs a comprehensive list of properties of compounds. Properties are listed under their chemical formulations, as can be seen from some of the tables included in this volume. However, the suitability of a material in practice depends on features that are marginal to the functional properties. The selection of a material is made not only on the basis of listed properties that have been determined in other ambients and under other conditions, but must also take into account the limits set by the manufacturer, which further restrict the freedom of choice (Section 9.1). For instance, a manufacturer of X-ray tubes observes an occasional unexpected dielectric breakdown at the surface of a ceramic alumina insulator of an anode under high voltages in vacuum. It is of no use to suggest replacing the alumina as an anode insulator by an insulator with a higher dielectric strength for at least two reasons:

1. The producer refuses to switch from alumina to, e.g., boron nitride because that would be fatal for production. The production technicians are experienced in making alumina and a change means unacceptable production delays. Alumina it is and alumina it stays, and to make it perform better it must be modified but not too thoroughly. This is the imposed constraint for the designer.
2. The property that needs improvement is extrinsic and is therefore manufacture-dependent: dielectric strength depends largely on microstructure. Flaws such as microcracks lead to local voltage peaks that may trigger dielectric breakdown. Moreover, the insulator in the X-ray tubes operates in a plasma that contains electrons and metal atoms that both interact strongly with the surface of the insulator. The values of the material properties that are listed in tables have not been determined in such an exceptional atmosphere and those values are therefore not very reliable selection criteria. The situation is similar to mechanical yield strength, which also depends on surface flaws.

In this case the design of material for anode insulators means a slight modification of alumina that eliminates erratic surface behavior. A thin surface deposition of an electrically insulating compound with better chemical behavior toward the species in the plasma or another surface finish could be the solution. Even a simple etch might do. A few directed experiments will be helpful with the selection.

The assembly of requirements is consulted before every decision in the entire design process. This list of requirements does not always point to only one material at the end and there may be alternatives. If there are several solutions the one that seems best (even perhaps on rather flimsy grounds) is chosen and developed further. As a result certain materials and processes are now well established that should have been replaced by much better or more efficient materials but are not. This positive feedback in development has earned industry a reputation of being conservative. Examples abound. At its start the internal combustion engine was preferred over the fuel cell although it was known to be much less efficient: the available materials for it happened to be slightly more suitable. Now the materials for fuel cells are there but replacing car engines has become a formidable problem. Another example is ceramic heat shield coatings on turbine parts in jet engines. These are deposited by means of flame spraying although now electron-beam PVD has been shown to produce materials that confer a much higher lifetime on critical engine parts. Operating costs might also be lowered if the deposition technique were to be changed but coating processes cannot be altered as the original procedure has been canonized in IAEA prescriptions and engine parts treated any other way are barred from use in civil aviation.

At the stage of setting up the requirements for the material the options can be very divergent as in the case of the traction engines mentioned above. Another example of two different materials for one purpose is found in the case of solar energy conversion. The goal is to convert photons to chemical energy efficiently. The subrequirements are: (a) efficient absorption of sunlight; (b) prevention of conversion of the absorbed energy to heat; (c) long lifetime; and (d) cheap production.

Solar energy conversion is a materials problem. What materials are there for it at this stage? That depends on the process chosen for it. It can be done with silicon or compound semiconductors, both crystalline and amorphous. That technology is based on *p–n*-junctions in semiconductors that induce an electric potential. Mobile charge carriers generated by light are separated by this built-in field and harvested by suitable chemicals to make fuel or by electrodes for electrical power generation. Another way is possible, which is analogous to the light–matter interaction in luminescent lamps. These two options would require entirely different materials!

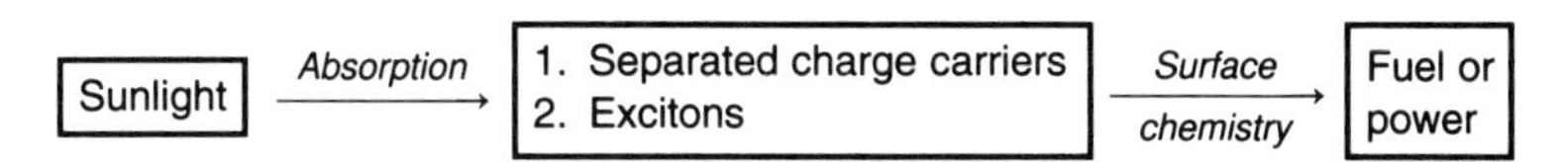


Basic to both options is the fact that a material medium is necessary to convert light into some intermediate form of energy since direct photochemistry to make fuel, e.g., hydrogen, from water with sunlight is as yet unknown. The material can be a semiconductor with a suitable bandgap and absorptivity. This represents the first option that uses charge separation. The second option is excitation without charge separation, i.e., the formation of excitons. These are known in pigments for luminescent lighting, where they are employed for making visible light from ultraviolet quanta, a different purpose. Both excitons and separated charges can in principle be made to react chemically at the surface to form fuel.

A major problem is recombination of the charge carriers or quenching of the excitons prior to forming fuel. This is an unwanted side reaction that converts the absorbed energy in the material to heat, which should be suppressed in order to increase the efficiency of the overall conversion of light to chemical energy. Quenching or premature relaxation is countered by eliminating certain impurities and traps in the crystallites or at their surfaces. Surface tailoring of the pigment particles by, e.g., a fluidized-bed treatment is essential for their performance. Apart from passivating surface sites that quench there is in some cases a second reason to change the surface. Fuel is made by molecular chemistry at the surface. If no single material can be found that has the right optoelectrical properties as well as the necessary catalytic surface chemistry, the surface of suitable semiconductors could be provided with the right chemical activity with a coating without lowering the original optoelectric performance of the interior.

In this section on the procedure for design of materials a tentative checklist was given that roughly shows what type of considerations lead to an answer to a demand for a new material. However, in design, general prescriptions such as the given checklist cannot be expected to be applicable to all cases in the same way, as can be seen from some of the illustrations above. Where expedient, the designer skips some of the steps on the list or adds others. The list of criteria is not static; some requirements may be dropped and novel ones added during the design procedure as a result of the inquiry itself. When the design procedure eventually comes up with a novel material it is usually not a unique solution. Many alternatives remain that have been rejected along the way for reasons that may become less convincing afterward as a result of novel materials research in other fields. Such alternatives could later become preferable to the initial solution and options are best kept open as long as possible.

## 9.3. Combination Properties of Composites

Simple materials or solid compounds are combined into one composite for several reasons:

1. The desirable properties of a material can be improved by combining it with several others in one composite material.
2. At the same time the unwanted characteristics of the components can in many cases be neutralized or eliminated in the composite.
3. Novel properties that do not exist in simple materials can be created and apparently conflicting properties can be combined in one composite.

Examples of synergy in composites are:

1. The simple compounds SiC and $HfB_2$ are both hard and highly inert. A very hard and refractory ceramic composite of fine-grained SiC and $HfB_2$ phases is even more corrosion-resistant at high temperatures in air than the components separately.
2. Hardness and toughness are conflicting properties in simple materials. A ceramic hardcoating on a metal substrate (see Chapters 6 and 8 for its fabrication) gives a hard, wear-resistant, and corrosion-resistant product that is also strong and tough. It combines the good bulk and surface properties of the components and eliminates their less desirable characteristics.

3. A magnetoelectric composite material reacts to an external magnetic field by forming an electric dipole moment. Such a material is composed of a ferromagnetic phase with magnetostriction and a piezoelectric component. Powders of the two components are well mixed and sintered together.



Properties of composites can be more easily adjusted within wide ranges than those of simple solid compounds because the behavior of composites depends on how much each component contributes and how the components are made to cooperate in the composite. These are fixed at the preparation stage. Two types of composite properties are recognized: sum properties and product properties.

### *9.3.1. Sum Properties*

Sum properties of composites are a function of the corresponding properties of the components. The relation between the sum property values and the property values of the components is not necessarily simple or linear, and it is definitely morphology-dependent. Four examples of sum properties are:

1. The dielectric constant ε of lithium aluminosilicate can be lowered from its bulk values of 5.6 to 2.1 by making a foam of the oxide, which has a porosity of 60%. This composite is an excellent insulator for microwave applications. The dielectric constant of the composite is some function of the dielectric constant of the components, the silicate, and air (Figure 9.1). The porosity should be of the closed type if impermeability for gas is also required.
2. Ferroelectrics such as $BaTiO_3$ usually have a high dielectric constant ε, which decreases with increasing high external electric fields because of saturation. A composite of this ferroelectric with antiferroelectric $NaNbO_3$ changes its saturation behavior and with a suitable composition and sintering process,

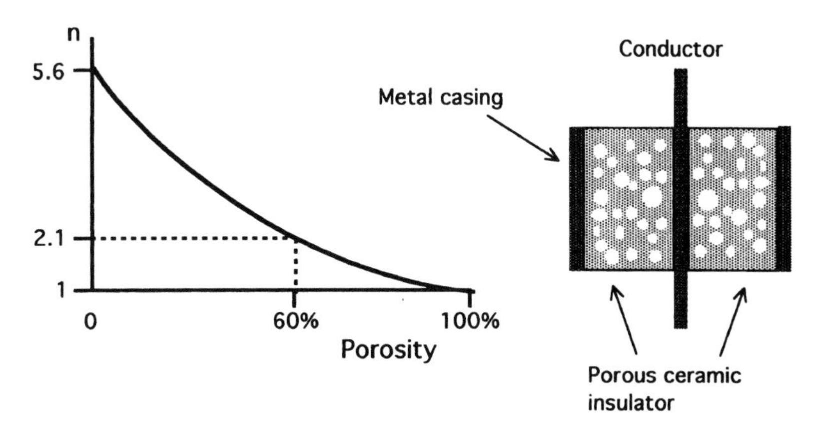


**Figure 9.1.** Microwave lead-through insulator of porous glass for a vacuum reactor.[7] The dielectric constant *n* can be made low by making a glass–air composite with closed porosity. Adapted with permission from the *Journal of Materials Education.*

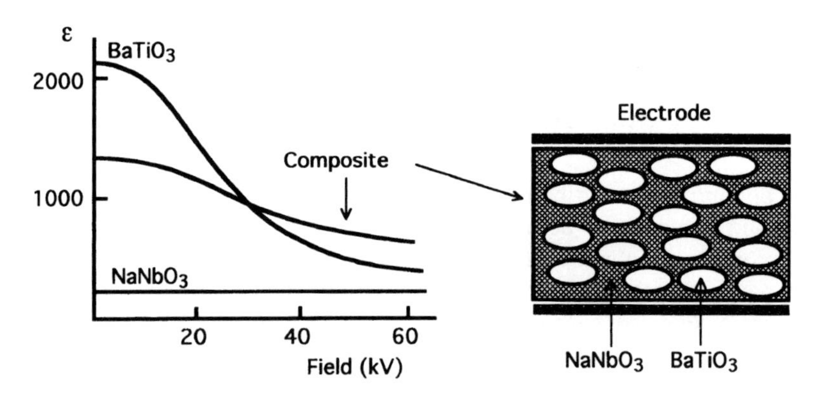


**Figure 9.2.** A ceramic composite with a high dielectric constant that is less easily saturated than its operative component.[7] Published with permission from the *Journal of Materials Education.*

the composite has a more constant ε (with electric field) than $BaTiO_3$ (Figure 9.2).

3. The dielectric constant of parallel plates of two dielectrics with different $\varepsilon_i$'s is a sum property of a composite. The value for the composite is strongly anisotropic: if the external field is parallel to the plates the resulting dielectric constant $\varepsilon = V_A\varepsilon_A + V_B\varepsilon_B$ is larger than if the field is perpendicular to the plates $1/\varepsilon = V_A/\varepsilon_A + V_A/\varepsilon_A$ (Figure 9.3) for the parallel case.
4. The anisotropic sound velocity in an oriented fiber-reinforced composite made of a matrix that has a low sound propagation velocity and fibers that are oriented in parallel that have a much higher sound velocity for transverse waves (Figure 9.4).

Thus sum properties are made up of the same properties of the individual components: the dielectric constant of the composite is some function of the dielectric constants of the components; the velocity of sound in the composite depends on the velocity of sound in the matrix and reinforcement phases. This is not true for product properties.

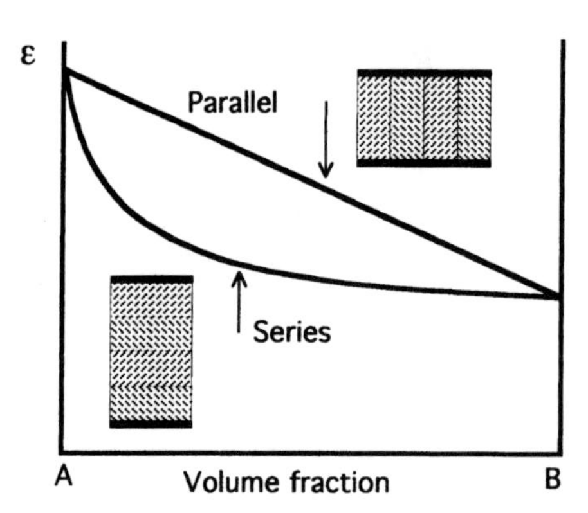


**Figure 9.3.** The effective dielectric constant of a composite consisting of parallel plates of materials with different dielectric constants.[7] Published with permission from the *Journal of Materials Education.*

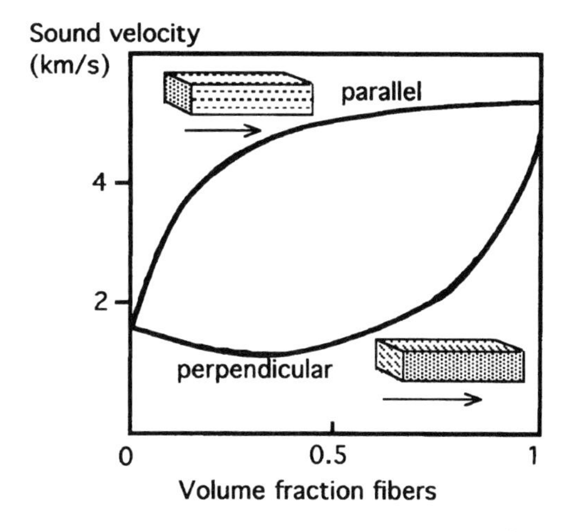


**Figure 9.4.** Direction dependence of the sound velocity in a composite of parallel fibers in a matrix.[7] Published with permission from the *Journal of Materials Education.*

### *9.3.2. Product Properties*

Product properties are entirely novel properties that are determined by other kinds of properties of the components and they can be completely absent in the individual components. Table 9.2[6] lists some examples.

Two examples of product properties are magnetoelectricity and thermomagnetism.[7] A magnetoelectric ceramic composite is made of grains of piezoelectric $BaTiO_3$ and cobalt- or titanium-doped $Fe_3O_4$ (a ferrimagnet that has a strong magnetostriction). When such a composite is put in a magnetic field the magnetostriction in the ferrimagnetic grains will induce a mechanical stress on the grains of the ferroelectric component that convert the stress to an electric dipole moment. This is easily measured: a magnetic field induces an electric feld and this product property (the components barium titanate and magnetite are not by themselves magnetoelectric) allows sensitive frequency-independent magnetic measurements.

The paramagnetic spinel $CdCr_2O_4$, when surrounding a superconducting wire shows a kind of electrothermal behavior: changes in current induce changes in temperature. A current in the wire has a magnetic field around it and the chromite

**Table 9.2. Examples of Combination Properties**

| Product property | Property of phase 1 | Property of phase 2 |
|---|---|---|
| Thermoresistance | Electrical resistivity | Thermal expansion |
| Thermoresistance | Electrical resistivity | Phase transition |
| Mechanoelectricity | Elasticity | Polarizability |
| Magnetoelectricity | Magnetostriction | Piezoelectricity |
| Magnetoresistance | Electrical conductivity | Hall effect |
| Thermoreflectance | Light absorption | Mott transition |
| Electrothermal effect | Superconductivity | Adiabatic demagnetization |
| Pyroelectricity | Piezoelectricity | Thermal expansion |

is in that field. The magnetic field orients the atomic magnetic moments in the chromite (the temperature is very low) and the paramagnetism is saturated. Now changes in current mean changes in magnetic field. If, for instance, the spin saturation decreases because the magnetic field becomes locally smaller owing to a lowering of the current, the spin temperature drops and the spin system takes heat from the lattice. This cools the chromite by so-called adiabatic demagnetization and there is negative feedback that reestablishes the status quo: if the wire locally becomes normal, the current drops; hence the surrounding chromite coat cools, which makes the wire superconducting again. Catastrophic local heating of the conductor is thus countered by an actively cooling insulator.

### 9.3.3. Morphology

The examples described above show clearly that the morphology of the composite strongly affects its properties. The connectivity of the phases in composites (how are they linked or oriented) can in many cases be expressed in the dimensions of the phases.[10] Figure 8.22 gives some idealized limiting cases of biphasic composites.

Percolation as discussed in Section 7.4 is a key concept for discussing properties of composites. Consider a composite that consists of a mixture of fine conducting and insulating grains with a low concentration of the first phase. If the concentration of the conducting particles is gradually increased, clusters of these particles start to form when they are no longer isolated from each other by the particles of the second phase and become neighbors. At a sufficiently high concentration the first phase is no longer zero-dimensional (isolated particles) but has formed agglomerates of dimension $x$ and the conducting component then has an $x$-dimensional distribution with $x$ somewhere between 1 and 2.5. If the first phase consists of holes, the cohering holes form pores that permit liquid or gas to flow through the composite. A composite with conducting particles in an insulating matrix allows current to flow through the composite if the conducting particles touch each other and if there is some connecting path for the charge to percolate from one side to the other. Below the percolation threshold there is no conductance; above the percolation threshold the conductance improves with concentration as the number of percolating paths increases with concentration. The graph of the conductivity of a composite of conducting particles in an insulating matrix (Figure 9.5) illustrates this. In a mixture of small conducting and insulating grains the value of the percolation limit depends on the difference in the particle size of the two phases. Smaller particles have a lower percolation threshold because they form the grain boundary of the large particles. Use is made of percolation in constructing polycrystalline sensors that are based on grain-boundary effects, e.g., in a composite NTC-PTC thermistor that has a conductance window.

Sum properties of disordered materials can be expressed in terms of compound properties and their volume fractions. As an example, the conductivity (electric or thermal) of a dispersion of a phase in a matrix is given in terms of the component conductivities. For inhomogeneous solids a general empirical equation is used to

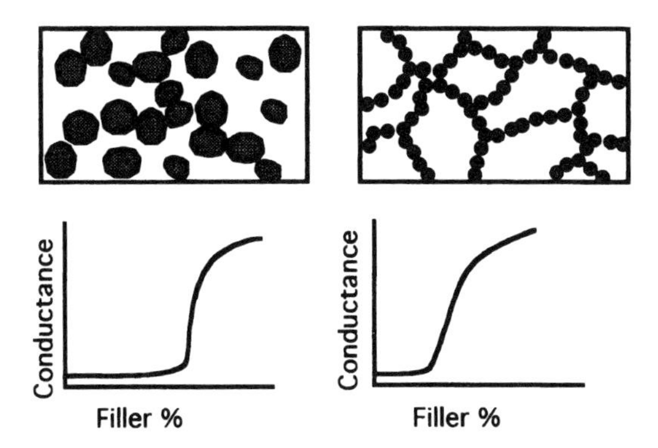


**Figure 9.5.** The resistivity of two 0–0 composites, one with large conducting particles (left) and one with small conducting particles, both with large nonconducting grains. The small–large mixture percolates at lower concentrations.[7] Published with permission from the *Journal of Materials Education.*

estimate the properties.[10] This equation also holds for dielectric polarization:

$$\frac{f(\sigma_l^{1/t}-\sigma_m^{1/t})}{\sigma_l^{1/t}+A\sigma_m^{1/t}}+\frac{(1-f)(\sigma_h^{1/t}-\sigma_m^{1/t})}{\sigma_l^{1/t}+A\sigma_m^{1/t}}=0$$

In which the constant $A$ is

$$A=\frac{1-\phi_C}{\phi_C}=\frac{f_C}{1-f_C}$$

where $f$ is the volume fraction of the poor-conductivity phase $(1-\phi)$; $f_C$ is the critical volume percolation fraction of the poor-conducting phase $(1-\phi_C)$; $\phi$ is the volume fraction of the high-conductivity phase $(1-f)$, $\phi_C$ is the critical volume fraction $(1-f_C)$ of the percolation of the high-conductivity phase; $\sigma_l$ is the conductivity of the low-conductivity phase, $\sigma_h$ is the conductivity of the high-conductivity phase, $\sigma_m$ is the conductivity of the composite (the resistivity is $\rho=1/\sigma$), and $t$ is the empirical exponent for the equation. The values of $t$ are normally in the range from 1.65 to 2.0, but when the conductivity of the particles in the matrix is very high, $t$ is found to be higher and values up to 4 have been measured.

This equation can be used to estimate electrical and thermal conductivities of binary composites as well as their static permeabilities when those of the components are known.

## 9.4. Functional Materials

Ceramics are irreplaceable as functional materials because of their unique properties. Functional ceramics are also key components of certain so-called active materials that are sometimes called smart materials.[11] Although in a sense structural, such novel materials are in large part functional ceramics, as will be explained below.

A growing application area of functional materials is the field of sensors and actuators. A sensor is made of a material that produces a signal when it is stimulated from the ambient. The size of the signal depends on the strength of the stimulus. The signal is usually an electric potential (a voltage): e.g., an acoustic sensor element in a microphone converts a mechanical stress to a proportional voltage. A thermocouple delivers a voltage that increases with a temperature difference. Such devices are made of sensor materials. Sensors are in increasing demand as a result of mechanized production; e.g., robots need sense organs to operate well. Actuator materials are solids that change or generate physical signals if triggered. Actuators are inverse sensors. Examples of actuators are piezoelectric buzzers that produce sound from an electric current (inverse microphones), ceramic "muscles" in scanning tunneling microscopes (inverse pickup elements), and a smart window that changes color under an electric voltage (an inverse light-sensitive cell). Sensors and actuators are transducers; they convert effects. Transducers change one action type to another, ferroelectrics change mechanical forces to voltages and vice versa, semiconductors in solar cells convert light to electrical energy, and fuel cells and batteries change chemical energy to electrical energy and vice versa.

Sensors are also increasingly needed for automatically monitoring air and water contamination and for measuring and controlling industrial processes. Most failures in automated operations (cars, aeroplanes) are due to wrong signals from sensors (a less developed area) and not to malfunction of processors. Two types of materials are used for chemical sensors: semiconductors and ionic conductors. Physical sensors measure heat, flow, displacement, and light and are usually ferroelectric ceramics or planar electronic devices made with standard electronic technology. There are also physical sensors (Taguchi-type sensors), based on the surface properties of semiconductors that change with gas adsorption. The degree of adsorption depends on the gas. The changes in surface conductivity are then related to gas concentrations in the ambient and the surface can be used as a material for a sensor. The construction is similar to that of a field effect transistor (FET), and so these sensors are called chemfets. Materials for chemfets are often oxides with a high adsorption coefficient for the molecules whose concentration is to be measured.

Chemical sensors that are based on ion conductors are of the Nernst type. They are concentration cells and the voltage developed by them is a measure of the concentration or activity difference of the active species at the two electrodes. One of the electrodes is exposed to the ambient and the other to a reference concentration. The materials used for electrolytes and electrodes in Nernst sensors are the same ones as those used for fuel cells and the requirements (high ionic conductivity and low electronic conduction) are the same. Examples of solid electrolytes having a good ionic conductivity are $LiAlSiO_4$ ($\beta$-eucryptite, a one-dimensional ion conductor), $Li_xTiS_2$, $Li_3N$, and Na $\beta$-aluminate (this compound conducts ions in crystal planes), and three-dimensional cation conductors such as $Na_3Zr_2PSi_2O_{12}$ (nasicon or nasiglass), $Li_4GeO_4$, and $NaZr_2(PO_4)_3$. Zirconium dioxide ($ZrO_2$), stabilized with $Ca^{2+}$ or $Y^{3+}$ ions, is a good oxide ion conductor at elevated temperatures and is widely used as the electrolyte in solid oxide fuel cells. Lanthanum chromite ($LnCrO_3$) is a mixed conductor, of both electrons and ions. In many of these cases of solid electrolytes the grain boundaries strongly affect the conductivity. The microstructure is thus determinitive for these applications.

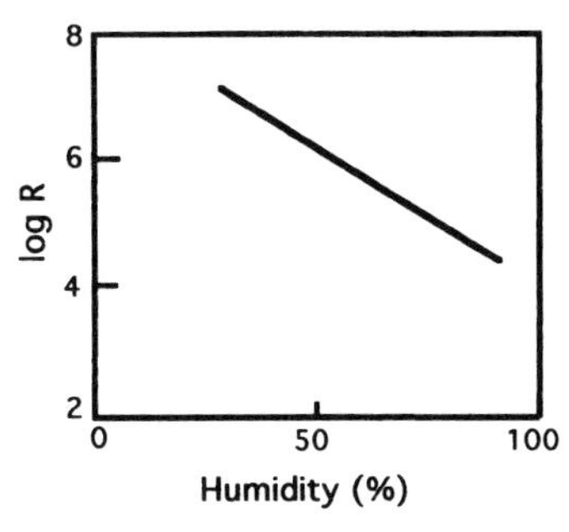


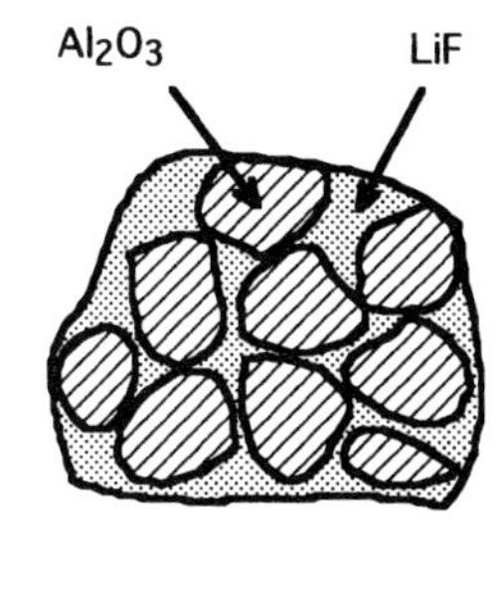


**Figure 9.6.** A composite of alumina particles in a LiF matrix as a humidity sensor.[7] The graph shows the response of the resistance to the relative humidity. Published with permission from the *Journal of Materials Education.*

The grains of the component phases in a composite and their surface or the intergrain material have different properties and therefore contribute to a different extent to the behavior of a composite. The interior of the grains determines the operation of solid oxide ion electrolytes, e.g., in zirconia or the operation of the electronic conductivity in NTC thermistors. The grain boundaries determine the performance of humidity sensors, PTC thermistors, and varistors. Humidity sensors for controlling, e.g., the electronic ignition in automobile engines are porous, fine-grained, metal oxides such as ZnO, $Fe_2O_3$, $TiO_2$, or $Al_2O_3$, impregnated with LiF or apatite (Figure 9.6). Adsorbed water molecules on the grain surfaces are good conductors of protons because the protons are easily passed on through a chain of neighboring water molecules (the Grotthus mechanism). The resistance of the porous composite changes by many orders of magnitude with the changes in air humidity, which changes the quantity of water molecules adsorbed at the interface.

Specific examples of these generalities are thermistors and varistor materials described below. They are exemplary composites that illustrate some basic features in passive composites. The design of a thermo-optic active material is also given as an example.

### *9.4.1. Thermistors*

Thermistors are resistors with a temperature-dependent value of their resistance. There are three types of thermistors: CTT (critical temperature thermistors), NTC (negative temperature coefficient), and PTC (positive temperature coefficient) thermistors. Their thermal behavior is shown in Figure 9.7.

The operation of CTT thermistors is based on the occurrence of a Mott transition in the oxide. The electrical conductivity is a step function. Below the critical temperature the material is a semiconductor or an insulator; above this transition temperature the material is a metallic conductor. $VO_2$ is used for this type of thermistor.

NTC thermistors are doped polycrystalline oxides that have the temperature-dependent conductivity of a semiconductor. NiO doped with $Li^+$ is a *p*-type semiconductor with hopping electron holes (see Chapter 10) and $Fe_2O_3$ doped with $Ti^{4+}$ an *n*-type semiconductor that has a conductivity that strongly depends on the

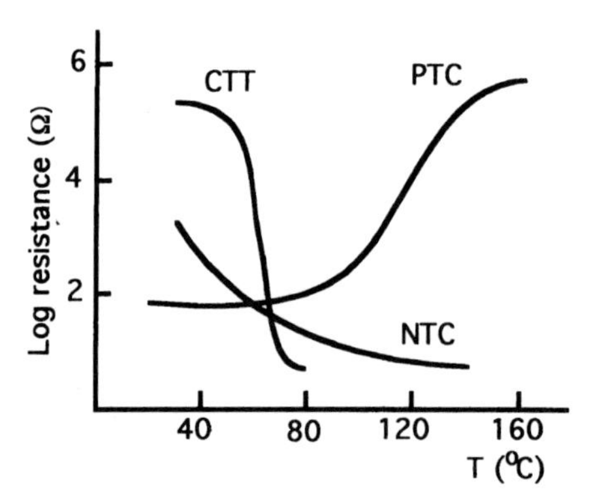


**Figure 9.7.** The temperature dependence of three types of thermistors.[6] Published with permission from the *Journal of Materials Education.*

degree of doping (Figure 9.8). The dissolution reaction of titania in iron oxide is

$$2TiO_2 \rightarrow 2Ti^{\bullet}_{Fe} + 3O^x_O + \tfrac{1}{2}O_2(g) + 2e'$$

This equation describes what happens when $TiO_2$ is dissolved in $Fe_2O_3$. Some trivalent iron ions are substituted by four-valent titanium ions and some oxygen gas escapes, leaving electrons behind in the lattice to compensate for the net positive charge on the substituted titanium. A small amount of titanium oxide dissolved in iron oxide makes the host solid an *n*-type semiconductor. The symbols in this equation will be more fully discussed in Chapter 10.

Pure NiO has a large bandgap and is an insulator. The apparent bandgap of NiO doped with 5% lithium ions is 0.15 eV because of the formation of clusters of defects $\{Li'_{Ni}, h^{\bullet}\}$. This material has a conductivity $\sigma = nq\mu$, in which $n$ is the concentration of charge carriers in the lattice (electrons or in this particular case electron holes), $q$ is the elementary charge, and $\mu$ is the electric mobility of the charge carrier in the lattice. Both $n$ and $\mu$ depend exponentially on the temperature as $\exp(-E/RT)$, where $E$ is the sum of the activation energy for displacement of the charge carriers (the diffusion activation energy) and the formation energy of these charge carriers (bandgap). NTC resistors are fabricated with resistances of 1–10,000 Ω with a temperature coefficient $R^{-1}(dR/dT)$ of 4% per degree C ($R$ decreases

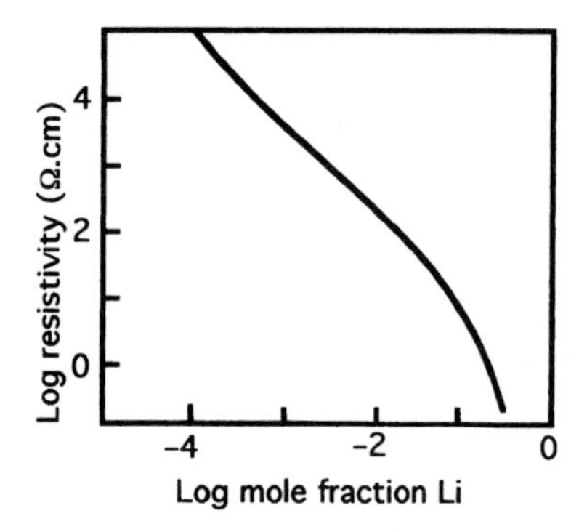


**Figure 9.8.** The resistivity of lithium-doped nickel oxide.[6] Published with permission from the *Journal of Materials Education.*

with increasing temperature). NTC resistors are used in gas flow sensors and in devices that protect against initial overheating when the current is switched on and the resistance of cold filaments is still too low.

In PTC thermistors the resistance increases strongly within a narrow temperature interval with increasing temperature and less rapidly at higher temperatures. To obtain this effect use is made of a ferroelectric phase transition in $BaTiO_3$. This oxide is made $n$-type semiconducting by doping with $La^{3+}$ or $Ce^{3+}$ ions in the barium sublattice or substituting $Nb^{5+}$ for $Ti^{4+}$. After sintering oxygen adsorbs on the grain surfaces, which react with lattice electrons to form some excess of oxide ions fixed in the grain boundaries. These negative charges make the material less conductive (they repel approaching electrons) and the comparatively conducting grains are surrounded by a thin insulating boundary. The resistance is inversely proportional to the grain size: electrons traversing a solid having small grains have more insulating barriers to cross.

The temperature dependence is the result of a ferroelectric phase transition in $BaTiO_3$ from the tetragonal to the cubic modification at the Curie temperature of 130°C. The dielectric constant $\varepsilon$ has a maximum at this temperature and decreases hyperbolically with $T$ as $\varepsilon \approx 10^5/(T - T_C)$. At temperatures below $T_C$ the positive ends of the electric dipole moments in the grains compensate for the negative charges at the grain boundaries sufficiently to make conductance possible and the grains are in conductive contact. At these temperatures, the resistance is rather low. Above the Curie temperature the dipole moments of the ferroelectric grains disappear, the charges are no longer compensated, and the resistance can increase (Figure 9.9). First it increases only slightly because the dielectric constant $\varepsilon$ is so high around $T_C$. A high $\varepsilon$ means efficient screening of obstructive charges and the electrons are free to cross the thin boundaries to the neighboring grains. At increasing $T > T_C$, $\varepsilon$ strongly decreases, the negative boundary charges are increasingly able to repel the conductance electrons, the grain boundaries become more insulating with increasing temperature, and the resistance increases. PTC resistors are used to protect circuits against short-circuiting accidents in thermostats and in liquid-level indicators.

### 9.4.2. Varistors

Varistors are semiconductors that are used as resistors having a nonohmic behavior. At low voltage differences over the resistor, these varistors behave as weakly temperature-dependent (NTC) resistors but above a critical voltage difference $V_C$ over the material the resistance drops sharply (Figure 9.10). This is a

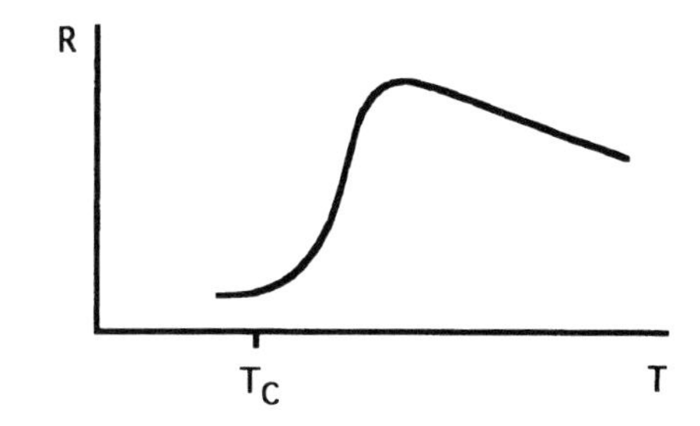


**Figure 9.9.** The temperature dependence of a PTC thermistor.[6] Published with permission from the *Journal of Materials Education.*

reversible effect that can be adjusted with the microstructure that is formed during sintering. The varistor effect is the result of tunneling of conductance electrons through the grain boundaries of zinc oxide. To make a varistor material ZnO powder is sintered in air at 1350°C after adding 3–10% $Bi_2O_3$ powder. The result is microcrystalline, *n*-type, bismuth-doped ZnO, while the grain boundaries are insulating zinc-doped bismuth oxide layers, 3 nm thick. The situation is somewhat similar to the PTC resistor based on ferroelectric grains above the Curie temperature. The band scheme of two neighboring grains in contact is like that of an *npn*-transistor (Figure 9.11). Between the grains there is a barrier of 0.8 eV, which is the reason for the expected NTC behavior at low voltage gradients. At high potential gradients the space charge layer in $Bi_2O_3$ becomes so thin that electrons can tunnel through it more freely, which means a strong decrease in the resistance. Varistors are used for protection against damage by lightning.

In designing composites there often remain conflicting requirements that cannot easily be solved with a proper combination of phases. Materials for fuel cell electrodes are an example. They must simultaneously be porous for gases, catalytically active, thermally inert, excellent electrical conductors, and have a large interface area with the electrolyte phase. To obtain an acceptable compromise for these properties, percolation of electronic and ionic conductors, open porosity, and interface "wetting" are imposed on the composite by choosing suitable components and fabrication processes.

### 9.4.3. Active Materials

An active material is a functional solid that has both sensor and actuator behavior. They are coupled to give an integrated material that responds with its actuator behavior to some stimulus that triggers the sensor moiety. The signal of the sensor component is directly converted by the actuator component into a physical or a chemical effect. Some examples of active materials are:

1. A membrane that becomes more or less porous depending on the size of the pressure on it (a valve action).
2. A material that changes color reversibly if the acidity of the ambient changes.
3. A material that converts solar energy to chemical energy in fuel by splitting water into hydrogen and oxygen.

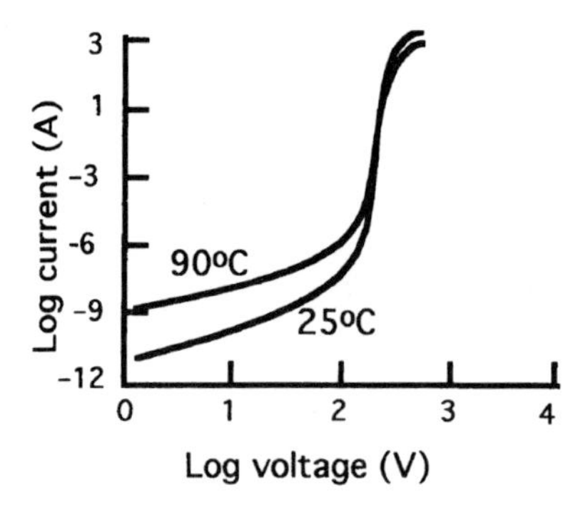


**Figure 9.10.** The voltage-dependent conductance of a zinc oxide varistor.[6] Published with permission from the *Journal of Materials Education.*

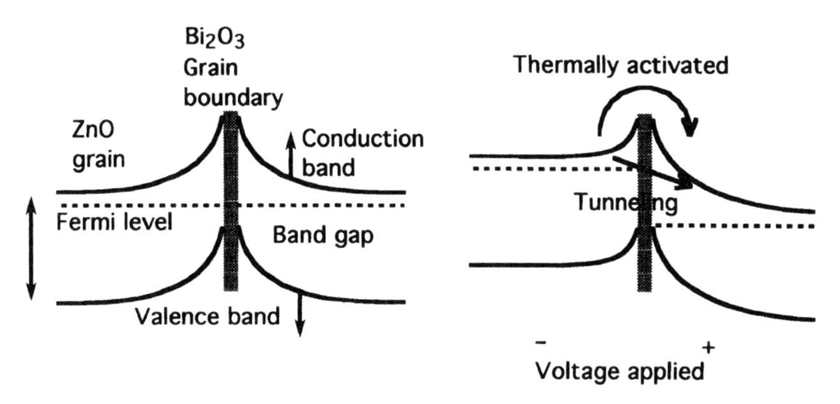


**Figure 9.11.** Band scheme of a $ZnO/Bi_2O_3$ varistor. The negatively charged grain boundaries of the insulating $Bi_2O_3$ are barriers for electron transport between the ZnO grains.[6] If the potential difference is high enough the barrier decreases and electrons start to flow. Published with permission from the *Journal of Materials Education.*

4. A material that has a catalytic activity that depends on some (electric) signal caused by some agent such as light or the presence of a chemical species that is not involved in the catalytic reaction.

Active materials are often composites in which one phase has a sensor action and the other component, coupled to it, has the actuator action. Other active materials have these functions divided between the interior and the surface of the material.

A smart material is a hypothetic active material that in addition to integrated sensor and actuator action has built-in memory or learning functions. These learning functions could in principle be realized with chemical changes invoked by repeated stimulation. A smart material has a memory and reacts to the circumstances in a complex way determined by the composition and prior stimulations of the material. Materials that are smart in this sense do not yet exist but the possibilities of making them are being explored. Living biomaterials are not smart because their intelligence is for a large part external and not stored in the material itself.

The coupling between the sensor and actuator components strongly affects the performance of an active composite. The sensor signal must be passed on to the actuator in an effective way. The contact is established during synthesis. It may happen that not all materials that are in principle suitable are actually also compatible in the synthesis process, and process stages must be found to combine them effectively.

One example of an optothermal active material is sketched below. The design rules for an active material are as follows:

1. Make up the list of requirements. What are the functions of the material, what should its properties be, and what should it not do? What are the sensor and actuator moieties supposed to do?
2. Make a list as complete as possible of the sensor behavior that the active material must display and list the materials that have all or some of this behavior by themselves. Do not limit the list with respect to the nature of the sensor signal; it can be anything from an electric signal to a color or strain

change. Include materials that partially fit the requirements but would be unsuitable for some reason.

3. Make a list of existing materials that might be suitable for application as an actuator. At this stage the list should not be limited by the nature of the physical or chemical effects or signals as these may be changed by some third agency.
4. Select all combinations that in principle could work in view of the signals from the sensor to the actuator moieties. This stage is the core of the design process for an active material.
5. Search for a fabrication process for every one of these combinations. Thermodynamic incompatibilities should be ignored here as in the other stages. Not all combinations of process 3 can be fabricated and the practical limitations further restrict the number of possibilities. In novel designs the relevant literature is rather insufficient and some experimental work might be called for at this stage.
6. Finally select from the surviving alternatives those combinations that need abundantly available cheap resources and develop a cost-effective fabrication process without waste.

In order to make these general procedures more concrete a design is described here of an optothermal active material that might be used as a photochromic cover of a solar collector with automatic feedback. It could also be used as an active window material for interior climate control.

The instruction is to design a material that reflects all incident light when it has a sufficiently high temperature ($T > T_C$). Below that threshold temperature ($T < T_C$) it must reflect light to a degree that increases with the incident light intensity. This behavior is somewhat similar to that of a photographic film (it becomes darker if illuminated further) but film is not a material. The sought for behavior is shown in Figure 9.12.

RULE 1. The list of materials with sensor properties (reaction to light and heat) is:

- Dyes and pigments
- Photochromic transition metal complexes
- Silver and lead halides that show photochromic activity
- Thermochromic and pyroelectric compounds
- Semiconductors that have a photovoltaic effect when combined
- Solids that have a phase transition between colored/noncolored, transparent/opaque, or amorphous/polycrystalline phases.

RULE 2. The list of solids with actuator action (light reflection, interaction with electric, mechanic, or other signals) is:

- Photoconductors, varistors, thermistors
- Polarization filters
- Semiconductors with a high Peltier effect
- Electroluminescent solids, liquid crystals
- Reflecting metals
- Compounds that have a Mott transition

RULE 3. Which of the possible couples could do the job? At the sensor side a selection from the possibilities by elimination of the impossible or difficult comb-

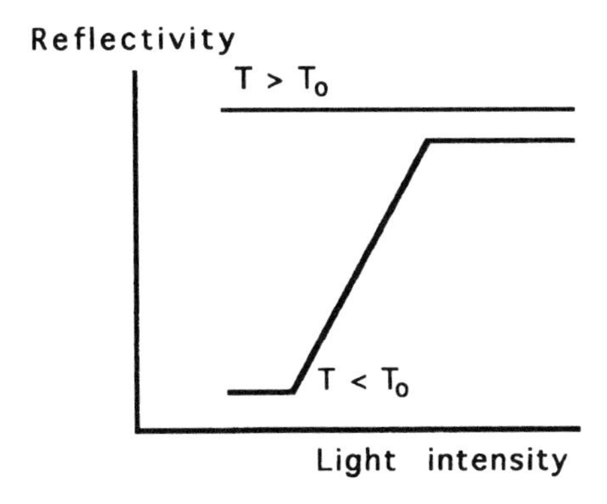


**Figure 9.12.** Optothermal behavior of a composite.

inations does not seem to be obvious. The actuator part seems to be more restrictive. First the temperature demand seems to point to a phase transition between a fully reflecting modification (metallic perhaps?) at the high-temperature side and some phase with a different behavior (light-intensity-dependent reflection) at lower temperatures. A simple solid that shows precisely that effect is not known, so it must be designed. The phase transitions that is closest is the Mott transition between an insulator and a metal. What can be done with it? A part of the problem (the behavior at high temperatures) is solved by this choice at this stage. The next question is how an intensity-dependent reflection can be imposed on a Mott insulator and what sensor materials are capable of that? One possibility might be that light triggers a temporary and local transition to the metallic state in the insulator so that it reflects partially (only where it is converted). If light is to trigger anything, it must first be absorbed and a photovoltaic or photochemically active phase is needed. In the Mott insulator pigments or dye molecules (chromophores) can be dispersed that convert the absorbed light to energy. This energy renders the solid locally and temporarily conducting so that it reflects at those sites. In order to be able to do that the chromophores must transfer the absorbed energy efficiently to the actuator matrix. The ease of excitation-energy transfer from chromophore to matrix is a spectroscopic requirement that limits the number of suitable pigments that can act as sensors. One possible combination is copper phthalocyanine (Cupc) in a vanadium oxide matrix. The values of the optothermal behavior (how large $T_c$ is, what the threshold intensities are, what the sensitivities are) can be adjusted with the microstructure and the concentrations. This active composite has two components, the pigment that absorbs the light and acts as sensor atomically dispersed in the thermochromic vanadium oxide matrix that is the actuator.

RULE 4. How to synthesize it. A thermodynamically schooled engineer would object to the combination described above on the grounds that copper phthalocyanine does not dissolve in vanadium oxide even if it would be stable in the melt and that therefore the combination is unsuitable. Thermodynamic impossibilities are unnecessarily restrictive and should be ignored at any stage in the design process as allegedly incompatible materials can be combined in practice by proper choice of process conditions. In the present case the sol-gel method (Chapter 8) for making this composite would be one possibility.

RULE 5. Production optimization. This can be developed only after the previous stages have been completed. Sometimes entirely new fabrication technologies have to be developed for a novel material but in this case that does not appear to be necessary. Established sol-gel methods for vanadium oxide can be easily scaled up and pigments can be dispersed in it by making suitable particles in the gas phase with a coating that improves adhesion if necessary.

## 9.5. Fabrication of Composites

The methods of synthesis of materials in Chapter 8 were described only for simple compounds. The same methods suitably combined can be used for preparing composites. Sometimes they can be combined in one process or in closely coupled processes. Many combinations are possible and an enumeration would need a book itself. Instead a few representative examples are given (the numbers indicate the dimensions of the components).

- 0–0 composites are made with sol-gel methods if they are oxides or by PP-CVD if they are not. Just mixing powders and heating (ceramics) works in many cases. Particles can easily be made so that one component covers the other: spinel is made efficiently starting from alumina particles covered with a silica layer. Particles can be covered in fluidized beds with another material to make them compatible with a third material when processed.
- 0–1 composites are made with sol-infiltration of whiskers or fibers and sintering; another possibility is melt-infiltrating a felt or a mat of woven fibers.
- 1–1 is a combination of twined strands or a mixed whisker felt. A whisker felt can be deposited from the gas phase by VLS. Subsequently this can be infiltrated with fibrous (VLS) material also from the gas phase in the second half of the process.
- 2–2 composites are made with time-modulated deposition conditions or eutectic solidification. Laminar composites can be made by alternating the deposition fluid from which the material is deposited on the surface a sufficient number of times. Multiple optical interference filters are made on silica lamp bulbs in that manner.
- 2–1 combinations in a material are made by alternating whisker growth (by VLS) with layer deposition, both from the gas phase.
- 0–2 is a connectivity made like the 2–2 composite but one of the two process phases deposits powder.
- 0–3 materials are particle reinforced matrices made by nucleation of particles in a crystalline or glassy solid phase or by dispersing particles in a melt. An example is partially stabilized zirconia.
- 1–3 composites are made by CVI or infiltrating whiskers or fibers by a liquid matrix. Examples are fiber-reinforced glass ceramics, carbon–silicon carbide composites, and ceramics-reinforced metal matrix hybrids. An alternative method for 1–3 composites is eutectic solidification, but that is less universally applicable than oscillating CVD as it depends on the existence of a suitable eutectic in the system.

- 2–3 materials are made by filling up a mass of loosely stacked platelets with a matrix from the gas phase or from a liquid phase.
- 3–3 combinations are the result of spinodal decomposition in glass or sintering a mixture of component powders that do not react with each other.

The central problem in fabrication is keeping control of the interface properties without at the same time having the intrinsic bulk properties deteriorate. The interface between crystallites is essential for coupling the components in active materials. There are three general basic guidelines for fabrication of composites:

1. Materials must be prevented from reacting with each other destructively during synthesis or during operation. The compounds must be compatible with each other at the temperature of fabrication and once they are present they should not form unwanted compounds. If necessary a diffusion barrier is deposited on one (e.g., reinforcing) phase to separate it from the other. One example is a layer of $TiB_2$ on steel in order to protect it against a carburating ambient in coal gasification equipment. $TiB_2$ has to be deposited by means of CVD at high temperatures. Steel cannot be directly covered because the deposition atmosphere corrodes the steel substrate, so it has to be protected by a thin diffusion barrier of TiN on which the boride can be grown. Another example is making composites with metallic aluminum. Molten aluminum is very reactive and $B_4C$ cannot be infiltrated with it to form a hybrid composite without extensive reaction. $TiB_2$ on the other hand, is inert toward aluminum and can be used to make a cermet if the surface does not contain BN, which prevents wetting and precludes the capillary action that is necessary for infiltration with molten aluminum. Obviously materials should not form unwanted side products or decompose at the temperatures needed for the other components.

2. Conversely, materials should not fail to react somewhat at their surface and they should not form porous products or unconnected product phases. The reason for the better interphase adhesion in Frank–van der Merwe type growth is the fast reaction rate, which precludes the occlusions and pores that are more likely to occur in Volmer–Weber type growth. One example is the combination of TiN with either steel or alumina ceramics as described in Chapter 6: some increase in chemical activation at the interface is necessary for good adhesion between the phases, which implies a high surface reactivity during preparation. As described in that chapter, coupling agents connect functional molecules (such as dyes) chemically with inorganic surfaces for good electron transfer or for an improved adhesion with an organic matrix.

3. Composite materials should combine and enhance the favorable properties of the components and at the same time eliminate their unwanted properties. The less desirable characteristics are eliminated by spatially arranging the components during fabrication with time modulation of the fabrication conditions. The clearest example is a ceramic coating deposited on a metal substrate or making a cermet with ceramic techniques as in the case of WC/Co hardmetal: the toughness of the metal is combined here with the wear and corrosion resistance of ceramics. Thermal stresses that develop in a composite during fabrication or use should be taken care of in the design.[12,13]

Some synthetic processes can be easily scaled up, but others cannot. Powder synthesis, single-crystal fabrication, and many CVD methods usually mean more of the same when scaling up to industrial level. Sometimes, however (for example in fluidized bed reactors), the scale of production is essential to the nature of the process. The treatises for chemical engineers on unit operations in the physical technology of bulk powders as mentioned in Chapter 8 should be consulted for upscaling. Materials technologists use more types of unit operations than chemical engineers. Casting, rolling, drawing, forging, sintering, and surface-hardening of metals can be found in the standard volumes on materials processing. The processing of the different types of materials is specific. Plastic deformation methods are used for metals and glasses and powder technologies for ceramics. Finally, the methods for so-called powderless processing (sol-gel, CVD, and conversion) in materials technology should be mentioned. These powderless alternatives to conventional ceramic procedures are less expensive and permit better control over the morphology of the end product than conventional powder technology. Such methods are therefore increasingly used in making advanced inorganic composites.

## Exercises

Exercises in design are by their nature always open-ended. Answers cannot be looked up and cannot be judged right or wrong. They are discussions. Three examples of such questions with comments (not answers) follow:

1. Is self-repair possible in inorganic materials as it is in organisms? If not, why not? If so, how and to what extent could that be built into materials?

   Self-repair in biological tissues is not an acquired property that is merely added to other characteristics but is inherent in the permanent state of growth and overhaul of living biological matter. Synthetic materials are basically different; the name materials for both implies too much similarity. That does not mean that localized repair of damage in nonliving materials is impossible. After all, its converse, stress corrosion, exists—a local corrosion at sites where stress concentrations such as at crack tips have increased chemical reactivity. Now enhanced local etching like its reverse, enhanced local growth, exists by virtue of the enhanced chemical reactivity at those sites. Therefore cracks in materials can be repaired if use is made of that local reactivity at these sites. It is sufficient to expose them to the proper ambient for the cracks to fill up as long as the reactivity enhancement lasts. For example, piezoelectric materials have excess surface charges under stress and therefore an altered reactivity that can be used to deposit solids with the right reactants.

2. Hierarchy in materials[14] has been shown to contribute to the extraordinary properties of biological tissues. Give a few examples of hierarchy in inorganic materials introduced for performance enhancement.

   Hierarchy can be described in analogy to rope (stretched polymer molecules in domains that make up nanofibers, combined to microwhiskers, bundled into fibers that are spun into yarn that is twined to make up the rope). Wood and tendon are biological examples that have six or more hierarchical levels. Compared to these, fiber-reinforced matrix composites made up of simple massive fibers embedded in a metallic, ceramic, or polymer matrix are primitive. Hierarchical inorganic materials, as discussed in Chapter 7, can be made with processes for fractal-like solid products: spinodal decomposition, diffusion-limited growth, particle precipitation from the vapor, and percolation. Fractal-like solids have holes and clusters of all sizes and are therefore hierarchical if the interactions between the composite phases are carefully controlled. Thus toughening a material by introducing a hierarchical morphology means deflecting a microcrack along a fractal interface. The higher the dimension of the crack, the tougher the material.

3. A glass is often a better ion conductor than the corresponding crystalline compound if the crystalline modification is structurally a bad conductor. Yet many glasses are not used as solid electrolytes because they crystallize during use. Glass is much more convenient for designers than ceramics. How could one adapt a glass as a solid electrolyte for a rechargeable battery so that it does not crystallize under charge/discharge cycles and by doing so significantly increase the lifetime and efficiency of the battery?

   The problem is how to prevent nucleation and, more importantly, crystallite growth from the glass during operation. The ion movement in the glassy electrolyte means a high atomic mobility and in general that helps crystallization. Which additions to the glass prevent the transition to the crystalline modification by poisoning incidental nuclei? To answer that three things should be realized, which might hold the answer for a specific electrolyte: (a) In ceramics extended use is made of densification additives, which are surface-active inorganic impurities added to lower rates of crystallite growth during sintering. Growth inhibitors improve the final density of the sintered ceramic product. (b) In the technology of glass ceramics certain inorganic additives are used to control nucleation and crystallite growth in order to make the right microstructure. (c) In solution and in the gas phase nucleation of a solute can be shifted to temperatures several hundred degrees higher by the right impurities. Hence as everywhere, the answer is specific and depends on the nature of the glass electrolyte. Inspiration for ion-conducting silicate glasses can be found in ceramics.

Some other exercises:

4. Give the defect reactions when doping material for PTC thermistors.
5. Indicate how the microstructure of a ceramic can affect the ion conductivity.
6. Choose a known material and discuss how the properties and the fabrication methods should be altered to make it suitable for a novel application. A functional example would be fluorescent pigments—actuators and a structural example porcelain—ship hulls.
7. Give an analytic expression for the ohmic resistance of the following composites in terms of the resistances of the separate components: (a) connectivity 2–2 and direction of current parallel and perpendicular to the planes; (b) connectivity 1–3 and direction of current parallel and perpendicular to the fibers; and (c) connectivity 0–3.
8. What if any is the effect of connectivity on the behavior of the photochromic composite described in Section 9.4 in the following cases? (a) a 0–3 composite: isolated sensor particles in a three-dimensional actuator matrix; (b) a 1–3 composite: sensor fibers or whiskers in an actuator matrix; and (c) a 2–2 composite: sensor layers alternated with actuator layers.
9. Design a composite ceramic material (give the component materials and the morphology) that obeys certain requirements as indicated below. Also indicate how this material is to be made and give the fabrication process conditions. If the answer to a question is "that is impossible" then give the arguments:

   1. A plate that responds with an electric dipole moment to a bending stress.
   2. A membrane that has an electrically controlled pore size.
   3. A membrane that has a thermally controlled pore size.
   4. A solid that can be heated very fast (300 ms), without overshoot, to a precise temperature that depends on the material (e.g., from 20 to 753°C).
   5. A photochromic coating in which the absorption coefficient for light of a certain wavelength increases with light intensity.

6. A ceramic "muscle" that is powdered by light.
7. An antiradar coating.
8. A material that shows temperature-dependent color and light-scattering.
9. A material that has a thermal conductivity that can be controlled with the amount of incoming light.
10. A gas sensor that has a strong response within a narrow redox interval and no response outside of that interval.
11. An optical low-pass filter that has an electrically adjustable threshold.
12. A temperature sensor for use at 2500°C under strongly carburating conditions (any refractory metal fails then).
13. A material that absorbs light under static mechanical stress; the wavelength of the absorbed light should depend on the stress level.
14. A material that emits light when exposed to a dynamic (alternating) mechanical stress such as sound. The wavelength of the emitted light depends on the dissipated mechanical energy.
15. A resistor that is a thermistor within a given window of temperatures.
16. A ferroelectric with a temperature-dependent permanent electric dipole moment.
17. A biocompatible ceramic that mimics bone growth when implanted.
18. A transparent material for laser protection goggles that becomes opaque instantly as soon as laser radiation (coherent light) hits it.
19. A red pigment to replace toxic elements[15] in the usual ones.

Combine the above requirements with some added constraints and see how they affect the design.

## References


1. C. Newey and G. Weaver (eds.). *Materials, Principles and Practice:* Materials in Action Series of the Open University. Butterworth-Heinemann, Oxford (1991).
2. D. L. Cocke and A. Clearfield. *Design of New Materials.* Plenum, New York (1987).
3. D. W. Bruce and D. O'Hare. *Inorganic Materials.* Wiley, Chichester (1992).
4. M. Diani. *The Immaterial Society.* Prentice Hall, Englewood Cliffs (1992).
5. N. F. M. Roozenburg and J. Eekels. *Product Design: Fundamentals and Methods.* Wiley, Chichester (1995).
6. R. E. Newnham. Structure-property relations in electronic ceramics. *J. Mater. Educ.* **6**, 807 (1984).
7. R. E. Newnham. Composite electroceramics. *J. Mater. Educ.* **7**, 601 (1985).
8. F. E. Fujita (ed.). *Physics of New Materials.* Springer-Verlag, Berlin (1994).
9. P. Calvert. Biomimetic ceramics. *Mat. Res. Soc. Symp. Proc.* **180**, 619 (1990).
10. D. S. McLachlan, M. Blaskiewicz, and R. E. Newnham. Electrical resistivity of solids. *J. Am. Ceram. Soc.* **73**, 2187 (1990).
11. R. E. Newnham and G. R. Ruschau. Smart electroceramics. *J. Am. Ceram. Soc.* **74**, 463 (1991).
12. G. Grimvall. *Thermophysical Properties of Materials.* North-Holland, Amsterdam (1986).
13. D. P. H. Hasselman and J. P. Singh. Criteria for the thermal stress failure of brittle structural ceramics. In: R. B. Hetnarski (ed.) *Thermal Stresses.* North-Holland, Amsterdam (1986), p. 263.
14. NMAB, National Materials Advisory Board. *Hierarchical Structures in Biology as a Guide for New Materials Technology.* National Academy Press, Washington D.C (1994).
15. G. Buxbaum (ed). *Industrial Inorganic Pigments.* VCH, Weinheim (1993).

*Chapter 10*

# INORGANIC PHYSICAL CHEMISTRY

*Toute règle qui fonctionne mérite d'être prise en considération.*

*J. M. Guéhenno, 1993*

## 10.1. Introduction

This chapter concerns those parts of physical chemistry that are of some use for materials chemists. It has been placed at the end in the book rather than at the beginning, which is unconventional. For sound didactic reasons, textbooks generally start out with physical models for chemical processes. After that, facts are given to support the models while the facts that do not fit well with the models are left out. The customary hierarchy is reversed here (with the exception of the chemical bond) because for actual practice, facts come first and theories second. Also selection is necessary: the parts of physical chemistry that are most relevant for the synthesis of inorganic materials are chemical equilibrium thermodynamics, phase diagrams, driving forces, and homogeneous kinetics based on averages. There are some other aspects that are occasionally invoked in materials chemistry but are not discussed here, e.g., the deviations from thermodynamic ideality, local equilibrium, surface tensions, and misfit terms in driving forces during materials formation. These are omitted because they are refinements of models that themselves are not of first importance for materials and remain so after the efforts to improve them.

This chapter can be seen as theoretical appendix that to some degree links the subjects of this book with scientific chemistry. Physical chemistry can further or inhibit progress in materials technology, as any academic subject can. The generalities in this chapter may be of use in synthetic inorganic chemistry because they give a semiquantitative impression of the role of the driving forces in chemical processes. Physical chemistry covers that part of chemistry that can be generalized and simplified. That is a great deal but a large part of materials chemistry is specific and complex. Like weather, chemistry cannot easily be dealt with by a physical approach, which is so successful for simple systems (Section 1.6). The label *physical chemistry* is an oxymoron in a sense but like several branches of physics, parts of it are definitely useful. Some physical properties of inorganic compounds were discussed in Chapter 4. This chapter will deal with the chemical consequences of the thermodynamic driving forces.

A caveat signals the limited role that scientific models in general have for innovation:[1]

> Invention has in common with discovery that it can claim to be what it is only if it is surprising. It must be separated from its antecedents by a considerable logical gap. The width (... of the gap...) measures the ingenuity of the invention..."

## 10.2. Equilibrium Thermodynamics

Thermodynamics has been described as "... the science that deals with energy, matter, and the laws governing their interaction,"[2] or "Thermodynamics may be considered to be the science of the possible."[3] If those claims are taken literally, thermodynamics is the long-sought theory of everything. It includes all of chemistry, physics, and materials science; it encompasses catalysis, electrolysis, corrosion, spectroscopy, chemical reactions, chemical bonding, and elementary particles. That would be quite remarkable for a theory that does not even recognize the existence of time or atoms. According to these definitions, the subject would also include chemical synthesis. Some people actually agree to that. As a recent handbook on thin-film processes[4] advises: "The first step in the design of a CVD experiment should be the calculation of the thermodynamic equilibrium...," although "Such calculations cannot, as a general rule, predict *exactly* what will happen in a vapor growth experiment...." It continues notwithstanding the facts: "Nevertheless... it is possible to predict both the feasibility of the process and nature and amount of the solid and gaseous species." Some reservation is added: "The vast majority of practical CVD processes proceed under nonequilibrium conditions; therefore overall analysis of CVD processes *also* includes, beside the equilibrium predictions, the chemical kinetics and mass transport analysis." Italics are mine. Although it is commonly known that equilibrium thermodynamics is not appropriate for nonequilibrium processes, too many authors continue to use it freely without any reservations. This has caused a critical thermodynamicist to remark on his subject: "While its claims are often grandiose, its applications are usually trivial."[5] The same physicist criticized the usual way in which chemical engineers deal (on paper) with liquids in a perfectly stirred flow reactor:

> The numerical values, irrespective of the size of the vessel and of what be the reactants, seem intended to help us keep our feet firmly on the ground of empirical science. I cannot help wishing to see some of my classically thermodynamic friends try the experiment with asphalt as one ingredient and nitroglycerin as the other.

What can this branch of physical chemistry offer materials chemists? Thermodynamic considerations can estimate the driving forces for simple chemical processes, which can be useful for those cases where the atoms have a high mobility and reactivity because then the end state can sometimes be described correctly. Driving forces for processes in the solid are discussed below. The equilibrium thermodynamics of processes in liquid phases can be found in undergraduate texts[3,6] if needed and are not included here. Some thermodynamics of chemical processes in solids that are of interest for inorganic materials are summarized below.

Often reaction rates are seen to change when the driving forces for the process are altered. The observed variations in the process rates are duly assigned to thermodynamics, and this explains the habit of adding thermodynamic calculations to discuss observed chemistry. However, changes in process conditions that alter the driving forces also change the reaction resistances or, some might say, the activation energies. An example for the case of reactions between molecules was given in Chapter 3. It is easy to ascribe rate changes to changes in thermodynamic potential differences. However, if the thermodynamic potential for the reaction is high with respect to the mean kinetic energy of the molecules (the measure of which is $RT$, with $R$ the gas constant and $T$ the absolute temperature), the reaction rates are not related to the driving force but to the reaction resistance as discussed in Section 1.4. The analogy of Ohm's law for chemical reactions would be convenient but does not hold[7] in systems which cannot be said to be in "local" equilibrium.

Thus although thermodynamic calculations permit estimation of the driving force for solid state reactions, they cannot predict rates of production and the nature of the reaction products. Contributions of the microstructure to the Gibbs energy of the formation reaction (or to the activation energy for that matter) can be large, surface energies are defined only for mobile macroscopic systems (liquids), growth of solid nuclei can be strongly affected by impurities, and the reactions may stop at metastable products and not continue to the thermodynamic end. In spite of all this, the use of $T$–$x$ phase diagrams can help to interpret observed solid state reactions if they are used with caution. Such diagrams summarize observations even if collected under near-equilibrium conditions.

The bonding energies in the compounds determine the phase diagram ($T$–$x$) of the system. Demixing occurs if the atoms of the components of the diagram prefer to surround themselves by their own sort while alloys or compounds occur if the atoms have a higher affinity for the other atoms than for their own sort. This is an equilibrium argument. Considered as a description of what will happen under certain circumstances, it assumes that equilibrium will be attained.

The stronger the affinity between the components, the larger the number of different stable compounds that the components in the system can form. Phase diagrams summarize a lot of systematic chemistry and are always consulted when designing thermal solid state reactions. They sometimes do not give very reliable guidelines because the metastable compounds that may also occur are usually omitted from the diagrams. These phase diagrams are empirical; they do summarize which phases have been found and what their compositions can be but it should be remembered that the equilibrium situation is aimed at in making them. Before determination of the composition of the phases that are present, the system is left for a long time under the conditions of the experiment in order to make sure that the composition does not change any further. There is no time for that in practice and phase diagrams must be used with this in mind.

There are existence areas in phase diagrams in which one phase is stable and coexistence areas where two phases are in equilibrium.[8–10] Line compounds (very narrow existence regions) are phases that have very low solubility for the separate

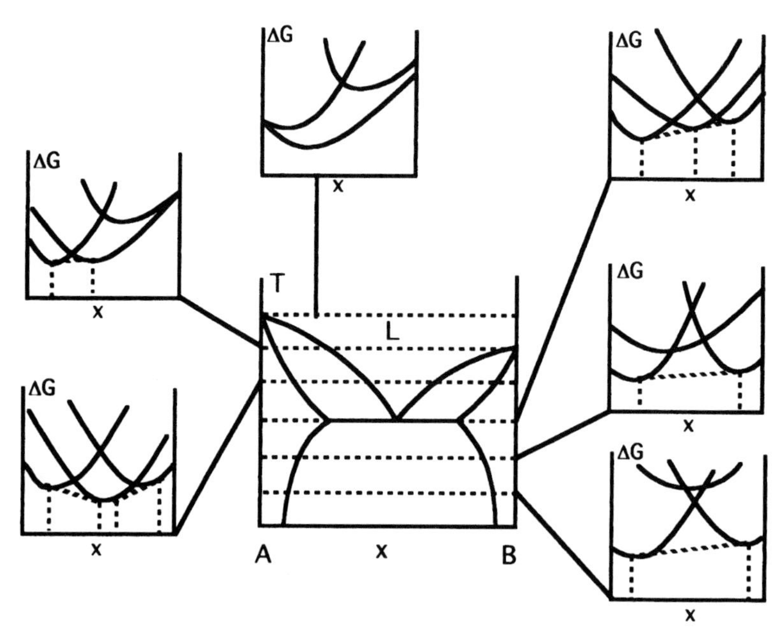


**Figure 10.1.** A model phase diagram of two compounds A and B that do not mix but dissolve somewhat in each other. The corresponding $\Delta G/x$ plots are given for six different temperatures. The three curves in each of these six plots correspond to the phases A, B, and the liquid phase.

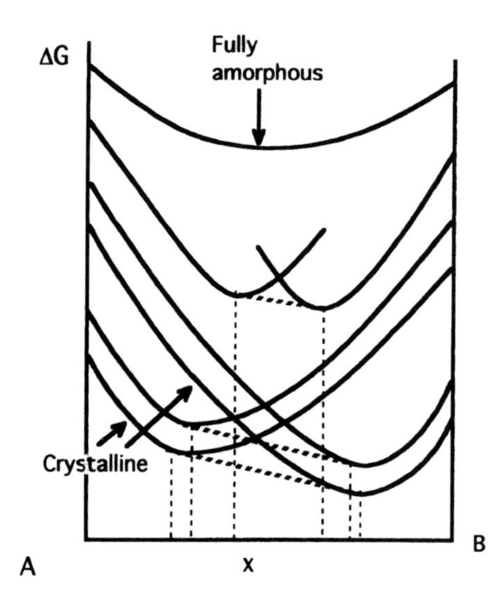


**Figure 10.2.** The $\Delta G/x$ curves for a binary system with some mutual solubility that increases with defect concentration. Introducing more defects into the lattice raises the $\Delta G/x$ curves toward the amorphous modification and that increases the mutual solubility as shown in the graph.

components of the diagram. Any possible nonstoichiometry can be read from the phase diagrams. The borides, carbides, pnictides, and chalcogenides show nonstoichiometry to a different extent. The widths of the existence areas of compounds, or the solubilities in them, tend to increase with increasing temperature because the entropy term in the Gibbs energy increases with the temperature (shown below). The

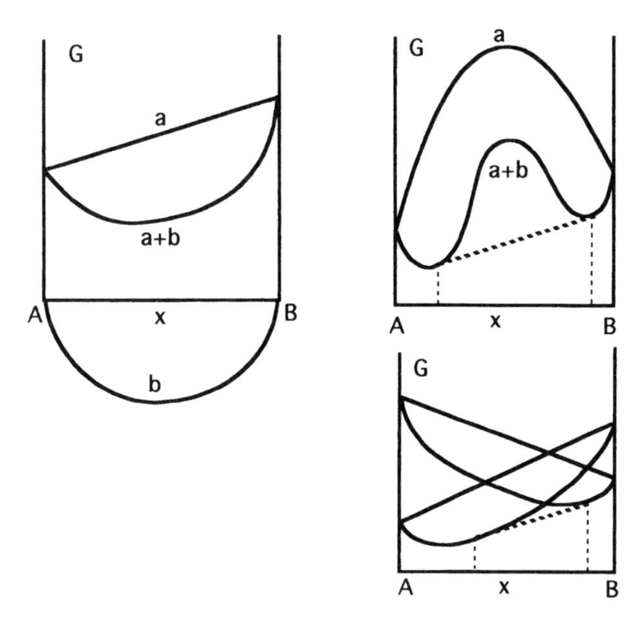


**Figure 10.3.** The $\Delta G/x$ plots of a few imaginary mixtures to show the contributions of enthalpy (a) and entropy (b) to binary phase diagrams. Top left: an ideal mixture ($\Delta H = 0$) at some temperature ($T > 0$). Top right: a nonideal mixture ($\Delta H > 0$) also with an entropy term. Bottom right: an ideal mixture at nonzero temperature, where A and B have a different crystal structure.[8]

phase diagrams $T$–$x$ for mixtures are related to the Gibbs energy $G(x)$ at different compositions. Figure 10.1 shows this relationship for several temperatures. The solubility is also increased by externally imposed disorder (more entropy) as Figure 10.2 shows for amorphous phases.

The argument for the coexistence areas in nonideal mixtures is the occurrence of a double minimum in the $G(x)/x$ curve at sufficiently low temperatures. Figure 10.3 illustrates the way in which the $G(T)/x$ graphs form the binodal in the $T/x$ diagram. It can also be seen that in the spinodal region the mixtures are unstable in the face of any fluctuation in the composition.

The thermodynamics of mixtures are expressed as follows. The Gibbs free energy $G_{PQ}$ of a mixture of the components $P$ and $Q$ is a sum of terms that the components $P$ and $Q$ contribute to the mixture $PQ$; it also has a mixing term, $G_{\text{mix}}$:

$$G_{PQ} = x_P G_P + x_Q G_Q + G_{\text{mix}}$$

In this equation $G_P$ and $G_Q$ are the Gibbs energies of the two components $P$ and $Q$ and $x_P$ and $x_Q$ are the concentrations of the components in the mixture. This equation can be rewritten by expressing the mixing term $G_{\text{mix}}$ in the mixing enthalpy (zero in ideal mixtures) $H_{\text{mix}}$ and the mixing entropy $S_{\text{mix}}$ ($G = H - TS$ is used). For the entropy $S_{\text{mix}}$ its expression in terms of the concentration $x$ in an ideal gas mixture is substituted in the equation. For the present purpose this is an acceptable approximation. The atomic concentration $x_P$ is called $x$ and for a binary system the following equation results:

$$\underbrace{G_{PQ} = (1-x)G_P + xG_Q + H_{\text{mix}}}_{a} + \underbrace{RT\{(1-x)\ln(1-x) + x\ln x\}}_{b}$$

It is convenient for the discussion to collect the terms into two groups, $a$ and $b$ (Figure 10.3).

For a so-called ideal mixture, $H_{\text{mix}} = 0$. The entropy of mixing at $T = 0\,\text{K}$ is also zero. Hence at that very low temperature, ideal mixtures have a straight $G_{PQ}(x)$ line between the $G$-values of the components $P(x = 1)$ and $Q(x = 0)$. Now, if $T > 0$, the second term $b$ (which is less than 0) starts to contribute and the $G_{PQ}(x)$ curve, even of an ideal mixture, is no longer linear and touches the vertical axes at $x = 0$ and $x = 1$. In a nonideal mixture at 0 K, $H_{\text{mix}} \neq 0$ and the first term $a$ is no longer linear but parabolic in $x$. The value of $H_{\text{mix}}$ can be positive or negative, depending on whether the components $P$ and $Q$ prefer to coordinate their own atoms or the others. The size of the mixing enthalpy can be estimated from the atomic parameters from bond models such as the Miedema model (see Section 2.6).

The simple equation for the Gibbs energy of a mixture describes the widening of the existence regions and solubilities at increasing temperatures. The mixing entropy term $b$ becomes larger at higher temperatures. At a temperature of 0 K the existence region of the compound $PQ$ is a line and the compound is stoichiometric. The solubilities and the nonstoichiometry also depend on the free energy of formation $\Delta G_{PQ}$ (other terms being kept the same) as can be seen in Figure 10.4.

If the components $P$ and $Q$ have the same crystal structure, the $G(x)$ curve is continuous over the whole region $x$. If $P$ and $Q$ have different structures, the $G(x)$ curve consists of two parts that intersect. Each part belongs to one structure, the structure of either of the components. The $G(x)$ part of component $P$ runs from $G_P$ at $x = 1$ to the fictitious $G$-value that $Q$ would have if it had the same structure as $P$. And conversely for the component $G$ curve for $Q$. This could lead to a double minimum that is a feature of demixing.

A certain affinity of atoms M and N for each other, or in other words a sufficiently negative mixing enthalpy, means that a number of compounds $M_iN_j$

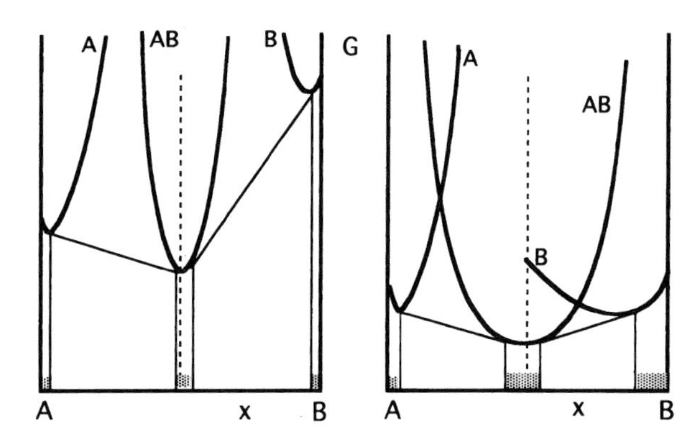


**Figure 10.4.** Schematic $G/x$ diagram for a system A/B with one compound, AB. The existence and solubility regions increase with temperature and also depend on reaction heat.[8]

exist. However, the converse, that if ordered compounds exist it means that there is some attraction between the atoms, is not true. Ideal mixtures of hard spherical atoms without attraction between them can form crystalline compounds that are "entropy driven." If the ratio of the atomic radii of M and N is 0.58, ordered lattices of $MN_2$ and $MN_{13}$ are formed at sufficiently high density although there is no attraction at all between M and N. This effect has been observed in colloidal systems in which the particles behave as hard spheres that repel each other only if they come too close. The fact that stoichiometric compounds and crystalline order are possible in the absence of bonding may be a surprise for anyone who is used to the idea that compound formation occurs if the attraction is high enough and assumes that the reverse also holds true.

Phase transitions (considered to be a part of physics) between one solid phase and another are similar to chemical solid state reactions. They have similar kinetics that are measured in the same way and in both cases chemical bonds are broken and remade. Polymorphic phase transitions are observed in many ceramic materials on heating, during which the crystal structures and properties change abruptly (not the composition).

There are two types of phase transitions in solids: first-order and continuous-phase. According to Ehrenfest the *order* of the phase transition is equal to the lowest order of the derivative of the Gibbs energy, which is discontinuous at the transition.

First-order phase transitions have discontinuous first partial derivatives of the Gibbs energy at the transformation temperature:

$$V = \left(\frac{\partial G}{\partial p}\right)_T \qquad S = -\left(\frac{\partial G}{\partial T}\right)_p$$

The symbols have their usual meaning. Often hysteresis is observed in first-order transitions. On cooling a melt in a deep eutectic or at a rapid rate, the solidification may be so much delayed that a glass forms instead of a crystalline solid.

Phase transitions that have continuous first-order transitions at the transition temperature are continuous. While the first-order partial derivatives do not jump at such phase transitions, the higher-order derivatives can change there. The relevant higher partial derivatives are:

$$\left(\frac{\partial^2 G}{\partial p^2}\right)_T = \left(\frac{\partial V}{\partial p}\right)_T = -V_m\chi$$

$$\left(\frac{\partial^2 G}{\partial p \partial T}\right) = \left(\frac{\partial V}{\partial T}\right)_p = V_m\alpha$$

$$\left(\frac{\partial^2 G}{\partial T^2}\right)_p = -\left(\frac{\partial S}{\partial T}\right)_p = -\frac{c_p}{T}$$

In these equations $V_m$ is the molar volume, $\chi$ is the compressibility, $\alpha$ the thermal expansion coefficient, and $c_p$ the specific heat at constant pressure. In a continuous-phase transition, the specific heat, the thermal expansivity, and the compressibility are all discontinuous.

During polymorphic phase transitions the structures and properties of single-phase materials change discontinuously, which is of considerable interest in the subject of structure–property relationships. When a compound has a displacive phase transition at a certain transition temperature there are symmetry relations between the modifications that are stable below and above the transition temperature. The modification on the higher-temperature side has the highest symmetry. The symmetry group below the transition point is a subgroup (fewer symmetry elements) of the symmetry group of the higher-temperature phase. If the coordination number changes in a phase transition the latent heat is usually high.

There are several types of transitions between solid phases and latent heats and the kinetics of the transformation both depend strongly on the particular type of transformation.

1. *Reconstructive* phase transitions occur between markedly different crystal structures. The two phases have no symmetry relationships. Bonds between the atoms are broken and remade in the transition. Large latent heats obtain if the coordination number changes during constructive transformations. Transition rates are low.
2. In a *displacive* transformation the atomic positions in the two phases differ by only small displacements and bonds are slightly modified but not broken. Usually the high-temperature phase has the higher symmetry. Transition rates of displacive transformations are usually high.
3. A *Dilatational* phase transition is a special displacive transformation in which there is a coordination change. The symmetry is not necessarily increased at the higher-temperature side of the transition point. The transition of the CsCl structure to the NaCl structure is one example. These transitions are rapid and may have high latent heats.
4. *Bond type* transformations are characterized by a change in the type of interatomic bonding. Examples are the transitions between allotropes of tin and carbon. These transformations, like reconstructive transitions, are usually very sluggish.
5. *Order–disorder* transitions are continuous transformations that are characterized by an order parameter that changes continuously from 1 at very low temperatures to 0 at the transition temperature. An ordered solution (in other words a compound) becomes a disordered alloy at temperatures above the transition point. Examples of materials having order–disorder transitions are ionic conductors and ferromagnetic and ferroelectric compounds. Substitutional order transitions involve diffusion and are sluggish; those involving rotational disorder are rapid.
6. *Martensitic* transitions are characterized by a cooperative, diffusionless, shearing displacement in the lattice. There is no activation energy and martensitic transitions occur in a temperature range with hysteresis. One example is the monoclinic to tetragonal transformation of undoped zirconium oxide ($ZrO_2$). The transition occurs at 1170°C when the temperature is increased but when cooling the tetragonal form from a high temperature the transition from the tetragonal to the monoclinic modification takes place between 850 and 1000°C.

Melting is an example of a first-order phase transition. There is latent heat and the molar volume is discontinuous at the melting point. One model of melting

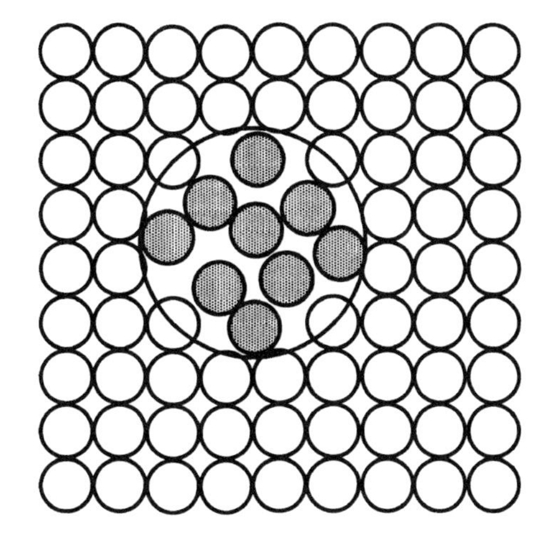

**Figure 10.5.** An aggregate of three vacancies in a schematic lattice. The cluster causes disorder in a domain that is much larger than the total of the separate volumes of the three neighboring vacancies themselves. Below the melting point such disordered domains are transient: if the vacancies separate again the lattice returns to its previous state. At high temperatures when the concentration of vacancy clusters is so high that they percolate the lattice, the solid melts. From A. L. G. Rees. Defect aggregation in solid state chemistry. In: E. A. V. Ebsworth, A. G. Maddox, and A. G. Sharpe (eds.). *New Pathways in Inorganic Chemistry*. With kind permission from Cambridge University Press.

indicates that the local disorder on vacancy clustering at high vacancy concentrations extends over a domain size several times the size of the cluster itself (Figure 10.5). The melting point can be seen as the percolation limit (Chapter 7) of disordering point defect agglomerations as follows. At increasing temperature when the melting point is approached, the vacancy concentration becomes high and defect agglomerates occur. The defects are mobile and the agglomerates have a fluctuating existence, being formed and after a while dissolving again. They disorder the lattice over a larger area than the actual size of the defect aggregate. Such disordered domains fluctuate, their number and average size increasing with temperature. If these disordered defect domains become connected, the domains percolate the lattice, which breaks down, and the solid melts. Below the melting point of a solid there is already considerable disorder in the lattice while slightly above the melting point there is still a lot of short-range order in the liquid.

Thermodynamic data are now sufficiently reliable to permit calculation of useful stability diagrams. These diagrams summarize equilibrium conditions (partial pressures and temperatures) under which particular solids are thermodynamically stable. Thermodynamic calculations can be useful as they permit estimation of the driving forces necessary for wanted processes as well as for unwanted side reactions or corrosion processes that may occur under the given conditions.

The thermodynamic stability of ceramic oxides at high temperatures can be read from an Ellingham diagram (see Figure 8.1), a graph of the Gibbs energy for the reaction $xM + O_2 \rightleftharpoons M_xO_2$ or some other metal oxide consistent with the valency of the metal M. It shows at a glance which metal oxides are stable in the presence of other metals or carbon. Most published Ellingham diagrams show (in the margin of the figure) which partial oxygen pressure is in equilibrium with a mixture of a metal and its oxide at a specified temperature. To read this pressure from the Ellingham diagram connect the point (0, 0) to the point corresponding to the value of $G(T)$ of the metal oxide and extrapolate to the margin to find the partial oxygen pressure in thermodynamic equilibrium with both metal and metal oxide at the chosen temperature. These pressures are calculated ones and in most cases have little chemical significance.

## 10.3. Defect Chemistry

In a crystal lattice there is translation symmetry but in a polycrystalline solid it exists only approximately within one grain. Similar to the outer surfaces of the crystallites, the grain boundaries are two-dimensional defects; the crystal lattice stops there. Dislocations are one-dimensional defects and pores are defects in solids having a dimension that is usually three but can be lower. Such higher-dimensional defects (Table 10.1) determine many properties: the dislocations in metals affect plasticity and the porosity, if open, determines gas permeability.

Point defects are zero-dimensional (Figure 10.6) and they are the only defects that are thermodynamically stable.[11–16] Line and plane defects are not thermodynamically stable and do not occur in equilibrium states. Point defects determine the extrinsic physical properties of solids such as electrical conductivity, work function, and color as well as the chemical properties such as diffusivity, stoichiometry, and sinter rate. Some examples of point defects are: (a) vacancies, where atoms or ions that should be on lattice sites are missing; (b) interstitials which are atoms or ions between the regular lattice sites of a solid; (c) foreign atoms or

**Table 10.1. Types of Defects in Solids**

| Defect | Description | Dimension |
|---|---|---|
| Substitutional dopants | Dissolved foreign atoms or ions replacing lattice atoms. If aliovalent they may introduce vacancies | 0 |
| Interstitial dopants | Dissolved atoms or ions in interstitial sites of the host lattice | 0 |
| Electronic charge carriers | Free or trapped electrons and holes in *n*-type and *p*-type compounds such as nonstoichiometric oxides | 0 |
| Antisite disorder | Exchange of two atoms from their regular site to that of the other in the lattice | 0 |
| Vacancy | Absence of an atom or ion from its regular site in the complete lattice | 0 |
| Color center | Electron or electron hole trapped in a vacancy | 0 |
| Vacancy clusters | Aggregation of vacancies; in a mobile lattice the onset of melting | 0–2.5 |
| Defect pairs | Schottky: pairs of anion and cation vacancies<br>Frenkel: pairs of a cation vacancy and the same cation as an interstitial<br>Anti-Frenkel: the same as Frenkel but for anions | 0 |
| Dislocations | Misfit of lines of atoms in a lattice | 1 |
| Shear planes | Misfit in lattice planes | 2 |
| Grain boundaries | End of a small piece of solid with one lattice orientation | 1.5–2.5 |
| Surfaces | Grain boundary between solid and vacuum | 1.5–2.5 |

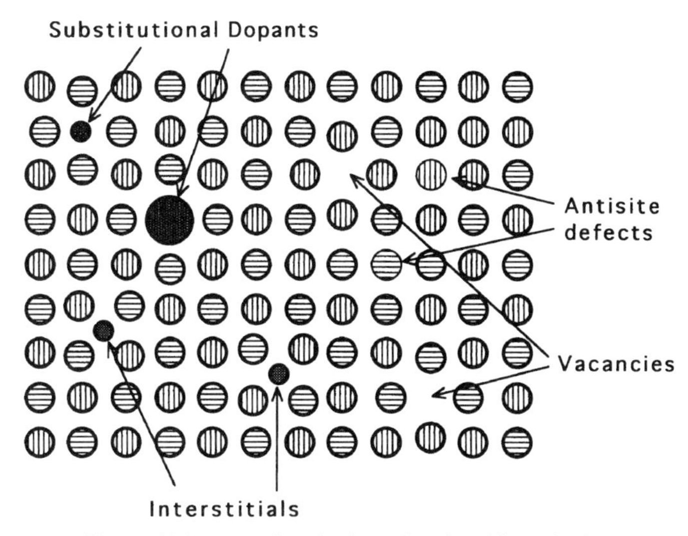


**Figure 10.6.** Examples of point defects in a binary lattice.

ions that have replaced regular lattice atoms or ions in a crystalline solid; and (d) extra electrons or electron holes in vacancies or on the lattice atoms.

Chemical reactions in and between solids during sintering are possible if there are point defects and if those defects are mobile. This is the only type of defect discussed in this section. Defect chemistry, which describes the way in which defect concentrations and reactivities can be controlled, is basic for the entire subject of solid state chemistry. However, only oxides are treated here because they are representative for all ionic compounds and technologically among the most important solids.

The reaction between defects is described with a characteristic chemical notation, the Kröger–Vink notation, which indicates defects as changes in the background of the perfect lattice, where every atom is in its place. The symbol for a defect species in a solid is $D_R^z$ in which $D$ gives the nature of the defect, $R$ is the sublattice that has the defect, and $z$ is the formal charge change in the background because of the presence of the defect. The "sublattice" $R$ can also mean interstitial space and is then the subscript $i$. Only changes in the background of the perfect lattice are defects in that lattice. Such a change may be a vacancy or a different atom (a dopant) that replaces an originally present lattice atom. Another change in the host lattice occurs when an atom has a valency other than it had initially in the host lattice. Interstitials or atoms that are not on lattice sites but between them are another example of changes in the original situation. Interstitial ions keep their own ionic charge as defect charge because the original empty interstitial has no charge of its own. The formal charge $z$ of the defect is indicated with a dot ($^\bullet$) for a positive charge and a prime ($'$) for a negative charge. A neutral defect is often given the superscript $x$. Examples of defects are:

- *Vacancies* (lattice sites that have no atoms or ions): $V$ (or perhaps $\nabla$ if the compound has vanadium to avoid confusion). In nickel oxide there are the vacancies $V_{Ni}''$ and $V_O^{\bullet\bullet}$. NaCl has vacancies $V_{Cl}^{\bullet}$ and $V_{Na}'$; AgBr has $V_{Ag}'$.

- *Interstitials* (atoms or ions between lattice sites): subscript *i*. In AgBr: $Ag_i^{\bullet}$. In zirconia ($ZrO_2$): $O_i''$.
- *Dopants* (foreign atoms or ions on lattice sites replacing the original atoms): (i) boron substituting silicon atoms in silicon ($B_{Si}$ or $B_{Si}'$); (ii) CaS dissolved in NaCl ($Ca_{Na}^{\bullet}$ and $S_{Cl}'$); (iii) $Y^{3+}$ ions in $ZrO_2$ ($Y_{Zr}'$); (iv) $Ti^{4+}$ in $CeO_2$ ($Ti_{Ce}^{x}$).
- Electrons or electron holes in the valence band or in the conduction band: $e'$ and $h^{\bullet}$. The site or sublattice of these defects is not indicated in their notation.

The defect concentrations are not independent of each other, and there are a few relationships among them:

1. Electroneutrality requirement: the number of formally positively charged defects equals the number of negative defects; each positive defect has to be compensated by a negative one.
2. Mass balance: chemical reactions do not create or annihilate matter but only displace it.
3. Site balance: the number of available lattice sites is not usually changed by a reaction.

The *intrinsic* defect concentration (at equilibrium) is calculated from the enthalpies and entropies of the defect formation reactions and the concentrations are strongly temperature-dependent. The *extrinsic* defect concentrations are strongly dependent on the quantity of dopant or impurity dissolved in the lattice and not by the temperature.

### *10.3.1. Thermal Disorder: Intrinsic Defect Concentrations*

The formation of lattice defects requires energy, yet their presence means a certain entropy. Therefore according to equilibrium thermodynamics at all temperatures above absolute zero there are a certain number of defects. The concentration of defects $[n]$ at equilibrium can be derived from the equilibrium constants and has the usual exponential temperature dependence:

$$n \propto \exp(\Delta H_f/RT)$$

in which $\Delta H_f$ is the formation enthalpy of a mole of defects. The equilibrium constants for the formation reaction of defects are exponentially dependent on the Gibbs free energy:

$$K = \exp(-\Delta G_f/RT)$$

in which $\Delta G_f$ is the change in free energy of the system during the formation.

Thermal disorder results in defects such as vacancies or interstitials. Such thermal defects are often formed in pairs as explained below. As long as there is no matter exchange with the surroundings of the system the stoichiometry of the compound does not change by heat-induced disorder.

There are different types of formation reactions and equilibria, depending on the type of lattice and the type of defect. The types of disorders are known as Schottky, Frenkel, and anti-Frenkel.

1. SCHOTTKY EQUILIBRIA IN CATION AND ANION SUBLATTICES: In NaCl the reaction is $0 \rightleftharpoons V'_{Na} + V^{\bullet}_{Cl}$; $K_S = [V'_{Na}][V^{\bullet}_{Cl}]$ and in NiO; $0 \rightleftharpoons V''_{Ni} + V^{\bullet\bullet}_{O}$; $K_S = [V''_{Ni}][V^{\bullet\bullet}_{O}]$. Schottky equilibrium leads to vacancy pairs and this type of disorder dominates in alkaline halides and in oxides with a halite (NaCl) structure.

2. FRENKEL DISORDER: a pair of defects involving only cations, with the pair consisting of a cation vacancy and an interstitial cation; one example of Frenkel-type disorder is found in AgBr, where the reaction is $0 \rightleftharpoons V'_{Ag} + Ag^{\bullet}_{i}$, and the Frenkel equilibrium constant $K_F = [V'_{Ag}][Ag^{\bullet}_{i}]$. This type of disorder prevails in silver halides and in cation conductors such as Na $\beta$-alumina and $Li_3N$. The cations are monovalent.

3. ANTI-FRENKEL DISORDER: similar to Frenkel disorder except that the interstitials are anions and vacancies are therefore in the anion sublattice. In $ZrO_2$ the reaction is $0 \rightleftharpoons V^{\bullet\bullet}_{O} + O'_{i}$ and the anti-Frenkel equilibrium constant is $K_{AF} = [V^{\bullet\bullet}_{O}][O''_{i}]$. This type of thermal defect is found in lattices that have a fluorite structure ($CaF_2$, $ZrO_2$), which means that there are many large interstitial sites where the anions can be accommodated, but not the cations because their charge is larger, and they are less well screened from each other.

4. ELECTRONIC DISORDER: thermal excitation of electrons from the fully filled valence band to the empty conduction band, which means that electronic conduction by electrons and holes becomes possible. The thermal equilibrium is $0 \rightleftharpoons e' + h^{\bullet}$ with $K_i = [e'][h^{\bullet}]$. The formation enthalpy of the electronic defect pairs is proportional to the bandgap and the higher the bandgap, the smaller the equilibrium constant. The charge carriers are on the atoms or ions in the orbital model.

The defect concentrations that are the result of thermal disorder are small in most oxides. The formation enthalpy of vacancy pairs in MgO is 7 eV, which gives a vacancy concentration of $10^{-6}$ ppm at 1000°C. In most oxides the bandgap is also large (>4 eV) and at 1000°C the charge carrier concentration is lower than $10^{-4}$ ppm. Now, oxides can be made with an impurity concentration of at best 10–100 ppm. The concentration of impurities contributes much more to the defect concentration than the thermal disorder at these low formation equilibrium constants and the thermal (intrinsic) contribution to the defect concentration can usually be disregarded.

### 10.3.2. Doping: Extrinsic Defect Concentrations

In solid solutions or alloys the atoms on the lattice sites are replaced by the atoms of the dissolved species, the solute. Apart from these substitutional solutions, interstitial solutions can also occur if the atoms or ions of the solute are small and can be accommodated in the interstices of the host lattice. The solubilities vary. The phase diagrams of spinel, lithium aluminum silicate (LAS), Ta/C, and Ti/N show to what extent dissolution is possible. Apart from having the dopants in the lattice, these alloys often have other defects that are the result of a difference in charge between the ions being replaced in the lattice and those replacing them.

Doping or replacing atoms in a host crystal by foreign atoms with another valence, e.g., doping magnesium oxide by replacing $Mg^{2+}$ with $Li^{+}$ or with $Sc^{3+}$, is called aliovalent substitution. The dissolution reaction of lithium oxide in magnesium oxide is

$$Li_2O \rightarrow 2Li'_{Mg} + O_O^x + V_O^{\bullet\bullet}$$

The $Li'_{Mg}$ defects are charged and need positive counterions to maintain electroneutrality. This charge compensation is provided by the positive oxygen vacancies.

If scandium oxide is dissolved in MgO, $Sc^{3+}$ replaces $Mg^{2+}$ and the defect $Sc^{\bullet}_{Mg}$ needs a negative countercharge, which is a metal vacancy in this case. The dissolution reaction is

$$Sc_2O_3 \rightarrow 2Sc^{\bullet}_{Mg} + 3O_O^x + V''_{Mg}$$

Hence, doping scandium in magnesium oxide increases the magnesium vacancies.

Doping with aliovalent species affects the concentrations of the defects that are formed thermally in the intrinsic equilibrium. An increase in the concentration of one defect in a pair implies a decrease in the concentration of the other defect. The Schottky or Frenkel equilibrium constants $K_S$ or $K_F$, the products of the defect concentrations, are constants that depend only on temperature. This means that if one defect is added by doping, its partner in the thermal equilibrium disappears.

### *10.3.3. Nonstoichiometry: Gas Equilibria*

If one of the components of a binary solid is involved in exchange with the gas phase the stoichiometry depends on the partial vapor pressure of that component. Nonstoichiometric compounds can be recognized by their existence region in the $T$–$x$ diagram. If the vertical line that corresponds to the stoichiometric composition of that compound lies fully in the existence region, the compound can exist with an excess of either of the components. Sometimes the stoichiometric line in the phase diagram is not inside but slightly outside the existence region of a compound. FeO is an example: it does not exist in a 1:1 stoichiometry but has an excess of oxygen. The composition of the equilibrium product depends on the partial pressures and temperature during formation as the phase diagram prescribes.

In general oxides have stoichiometries that are not exactly integer owing to two oxygen exchange reactions between the solid and the gas phase. These reactions determine their stoichiometry and the possible defects that can occur.

The first reaction of a fictitious oxide MO is taking up of gaseous oxygen from the gas phase:

$$\tfrac{1}{2}O_2(g) \rightarrow O_O^x + 2h^{\bullet} + V''_M \qquad \text{with } K_a = [h^{\bullet}]^2[V''_M].p(O_2)^{-1/2}$$

The oxygen atom needs two electrons to form the oxide ion in the lattice. The two electrons are taken from the solid and leave two positive electron holes behind labeled $h^{\bullet}$. The equilibrium constant corresponding to oxygen absorption is $K_a$. The expression for the equilibrium constant shows that increasing the oxygen pressure makes the oxide more $p$-type conducting and oxygen-rich or metal-deficient. Its composition becomes $MO_{1+x}$ with $x > 0$.

In some oxides the second reaction, depletion of oxygen from the lattice, is easier, which means that the oxygen depletion reaction constant $K_d$ is much larger than the absorption constant $K_a$. This depletion rection is

$$O_O^x \rightarrow \tfrac{1}{2}O_2(g) + 2e' + V_O^{\bullet\bullet} \qquad \text{with } K_d = [e']^2[V_O^{\bullet\bullet}].p(O_2)^{1/2}$$

Oxides in which the second reaction prevails over the first or, in other words, oxides that for energy reasons prefer to give up oxygen and can more easily accommodate electrons than holes in their lattice are oxygen-deficient, *n*-type semiconductors, and they become less conducting and less oxygen deficient with increasing oxygen pressure, always assuming that there is equilibrium.

Oxides that have a crystal structure that prefers interstitial formation have Frenkel and anti-Frenkel disorder and obey the following defect formation equations.

INCORPORATION OF OXYGEN:

$$\tfrac{1}{2}O_2(g) \rightarrow O_i'' + 2h^{\bullet} \qquad \text{with } K_a = [h^{\bullet}]^2[O_i''].p(O_2)^{-1/2}$$

and an oxygen-rich *p*-type conducting oxide forms from an oxide MO with anti-Frenkel disorder.

DEPLETION OF OXYGEN:

$$M_M^x + O_O^x \rightarrow \tfrac{1}{2}O_2(g) + 2e' + M_i^{\bullet\bullet} \qquad \text{with } K_d = [e']^2[M_i^{\bullet\bullet}].p(O_2)^{1/2}$$

which means that an oxide with Frenkel disorder would form a metal-rich *n*-type semiconductor.

All these defect concentrations depend on each other through the equilibrium constants. If an oxide has Schottky-type thermal disorder the vacancy concentrations are inversely proportional to each other through $K_S$. The electron and electron hole concentrations are interrelated through the formation equilibrium constant $K_i$.

From the constants $K_a$, $K_d$, $K_i$, and $K_s$ or $K_F$ and use of the electroneutrality requirement, the dependence of the defect concentration on the oxygen partial pressure can be derived by simple elimination. Figure 10.7 shows his dependence for a fictitious pure oxide MO with Schottky disorder for two cases: $K_S \gg K_i$ and $K_s \ll K_i$.

In Figure 10.7 the center of the graph is at the oxygen partial pressure that is in equilibrium with the stoichiometric oxide $M_1O_1$. The value of the equilibrium constants $K_a$ and $K_d$ depends on the reaction heat for oxidation or reduction of the oxide. $K_i$ depends exponentially on the bandgap. This figure shows how oxygen pressures affect the electic behavior of the oxide:

- If $K_i \ll K_S$ or, in other words, if the oxide has a wide bandgap, the oxide is a pure ionic conductor under an oxygen pressure at the middle region near the stoichiometric composition. Electron or hole concentrations in the middle region are very low but to the left or to the right of this area the oxide is a *p*-type or an *n*-type semiconductor because there is an excess of holes or electrons as defects.
- If $K_i \gg K_S$, which holds for oxides that have narrow bandgaps, the oxide is a semiconductor at all oxygen pressures. At low oxygen pressures (in the left part) it is *n*-type, at high oxygen pressure it is *p*-type, and in the middle it is an intrinsic semiconductor.

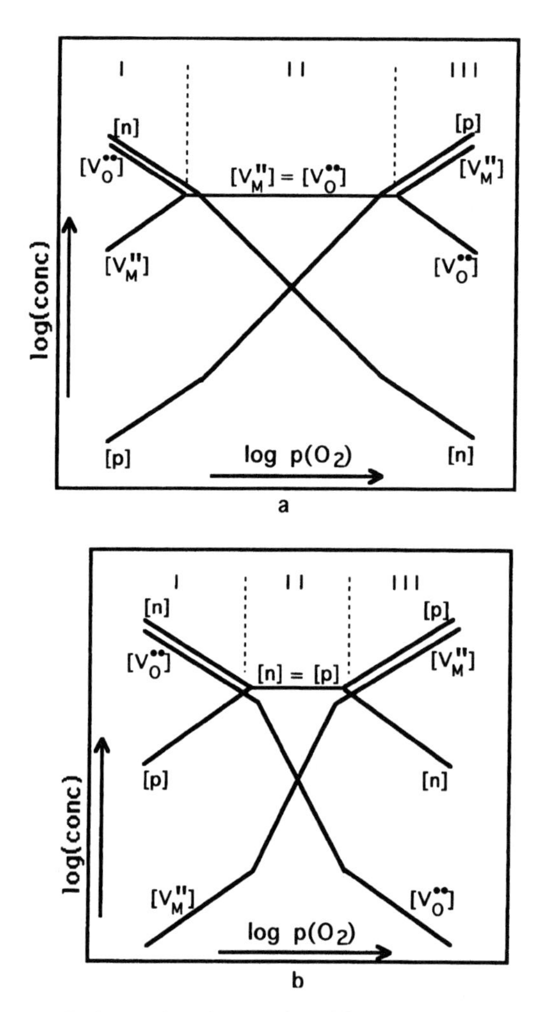


**Figure 10.7.** Calculated equilibrium defect diagrams for a binary oxide MO with Schottky defect pairs. In case (a) the equilibrium constant for vacancies is taken to be much larger than for electronic disorder ($K_S = 10^6 K_i$); case (b) gives the concentrations if the Schottky disorder is the smaller. The defect concentrations in regions I and III have a power dependence on the oxygen pressure with the exponent $\frac{1}{6}$. Region II in case (a), which includes the electrolytic domain has an exponent of the defect lines of $\frac{1}{4}$; in case (b) the exponent is $\frac{1}{2}$.

Transition metal oxides in which the metal is in a low stable valence state but can possibly reach a higher valence are, e.g., MnO, FeO, CoO, and NiO. These oxides tend to be oxygen-rich and have the composition $MO_{1+x}(x > 0)$. The maximum that $x$ can have decreases in this series because it is increasingly difficult to oxidize the two-valent metal to a three-valent state. The third ionization potential increases in this series from left to right. Such oxides are usually *p*-type semiconductors.

For oxides of metals in their highest valence state such as $Ta_2O_5$, $CeO_2$, and ZnO it is easier to get rid of some oxygen than to take up oxygen from the gas phase.

The metals cannot be oxidized more than they already are in these oxides but they can have lower valencies. The lattice can accommodate extra electrons better than electron holes, as is seen in the value of $K_a$ and $K_d$. Such oxides tend to be $n$-type semiconductors. Table 4.7 lists examples of both types.

The nonstoichiometry of an oxide strongly depends on the presence of aliovalent impurities and dopants affect the number of thermal defects in nonstoichiometric oxides and their electrical conductivity because the solutes have a valence other than the atoms they replace. This is illustrated in the case of lithium and chromium doping of nickel and zinc oxides in equilibrium with gaseous oxygen.

The defect formation reaction on dissolving $Li_2O$ in NiO is

$$Li_2O \rightarrow 2Li'_{Ni} + O^x_O + V^{\bullet\bullet}_O$$

Some excess oxygen can now be taken up by the oxygen vacancies:

$$\tfrac{1}{2}O_2(g) + V^{\bullet\bullet}_O \rightarrow O^x_O + 2h^\bullet$$

Combining these two equations yields

$$Li_2O + \tfrac{1}{2}O_2(g) \rightarrow 2Li'_{Ni} + 2O^x_O + 2h^\bullet$$

which shows that lithium doping increases the $p$-type conductivity of the nonstoichiometric NiO and lithium helps in further oxidation of NiO. The oxygen pressure and the values of the equilibrium constants determine to what extent the two contributing reactions occur: the compensation for the negative charge of $Li'_{Ni}$ can be half an oxygen vacancy or an electron hole. The fact that lithium-doped nickel oxide is strongly oxidized follows from the chlorine developed through its reaction with an aqueous solution of chlorides.

The opposite happens when nickel oxide is doped with a trivalent metal ion such as chromium. The defect formation reaction on dissolving $Cr_2O_3$ in NiO is

$$Cr_2O_3 \rightarrow 2Cr^{+}_{Ni} + 3O^x_O + V''_{Ni}$$

The excess oxygen cannot be dissolved in the lattice and there is a tendency to shed some of it. The dissolution reaction becomes

$$Cr_2O_3 \rightarrow 2Cr^{+}_{Ni} + 2O^x_O + 2e' + \tfrac{1}{2}O_2(g)$$

On doping with chromium, nickel oxide that is initially $p$-type becomes less $p$-type conducting. Whether this permits making NiO $n$-type in this way depends on the equilibrium constants. The electronic conductivity also depends on the capture reactions of the charge carriers by the localized stationary defects.

The two examples just described illustrate the way in which $p$-type oxides react to aliovalent doping. The effect of doping an $n$-type oxide is shown with the case of zinc oxide. $Li_2O$ dissolves in ZnO as

$$Li_2O \rightarrow 2Li'_{Zn} + O^x_O + V^{\bullet\bullet}_O$$

As in the case of NiO some oxygen can now be taken up in the vacancies as an oxide ion and the dissolution equation of $Li_2O$ in ZnO becomes

$$Li_2O + \tfrac{1}{2}O_2(g) + 2e' \rightarrow 2Li'_{Zn} + 2O_O^x$$

Lithium doping oxidizes ZnO and lowers its $n$-type conductivity by consuming the surplus electrons in the conduction band.

Doping ZnO with $Cr_2O_3$ that has too many oxide ions for ZnO evolves gaseous oxygen that leaves electrons behind:

$$Cr_2O_3 \rightarrow 2Cr^{\bullet}_{Zn} + 2O_O^x + 2e' + \tfrac{1}{2}O_2(g)$$

The dissolved chromium oxide enhances the $n$-type character of ZnO.

### 10.3.4. Defect Reactions

Many defects are charged, they diffuse through the lattice, if the temperature is high enough, and they can associate with each other to form defect clusters or color centers. Such associates strongly affect the properties of the solid and materials engineers control the defect concentrations in solids to obtain the wanted properties. The following are examples of reactions between defects.

IONIZATION REACTIONS AND TRAPPING OF CHARGES IN COLOR CENTERS:

$$V_O^{\bullet\bullet} + e' \rightarrow V_O^{\bullet} \qquad V_O^{\bullet} + e' \rightarrow V_O^x$$

$$Cr^{\bullet}_{Ni} + e' \rightarrow Cr^x_{Ni} \qquad Zn_i^{\bullet\bullet} + e' \rightarrow Zn_i^{\bullet}$$

The strength of the bond between the stationary atomic defect and the mobile electron is also called the depth of the electron trap. If a certain type of charge carrier, electron or electron hole, has sufficiently deep traps in a lattice, the electronic conductivity can be lower than expected because the mobile charge carriers are bound. If the above reactions are reversed the traps are ionized. Trapping and ionization reactions create equilibria.

CLUSTER FORMATION: $ZrO_2$ with CaO dissolved in it has two charged defects, $V_O^{\bullet\bullet}$ and $Ca''_{Zr}$. At high temperatures these diffuse and may associate on collision:

$$V_O^{\bullet\bullet} + Ca''_{Zr} \rightleftharpoons \{V_O Ca_{Zr}\}^x$$

This association may be the reason for an ionic conductivity that decreases on increasing the defect concentrations. Defect associates can be nuclei for a phase transition or a solid state reaction.

### 10.3.5. Applications

There are three ways to control the defect concentrations in solids.

1. The temperature exponentially determines the product of the defect pairs in the thermal disorder reactions, the Schottky or Frenkel equilibria. In pure compounds the concentrations of the two defects are equal.
2. Doping with aliovalent species makes the defect concentrations in the thermal disorder pair unequal.

3. A heat treatment in a gas with the proper partial pressure determines the stoichiometry of the solid and the concentrations of the defects involved in gas exchange.

The following applications show how defect chemistry is used in ionic materials.

### *Solid Oxide Fuel Cells*

Fuel cells are galvanic cells that have gases as electrodes rather than solids. These are fuel gas such as hydrogen and oxygen or air in separate electrodes. The electrodes are porous conductors that harvest and transfer the electrons. Like batteries, fuel cells convert the chemical energy from an oxidation reaction directly to electrical energy. As in conventional batteries, in a fuel cell there is ionic transport between the electrodes. The electrolyte must be impermeable for both of the gases. There are different types of fuel cells, depending on the electrolyte phase used. Liquid electrolytes can be phosphoric acid (in the PAFC) at room temperature or a mixture of molten carbonates (in the MCFC) at higher temperatures. A solid oxide fuel cell (SOFC) has a solid ion-conducting oxide as an electrolyte and operates at still higher temperatures. The advantage over liquid electrolyte cells is that SOFCs have fewer corrosion problems.

The operation of a fuel cell is shown schematically in Figure 10.8: the electrodes for oxygen and fuel gas are porous layers of electron-conducting mixed oxides or cermets on either side of a thin but gas-tight oxygen ion conductor. The solid

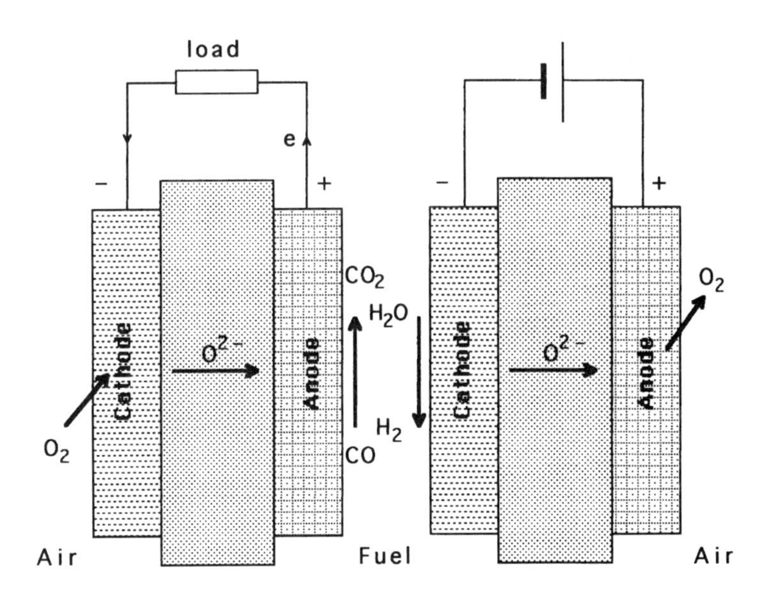


**Figure 10.8.** Schematic of a fuel cell for power generation (left) and its opposite, a reactor component for gas electrolysis (right). All components are solid: the electrolyte is a nonporous, gas-impermeable oxide ion conductor (stabilized zirconia), and the electrodes are porous electron conductors (oxides and cermets).

**Table 10.2. Some Ion Conductors**

| Compound | Ion ($T$, °C) | Conductivity $(\Omega m)^{-1}$ |
|---|---|---|
| $ZrO_2$ (12% CaO) | $O^{2-}$ (1000) | 0.8 |
| $CeO_2$ (12% CaO) | $O^{2-}$ (700) | 4 |
| $Li_{0.5}Zr_{1.5}Ta_{0.5}P_3O_{12}$ | $Li^+$ (200) | 0.1 |
| $Na_3Zr_2PSi_2O_{12}$ | $Na^+$ (300) | 20 |
| $NaAl_{11}O_{17}$ | $Na^+$ (300) | 35 |
| $K_{1.4}Fe_{11}O_{17}$ | $K^+$ (300) | 2 |

electrolytes are functional ceramics, which should easily conduct ions for a high rate of energy conversion. Therefore they must be thin, but in spite of being thin they should not conduct electrons or electron holes so as not to short-circuit the electrodes internally. Moreover they have to keep the gases separate and should not leak. This, of course, presents a nice materials design problem, the solution of which will involve a lot of chemical ingenuity. Table 10.2 lists several solid ion conductors that are used in energy technology. At the operating temperature and partial oxygen pressures in the two compartments the electron and electron hole concentrations in the solid oxide electrolyte must remain very low. $ThO_2$, $HfO_2$, and especially $Ca^{2+}$ or $Y^{3+}$ stabilized zirconia are suitable oxide ion conductors.

At the cathode, the oxygen electrode, there is a high oxygen pressure and oxygen is reduced by electrons from the cathode conductor to oxide ions, which are then taken up by the lattice of the electrolyte and diffuse to the other electrode, the anode, where they are used to burn the fuel. The oxide ion concentration gradient over the thickness of the electrolyte drives the oxide ions from the cathode to the anode. At the anode the oxide ions oxidize the fuel and leave the electrons on the cathode conductor. The potential over the cell (between the cell electrodes) is proportional to the difference between the oxygen potentials on the two electrodes:

$$\Delta G = \mu_O(\text{anode}) - \mu_O(\text{cathode}) = -2FE$$

in which $F$ is the Faraday number and $E$ the cell voltage.

The power that a fuel cell can produce is largely determined by the rate at which the oxygen ions can diffuse through the thin electrolyte layer. In addition to the defect concentration, the ion conductivity also determines the suitability of an oxide for use as an electrolyte in a fuel cell. In the case of an oxide that has a large bandgap and thus a small $K_i$, the stoichiometric oxide has a much higher concentration of thermally generated ionic defects such as vacancies or interstitials than electronic defects such as free electrons or holes (Figure 10.7a). In the middle area the solid is a purely ionic conductor. On both sides of this area in the $[n]/p(O_2)$ plot the electron and electron hole concentrations are comparable with that of the ionic defects and at low or high oxygen pressures the oxide has free electrons or electron holes and is a mixed conductor, i.e., it conducts ions as well as electrons. The mobility of the electrons or electron holes is usually higher than that of the ion defects and a mixed conductor at equal concentrations of electronic and ionic defects has a higher electronic conductivity than an ionic conductivity. Therefore near stoichiometry at an oxygen pressure at which the oxide conducts only ions and not electrons (or

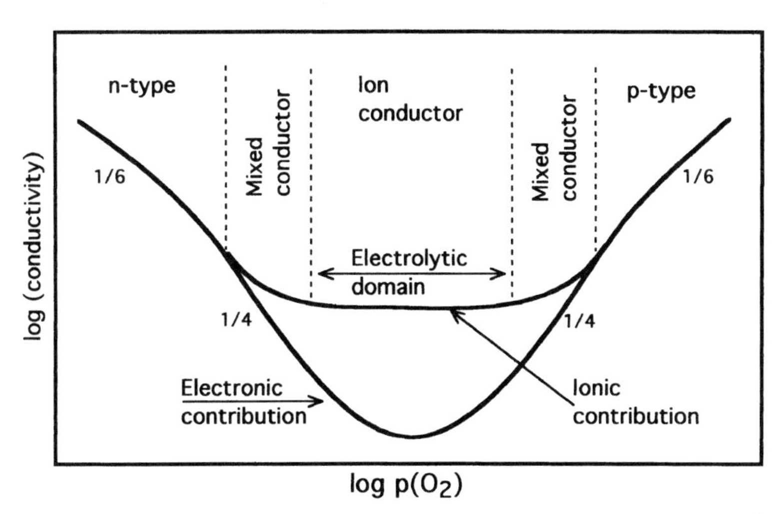


**Figure 10.9.** Schematic of the total electric conductivity at different oxygen pressures of an oxidic electrolyte [like case (a) in Figure 10.7]. At extremes in oxygen pressure the compound is an *n*-type or *p*-type semiconductor because the mobility of the electronic charge carriers is much higher than that of the ionic charges. When the concentrations of the electronic charge carriers drop below the ionic defect concentrations the compound becomes a mixed conductor. In the electrolytic domain there is no contribution of electrons to the conductivity. From Ø. Johannesen and P. Kofstad. Electrical conductivity in binary metal oxides. Part 2. *J. Mater. Educ.* **7**, 969 (1985) with permission from the *Journal of Materials Education.*

holes) is the electrolytic domain. This domain is somewhat narrower than the regime in which the ionic defect concentrations are larger than those of the electrons or holes (Figure 10.9). The electrolytic domain of zirconia is very wide and it remains an ion conductor at both electrodes even at very different oxygen pressures.

### *Sensors*

As described in Section 9.4 a Nernst-type sensor for measuring the gas concentration of oxygen or hydrogen is basically a small fuel cell. It is a small SOFC that has a reference electrode with a constant oxygen or hydrogen concentration and therefore a constant potential. The other electrode is the measuring electrode and the cell voltage is the result of a difference in gas concentration between the two electrodes. Chemical sensors of the Nernst type usually have a rather slow response (of the order of minutes) even at elevated temperatures. A new variant of this type of gas sensor is the kinetic sensor, which has its electrodes (made of different metals) exposed to the same mixture of oxidizing and reducing gases. It generates a potential that depends on the difference in catalytic activity of the two metal electrodes and on the composition of the gas. The signal from the sensor is the result of the difference between the surface reaction rates at the two electrodes.[17]

### *Image Formation in Photographic Film Emulsions*

Simply put, a photographic film consists of a gelatine layer in which a light-sensitive colloid of a silver halide (particle size $<1\ \mu m$) is dispersed. Incident light forms catalytically active nuclei in each photosensitive crystallite in the film and the particles with such nuclei form a latent image. An active nucleus in a silver halide grain allows the developing process to blacken the entire crystallite. These reduced black grains turn the negative film black at the places that have received a sufficient amount of light. A silver halide particle needs a minimum amount of light to form such a critical nucleus. Crystallites that have not absorbed sufficient light and therefore cannot be reduced to silver grains in the development process do not turn black but are removed from the film in the fixating process. Too much light may darken the film visibly even without developing, which is called printout. The formation of the critical nucleus is described by the defect chemistry in the silver halide as being due to the action of light.[18,19]

Silver bromide is a semiconductor. Absorbed light excites electrons in the conduction band and electron holes in the valence band. The holes oxidize bromide ions to traces of bromine that are dissolved in the gelatine layer. The free electrons reduce silver ions to silver atoms that form the catalytically active clusters in the halide crystallites. The reactions are:

$$h\nu \rightarrow e^- + h^+; \qquad 2Br^- + 2h^+ \rightarrow Br_2; \qquad Ag^+ + e^- \rightarrow Ag; \qquad nAg \rightarrow Ag_n$$

The shape of the AgBr crystallites affects the rate of formation of the latent image. When the colloid is made, the conversion reaction

$$AgNO_3 + NaBr \rightarrow AgBr + NaNO_3$$

yields submicron cubes bounded by $\{100\}$ faces if the reactants in the solution are mixed in stoichiometric amounts. A slight excess of silver nitrate in the mixture yields tetrahedral AgBr particles bounded by $\{111\}$ faces. Because charge compensation differs in the different planes, the photochemical properties depend on the particle shape.

Figure 10.10 gives a qualitative indication of how the blackening $D$ depends on the illumination $E$ (the amount of light is the product of the light flux $I_e$ and the

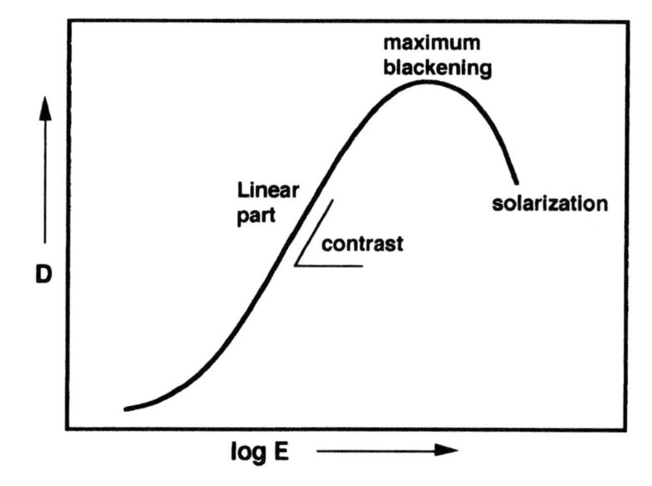


**Figure 10.10.** Dependence of photographic blackening $D$ on illumination intensity $E$. Film is normally used in the linear region. A curve shifted to the left means higher sensitivity and a film with a steeper curve has more contrast.

exposure time). The definitions are:

$$D = {}^{10}\log(I_0/I) \qquad \text{and} \qquad E = I_e t$$

in which $I_0$ is the intensity of the incident light, $I$ the intensity of the light that the developed film transmits, and $I_e$ the intensity of the incident light during exposure of the film.

The curve shows four domains:

1. The threshold or background haze: any blackening here is the result of reduction of nonexposed grains during development.
2. The linear part in which the film is used. The steepness of the curve is the gradation (hard films with strong contrast have steep curves). The slope is related to the grain size.
3. The shoulder in which the blackening on illumination is retarded because of recombination of excess free silver and bromine to AgBr.
4. The solarization domain, where the active nuclei that were formed previously are broken down by strong recombination in excess light; in this domain blackening decreases with increasing illumination.

In silver bromide, Frenkel-type defects predominate:

$$\mathrm{Ag}_{\mathrm{Ag}}^{x} + V_i^{x} \rightleftarrows \mathrm{Ag}_i^{\bullet} + V_{\mathrm{Ag}}' \qquad \text{with } K_F = 30\exp(-1.06\,\mathrm{eV}/kT)\,\mathrm{cm}^{-6}$$

At room temperature $[\mathrm{Ag}_i^{\bullet}]$ would be equal to $\sqrt{K_F}$ or $8.10^{14}\,\mathrm{cm}^{-3}$ in stoichiometric AgBr. The concentration of impurities (0.1–1 ppm) is much higher than the thermally generated point defect concentration and this means that the interstitial silver ion concentration ($\approx 10^{17}\,\mathrm{cm}^{-3}$) is much higher than the intrinsic value. To control the defect concentration in AgBr suitable dopands are introduced:

$$\mathrm{Ag_2S} \rightarrow \mathrm{Ag}_{\mathrm{Ag}}^{x} + \mathrm{Ag}_i^{\bullet} + S_{\mathrm{Br}}'; \ \mathrm{CdBr_2} \rightarrow \mathrm{Cd}_{\mathrm{Ag}}^{\bullet} + V_{\mathrm{Ag}}' + 2\mathrm{Br}_{\mathrm{Br}}^{x}; \ \mathrm{CuCl} \rightarrow \mathrm{Cu}_{\mathrm{Ag}}^{x} + \mathrm{Cl}_{\mathrm{Br}}^{x}$$

Hence aliovalent impurity concentrations affect the concentrations of interstitials and vacancies. Dopands can also act as sensitizers (they improve the formation of the latent image) or as efficient traps for defects: $\mathrm{Pb}_{\mathrm{Ag}}^{\bullet}$ for $e'$, $\mathrm{Cu}_{\mathrm{Ag}}^{x}$ for $h^{\bullet}$, and $\mathrm{Cu}_{\mathrm{Ag}}^{\bullet}$ for $V_{\mathrm{Ag}}'$.

The mobility of the interstitials is much higher than that of the vacancies at room temperature:

$$\mu(\mathrm{Ag}_i^{\bullet}) = \frac{34.8}{T}\exp\frac{-0.15\,\mathrm{eV}}{kT}\,\mathrm{cm^2/Vs}$$

$$\mu(V_{\mathrm{Ag}}') = \frac{620}{T}\exp\frac{-0.34\,\mathrm{eV}}{kT}\,\mathrm{cm^2/Vs}$$

The interstitial silver ions diffuse according to the so-called interstitialcy mechanism

$$\mathrm{Ag}_i^{\bullet}(1) + \mathrm{Ag}_{\mathrm{Ag}}^{x}(2) \rightarrow \mathrm{Ag}_{\mathrm{Ag}}^{x}(1) + \mathrm{Ag}_i^{\bullet}(2)$$

Similarly, the mobility of the photoelectrons in AgBr is considerably higher than the mobility of the holes:

$$\mu(e') = 49.5\,\mathrm{cm^2/Vs} \qquad \mu(h^{\bullet}) = 1\,\mathrm{cm^2/Vs}$$

The photogenerated electrons and holes are trapped by positive and negative lattice defects:

$$Ag_i^{\bullet} + e' \leftrightarrows Ag_i^x \quad \text{and} \quad V'_{Ag} + h^{\bullet} \leftrightarrows (V_{Ag}h)^x$$

A latent image is the result of a stepwise formation of nuclei that later may activate the reduction of the entire grain by the developing solution (a reducer). The photogenerated electron holes are rapidly trapped by the available vacancies. The photoelectrons, on the other hand, are not trapped by the silver interstitials in the interior of the silver bromide grain but only by the interstitials in suitable places on the surface. The activating silver metal nuclei are formed at the surface.

The surface of a AgBr crystallite is incomplete and negatively charged. On the (111)-plane, only three-sevenths of the stoichiometrically needed number of silver ions is present (in a hexagonal arrangement), which means a net surface charge density of $-4.6\,10^{13}\,C/cm^2$. The silver interstitials prefer the interior of the lattice and are present in a slight excess there. They form a positive space charge that compensates the negative surface charge.

The critical activating nucleus forms as follows: A lattice defect at the surface (e.g., a dislocation, a step, or a kink) can first trap an electron, then one or two interstitial silver ions, then one or two more electrons, and so on, reacting with electrons and ionic defects alternately:

$$(\text{Surface trap})^x + e' \rightarrow e'_s; \quad e'_s + Ag_i^{\bullet} \rightarrow Ag_s^x; \quad Ag_s^x + e' \rightarrow Ag'_s; \quad Ag'_s + Ag_i^{\bullet} \rightarrow Ag_{s2}^x;$$
$$Ag_{s2}^x + e' \rightarrow Ag'_{s2} \quad \text{or} \quad Ag_{s2}^x + Ag_i^{\bullet} \rightarrow Ag_{s3}^{\bullet} \quad \text{etc.}$$

The presence of one $Ag_{s3}^x$ cluster is sufficient to cause the grain to be fully reduced to metallic silver in the developing stage.

The electron holes that are initially trapped in silver vacancies move around and are finally removed as bromine on the surface:

$$(V_{Ag}h)^x + Br_{Br}^x(s) \rightarrow \tfrac{1}{2}Br_2(l)$$

The gelatine traps the bromine and prevents it from reoxidizing the silver nuclei on the AgBr crystallites.

## 10.4. Diffusion in Solids

Chemical reactions in a solid differ basically from reactions in a liquid solution because solids have atoms in lattices that interreact rather than molecules in solution and thus diffusion rates are much lower. Species can be dissolved in solids as they can in liquid solvents but a solid cannot be stirred and rates of reactions between two spatially separated species in solids are always limited by diffusion. Control of solid state reactions means control of the factors that determine diffusion. Measuring diffusion rates is central for solid state chemistry.

Diffusion in a solid is possible only if the lattice is not perfect but has a certain degree of disorder and contains vacancies and interstitials.[20] The main defects that help diffusion were listed in the previous section. Nonstoichiometric compounds (see Table 10.3 for examples) have high concentrations of defects such as ion vacancies and dopants and these concentrations strongly affect diffusion rates. Figure 10.11

**Table 10.3. Some Approximate Existence Boundaries of Nonstoichiometric Compounds**[a]

| Compound | Composition range of $x$: Lower boundary | Composition range of $x$: Upper boundary | Compound | Composition range of $x$: Lower boundary | Composition range of $x$: Upper boundary |
|---|---|---|---|---|---|
| $TiO_x$ | 0.65 | 1.25 | $In_xWO_3$ | 0.20 | 0.33 |
| | 1.998 | 2.000 | $TiS_x$ | 0.971 | 1.064 |
| $VO_x$ | 0.8 | 1.3 | | 1.112 | 1.205 |
| $CeO_x$ | 1.50 | 1.52 | | 1.282 | 1.300 |
| | 1.805 | 1.812 | | 1.370 | 1.587 |
| $ZrO_x$ | 1.700 | 2.004 | | 1.818 | 1.923 |
| $Fe_xO$ | 0.83 | 0.96 | $Nb_xS$ | 0.92 | 1.00 |
| $Co_xO$ | 0.988 | 1.000 | $Ta_xS_2$ | 1.0 | 1.35 |
| $Ni_xO$ | 0.999 | 1.000 | $Ni_xSe$ | 0.77 | 0.81 |
| $Li_xV_2O_5$ | 0.13 | 0.31 | $Ti_xC$ | 0.47 | 0.95 |
| $Li_xV_3O_8$ | 1.13 | 1.33 | $Ta_xC$ | 0.7 | 1.0 |
| $Li_xWO_3$ | 0 | 0.5 | $Ti_xN$ | 0.6 | 1.0 |
| $Ca_xWO_3$ | 0 | 0.12 | $Nb_xN$ | 0.75 | 1.06 |

[a]Existence ranges are temperature dependent and more precise values at different temperatures can be read from phase diagrams.

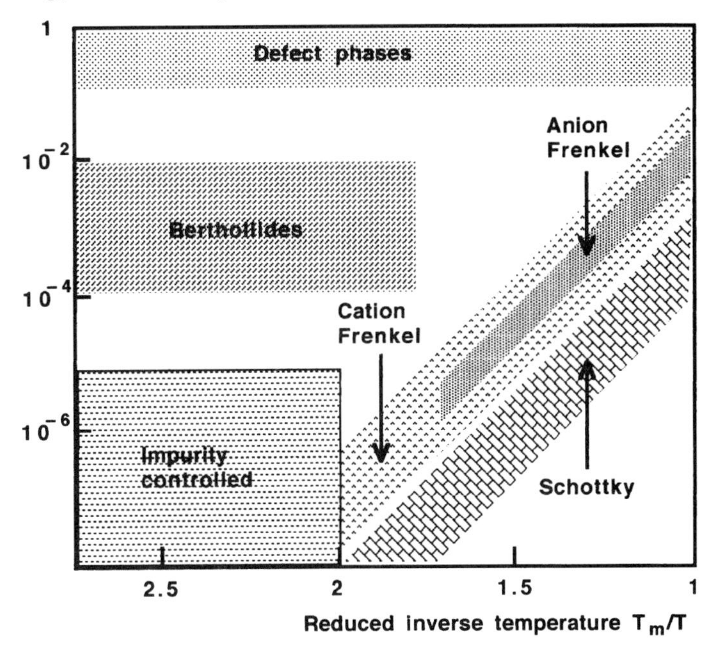


**Figure 10.11.** Fractional concentrations of defects. Thermal defect concentrations increase with temperature ($T_m$ is the melting temperature).

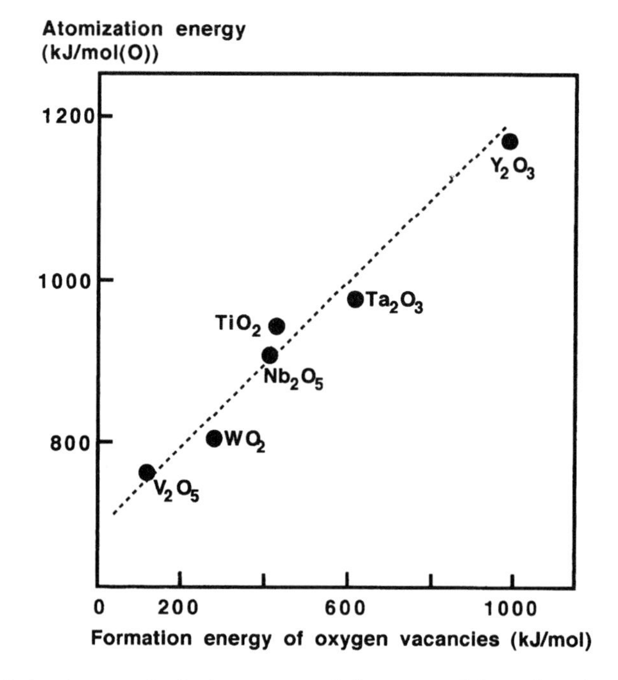


**Figure 10.12.** Relation between the lattice energy and the energy of formation of oxygen vacancies in some oxides.

gives orders of magnitude and shows how defect concentrations depend on the temperature. Figure 10.12 shows for a few $n$-type oxides that the energy of formation of thermally generated defects strongly correlates with lattice energy.

Diffusion in solids is possible if defects are present (and, again, if they have enough mobility). Direct or cyclic exchange of atoms in a complete lattice is of little significance because the activation energy for that is too high. The most important mechanisms are diffusion of vacancies, which means that neighboring atoms jump into them, and diffusion of atoms through interstitials if the lattice permits it.

Two of the laws of diffusion in solids are Fick's laws:

| Fick 1 | Fick 2 |
|---|---|
| $J = -\nabla c$ | $\dfrac{\partial c}{\partial t} = \nabla(D\nabla c)$ |

in which $J$ is the matter flux, $c$ the concentration, $\nabla$ the gradient operator, and $D$ the diffusion coefficient, which is temperature-dependent:

$$D = D_0 \exp(-Q/RT)$$

There is some correlation between the parameters $D_0$ and $Q$. The Meyer–Neldel rule says that $\ln D_0 = a + bQ$, with $a$ and $b$ positive numbers. $D_0$ compensates somewhat for the influence of $Q$. The higher the activation energy $Q$, the higher the preexponential factor $D_0$.

The preexponential factor $D_0$ depends on the crystal structure, the jumping distance, and the Debye frequency. The dependence will be shown for two simplified cases, without and with a driving force for ion transport.

Movement of an atom or ion from one position to another is a thermally activated process. The probability $p_0$ per mole that one atom will leave its position to jump to a neighboring available position is

$$p_0 = f \exp(-G_m/RT)$$

in which $f$ is a characteristic vibration frequency, for which the Debye frequency is often chosen; $G_m$ is the height of the barrier between two atomic positions.

If $n$ is the number of atoms per unit of surface and $\lambda$ the distance between lattice planes, then $n = \lambda c$. The net ion flux between the two planes 1 and 2 is

$$J = n_1 p_{12} - n_2 p_{21}$$

in which $p_{12} = p_{21} = \frac{1}{2} p_0$ (Figure 10.13) for the two planes labeled 1 and 2 if there is no driving force. The flux from surface 1 to surface 2 is then

$$J = \tfrac{1}{2} \lambda p_0 (c_1 - c_2)$$

By definition

$$\frac{c_1 - c_2}{\lambda} = -\frac{\partial c}{\partial z}$$

so that the flux becomes

$$J = \tfrac{1}{2} \lambda^2 p_0 \frac{\partial c}{\partial z}$$

If we consider six directions for the atoms to move in instead of two as in Figure 10.13 the factor $\frac{1}{2}$ becomes $\frac{1}{6}$, or more generally a factor $g$, which also contains the correlation factor or the structure-dependent chance that the atom after jumping returns to its original position. For the fcc lattice, $g = 0.78$ and for bcc, $g = 0.73$.

Using Fick's first law, we find that $D$ becomes

$$D = g\lambda^2 p_0 \qquad \text{or} \qquad D = g\lambda^2 f \exp(-G_m/RT)$$

For the representative values $\lambda \approx 0.2$ nm, $f \approx 10^{13}\,\text{s}^{-1}$, $G_m \approx 100$ kJ/mole and $T = 1200°\text{C}$, the value for $D$ is $3 \times 10^{-12}\,\text{m}^2/\text{s}$.

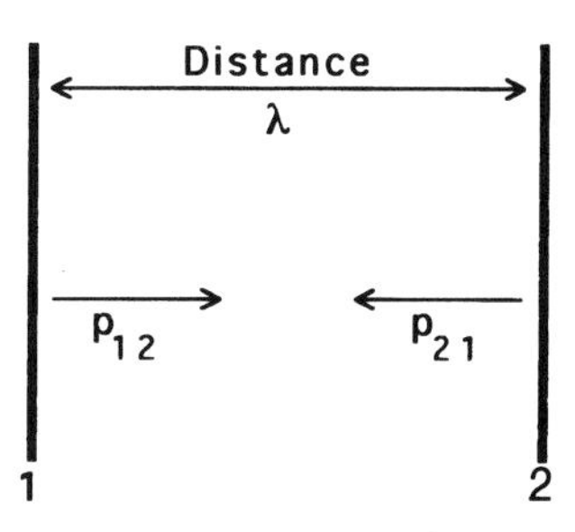


**Figure 10.13.** Two lattice planes with the probability $p$ of atom transfer between them.

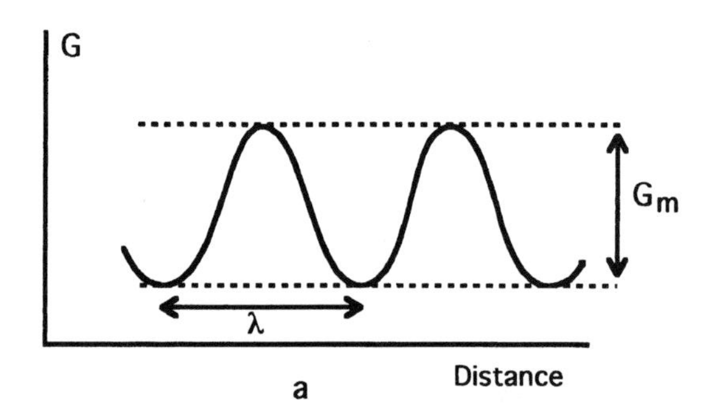


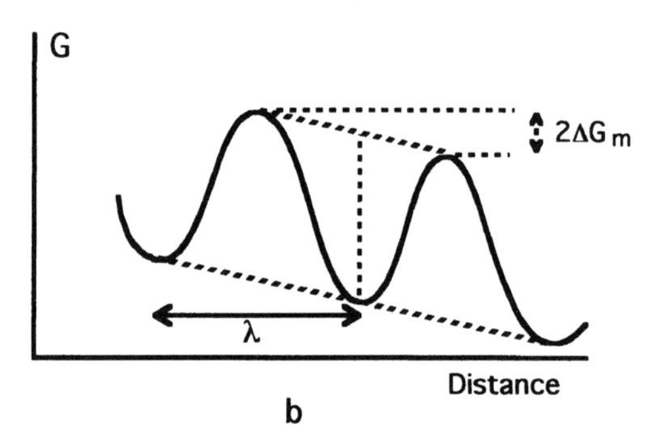


**Figure 10.14.** The activation energy of an atom moving from one site to the next over a distance $\lambda$ during diffusion.[20] If a directed potential is present the activation energy becomes direction-dependent. The extra driving force owing to the directed potential $\Delta G$ is small compared to $kT$.

If there is a driving force ($F = -\nabla\Phi$, where $\Phi$ is the thermodynamic potential) then the flux is proportional to it, provided the resistance remains the same. The activation energy is now no longer the same for the two directions (Figure 10.14), $\Delta G_m = -\frac{1}{2}\lambda F$. The probability $p_{\text{right}}$ that an atom or ion jumps to the right from a lattice plane is

$$p_{\text{right}} = \tfrac{1}{2} f \exp(-G_m + \Delta G_m/RT) = \tfrac{1}{2} p_0 \exp(\Delta G_m/RT) = \tfrac{1}{2}\left(1 + \frac{\Delta G_m}{RT}\right)$$

for $\Delta G_m \ll RT$.

Similarly

$$p_{\text{left}} = \tfrac{1}{2} p_0 \left(1 - \frac{\Delta G_m}{RT}\right)$$

The net chance of jumping $p$ is

$$p = p_{\text{right}} - p_{\text{left}} = \tfrac{1}{2} p_0 \frac{2\Delta G_m}{RT} = \tfrac{1}{2} p_0 \frac{\lambda F}{RT}$$

The drift velocity of the jumping atom is $v = \lambda p$; the atom flux is $J = cv$, where $c$ is the concentration; hence:

$$J = \tfrac{1}{2}p_0 \frac{F\lambda^2 c}{RT} = \tfrac{1}{2}f \frac{F\lambda^2 c}{RT} \exp(-G_m/RT)$$

Finally the mobility $B$ can be introduced, which is defined by the equation $J = BcF$. The mobility is then

$$B = \tfrac{1}{2}\frac{f\lambda^2}{RT} \exp(-G_m/RT)$$

in one direction, for instance, $z$. Hence the diffusion coefficient and the mobility are related: $D = RTB$.

As $G_m = H_m - TS_m$ ($m$ for mobility), the diffusion coefficient that corresponds to interstitial diffusion is

$$D = g\lambda^2 f \exp(S_m/R) \exp(-H_m/RT) \qquad \text{or} \qquad D = D_0 \exp(-Q/RT)$$

Table 10.4 gives a few representative values for $D_0$ and $Q$.

The equation above holds for diffusion of matter that is the result of moving interstitials. If atomic diffusion takes place through a diffusion of vacancies, the jumping frequency $p$ of the diffusing atom is proportional to the concentration of the vacancies $[V]$, which is also temperature-dependent. The expression for the diffusion coefficient of the vacancies has to be multiplied by this concentration to obtain the atomic diffusion coefficient. In atom diffusion in a compound with Schottky-type disorder (vacancies) both $D_0$ and $Q$ depend on the vacancy concentration, and the formation free energy $G_0$ as in the equation for interstitials must be replaced by $G_m + G_f$, where $G_f$ is the Gibbs free energy for the formation of vacancies.

**Table 10.4. Values of Diffusion Coefficients of Some Species in Selected Materials**

| Species | Solid | $D$ (530°C) ($m^2/s$) | $D_0$ ($m^2/s$) | $Q$ (kJ/mol) |
|---|---|---|---|---|
| H | α-Fe | $1.2 \times 10^{-8}$ | $8.8. \times 10^{-8}$ | 13 |
| C | α-Fe | $6.6 \times 10^{-12}$ | $2.0 \times 10^{-6}$ | 84 |
| Fe | α-Fe | $4.7 \times 10^{-20}$ | $1.9 \times 10^{-4}$ | 239 |
| H | Ag | $2.5 \times 10^{-9}$ | $2.8 \times 10^{-7}$ | 31 |
| Cu | Ag | $3.0 \times 10^{-17}$ | $1.2 \times 10^{-4}$ | 193 |
| Ag | Ag | $3.6 \times 10^{-17}$ | $4.0 \times 10^{-5}$ | 185 |
| Li | Ge | $1.5 \times 10^{-10}$ | $1.2 \times 10^{-7}$ | 45 |
| P | Ge | $6.2 \times 10^{-20}$ | $2.5 \times 10^{-4}$ | 239 |
| Ge | Ge | $1.4 \times 10^{-22}$ | $7.8 \times 10^{-4}$ | 287 |
| $Na^+$ | NaCl | $3.9 \times 10^{-15}$ | $5.0 \times 10^{-5}$ | 155 |
| $Cl^-$ | NaCl | $1.0 \times 10^{-16}$ | $1.1 \times 10^{-2}$ | 215 |
| $Ni^{2+}$ | NaCl | $1.3 \times 10^{-14}$ | $2.0 \times 10^{-6}$ | 126 |
| $Cu^+$ | $Cu_2O$ | $5.9 \times 10^{-16}$ | $4.3 \times 10^{-6}$ | 151 |
| $O^{2-}$ | $Cu_2O$ | $1.3 \times 10^{-17}$ | $6.5 \times 10^{-7}$ | 164 |
| $Ag^+$ | α-AgI | $4.1 \times 10^{-9}$ | $1.6 \times 10^{-8}$ | 9 |
| $I^-$ | α-AgI | $5.0 \times 10^{-13}$ | $4.8 \times 10^{-9}$ | 61 |

The vacancy concentration that affects $D$ can be extrinsic, in which case it does not depend on the temperature. If $Na_2O$ is dissolved in NaCl:

$$Na_2O \rightarrow 2Na_{Na}^x + O'_{Cl} + V_{Cl}^{\bullet}$$

At low temperatures when the compound is extrinsic, $[V_{Cl}^{\bullet}] = [Na_2O]$ and $D$ depends on $G_m$ and not on $G_f$. The vacancy concentration is determined by the impurity, not by the temperature. At sufficiently high temperatures, when the number of thermally generated defects is much larger than the impurity concentration, $D$ depends on both $G_m$ and $G_f$.

The diffusion of Ni in NiO, taking another example, depends on the oxygen pressure because this affects the nickel vacancy concentration, which in this case determines the diffusion coefficient:

$$\tfrac{1}{2}O_2(g) \rightleftharpoons O_C^x + V''_{Ni} + 2h^{\bullet}$$

with, say, the equilibrium constant $K_a$. Hence

$$[V''_{Ni}] = (\tfrac{1}{4}K_a)^{1/3} p(O_2)^{1/6}$$

Vacancies also affect the physical properties of solids. Four examples are:

1. Ionic conductivity: If $c$ is the concentration, $q$ the charge, and $\mu$ the mobility of charge carriers, the conductivity $\sigma = cq\mu$. Hence,

$$\mu = \frac{qD}{RT} = \frac{qD_0}{RT}\exp(-H_m/RT)$$

The ionic conductivity $\sigma$ depends on both the migration enthalpy and the formation enthalpy of vacancies. The electric mobility $\mu$ here is not the same as the mechanical mobility $B$ used above. The relation is $\mu = Bq = B|z|e$. Here $q$ is the charge per charge carrier, $e$ the elementary charge, and $z$ the number of elementary charges per charge carrier.

2. Density of the solid: As the number of dissolved vacancies increases the solid expands and this can be measured: on doping $CaCl_2$ in KCl the density decreases because the vacancies generated overcompensate for the mass increase on replacing K by Ca: $CaCl_2 \rightarrow Ca_K^{\bullet} + V'_K + 2Cl_{Cl}^x$.

3. Expansion coefficient: Owing to the formation of vacancies the linear coefficient of thermal expansion that is determined from the size of a crystal is larger than the value derived from the lattice constant increase as measured with X-ray diffraction.

4. Specific heat: The specific heat increases at higher temperatures because the formation of vacancies needs energy.

The resistances for atomic transport are rate-determining in solid state chemistry. The various diffusion rates also determine the morphology of the solid product of those reactions: different microstructures are formed when the rates are changed and the fastest-growing form overwhelms slower-growing forms when they compete in parallel. This rate-dependent morphology also holds for the opposite of growth, i.e., corrosion, which attacks grain boundaries first. There are several paths for atomic transport in a polycrystalline solid: grain boundary diffusion along the grain

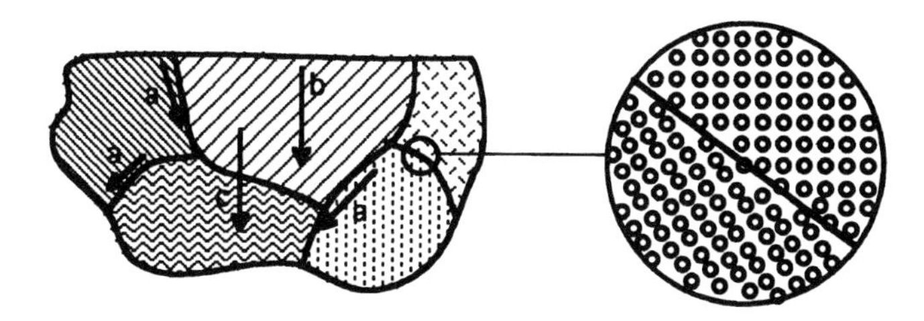


**Figure 10.15.** Diffusion through a polycrystalline solid. The three paths have different rates: (a) in the grain boundaries; (b) within the grains; (c) between the grains.

boundaries, bulk diffusion in the grains, and cross-boundary diffusion followed by bulk diffusion. In polycrystalline solids the diffusion resistance in the grain boundaries is often lower than in the interior of the crystallites (Figure 10.15). Obviously atomic transport and therefore chemical reaction is morphology-dependent. The microstructure, which is itself the result of the formation chemistry, strongly affects the chemical properties.

Figure 10.16 shows the Arrhenius plots with the temperature-dependent diffusion coefficients of several solids. The differences can be considerable. For some applications good ionic conductivity is required while other materials are chosen because they are good diffusion barriers.

Low diffusion resistances (high diffusion coefficients) for ions are necessary for electrolytes of solid state barriers, fuel cells, and Nerst-type sensors. High diffusion coefficients in solids often also entail high sinter rates, which may be a less desirable characteristic in devices. High resistances against solid state diffusion, on the other hand, is necessary in passivating coatings that are intended to protect against

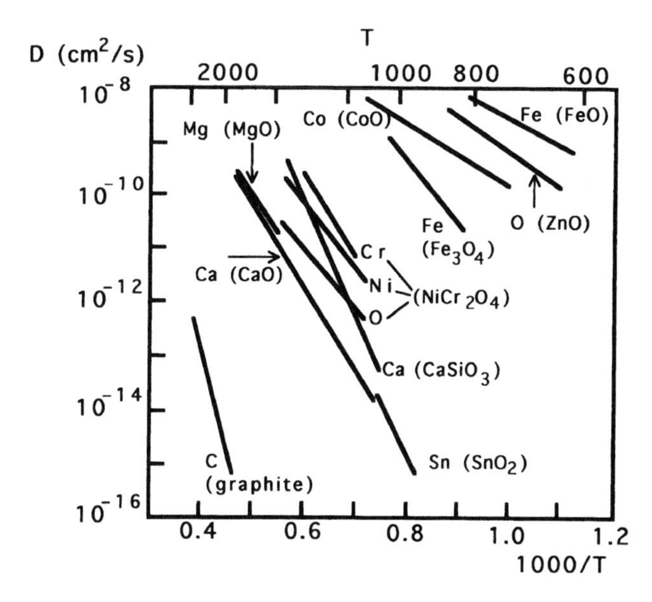


**Figure 10.16.** The temperature-dependence of the diffusion coefficients of atoms in oxides.

**Table 10.5. Diffusion Data for Selected Nonoxide Ceramics**

| Species | Solid | Temperature range (°C) | $D_0$ ($cm^2/s$) | $Q$ (eV) | $\log D$ at $0.6\,T_m$ ($cm^2/s$) |
|---|---|---|---|---|---|
| Ti | $TiC_{0.97}$ | 1800–2200 | $4.4 \times 10^4$ | 7.6 | −14.5 |
| C | $TiC_{0.97}$ | 1500–2700 | 7.0 | 4.1 | −9.6 |
| C | $TiC_{0.87}$ | 1800–2100 | 45.4 | 4.6 | −9.9 |
| Zr | $ZrC_{0.97}$ | 2300–2800 | $1 \times 10^3$ | 7.5 | −14.1 |
| C | $ZrC_{0.97}$ | 1300–2100 | $1.3 \times 10^2$ | 5.6 | −9.0 |
| Nb | $NbC_{0.87}$ | 2100–2400 | 4.5 | 6.1 | −12.5 |
| C | $NbC_{0.87}$ | 1600–2100 | 2.6 | 4.3 | −9.0 |
| Si | $\alpha$-SiC | 2000–2300 | $5 \times 10^2$ | 7.2 | −17.2 |
| C | $\alpha$-SiC | 1850–2200 | $8.6 \times 10^5$ | 7.4 | −14.5 |
| Si | $\beta$-SiC | 2000–2300 | $8.4 \times 10^7$ | 9.5 | −18.3 |
| C | $\beta$-SiC | 1900–2200 | $4.4 \times 10^7$ | 8.7 | −16.4 |
| C | WC | 2000–2400 | $1.9 \times 10^{-6}$ | 3.8 | −16.1 |
| N | $\alpha$-$Si_3N_4$ | 1200–1400 | $1.2 \times 10^{-12}$ | 2.4 | −20.1 |
| N | $\beta$-$Si_3N_4$ | 1300–1400 | $6 \times 10^6$ | 8.1 | −20.8 |

corrosion at high temperatures. Metal nitrides, carbides, and borides are rather inert and good diffusion barriers. Atomic transport rates differ with the elements but they also are understandably very dependent on the stoichiometry, which in this type of compound may be considerable. Table 10.5 gives the diffusion parameters of some carbides and nitrides. Higher vacancy concentrations imply lower diffusion resistances.

The concentration distribution of diffusing atoms from a flat surface into the bulk of an alloy can be estimated by solving Fick's second diffusion equation with the imposed boundary conditions. The equation in one spatial dimension $z$ perpendicular to the surface is

$$\frac{\partial c(z, t)}{\partial t} = D\frac{\partial^2 c(z, t)}{\partial z}$$

in which $c(z, t)$ is the concentration at time $t$ and depth $z$ measured from the surface. $D$ is assumed not to depend on $z$. The solution of this equation is given here for two boundary conditions.

1. There is an initial concentration at the surface of a species M that diffuses into the substrate. The boundary and initial conditions for this particular case are:

   $c(0, 0) = \mathrm{M}$ (initial concentration at the surface)

   $c(z, 0) = 0$ (initially the substrate does not contain the diffusing species)

   The solution of Fick's equation with these boundary conditions is a Gaussian distribution, which flattens in time:

$$c(z, t) = \frac{M}{(\pi D t)^{1/2}}\exp(-z^2/4Dt)$$

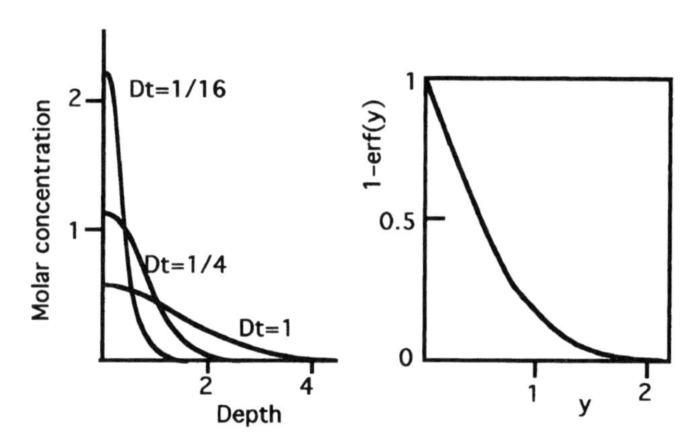


**Figure 10.17.** Two solutions of the Fick equation with different boundary conditions.[20] Left: a fixed total amount, initially present at the surface, diffuses into the solid. The curves give the concentration at different instants after the start. Right: the curve gives the concentration in the solid when the concentration at the surface is kept constant during the diffusion process.

2. The second limiting case involves a surface concentration that is kept constant during the process if, e.g., atoms diffuse into a solid from a flowing gas or a liquid with constant concentration at the solid surface. The initial and boundary conditions are:

$$c(0, t) = c_0 \text{ (constant surface concentration } c_0\text{)}$$

$$c(z, 0) = 0 \text{ (initially no diffusing atoms in the substrate}$$

The solution of Fick's equation is then

$$c(z, t) = c_0 \left\{ 1 - \operatorname{erf}\left( \frac{z}{2\sqrt{D t}} \right) \right\} \qquad \text{with } \operatorname{erf}(y) = \frac{2}{\sqrt{\pi}} \int_0^y \exp(-z^2)\, dz$$

Figure 10.17 shows how a concentration with an erf distribution drops with increasing depth. The "penetration depth" $z_d$ for this case is $z_d = \sqrt{(Dt)}$ and this equation explains the parabolic rates observed in many solid state reactions.

## 10.5. A Note on Diffusion Coefficients

There are different kinds of diffusion coefficients: those of defects, of individual atoms, and of ions that depend on the diffusion rates of other ions. Some simple relations follow.

1. Diffusion of defects: For a defect (an interstitial or a vacancy) that jumps from one site to another the equation for the diffusion coefficient $D_d$ was derived as

$$D_d = g\lambda^2 f \exp(-\Delta G_m/RT)$$

in which $g$ is the geometric factor, $\lambda$ the jump distance, $f$ the vibration frequency, and $\Delta G_m$ the Gibbs energy for migration of the defects. This equation is used to derive expressions for the other diffusion coefficients.

2. SELF-DIFFUSION OF INDIVIDUAL ATOMS OR IONS IN SOLIDS: The self-diffusion coefficient is proportional to the defect concentration and the diffusion coefficient of the defects that make the diffusion of species $i$ possible: $D_i = D_d c_d$, in which $c_d$ is the molar fraction of defects. In the intrinsic case, where the disorder is of thermal origin, the concentration of the defects is temperature-dependent:

$$c_d = \exp(-\Delta G_f/RT)$$

where $\Delta G_f$ is the free energy for formation of the defects. Not only the concentration but also the diffusion coefficient $D_d$ itself is temperature-dependent as shown above under 1. Hence in the product, the exponential temperature dependence of the diffusion coefficient for atoms contains the migration activation free energy $\Delta G_m$ as well as the defect formation free energy $\Delta G_f$. The self-diffusion coefficient $D_i$ then becomes

$$D_i = g\lambda^2 f \exp(-(\Delta G_m + \Delta G_f)RT)$$

If the solid has several types of diffusing atoms or ions that have different self-diffusion coefficients they are distinguished with the subscript $i$. $D_i$ is the self-diffusion coefficient of species $i$ in the solid under consideration.

Doping can increase or decrease this diffusion coefficient because it affects the concentrations of the defects as was discussed earlier. Doping with anions that have a higher valence than the original anions increases the anion vacancy concentration in the lattice. Increasing the oxygen pressure on oxides can increase the cation vacancy concentration. Thus certain impurities and ambients may affect the self-diffusion coefficients.

The method to determine defect concentrations, defect mobilities, and their influence on the diffusion is to use conductivity measurements at different temperatures and measuring frequencies. In a single experiment the activation energies for formation and displacement of the defects can be established if both the intrinsic and extrinsic domain can be covered. Some defects can be identified spectroscopically, e.g., with electron paramagnetic resonance (EPR or ESR) if the defect has unpaired electrons.

3. THE CHEMICAL DIFFUSION COEFFICIENT: In solid state reactions several atoms or ions may diffuse in various directions, each with a different self-diffusion coefficient. This can develop electrochemical potentials that affect the overall diffusion coefficients. The chemical diffusion coefficient $D$ of the atoms A and B in a compound $A_x B_y$ is a weighted mean of the self-diffusion coefficients $D_A$ and $D_B$ of the diffusing atoms or ions A and B. $D$ becomes

$$D = \frac{D_A D_B}{c_A D_A + c_B D_B}$$

This is seen as follows: The diffusion rates of the two species A and B are initially different and therefore they develop a potential $E$ in the lattice. This potential affects the drift velocity $\langle v_i \rangle$ of the atoms. Besides the Fick term the expression for the particle flux then has another term:

$$J_i = -D_i \frac{dc_i}{dx} + c_i \langle v_i \rangle$$

where $c_i$ is the mole fraction of species $i$ ($i$ = A or B). Now by definition of the chemical diffusion coefficient $D$:

$$J_i = -D\frac{dc_i}{dx}$$

In the reaction zone where A and B form the compound, $J_A = -J_B$ and $c_A + c_B = 1$, and it follows that

$$\langle v_A \rangle = \frac{qED_A}{RT}$$

with a similar expression for atom B. From these equations the drift velocities $\langle v_i \rangle$ can be eliminated. For $J_A$ this yields

$$J_A = -D_A\frac{dc_A}{dx} + c_A D_A \frac{(D_A - D_B)(dc_A/dx)}{c_A D_A + c_B D_B}$$

with a similar expression for $J_B$.

This results in the chemical diffusion coefficient

$$D = \frac{D_A D_B}{c_A D_A + c_B D_B}$$

This coefficient is determined by the self-diffusion coefficients $D_i$ of both species but is dominated by the atom or ion that has the highest resistance for diffusion.

## Exercises

1. In Chapter 6 a derivation was given for an expression of the Sherwood number in the case of a simple first-order reaction. It is basically incorrect to use the analogy of Ohm's law in the derivation. Why?
2. Derive the values of the three slopes ($\frac{1}{2}$, $\frac{1}{4}$, $\frac{1}{6}$) in Figure 10.7 from the four defect equilibria and the electoneutrality constraint (number of positive defects equals number of negative defects). Hint: use approximations in which some concentrations can be ignored in the limits of low, high, and stoichiometric oxygen pressures.
3. Give the Kröger–Vink notation of the following point defects. (a) In a spinel: an $Al^{3+}$ ion on an $Mg^{2+}$ site; an $Mg^{2+}$ ion on an $Al^{3+}$ site; and an $F^-$ ion on an $O^{2-}$ site. (b) In lead zirconate, the perovskite $PbZrO_3$: $Ti^{4+}$ on a $Zr^{4+}$ site; $La^{3+}$ on a $Pb^{2+}$ site; and $Nb^{5+}$ on a $Zr^{4+}$ site.
4. Do you expect the following defect concentrations to be strongly temperature-dependent? (a) $V_{Mn}''$, $V_O^{\bullet\bullet}$, $Co_{Mn}^x$, $Na_{Mn}'$, and $S_O^x$ in MnO and (b) $V_O^{\bullet\bullet}$, $V_{Co}''$, $Co_{Cr}'$, and $Ti_{Cr}^{\bullet}$ in the spinel $CoCr_2O_4$.
5. The electroneutrality principle requires that the number of positively charged defects equal the number of negatively charged ones. Give possible charge compensating defects for those given in Exercise 3.
6. The temperature dependence of the electron concentration in an intrinsic oxide semiconductor depends on the equilibrium constant for thermal generation of electrons and electron holes $K_i$ and on the bandgap $E_g$. What is this relationship?
7. Give the dissolution reaction of: (a) Zn in ZnO; (b) $O_2$ in FeO; $O_2$ in FeO that is doped with $Na_2O$ or $Sc_2O_3$; (c) ZnS in GaAs; $Zn_3P_2$ in GaAs; and (d) AlN in $Si_3N_4$; AlN in SiC.
8. What happens to the conductivity of the semiconductor BP (boron phosphide) on doping it with silicon?

9. Calcium-doped cerium oxide ($CeO_2$) is unsuitable as a solid electrolyte in a Nernst sensor for measuring the oxygen concentration and in a fuel cell. Why?
10. Name several solid reference electrode materials for a hydrogen sensor and for an oxygen sensor.
11. If an oxide has a sufficiently large bandgap (small $K_i$), the stoichiometric oxide MO (in the center of the electrolytic domain) is an ion conductor. Assume that $K_S/K_i = 10^6$ and that the Schottky defect formation enthalpy is 2 eV. What is the bandgap energy of the oxide?
12. Explain in thermodynamic terms why FeO is nonstoichiometric and always has an excess of oxygen.
13. Photographic films have light sensitivities and contrast that are controlled with the size of the silver halide particles among other tricks. How should the halide particle size be changed in order to increase the contrast? And how for more sensitivity? What do these changes mean for photochemical characteristics of the film? Increasing sensitivity, for instance, increases the graininess of the picture. Why?
14. What is the atomistic reason for the Meyer–Neldel rule? Design two ion conductors (for $Li^+$ and $O^{2-}$) that have a low value for $L$ and simultaneously a high $D_0$.

## References


1. M. Polanyi. *Personal Knowledge. Towards Post-Critical Philosophy.* Routledge and Kegan, London (1978).
2. F. E. Huang. *Engineering Thermodynamics: Fundamentals and Applications.* MacMillan, New York (1976).
3. N. E. Dasent. *Inorganic Energetics.* Cambridge University Press, Cambridge (1982).
4. D. A. Glocker and S. I. Shah. *Handbook of Thin Film Process Technology.* Institute of Physics, Bristol (1995).
5. C. Truesdell. *Rational Thermodynamics.* Springer-Verlag, New York (1984).
6. D. A. Johnson. *Some Thermodynamic Aspects of Inorganic Chemistry.* Cambridge University Press, Cambridge (1982).
7. P. Glansdorff and I. Prigogine. *Thermodynamic Theory of Structure, Stability and Fluctuations.* Wiley Interscience, New York (1977).
8. W. Albers. *Non-stoichiometry of Inorganic Solids.* In: *Current Topics in Materials Science*, Vol. 10, E. Kaldis (ed.). North-Holland, Amsterdam (1982), p. 191.
9. Alper, A. M. (ed.). *Refractory Materials*, Vol. 6: *Phase Diagrams.* Academic, New York (1995).
10. E. M. Levin, C. R. Robbins, and H. F. McMurdie (eds.). *Phase Diagrams for Ceramists.* American Ceramic Society, Columbus (1964).
11. R. F. Davis. Point (atomic) defects in stoichiometric ceramic materials. Part I: Solid solutions; Part II: Schottky and Frenkel defects. *J. Mater. Educ.* **2**, 809, 837 (1980).
12. W. Hayes and A. M. Stoneham. *Defects and Defect Processes in Nonmetallic Solids.* Wiley, New York (1985).
13. G. G. Libowitz. Defect Chemistry in Solids. In: *Treatise on Solid State Chemistry: I. The Chemical Structure of Solids*, N. N. Hannay (ed.). Plenum, New York (1973), p. 335.
14. R. J. D. Tilley. *Defect Crystal Chemistry and Its Applications.* Blackie, Glasgow (1987).
15. D. M. Smyth. The role of impurities in insulating transition metal oxides. *Progr. Sol. St. Chem.* **15**, 145 (1984).
16. W. L. Worrell. Oxide Solid Electrolytes. In: *Topics in Applied Physics, Vol. 21*, S. Geller (ed.). Springer, Berlin (1977), p. 143.
17. I. Riess, P. J. van der Put, and J. Schoonman. Solid oxide fuel cells operating on uniform mixtures of fuel and air. *Sol. St. Ionics* **82**, 1 (1995).
18. J. F. Hamilton. The photographic process. *Progr. Sol. St. Chem.* **8**, 167 (1973).
19. Y. T. Tan. Silver halides in photography. *Mater. Res. Soc. Bull.*, p. 13 (May 1989).
20. R. Metselaar. Diffusion in Solids I-III. *J. Mater. Educ.* **6**, 231 (1984) **7**, 655 (1985); **10**, 621 (1988).

# INDEX